Notice

The 2009 edition of the *Uniform Mechanical Code* is developed through a consensus standards development process approved by the American National Standards Institute. This process brings together volunteers representing varied viewpoints and interests to achieve consensus on mechanical issues. While the International Association of Plumbing and Mechanical Officials (IAPMO) administers the process and establishes rules to promote fairness in the development of consensus, it does not independently test, evaluate, or verify the accuracy of any information or the soundness of any judgments contained in its codes and standards.

IAPMO disclaims liability for any personal injury property or other damages of any nature whatsoever, whether special, indirect, consequential, or compensatory, directly or indirectly resulting from the publication, use of, or reliance on this document. IAPMO also makes no guaranty or warranty as to the accuracy or completeness of any information published herein.

In issuing and making this document available, IAPMO is not undertaking to render professional or other services for or on behalf of any person or entity. Nor is IAPMO undertaking to perform any duty owed by any person or entity to someone else. Anyone using this document should rely on his or her own independent judgment or, as appropriate, seek the advice of a competent professional in determining the exercise of reasonable care in any given circumstances.

An American National Standard

IAPMO/ANSI UMC 1-2009

2009 UNIFORM MECHANICAL CODE™

Approved American National Standard

ANSI

REVISION MARKINGS

Code changes from the 2006 edition are marked in the margins as follows:

→ An arrow denotes a deletion.

| A vertical denotes a change.

TIA TIA TIA TIA TIA TIA

TIA indicates that the revision is the result of a Tentative Interim Amendment
For further information on tentative interim amendments see Section 5 of the IAPMO Regulations Governing Committee Projects available at http://www.iapmo.org/pages/codedevelopmentprocedures.aspx

NFPA NFPA NFPA NFPA NFPA NFPA NFPA NFPA NFPA NFPA NFPA NFPA NFPA NFPA NFPA NFPA NFPA NFPA

A reference in brackets [ ] following a section or paragraph indicates material that has been extracted from another document. This reprinted material is not the complete and official position of the National Fire Protection Association on the reference subject which is represented by the standard in its entirety. Text which has been extracted pursuant to IAPMO's Extract Guidelines is denoted with the use of "NFPA" in the margin. This text has not been fully processed by IAPMO in accordance with ANSI's public announcement consensus requirements for an American National Standard (ANS) nor approved by ANSI's Board of Standards Review, but will be fully processed in accordance with those requirements as part of the next revision cycle for this document.

Information on referenced publications can be found in Chapter 17.

All pressures used in this code are gauge pressures unless otherwise indicated.

Twenty-Fifth Edition

First Printing, January 2009
Second Printing, April 2009

ISSN 0733-2335
Published by the International Association of Plumbing and Mechanical Officials
5001 E. Philadelphia Street • Ontario, CA 91761-2816 – USA
Main Phone: (909) 472-4100 • Main Fax: (909) 472-4150

GCIU 897-M IBT 853 PRINTED WITH SOY INK

2009 UMC Foreword

The *Uniform Mechanical Code*™ (UMC™) provides complete requirements for the installation and maintenance of heating, ventilating, cooling, and refrigeration systems, while at the same time allowing latitude for innovation and new technologies.

This code was first published in 1967. With the publication of the 2003 edition of the *Uniform Mechanical Code*, another significant milestone was reached. For the first time in the history of the United States, a mechanical code was developed through a true consensus process. The 2009 edition represents the most current approaches in the mechanical field and is the third edition developed under the ANSI Consensus process. Contributions to the content of this code were made by every segment of the built industry, including such diverse interests as consumers, enforcing authorities, installers/maintainers, labor, manufacturers, research/standards/testing laboratories, special experts, and users.

The public at large is encouraged and invited to participate in IAPMO's open consensus code development process. This code is updated every three years. A code development timeline and other relevant information is available at IAPMO's website at www.iapmo.org.

The *Uniform Mechanical Code* is dedicated to all those who, in working to achieve "the ultimate mechanical code," have unselfishly devoted their time, effort, and personal funds to create and maintain this, the finest mechanical code in existence today.

The 2009 *Uniform Mechanical Code* is supported by the American Society of Sanitary Engineering (ASSE), Mechanical Contractors Association of America (MCAA), Plumbing-Heating-Cooling Contractors National Association (PHCC-NA), the United Association (UA), and the World Plumbing Council (WPC). The presence of these logos, while reflecting support, does not imply any ownership of the copyright to the UMC which is held exclusively by IAPMO. Further, the logos of these associations indicates the support of IAPMO's open, consensus process being used to develop IAPMO's codes and standards.

The addresses of the organizations are as follows:

ASSE – 901 Canterbury Road, Suite A • Westlake, OH 44145-7201 • (440) 835-3040

MCAA – 1385 Piccard Drive • Rockville, MD 20850 • (301) 869-5800

PHCC-NA – PO Box 6808 • Falls Church, VA 22046 • (800) 533-7694

UA – 901 Massachusetts Avenue NW • Washington, DC 20001 • (202) 628-5823

WPC – WPC Secretary • c/o Mechanical Contractors Association of Alberta

#204 2725-12 St. NE Calgary • Alberta T2E 7J2 Canada • +1-40-325-072-37

P.O. Box 2005 • Marmion, Western Australia 6020 • +61-8-9448-0430

Tentative Interim Amendment (TIA)

A Tentative Interim Amendment (TIA) to any Document may be processed if the Council Secretary determines, after a preliminary review, and consultation with the appropriate Chair, that the Amendment appears to be of an emergency nature requiring prompt action and has the endorsement of a Member of the involved Technical Committee.

Sections 311.1 , 311.3 - 311.5 – This Tentative Interim Amendment (TIA) was issued on November 14, 2008.
Chapter 4, Table 4-4 – This Tentative Interim Amendment (TIA) was issued on March 31, 2008.

For further information on TIAs see Section 5 of the IAPMO Rules Governing Committee Projects – www.iapmo.org

COMMITTEE ON UNIFORM MECHANICAL CODE

These lists represent the membership at the time the Committee was balloted on the final text of this edition. Since that time, changes in the membership may have occurred.

IAPMO Standards Council

J. Richard Wagner, *Chairman*
The Poole & Kent Company, [SE]

Kevin Cotter, UA Local 12 [L]
Bob Courtnier, State of California [U]
Jim Crawford, City of Vancouver [U]
John Garvelink, Commercial Design Engineering Ltd. [I/M]
Ron George, Ron George Design & Consulting Services [SE]
Dave Levanger, Carbon County, Utah [E]
Phil Ribbs, PHR Consultants [SE]
Terry Swisher, State of Oregon [E]

Nonvoting

Gabriella M. Davis, Secretary
International Association of Plumbing and Mechanical Officials

Lynne Simnick, Recording Secretary
International Association of Plumbing and Mechanical Officials

IAPMO Uniform Mechanical Code Technical Committee

April Trafton, *Chairperson*
American Society of Plumbing Engineers [SE]

Bob Adler, City of San Jose, California [E]
Denise Beach, National Propane Gas Association [U]
DJ Berger, New Orleans Pipe Trades [L]
Paul Cabot, American Gas Association [U]
Sid Cavanaugh, Cavanaugh Consulting [SE]
Ian Chang, Intertek Testing Services [R/S/T]
Richard Church, Plastic Pipe & Fittings Association [M]
Harvey Kreitenberg, Harvey Kreitenberg & Associates [I/M]
Theodore Lemoff, NFPA International [R/S/T]
Ted Lowe, Radiant Panel Association [I/M]
David Mann, Plumbers & Steamfitters Local [L]
Nasser Nikravi, State of California [E]
Tim Orris, Air Movement and Control Association International Inc. [M]
James Pavesic, United Association, [L]
Anthony Scarano, Plastic Piping Consultant [SE]
John D. Sedine, Air Conditioning Contractors of America [I/M]
Amir Tabakh, City of Los Angeles [E]
John Taecker, Underwriters Laboratories [R/S/T]
Steven Taylor, Taylor Engineering LLC
Rep. American Society of Heating, Refrigerating, and Air-Conditioning Engineers (ASHRAE) [U]
Richard W. Vrana, City of Houston [E]

Alternates

Michael Cudahy, Plastic Pipe & Fittings Association [M]
Wesley Davis, Air Conditioning Contractors of America [I/M]
James Dingman, Underwriters Laboratories [R/S/T]
TJ Dodd, Austin Area Plumbing & Pipe Fitters JATC [L]
Scott Hamilton, United Association [L]
Vickie Lovell, Air Movement and Control Association International Inc. [M]
Michael Tharpe, City of Los Angeles [E]
Mike Wynne, Plumbers & Steamfitters Local [L]

Nonvoting

Ron Rice, Ex -Officio IAPMO [E]
Adam Muliawan, IAPMO Staff Liaison

COMMITTEE MEMBERSHIP CLASSIFICATION ABBREVIATIONS

These classifications apply to Technical Committee members and represent their principal interest in the activity of a committee.

M *Manufacturer:* A representative of a maker or marketer of a product, assembly or system, or portion thereof, that is affected by the standard.

U *User:* A representative of an entity that is subject to the provisions of the standard or that voluntarily uses the standard.

I/M *Installer/Maintainer:* A representative of an entity that is in the business of installing or maintaining a product, assembly or system affected by the standard.

L *Labor:* A labor representative or employee concerned with safety in the workplace.

R/S/T *Research/Standards/Testing Laboratory:* A representative of an independent research organization; an organization that develops codes, standards or other similar documents; or an independent testing laboratory.

E *Enforcing Authority:* A reperesentative or an agency or an organization that promulgates and/or enforces standards.

C *Consumer:* A person who is, or represents, the ultimate purchaser of a product, system, or service affected by the standard, but who is not included in the *User* classification.

SE *Special Expert:* A person not representing any of the previous classifications, but who has special expertise in the scope of the standard or portion thereof.

FORM FOR PROPOSALS ON IAPMO UPC/UMC COMMITTEE DOCUMENTS

NOTE: All proposals MUST be received by 5:00 PM PST/PDST on February 1, 2010.

For further information on the standards-making process, please contact Codes and Standards Administration at 909-472-4110. For technical assistance, please call IAPMO at 909-472-4111.	**FOR OFFICE USE ONLY** **LOG # :** ______ **DATE REC'D:** ______

PLEASE USE SEPARATE FORM FOR EACH PROPOSAL

Please indicate in which format you wish to receive your ROP/ROC: ☐ **Paper** ☐ **Download*** ☐ **CD-ROM**

***Note: In choosing the download option, you intend to view the ROP/ROC from our website. NO copy will be sent to you.**

Date: ______ Name: ______ Telephone #: ______

Company: ______

Street Address: ______ City: ______ State: ______ Zip: ______

Please Indicate Organization Represented (if any): ______

1. **IAPMO Document Title** ______ **IAPMO NO. & Year** ______

   Section/Paragraph ______

2. **Proposal Recommends** (check one): ☐ New Text ☐ Revise Text ☐ Delete Text

3. **Proposal** (Include proposed new wording, or identification of wording to be deleted): [Note: Proposed text should be in legislative format: i.e., use underscore to denote wording to be inserted (inserted wording) and strike-through to denote wording to be deleted (~~deleted wording~~).] Please note if you are referencing a standard or other publication, please provide two copies.

4. **Statement of Problem and Substantiation for Proposal**: [Note: State the problem that will be resolved by your recommendations; give the specific reason for your proposal, including copies of tests, research papers, etc. If more than 200 words, it may be abstracted for publication.]

5. ☐ **This proposal is original material.** [Note: Original material is considered to be the submitter's own idea based on or as a result of his/her own experience, thought, or research and, to the best of his/her knowledge, is not copied from another source.]

   ☐ **This proposal is not original material, its source (if known) is as follows:** ______

I hereby grant the IAPMO all and full rights in copyright, in this proposal, and I understand that I acquire no rights in any publication of IAPMO in which this proposal in this or another similar or analogous form is used.

Signature (*Required*): ______

IAPMO CODES DEPARTMENT FAX • (909) 472-4198 or (877) 852-6337
Mail to: Code Development • IAPMO • 5001 E Philadelphia Street • Ontario • CA • 91761-2816

INSTRUCTIONS FOR SUBMITTING PROPOSALS
– PLEASE READ CAREFULLY –

1. Type or print in BLACK ink.

2. Indicate the title of the document and the document year. Also indicate the specific section or paragraph that the proposed amendment applies to.

3. Check the appropriate box to indicate whether this proposal recommends adding new text, revising existing text, or deleting text.

4. In the space identified as "Proposal", indicate the exact wording you propose as new or revised text, or the text you propose be deleted.

5. In the space titled "Statement of Problem and Substantiation for Proposal", state the problem that will be resolved by your recommendation and give the specific reason for your proposal. Include copies of test results, research papers, fire experience, or other materials that substantiate your recommendation. [See note below, item (f).]

6. Check the appropriate box to indicate whether or not this proposal is original material, and, if it is not, indicate the source of the material.

7. Sign the proposal.

If supplementary material (photographs, diagrams, reports, etc.) is included, you may be required to submit sufficient copies for all members and alternates of the technical committee. For publication in the *Report on Proposals,* the technical committee is authorized to abstract the "Statement of Problem and Substantiation for Proposal" if it exceeds 200 words.

NOTE: The IAPMO Regulations Governing Committee Projects in Paragraph 4-3.3 state: Each proposal shall be submitted to the Council Secretary and shall include:

(a) identification of the submitter and his or her affiliation (i.e., technical committee, organization, company), where appropriate;
(b) identification of the document, edition of the document, and paragraph of the document to which the proposal is directed;
(c) the proposed text of the proposal, including the wording to be added, revised (and how revised), or deleted;
(d) a statement of the problem and substantiation for proposal;
(e) the signature of the submitter; and
(f) two copies of any document(s) (other than an IAPMO document) being proposed as a reference standard or publication (see 3-3.7).

TABLE OF CONTENTS

CHAPTER 1

ADMINISTRATION

Part I — General.

101.0 Title.

These regulations shall be known as the Uniform Mechanical Code, may be cited as such, and will be referred to herein as "this code."

102.0 Purpose.

The purpose of this code is to provide minimum standards to safeguard life or limb, health, property, and public welfare by regulating and controlling the design, construction, installation, quality of materials, location, operation, and maintenance or use of heating, ventilating, cooling, and refrigeration systems; incinerators; and other miscellaneous heat-producing appliances within this jurisdiction.

The purpose of this code is not to create or otherwise establish or designate any particular class or group of persons who will or should be especially protected or benefited by the terms of this code.

103.0 Scope.

103.1 General. The provisions of this code shall apply to the addition to or erection, installation, alteration, repair, relocation, replacement, use, or maintenance of any heating, ventilating, cooling, refrigeration systems; incinerators; or other miscellaneous heat-producing appliances within this jurisdiction.

Additions, alterations, repairs to, and replacement of equipment or systems shall comply with the provisions for new equipment and systems, except as otherwise provided in Section 104.0 of this code.

103.2 Conflicts. Where, in any specific case, different sections of this code or referenced standards specify different materials, methods of construction, or other requirements, the most restrictive shall govern as determined by the Authority Having Jurisdiction. When there is a conflict between a general requirement and a specific requirement, the specific requirement shall be applicable.

103.3 Design and Testing. The design and testing of equipment regulated by this code shall be subject to the approval of the Authority Having Jurisdiction.

103.4 Appendices. The standards contained in Appendix A shall be considered as part of this code. Appendices B and C contain recommended practices that shall not apply unless specifically adopted. Appendix D contains conversion tables and a table for determining the approximate minimum thickness for carbon sheet steel.

104.0 Application to Existing Mechanical Systems.

104.1 Additions, Alterations, or Repairs. Additions, alterations, or repairs shall be permitted to be made to any mechanical system without requiring the existing mechanical system to comply with all the requirements of this code, provided the addition, alteration, or repair conforms to that required for a new mechanical system. Additions, alterations, or repairs shall not cause an existing system to become unsafe or create unhealthy or overloaded conditions.

Minor additions, alterations, and repairs to existing mechanical systems shall be permitted to be installed in accordance with the law in effect at the time the original installation was made, when approved by the Authority Having Jurisdiction.

104.2 Existing Installations. Mechanical systems lawfully in existence at the time of the adoption of this code shall be permitted to have their use, maintenance, or repair continued if the use, maintenance, or repair is in accordance with the original design and location and no hazard to life, health, or property has been created by such mechanical systems.

104.3 Changes in Building Occupancy. Mechanical systems that are a part of any building or structure undergoing a change in use or occupancy, as defined in the Building Code, shall comply with the requirements of this code that may be applicable to the new use or occupancy.

104.4 Maintenance. Mechanical systems, materials, and appurtenances, both existing and new, and parts thereof shall be maintained in proper operating condition in accordance with the original design and in a safe and hazard-free condition. Devices or safeguards that are required by this code shall be maintained in conformance with the code edition under which installed. The owner or the owner's designated agent shall be responsible for maintenance of mechanical systems and equipment. To determine compliance with this subsection, the Authority Having Jurisdiction shall be permitted to cause a mechanical system or equipment to be reinspected.

104.5 Moved Buildings. Mechanical systems or equipment that is a part of buildings or structures moved into or within this jurisdiction shall comply with the provisions of this code for new installations.

105.0 Alternate Materials and Methods of Construction Equivalency.

Nothing in this code is intended to prevent the use of systems, methods, or devices of equivalent or superior quality, strength, fire resistance, effectiveness, durability, and safety over those prescribed by this code. Technical documentation shall be submitted to the Authority Having Jurisdiction to demonstrate equivalency. The Authority Having Jurisdiction shall have the authority to approve or disapprove the system, method or device for the intended purpose.

106.0 Intent.

However, the exercise of this discretionary approval by the Authority Having Jurisdiction shall have no effect beyond the jurisdictional boundaries of said Authority Having Jurisdiction. Any alternate material or method of construction so approved shall not be considered as conforming to the requirements and/or intent of this code for any purpose other than that granted by the Authority Having Jurisdiction.

107.0 Testing.

The Authority Having Jurisdiction shall have authority to require tests, as proof of equivalency.

107.1 Tests shall be made in accordance with approved testing standards by an approved testing agency at the expense of the applicant. In the absence of such standards, the Authority Having Jurisdiction shall have the authority to specify the test procedure.

107.2 The Authority Having Jurisdiction shall have the authority to require tests to be made or repeated if, at any time, there is reason to believe that any material or device no longer conforms to the requirements on which its approval was based.

Part II – Organization and Enforcement.

108.0 Powers and Duties of the Authority Having Jurisdiction.

108.1 General. The Authority Having Jurisdiction is hereby authorized and directed to enforce the provisions of this code. For such purposes, the Authority Having Jurisdiction shall have the powers of a law enforcement officer.

The Authority Having Jurisdiction shall have the power to render interpretations of this code and to adopt and enforce rules and regulations supplemental to this code as may be deemed necessary in order to clarify the application of the provisions of this code. Such interpretations, rules, and regulations shall be in accordance with the intent and purpose of this code.

108.2 Deputies. In accordance with the prescribed procedures and with the approval of the appointing authority, the Authority Having Jurisdiction shall be permitted to appoint such number of technical officers and inspectors and other employees as shall be authorized from time to time. The Authority Having Jurisdiction shall be permitted to deputize such inspectors or employees as may be necessary to carry out the functions of the code enforcement agency.

108.3 Right of Entry. When it is necessary to make an inspection to enforce the provisions of this code, or when the Authority Having Jurisdiction has reasonable cause to believe that there exists in a building or upon a premises a condition that is contrary to or in violation of this code that makes the building or premises unsafe, dangerous, or hazardous, the Authority Having Jurisdiction shall be permitted to enter the building or premises at reasonable times to inspect or to perform the duties imposed by this code, provided that if such building or premises be occupied that credentials were presented to the occupant and entry requested. If such building or premises be unoccupied, the Authority Having Jurisdiction shall first make a reasonable effort to locate the owner or other person having charge or control of the building or premises and request entry. If entry is refused, the Authority Having Jurisdiction shall have recourse to the remedies provided by law to secure entry.

108.4 Stop Orders. When any work is being done contrary to the provisions of this code, the Authority Having Jurisdiction shall have the authority to order the work stopped by notice in writing served on any persons engaged in doing or causing such work to be done, and such persons shall forthwith stop work until authorized by the Authority Having Jurisdiction to proceed with the work.

108.5 Authority to Disconnect Utilities in Emergencies. The Authority Having Jurisdiction or authorized representative shall be permitted to disconnect fuel gas utility service or energy supplies to a building, structure, premises, or equipment regulated by this code in case of emergency where necessary to eliminate an immediate hazard to life or property. The Authority Having Jurisdiction shall, whenever possible, notify the serving utility, the owner, and the occupant of the building, structure, or premises of the decision to disconnect prior to taking such action, and shall notify such serving utility, owner, and occupant of the building, structure, or premises in writing of such disconnection immediately thereafter.

108.6 Authority to Condemn Equipment. When the Authority Having Jurisdiction ascertains that any equipment, or portion thereof, regulated by this code has become hazardous to life, health, or property, it shall order in writing that the equipment either be removed or restored to a safe or sanitary condition, as appropriate. The written notice shall contain a fixed time limit for compliance with such order. Persons shall not use or maintain defective equipment after receiving a notice.

When equipment or an installation is to be disconnected, written notice of the disconnection and causes therefore shall be given within twenty-four (24) hours to the serving utility, owner, and occupant of the building, structure, or premises. When any equipment is maintained in violation of this code, and in violation of a notice issued pursuant to the provisions of this section, the Authority Having Jurisdiction shall institute an appropriate action to prevent, restrain, correct, or abate the violation.

108.7 Connection After Order to Disconnect. Persons shall not make connections from an energy, fuel, or power supply nor supply energy or fuel to any equipment regulated by this code that has been disconnected or ordered to be disconnected by the Authority Having Jurisdiction until it has authorized the reconnection and use of such equipment.

108.8 Liability. The Authority Having Jurisdiction charged with the enforcement of this code acting in good faith and without malice in the discharge of the duties required by this code or other pertinent law or ordinance shall not thereby be rendered personally liable for damages that may accrue to persons or property as a result of an act or by reason of an act or omission in the discharge of such duties. A suit brought against the Authority Having Jurisdiction or employee because of such act or omission performed by the Authority Having Jurisdiction or employee in the enforcement of any provision of such codes or other pertinent laws or ordinances implemented through the enforcement of this code or enforced by the code enforcement agency shall be defended by this jurisdiction until final termination of such proceedings, and any judgment resulting therefrom shall be assumed by this jurisdiction.

This code shall not be construed to relieve from or lessen the responsibility of any person owning, operating, or controlling any equipment regulated herein for damages to persons or property caused by defects, nor shall the code enforcement agency or its parent jurisdiction be held as assuming any such liability by reason of the inspections authorized by this code or any permits or certificates issued under this code.

108.9 Cooperation of Other Officials and Officers. The Authority Having Jurisdiction shall be permitted to request, and shall receive, the assistance and cooperation of other officials of this jurisdiction so far as required in the discharge of the duties required by this code or other pertinent law or ordinance.

109.0 Unsafe Equipment.

Equipment regulated by this code that is unsafe or that constitutes a fire or health hazard or is otherwise dangerous to human life is, for the purpose of this section, unsafe. Use of equipment regulated by this code constituting a hazard to safety, health, or public welfare by reason of inadequate maintenance, dilapidation, obsolescence, fire hazard, disaster, damage, or abandonment is, for the purpose of this section, an unsafe use. Unsafe equipment is hereby declared to be a public nuisance and shall be abated by repair, rehabilitation, demolition, or removal in accordance with procedures as may be adopted by this jurisdiction. As an alternative, the Authority Having Jurisdiction or other employee or official of this jurisdiction, as designated by the governing body, shall be permitted to institute other appropriate action to prevent, restrain, correct, or abate the violation.

110.0 Board of Appeals.

110.1 General. In order to hear and decide appeals of orders, decisions, or determinations made by the Authority Having Jurisdiction relative to the application and interpretations of this code, there shall be and is hereby created a Board of Appeals consisting of members who are qualified by experience and training to pass upon matters pertaining to mechanical design, construction, and maintenance and the public health aspects of mechanical systems and who are not employees of the jurisdiction. The Authority Having Jurisdiction shall be an ex-officio member and shall act as secretary to said board but shall have no vote upon any matter before the board. The Board of Appeals shall be appointed by the governing body and shall hold office at its pleasure. The board shall adopt rules of procedure for conducting its business and shall render all decisions and findings in writing to the appellant with a duplicate copy to the Authority Having Jurisdiction.

110.2 Limitations of Authority. The Board of Appeals shall have no authority relative to interpretation of the administrative provisions of this code, nor shall the board be empowered to waive requirements of this code.

111.0 Violations.

It shall be unlawful for a person, firm, or corporation to erect, construct, enlarge, alter, repair, move, improve, remove, convert or demolish, equip, use, or maintain mechanical systems or equipment or cause or permit the same to be done in violation of this code.

Part III – Permits and Inspections

112.0 Permits.

112.1 Permits Required. Except as permitted in Section 112.2, no mechanical system regulated by this code shall be installed, altered, repaired, replaced, or remodeled unless a separate mechanical permit for each separate building or structure has first been obtained.

112.2 Exempt Work. A mechanical permit shall not be required for the following:

112.2.1 A portable heating appliance, portable ventilating equipment, a portable cooling unit, or a portable evaporative cooler.

112.2.2 A closed system of steam, hot, or chilled water piping within heating or cooling equipment regulated by this code.

112.2.3 Replacement of any component part or assembly of an appliance that does not alter its original approval and complies with other applicable requirements of this code.

112.2.4 Refrigerating equipment that is part of the equipment for which a permit has been issued pursuant to the requirements of this code.

112.2.5 A unit refrigerating system. Exemption from the permit requirements of this code shall not be deemed to grant authorization for work to be done in violation of the provisions of this code or other laws or ordinances of this jurisdiction.

113.0 Application for Permit.

113.1 Application. To obtain a permit, the applicant shall first file an application in writing on a form furnished by the code enforcement agency for that purpose. Every such application shall:

113.1.1 Identify and describe the work to be covered by the permit for which application is made.

113.1.2 Describe the land on which the proposed work is to be done by legal description, street address, or similar description that will readily identify and definitely locate the proposed building or work.

113.1.3 Indicate the use or occupancy for which the proposed work is intended.

113.1.4 Be accompanied by plans, diagrams, computations, and specifications and other data as required in Section 113.2.

113.1.5 Be signed by the applicant or an authorized agent of the applicant.

113.1.6 Give such other data and information as may be required by the Authority Having Jurisdiction.

113.2 Plans and Specifications. Plans, engineering calculations, diagrams, and other data shall be submitted in one (1) or more sets with each application for a permit. When such plans are not prepared by an architect or engineer, the Authority Having Jurisdiction shall be permitted to require any applicant submitting such plans or other data to demonstrate that state law does not require that the plans be prepared by an architect or engineer. The Authority Having Jurisdiction shall also be permitted to require plans, computations, and specifications to be prepared and designed by an engineer or architect licensed by the state to practice as such even if not required by state law.

Exception: The Authority Having Jurisdiction shall be permitted to waive the submission of plans, calculations, or other data if it is found that the nature of the work applied for is such that review of plans is not necessary to obtain compliance with this code.

113.3 Information on Plans and Specifications. Plans and specifications shall be drawn to scale upon substantial paper or cloth and shall be of sufficient clarity to indicate the location, nature, and extent of the work proposed and show in detail that it will conform to the provisions of this code and relevant laws, ordinances, rules, and regulations.

Plans for buildings exceeding two (2) stories in height of other than Group R, Division 3 and Group U Occupancies shall indicate how required structural and fire-resistive integrity will be maintained where a penetration will be made for electrical, mechanical, plumbing, and communication conduits, pipes, and similar systems.

The installer shall submit plans showing the proposed installation, indicating the location of the heater and such accessories as may be required to ensure the proper and safe performance of its function.

114.0 Permit Issuance.

114.1 Issuance. If the Authority Having Jurisdiction finds that the work described in an application for a permit and the plans, specifications, and other data filed are in accordance with the requirements of this code and other pertinent laws and ordinances and that the fees specified in Section 115.0 have been paid, the Authority Having Jurisdiction shall issue a permit to the applicant.

When issuing a permit where plans are required, the Authority Having Jurisdiction shall endorse in writing or stamp the plans and specifications "APPROVED." Such approved plans and specifications shall not be changed, modified, or altered without approval from the Authority Having Jurisdiction, and all work regulated by this code shall be done in accordance with the approved plans.

The Authority Having Jurisdiction has the authority to issue a permit for the construction of a part of a mechanical system before the entire plans and specifications for the whole system have been submitted or approved, provided adequate information and detailed statements have been filed complying with all pertinent requirements of this code.

The holder of a partial permit shall be permitted to proceed without assurance that the permit for the entire building, structure, or mechanical system will be granted.

114.2 Retention of Plans. One (1) set of approved plans, specifications, and computations shall be retained by the Authority Having Jurisdiction until final approval of the work is covered therein. One (1) set of approved plans and specifications shall be returned to the applicant, and shall be kept on the site of the building or work at all times during which the work authorized thereby is in progress.

114.3 Validity of Permit. The issuance of a permit or approval of plans, specifications, and computations shall not be construed to be a permit for, or an approval of, any violation of the provisions of this code or of other ordinances of the jurisdiction. Permits presuming to give authority to violate or cancel the provisions of this code or of other ordinances of the jurisdiction shall not be valid.

The issuance of a permit based upon plans, specifications, computations, and other data shall not prevent the Authority Having Jurisdiction from thereafter requiring the correction of errors in said plans, specifications, and other data or from preventing building operations being carried on thereunder when in violation of this code or of other ordinances of this jurisdiction.

114.4 Expiration. Every permit issued under the provisions of this code shall expire by limitation and become null and void if the work authorized by such permit is not commenced within one-hundred and eighty (180) days from the date of such permit, or if the work authorized by such permit is suspended or abandoned at any time after the work is commenced for a period of one-hundred and eighty (180) days. Before such work can be recommenced, a new permit shall be first obtained, and the fee therefore shall be one-half (1/2) the amount required for a new permit for such work, provided no changes have been made or will be made in the original plans and specifications for such work and provided further that such suspension or abandonment has not exceeded one (1) year. No permit shall be extended more than once. To renew action on a permit after expiration, the permittee shall pay a new full permit fee.

A permittee holding an unexpired permit shall be permitted to apply for an extension of the time within which work may be commenced under that permit when the permittee is unable to commence work within the time required by this section for good and satisfactory reasons. The Authority Having Jurisdiction shall have the authority to extend the time for action by the permittee for a period not exceeding one-hundred and eighty (180) days upon written request by the permittee showing that circumstances beyond the control of said permittee have prevented action from being taken.

114.5 Suspension or Revocation. The Authority Having Jurisdiction shall have the authority to suspend or revoke a permit issued under the provisions of this code whenever the permit is issued in error or on the basis of incorrect information supplied or in violation of other ordinances or regulations of the jurisdiction.

115.0 Fees.

115.1 General. Fees shall be assessed in accordance with the provisions of this section and as set forth in the fee schedule, Table 1-1. The fees are to be determined and adopted by this jurisdiction.

115.2 Permit Fees. The fee for each permit shall be as set forth in Table 1-1.

115.3 Plan Review Fees. When plans or other data are required to be submitted by Section 113.2, a plan review fee shall be paid at the time of submitting plans and specifications for review. The plan review fees for mechanical work shall be determined and adopted by this jurisdiction.

The plan review fees specified in this subsection are separate fees from the permit fees specified in Section 115.2 and are in addition to the permit fees.

When plans are incomplete or changed so as to require additional plan review, an additional plan review fee shall be charged at the rate shown in Table 1-1.

115.4 Expiration of Plan Review. Applications for which no permit is issued within one-hundred and eighty (180) days following the date of application shall expire by limitation, and plans and other data submitted for review shall be permitted to be returned to the applicant or destroyed by the Authority Having Jurisdiction. The Authority Having Jurisdiction shall be permitted to extend the time for action by the applicant for a period not exceeding one-hundred and eighty (180) days upon request by the applicant showing that circumstances beyond the control of the applicant have prevented action from being taken. No application shall be extended more than once. In order to renew action on an application after expiration, the applicant shall resubmit plans and pay a new plan review fee.

115.5 Investigation Fees – Work Without a Permit.

115.5.1 Investigation. Whenever any work for which a permit is required by this code has been commenced without first obtaining said permit, a special investigation shall be made before a permit shall be permitted to be issued for such work.

115.5.2 Fee. An investigation fee, in addition to the permit fee, shall be collected whether or not a permit is then or subsequently issued. The investigation fee shall be equal to the amount of the permit fee that would be required by this code if a permit were to be issued. The payment of such investigation fee shall not exempt any person from compliance with other provisions of this code or from any penalty prescribed by law.

115.6 Fee Refunds. The Authority Having Jurisdiction shall be permitted to authorize the refunding of any fee paid hereunder that has been erroneously paid or collected.

115.6.1 The Authority Having Jurisdiction shall be permitted to authorize a refund not exceeding a percentage, as determined by this jurisdiction, when no work has been done under a permit issued in accordance with this code.

115.6.2 The Authority Having Jurisdiction shall be permitted to authorize refunding not exceeding a percentage, as determined by this jurisdiction, when an application for a permit for which a plan review fee has been paid is withdrawn or canceled before any plan review effort has been expended.

115.6.3 The Authority Having Jurisdiction shall not authorize refunding of any fee paid except upon written application filed by the original permittee not exceeding one-hundred and eighty (180) days after the date of fee payment.

116.0 Inspections.

116.1 General. Mechanical systems for which a permit is required by this code shall be subject to inspection and such mechanical systems shall remain accessible and exposed for inspection purposes until approved by the Authority Having Jurisdiction.

It shall be the duty of the permit applicant to cause the mechanical systems to remain accessible and exposed for inspection purposes. The Authority Having Jurisdiction shall not be liable for expense entailed in the removal or replacement of any material required to permit inspection. When the installation of a mechanical system is complete, an additional and final inspection shall be made. Mechanical systems regulated by this code shall not be connected to the energy fuel supply lines until authorized by the Authority Having Jurisdiction.

Approval, as a result of an inspection, shall not be construed to be an approval of a violation of the provisions of this code or of other ordinances of the jurisdiction. Inspections presuming to give authority to violate or cancel the provisions of this code or of other ordinances of the jurisdiction shall not be valid.

116.2 Operation of Mechanical Equipment. The requirements of this section shall not be considered to prohibit the operation of mechanical systems installed to replace existing equipment or fixtures serving an occupied portion of the building in the event a request for inspection of such equipment or fixture has been filed with the Authority Having Jurisdiction not exceeding forty-eight (48) hours after such replacement work is completed and before any portion of such mechanical system is concealed by any permanent portion of the building.

116.3 Testing of Equipment. Refrigeration equipment regulated by this code shall be tested and approved as required by Section 1124.0 of this code.

Steam and hot-water boilers and piping shall be tested and approved as required by Sections 1022, 1201.2.8, and 1207.0 of this code.

Where applicable (see Section 103.0), fuel gas piping shall be tested and approved as required by Section 1304.0 of this code.

116.4 Inspection Requests. It shall be the duty of the person doing the work authorized by a permit to notify the Authority Having Jurisdiction that such work is ready for inspection. The request for inspection must be filed not less than one (1) working day before such inspection is desired.

It shall be the duty of the person requesting inspections required by this code to provide access to and means for inspection of such work.

116.5 Other Inspections. In addition to the called inspections required by this code, the Authority Having Jurisdiction shall be permitted to make or require other inspections of mechanical work to ascertain compliance with the provisions of this code and other laws that are enforced by the code enforcement agency.

116.6 Reinspections. The Authority Having Jurisdiction shall be permitted to assess a reinspection fee for each inspection or reinspection when such portion of work for which inspection is requested is not complete or when required corrections have not been made.

This provision shall not be interpreted as requiring reinspection fees the first time a job is rejected for failure to comply with the requirements of this code, but as controlling the practice of calling for inspections before the job is ready for inspection or reinspection.

To obtain reinspection, the applicant shall file an application therefore in writing upon a form furnished for that purpose and pay the reinspection fee in accordance with Table 1-1 or as set forth in the fee schedule adopted by the jurisdiction.

In instances when reinspection fees have been assessed, no additional inspection of the work will be performed until the required fees have been paid.

117.0 Connection Approval.

117.1 Energy Connections. No person shall make connections from a source of energy or fuel to any mechanical system or equipment regulated by this code and for which a permit is required until approved by the Authority Having Jurisdiction.

117.2 Temporary Connections. The Authority Having Jurisdiction shall be permitted to authorize temporary connection of the mechanical equipment to the source of energy or fuel for the purpose of testing the equipment or for use under a temporary certificate of occupancy.

TABLE 1-1
Mechanical Permit Fees

Permit Issuance

1. For the issuance of each permit...______*
2. For issuing each supplemental permit for which the original permit has not expired or been canceled or finalized ....______*

Unit Fee Schedule

Note: The following do not include permit-issuing fees.

1. **Furnaces**

   For the installation or relocation of each forced-air or gravity-type furnace or burner, including ducts and vents attached to such appliance, up to and including 100,000 Btu/h (29.3 kW)........................______*

   For the installation or relocation of each forced-air or gravity-type furnace or burner, including ducts and vents attached to such appliance, over 100,000 Btu/h (29.3 kW)........................______*

   For the installation or relocation of each floor furnace, including vent........................______*

   For the installation or relocation of each suspended heater, recessed wall heater, or floor-mounted unit heater........................______*

2. **Appliance Vents**

   For the installation, relocation, or replacement of each appliance vent installed and not included in an appliance permit........................______*

3. **Repairs or Additions**

   For the repair of, alteration of, or addition to each heating appliance, refrigeration unit, cooling unit, absorption unit, or each heating, cooling, absorption, or evaporative cooling system, including installation of controls regulated by this code........................______*

4. **Boilers, Compressors, and Absorption Systems**

   For the installation or relocation of each boiler or compressor, up to and including three (3) horsepower, or each absorption system up to and including 100,000 Btu/h (29.3 kW)........................______*

   For the installation or relocation of each boiler or compressor exceeding three (3) horsepower (2.24 kW), up to and including 15 horsepower (11.19 kW), or each absorption system exceeding 100,000 Btu/h (29.3 kW) and including 500,000 Btu/h (146.48 kW)........................______*

   For the installation or relocation of each boiler or compressor exceeding 15 horsepower (11.19 kW), up to and including thirty (30) horsepower (22.37 kW), or each absorption system exceeding 500,000 Btu/h (146.48 kW), up to and including 1,000,000 Btu/h (29.3 kW)........................______*

   For the installation or relocation of each boiler or compressor exceeding thirty (30) horsepower (22.37 kW), up to and including fifty (50) horsepower (37.3 kW), or for each absorption system exceeding 1,000,000 Btu/h (292.95 kW), up to and including 1,750,000 Btu/h (512.66 kW)........................______*

   For the installation or relocation of each boiler or compressor exceeding fifty (50) horsepower (37.3 kW), or each absorption system exceeding 1,750,000 Btu/h (512.66 kW)........................______*

5. **Air Handlers**

   For each air-handling unit up to and including 10,000 cfm (4.72 m^3/S), including ducts attached thereto..______*

   Note: This fee shall not apply to an air-handling unit that is a portion of a factory-assembled appliance, cooling unit, evaporative cooler, or absorption unit for which a permit is required elsewhere in this code.

6. **Evaporative Coolers**

   For each air-handling unit exceeding 10,000 cfm (4.72 m^3/S) ....______*

   For each evaporative cooler other than portable type........................______*

TABLE 1-1 (continued)
Mechanical Permit Fees

7. **Ventilation and Exhaust**
   For each ventilation fan connected to a single duct .... _____*
   For each ventilation system that is not a portion of any heating or air-conditioning system authorized by a permit .... _____*
   For the installation of each hood that is served by mechanical exhaust, including the ducts for such hood .... _____*
8. **Incinerators**
   For the installation or relocation of each domestic-type incinerator .... _____*
   For the installation or relocation of each commercial or industrial-type incinerator .... _____*
9. **Miscellaneous**
   For each appliance or piece of equipment regulated by this code, but not classed in other appliance categories, or for which no other fee is listed in this table .... _____*
10. When Chapter 13 or Appendix B is applicable (see Section 103.0), permit fees for fuel-gas piping shall be as follows:
   For each gas piping system of one (1) to five (5) outlets .... _____*
   For each additional gas piping system, per outlet .... _____*
11. Process Piping:
   For each hazardous process piping system (HPP) of one (1) to four (4) outlets .... _____*
   For each HPP piping system of five (5) or more outlets, per outlet .... _____*
   For each nonhazardous process piping system (NPP) of one (1) to four (4) outlets .... _____*
   For each NPP piping system of five (5) or more outlets, per outlet .... _____*

Other Inspections and Fees

1. Inspections outside of normal business hours, per hour (minimum charge – two (2) hours) .... _____*
2. Reinspection fees assessed under provisions of Section 116.6, per inspection .... _____*
3. Inspections for which no fee is specifically indicated, per hour (minimum charge – one-half (1/2) hour) . _____*
4. Additional plan review required by changes, additions, or revisions to plans or to plans for which an initial review has been completed, per hour (minimum charge – one-half (1/2) hour) .... _____*

* Jurisdiction will indicate their fees here.

CHAPTER 2

DEFINITIONS

201.0 General.

For the purpose of this code, certain abbreviations, terms, phrases, words, and their derivatives shall be construed as specified in this chapter. Words used in the singular include the plural, and the plural the singular. Words used in the masculine gender include the feminine, and the feminine the masculine.

202.0 Accepted Meanings.

Except as defined in this chapter or elsewhere in this code, the interpretation of words used in this code shall be in accordance with the meanings defined in the applicable building, electric, fire, or plumbing code and *Webster's Third New International Dictionary of the English Language, Unabridged*, copyright 1993.

203.0 –A–

ABSORPTION UNIT – An absorption refrigeration system that has been factory-assembled and tested prior to its installation.

ACCESSIBLE – Having access to but which first shall be permitted to require the removal of an access panel, door, or similar obstruction covering the item described.

ACCESSIBLE, READILY – Capable of being reached safely and quickly for operation, repair, or inspection without requiring those to whom ready access is requisite to climb over or remove obstacles, or to resort to the use of portable access equipment.

AIR, COMBUSTION – See **Combustion Air**.

AIR, CONDITIONED – Air that has been treated to achieve a desired level of temperature, humidity, or cleanliness.

AIR, EXHAUST – Air being removed from any space or piece of equipment and conveyed directly to the atmosphere by means of openings or ducts.

AIR-HANDLING UNIT – A blower or fan used for the purpose of distributing supply air to a room, space, or area.

AIR, MAKEUP – Air that is provided to replace air being exhausted.

AIR-MOVING SYSTEM – A system designed to provide heating, cooling, or ventilation in which one or more air-handling units are used to supply air to a common space or are drawing air from a common plenum or space.

AIR, OUTSIDE – Air from outside the building intentionally conveyed by openings or ducts to rooms or to conditioning equipment.

AIR, RETURN – Air from the conditioned area that is returned to the conditioning equipment for reconditioning.

AIR, SUPPLY – Air being conveyed to a conditioned area through ducts or plenums from a heat exchanger of a heating, cooling, absorption, or evaporative cooling system.

APPLIANCE – A device that utilizes fuel or other forms of energy to produce light, heat, power, refrigeration, or air-conditioning. This definition also shall include a vented decorative appliance.

APPROVED – Acceptable to the Authority Having Jurisdiction.

APPROVED TESTING AGENCY – An organization primarily established for purposes of testing to approved standards and approved by the Authority Having Jurisdiction. [UPC:203.0]

ASSEMBLY BUILDING – A building or a portion of a building used for the gathering together of fifty (50) or more persons for such purposes as deliberation, education, instruction, worship, entertainment, amusement, drinking, dining, or awaiting transportation.

AUTHORITY HAVING JURISDICTION – The organization, office, or individual responsible for enforcing the requirements of a code or standard, or for approving equipment, materials, installations, or procedures. The Authority Having Jurisdiction shall be a federal, state, local, or other regional department or an individual such as a plumbing official, mechanical official, labor department official, health department official, building official, or others having statutory authority. In the absence of a statutory authority, the Authority Having Jurisdiction may be some other responsible party. This definition shall include the Authority Having Jurisdiction's duly authorized representative.

AUTOMATIC – That which provides a function without the necessity of human intervention. [NFPA 96:3.3.6]

AZEOTROPE – A refrigerant blend comprising multiple components of different volatiles that, when used in refrigeration cycles, do not change volumetric composition or saturation temperature as they evaporate or condense at constant pressure.

204.0 –B–

BOILER, HIGH-PRESSURE – A boiler furnishing steam at pressures in excess of fifteen (15) pounds per square inch (103.4 kPa) or hot water at temperatures in excess of 250°F (121°C) or at pressures in excess of one-hundred and sixty (160) pounds per square inch (1,102.4 kPa).

BOILER ROOM – Any room containing a steam or hot water boiler.

BONDING JUMPER – A reliable conductor to ensure the required electrical conductivity between metal parts required to be electrically connected. [NFPA 70:100.1]

BREATHING ZONE – The region within an occupied space between three (3) planes three (3) and seventy-two (72) inches (75 and 1,800 mm) above the floor and more than two (2) feet (600 mm) from the walls or fixed air-conditioning equipment. [ASHRAE 62.1:3]

BREATHING ZONE OUTDOOR AIRFLOW (*Vbz*) – The rate of outdoor airflow required in the breathing zone of an occupiable space.

BREECHING – A metal connector for medium- and high-heat appliances.

BTU/H – The listed maximum capacity of any appliance, absorption unit, or burner expressed in British thermal units input per hour, unless otherwise noted.

BUILDING CODE – The building code that is adopted by this jurisdiction.

BUILDING OFFICIAL – See **Authority Having Jurisdiction.**

205.0 –C–

CAS NUMBER – The Chemical Abstract System registry number.

CENTRAL HEATING PLANT or HEATING PLANT – Environmental heating equipment installed in a manner to supply heat by means of ducts or pipes to areas other than the room or space in which the equipment is located.

CERTIFIED – A formally stated recognition and approval of an acceptable level of competency, acceptable to the Authority Having Jurisdiction. [NFPA 96:3.3.9]

CHIMNEY – A vertical shaft enclosing one or more flues for conveying flue gases to the outside atmosphere.

Factory-Built Chimney – A listed chimney.

Masonry Chimney – A chimney of solid masonry units, bricks, stones, listed masonry units, or reinforced concrete, lined with suitable flue liners.

Metal Chimney – A chimney constructed of metal with a minimum thickness not less than one-hundred and twenty-seven one-hundredths (0.127) inch (3.23 mm) (No. 10 manufacturer's standard gauge) steel sheet.

CHIMNEY CLASSIFICATIONS:

Chimney, Residential Appliance-Type – A factory-built or masonry chimney suitable for removing products of combustion from residential-type appliances producing combustion gases not in excess of 1,000°F (538°C), measured at the appliance flue outlet. Factory-built Type HT chimneys have high-temperature thermal shock resistance.

Chimney, Low-Heat Appliance-Type – A factory-built, masonry, or metal chimney suitable for removing the products of combustion from fuel-burning low-heat appliances producing combustion gases not in excess of 1,000°F (538°C) under normal operating conditions, but capable of producing combustion gases of 1,400°F (760°C) during intermittent forced firing for periods up to one hour. All temperatures are measured at the appliance flue outlet.

Chimney, Medium-Heat Appliance-Type – A factory-built, masonry, or metal chimney suitable for removing the products of combustion from fuel-burning medium-heat appliances producing combustion gases not in excess of 2,000°F (1,093°C), measured at the appliance flue outlet.

Chimney, High-Heat Appliance-Type – A factory-built, masonry, or metal chimney suitable for removing the products of combustion from fuel-burning high-heat appliances producing combustion gases in excess of 2,000°F (1,093°C), measured at the appliance flue outlet.

CHIMNEY CONNECTOR – The pipe that connects a fuel-burning appliance to a chimney.

CLOSED COMBUSTION SOLID-FUEL-BURNING APPLIANCE – A heat-producing appliance that employs a combustion chamber that has no openings other than the flue collar, fuel-charging door, and adjustable openings provided to control the amount of combustion air that enters the combustion chamber.

CLOSET – See **Confined Space.**

CLOTHES DRYER – An appliance used to dry wet laundry by means of heat derived from the combustion of fuel gases or electric heating elements.

CLOTHES DRYER, TYPE 1 – Primarily used in family living environment. May or may not be coin operated for public use. [NFPA 54:3.3.18.1]

CLOTHES DRYER, TYPE 2 – Used in business with direct intercourse of the function with the public. May or may not be operated by public or hired attendant. May or may not be coin operated. [NFPA 54:3.3.18.2]

CODE – A standard that is an extensive compilation of provisions covering broad subject matter or that is suitable for adoption into law independently of other codes and standards.

COMBUSTION AIR – The total amount of air provided to the space that contains fuel-burning equipment. Includes air for fuel combustion, draft hood dilution, and ventilation of the equipment enclosure.

COMPRESSOR, POSITIVE DISPLACEMENT – A compressor in which increase in pressure is attained by changing the internal volume of the compression chamber.

COMPRESSOR, REFRIGERANT – A machine, with or without accessories, for compressing a refrigerant vapor.

CONDENSER – The part of the system designed to liquefy refrigerant vapor by removal of heat.

CONDENSING APPLIANCE – An appliance that condenses part of the water vapor generated by the burning of hydrogen in fuels.

CONDENSING UNIT – A mechanical refrigeration system, consisting of one or more power-driven compressors, condensers, liquid receivers if provided, and the regularly furnished accessories that have been factory assembled and tested prior to its installation.

CONDITIONED SPACE – An area, room, or space normally occupied and being heated or cooled for human habitation by any equipment.

CONFINED SPACE – A room or space having a volume less than fifty (50) cubic feet (1.42 m^3) per 1,000 Btu/h (293 W) of the aggregate input rating of all fuel-burning appliances installed in that space.

COOLING – Air cooling to provide a room or space temperature of 68°F (20°C) or above.

COOLING SYSTEM – All of that equipment, including associated refrigeration, intended or installed for the purpose of cooling air by mechanical means and discharging such air into any room or space. This definition shall not include any evaporative cooler.

COOLING UNIT – A self-contained refrigeration system that has been factory assembled, tested, and installed with or without conditioned air and ducts, without connecting any refrigerant-containing parts. This definition shall not include a portable cooling unit or an absorption unit.

CRAWL SPACE – In a building, an area accessible by crawling, having a clearance less than human height, for access to plumbing or wiring, storage, etc.

CSST – An acronym for corrugated stainless steel tubing.

206.0 –D–

DAMPERS – shall be defined as follows:

> **Fire Damper** – An automatic-closing metal assembly consisting of one or more louvers, blades, slats, or vanes that closes upon detection of heat so as to restrict the passage of flame and is listed to the applicable recognized standards.
>
> **Smoke Damper** – A damper arranged to seal off airflow automatically through a part of an air duct system so as to restrict the passage of smoke and is listed to the applicable recognized standard.
>
> **Volume Damper** – Any device that, when installed, will restrict, retard, or direct the flow of air in any duct, or the products of combustion in any heat-producing equipment, its vent connector, vent, or chimney.

DESIGN FLOOD ELEVATION – The elevation of the "design flood," including wave height, relative to the datum specified on the community's legally designated flood hazard map.

DIRECT GAS-FIRED MAKEUP AIR HEATER – A heater in which all the products of combustion generated by the gas-burning device are released into the outside airstream being heated.

DIRECT VENT APPLIANCE – An appliance that is constructed and installed so that all air for combustion is derived from the outside atmosphere and all flue gases are discharged to the outside atmosphere.

DISTRICT HEATING PLANT – A power boiler plant designed to distribute hot water or steam to users located off the premises.

DRAFT HOOD – A device built into an appliance or made a part of the vent connector from an appliance, that is designed to:

(1) Assure the ready escape of the flue gases in the event of no draft, back draft, or stoppage beyond the draft hood.

(2) Prevent a back draft from entering the appliance.

(3) Neutralize the effect of stack action of the chimney or gas vent upon the operation of the appliance.

DUCT – Any tube or conduit for transmission of air, fumes, vapors, or dusts. This definition shall not include:

(1) A vent, vent connector, or chimney connector.

(2) Any tube or conduit wherein the pressure of the air exceeds one (1) pound per square inch.

(3) The air passages of listed self-contained systems.

DUCT FURNACE – A warm-air furnace normally installed in an air-distribution duct to supply warm air for heating. This definition shall apply only to a warm-air heating appliance that depends for air circulation on a blower not furnished as part of the furnace.

DUCT SYSTEM – Includes all ducts, duct fittings, plenums, and fans assembled to form a continuous passageway for the distribution of air.

DWELLING – A building or portion thereof that contains not more than two dwelling units.

DWELLING UNIT – A building or portion thereof that contains living facilities, including provisions for sleeping, eating, cooking, and sanitation, as required by this code, for not more than one family.

207.0 –E–

ELECTRIC HEATING APPLIANCE – A device that produces heat energy to create a warm environment by the application of electric power to resistance elements, refrigerant compressors, or dissimilar material junctions.

ELECTRICAL CODE – The National Electrical Code promulgated by the National Fire Protection Association, as adopted by this jurisdiction.

EQUIPMENT – A general term including materials, fittings, devices, appliances, and apparatus used as part of or in connection with installations regulated by this code.

EVAPORATIVE COOLER – A device used for reducing the sensible heat of air for cooling by the process of evaporation of water into an airstream.

EVAPORATIVE COOLING SYSTEM – All of that equipment intended or installed for the purpose of environmental cooling by an evaporative cooler from which the conditioned air is distributed through ducts or plenums to the conditioned area.

EVAPORATOR – That part of a refrigeration system in which liquid refrigerant is vaporized to produce refrigeration.

EXCESS FLOW VALVE (EFV) – A valve designed to activate when the fuel gas passing through it exceeds a prescribed flow rate.

208.0 –F–

FIRE CODE – The fire code adopted by this jurisdiction.

FIREPLACE STOVE – A chimney-connected, solid-fuel-burning stove (appliance) having part of its fire chamber open to the room.

FIRE-RESISTIVE CONSTRUCTION – Construction complying with the requirements of the Building Code for the time period specified.

FLOOD HAZARD AREA – The greater of the following two areas:

(1) The area within a floodplain subject to a 1 percent or greater chance of flooding in any given year.

(2) The area designated as a flood hazard area on a community's flood hazard map, or otherwise legally designated.

FLOOD HAZARD AREA SUBJECT TO HIGH VELOCITY WAVE ACTION – Area within the flood hazard area that is subject to high-velocity wave action, and shown on a Flood Insurance Rate Map or other flood hazard map as Zone V, VO, VE or V1-30.

FLOOR FURNACE – A completely self-contained furnace suspended from the floor of the space being heated, taking air for combustion from outside such space and with means for observing flames and lighting the appliance from such space.

FORCED-AIR-TYPE CENTRAL FURNACE – A central furnace equipped with a fan or blower that provides the primary means for circulation of air.

Horizontal-Type Central Furnace – A furnace designed for low-headroom installations with airflow through the appliance in a horizontal path.

Upflow-Type Central Furnace – A furnace designed with airflow essentially in a vertical path, discharging air at or near the top of the furnace.

Downflow-Type Central Furnace – A furnace designed with airflow essentially in a vertical path, discharging air at or near the bottom of the furnace.

Enclosed Furnace – A specific heating or heating and ventilating furnace incorporating an integral total enclosure and using only outside air for combustion.

FRACTIONATION – A change in composition of a blend by preferential evaporation of the more volatile component or condensation of the less volatile component.

FUSIBLE PLUG – A device arranged to relieve pressure by operation of a fusible member at a predetermined temperature.

209.0 –G–

GALVANIZED STEEL – Any steel conforming to the requirements of UMC Standard No. 2-2.

GENERATOR – A device equipped with a means of heating used in an absorption system to drive refrigerant out of solution.

GRAVITY HEATING SYSTEM – Any heating system consisting of a gravity-type warm-air furnace, together with all air ducts or pipes and accessory apparatus installed in connection therewith.

GRAVITY-TYPE WARM-AIR FURNACE – A warm-air furnace depending primarily on circulation of air through the furnace by gravity. This definition also shall include any furnace approved with a booster-type fan that does not materially restrict free circulation of air through the furnace when the fan is not in operation.

GROUNDING ELECTRODE – A conducting object through which a direct connection to earth is established. [NFPA 70:100.1]

210.0 –H–

HAZARDOUS LOCATION – Any area or space where combustible dust, ignitable fibers or flammable, volatile liquids, gases, vapors or mixtures are or may be present in the air in quantities sufficient to produce explosive or ignitable mixtures.

HEAT (ENERGY) RECOVERY VENTILATOR – A device intended to remove air from buildings, replace it with outside air, and in the process transfer heat from the warmer to the colder airstreams.

HEATING DEGREE DAY – A unit, based upon temperature difference and time, used in estimating fuel consumption and specifying nominal annual heating load of a building. For any one day when the mean temperature is less than 65°F (18°C), there exist as many degree days as there are Fahrenheit degrees difference in temperature between mean temperature for the day and 65°F (18°C).

HEATING EQUIPMENT – Includes all warm-air furnaces, warm-air heaters, combustion products vents, heating air-distribution ducts and fans, and all steam and hot water piping, together with all control devices and accessories installed as part of, or in connection with, any environmental heating system or appliance regulated by this code.

HEATING SYSTEM – A warm air heating plant consisting of a heat exchanger enclosed in a casing, from which the heated air is distributed through ducts to various rooms and areas. A heating system includes the outside air, return air and supply air system, and all accessory apparatus and equipment installed in connection therewith.

HEAT PUMP – A refrigeration system that extracts heat from one substance and transfers it to another portion of the same substance or to a second substance at a higher temperature for a beneficial purpose.

HIGH SIDE – The parts of a refrigeration system subjected to approximately condenser pressure.

HOOD – An air-intake device connected to a mechanical exhaust system for collecting vapors, fumes, smoke, dust, steam, heat or odors from, at, or near the equipment, place, or area where generated, produced, or released.

211.0 –I–

IDLH (Immediately Dangerous to Life and Health) – A concentration of airborne contaminant's, normally expressed in parts per million (ppm) or milligrams per cubic meter (mg/m^3), that represents the maximum level from which one could escape within thirty (30) minutes without any escape-impairing symptoms or irreversible health effects. This level is established by the National Institute of Occupational Safety and Health (NIOSH).

INDUSTRIAL HEATING EQUIPMENT – Includes appliances, devices, or equipment used, or intended to be used, in an industrial, manufacturing, or commercial occupancy for applying heat to any material being processed, but shall not include water heaters, boilers, or portable equipment used by artisans in pursuit of a trade.

INSANITARY LOCATION – An area, space, or room where the air is unfit or undesirable for circulation to occupied parts of a building.

212.0 –J–

JOINT, BRAZED – Any joint obtained by joining of metal parts with alloys that melt at temperatures higher than 840°F (449°C) but lower than the melting temperature of the parts being joined.

JOINT, COMPRESSION – Any multipiece joint with cup-shaped threaded nuts that, when tightened, compress tapered sleeves so that they form a tight joint on the periphery of the tubing they connect.

JOINT, FLANGED – One made by bolting together a pair of flanged ends.

JOINT, FLARED – Any metal-to-metal compression joint in which a conical spread is made on the end of a tube that is compressed by a flare nut against a mating flare.

JOINT, MECHANICAL – General form for gastight joints obtained by the joining of metal parts through a positive holding mechanical construction (such as flanged joint, screwed joint, or flared joint).

JOINT, SOLDERED – A joint made by the joining of metal parts with metallic mixtures or alloys that melt at a temperature up to and including 840°F (449°C).

213.0 –K–

No definitions.

214.0 –L–

LABELED – Equipment or materials bearing a label of a listing agency (accredited conformity assessment body). See **Listed (Third Party Certified).** [UPC:214.0]

LEL (Lower Explosive Limit) – See **LFL.**

LFL (Lower Flammable Limit or Lower Limit of Flammability) – The minimum concentration of a combustible substance that is capable of propagating a flame through homogeneous mixture of the combustible and a gaseous oxidizer under the specified condition of test. The LFL is sometimes referred to as LEL (Lower Explosive Limit). For the purposes of this definition, LFL and LEL are identical.

LINE CONTACT INSTALLATION – An installation in which a furnace is installed so that building joists, studs, or framing are contacted by the furnace jacket upon the lines formed by the intersection of the jacket sides with the top surface.

LIQUEFIED PETROLEUM GAS or LPG (LP Gas) – Means and includes a material composed pre-dominantly of any of the following hydrocarbons or mixtures of them: propane, propylene, butane's (normal butane or isobutane), and butylenes. When reference is made to liquefied petroleum gas in this code, it shall refer to liquefied petroleum gases in either the liquid or gaseous state.

LISTED (THIRD PARTY CERTIFIED) – Equipment or materials included in a list published by a listing agency (accredited conformity assessment body) that maintains periodic inspection on current production of listed equipment or materials and whose listing states either that the equipment or material complies with approved standards or has been tested and found suitable for use in a specified manner. [UPC:214.0]

LISTING AGENCY – An agency accredited by an independent and authoritative conformity assessment body to operate a material and product listing and labeling (certification) system and that is accepted by the Authority Having Jurisdiction, which is in the business of listing or labeling. The system includes initial and ongoing product testing, a periodic inspection on current production of listed (certified) products, and that makes available a published report of such listing in which specific information is included that the material or product conforms to applicable standards and found safe for use in a specific manner. [UPC:214.0]

LOW-PRESSURE HOT-WATER-HEATING BOILER – A boiler furnishing hot water at pressures not exceeding 160 pounds per square inch (1,102.4 kPa) and at temperatures not exceeding 250°F (121°C).

LOW-PRESSURE STEAM-HEATING BOILER – A boiler furnishing steam at pressures not exceeding fifteen (15) pounds per square inch (103.4 kPa).

LOW SIDE – Refers to the parts of a refrigeration system subjected to approximate evaporator pressure.

215.0 –M–

MACHINERY – The refrigeration equipment forming a part of the refrigeration system, including, but not limited to, the following: compressors, condensers, liquid receivers, evaporators, and connecting piping.

MANUFACTURER – The company or organization that evidences its responsibility by affixing its name, trademark, or trade name to equipment or devices.

MANUFACTURER'S INSTALLATION INSTRUCTIONS – Printed instructions included with equipment or devices for the purpose of providing information regarding safe and proper installation and use whether or not as part of the conditions of listing.

216.0 –N–

NONCOMBUSTIBLE – As applied to building construction material, means a material that in the form in which it is used is either one of the following:

(1) A material that, in the form in which it is used and under the conditions anticipated, will not ignite, burn, support combustion, or release flammable vapors when subjected to fire or heat. Materials that are reported as passing ASTM E 136, *Standard Test Method for Behavior of Materials* in a vertical Tube Furnace at 750°, are considered noncombustible material. [NFPA 220:3.3.4]

(2) Material having a structural base of noncombustible material as defined in 1 above, with a surfacing material not over one-eighth (1/8) inch (3.2 mm) thick that has a flame-spread index not higher than fifty (50).

Noncombustible does not apply to surface finish materials. Material required to be noncombustible for reduced clearances to flues, heating appliances, or other sources of high temperature shall refer to material conforming to 1 above. No material shall be classed as noncombustible that is subject to increase in combustibility or flame-spread index beyond the limits herein established, through the effects of age, moisture, or other atmospheric condition.

217.0 –O–

OCCUPANCY – The purpose for which a building or part thereof is used or intended to be used.

OCCUPANCY CLASSIFICATION – Classifications are defined in the Building Code.

OCCUPIED SPACE – An enclosed space intended for human activities excluding those spaces intended primarily for other purposes such as storage rooms and equipment rooms that are only occupied occasionally and for short periods of time.

218.0 –P–

PEL (Permissible Exposure Limit) – The maximum permitted eight-hour time-weighted average concentration of an airborne contaminant. The maximum permitted time-weighted average exposures to be utilized are those published in 29 CFR 1910.1000.

PEX – An acronym for cross-linked polyethylene.

PIPING – The pipe or tube mains for interconnecting the various parts of a system. Piping includes pipe, tube, flanges, bolting, gaskets, valves, fittings the pressure-containing parts of other components such as expansion joints, strainers, and devices that serve such purposes as mixing, separating, snubbing, distributing, metering, or controlling flow pipe-supporting fixtures and structural attachments.

PLENUM – An air compartment or chamber including uninhabited crawl spaces areas above a ceiling or below a floor, including air spaces below raised floors of computer/data processing centers or attic spaces, to that one or more ducts are connected and that forms part of either the supply-air, return-air or exhaust-air system, other than the occupied space being conditioned.

PLUMBING CODE – The Uniform Plumbing Code promulgated by the International Association of Plumbing and Mechanical Officials, as adopted by this jurisdiction.

PORTABLE COOLING UNIT – A self-contained refrigerating system, not over 3 horsepower rating, that has been factory assembled and tested, installed without supply-air ducts and without connecting any refrigerant-containing parts. This definition shall not include an absorption unit.

PORTABLE EVAPORATIVE COOLER – An evaporative cooler that discharges the conditioned air directly into the conditioned area without the use of ducts and can be readily transported from place to place without dismantling any portion thereof.

PORTABLE HEATING APPLIANCE – A heating appliance designed for environmental heating that may have a self-contained fuel supply and is not secured or attached to a building by any means other than by a factory-installed power supply cord.

PORTABLE VENTILATING EQUIPMENT – Ventilating equipment that can be readily transported from place to place without dismantling a portion thereof and that is not connected to a duct.

POSITIVE DISPLACEMENT COMPRESSOR – A compressor in which increase in pressure is attained by changing the internal volume of the compression chamber.

POWER BOILER PLANT – One or more power steam boilers or power hot water boilers and connecting piping and vessels within the same premises.

PP – An acronym for polypropylene.

PRESSURE, DESIGN – The maximum working pressure for which a specific part of a refrigeration system is designed.

PRESSURE, FIELD TEST – A test performed in the field to prove system tightness.

PRESSURE-IMPOSING ELEMENT – A device or portion of the equipment used for the purpose of increasing the pressure of the refrigerant vapor.

PRESSURE-LIMITING DEVICE – A pressure-responsive mechanism designed to automatically stop the operation of the pressure-imposing element at a predetermined pressure.

PRESSURE-RELIEF DEVICE – A pressure-actuated valve or rupture member or fusible plug designed to automatically relieve excessive pressure.

PRESSURE TEST – The minimum gauge pressure to which a specific system component is subjected under test condition.

PRESSURE VESSEL – REFRIGERANT – A refrigerant-containing receptacle that is a portion of a refrigeration system, but shall not include evaporators, headers, or piping of certain limited size and capacity.

219.0 –Q–

QUALIFIED – A competent and capable person or company that has met the requirements and training for a given field acceptable to the Authority Having Jurisdiction. [NFPA 96:3.3.38]

220.0 –R–

RADIANT HEATER – A heater designed to transfer heat primarily by direct radiation.

RECEIVER, LIQUID – A vessel permanently connected to a refrigeration system by inlet and outlet pipes for storage of liquid.

REFRIGERANT SAFETY CLASSIFICATIONS – Made up of a letter (A or B), that indicates the toxicity class, followed by a number (1, 2, or 3), that indicates the flammability class. Refrigerant blends are similarly classified, based on the compositions at their worst cases of fractionation, as separately determined for toxicity and flammability. In some cases, the worst case of fractionation is the original formulation.

TOXICITY – Classes A and B signify refrigerants with "lower toxicity" and "higher toxicity," respectively, based on prescribed measures of chronic (long-term, repeated exposures) toxicity.

FLAMMABILITY – Class 1 indicates refrigerants that do not show flame propagation in air when tested by prescribed methods at specified conditions. Classes 2 and 3 signify refrigerants with "lower flammability" and "higher flammability," respectively. The distinction depends on both the Lower Flammability Limit (LFL) and heat of combustion.

REFRIGERATION MACHINERY ROOM – A space designed to safely house compressors and pressure vessels.

REFRIGERATION ROOM or SPACE – A room or space in which an evaporator or brine coil is located for the purpose of reducing or controlling the temperature within the room or space to below 68°F (20°C).

REFRIGERATION SYSTEM, ABSORPTION – A heat-operated closed-refrigeration cycle in which a secondary fluid, the absorbent, absorbs a primary fluid, the refrigerant, that has been vaporized in the evaporator.

REFRIGERATION SYSTEM, MECHANICAL – A combination of interconnected refrigerant-containing parts constituting one closed refrigerant circuit in which a refrigerant is circulated for the purpose of extracting heat and in which a compressor is used for compressing the refrigerant vapor.

REFRIGERATION SYSTEM, SELF-CONTAINED – A complete factory-assembled and tested system that is shipped in one or more sections and has no refrigerant-containing parts that are joined in the field by other than companion or block valves.

RESIDENTIAL BUILDING – A building or portion thereof designed or used for human habitation.

RISER HEAT PIPE – A duct that extends at an angle of more than forty-five (45) degrees (0.79 rad) from the horizontal. This definition shall not include any boot connection.

ROOM HEATER – A freestanding, nonrecessed, environmental heating appliance installed in the space being heated and not connected to ducts.

ROOM HEATER, UNVENTED – An unvented, self-contained, freestanding, nonrecessed, fuel-gas burning appliance for furnishing warm air by gravity or fan circulation to the space in which installed, directly from the heater without duct connection. [NFPA 54:3.3.55.6]

ROOM LARGE IN COMPARISON WITH SIZE OF EQUIPMENT – A room having a volume of at least twelve (12) times the total volume of a furnace or air-conditioning appliance and at least sixteen (16) times the total volume of a boiler. Total volume of the appliance is determined from exterior dimensions and includes fan compartments and burner vestibules when used. When the actual ceiling height of a room is greater than eight (8) feet (2,438 mm), the volume of the room is figured on the basis of a ceiling height of eight (8) feet (2,438 mm).

RUPTURE MEMBER – A pressure-relief device that operates by the rupture of a diaphragm within the device on a rise to a predetermined pressure.

221.0 –S–

SEAM, WELDED – See **Joint, Welded.**

SELF-CONTAINED – Having all essential working parts, except energy and control connections, so contained in a case or framework that they do not depend on appliances or fastenings outside of the machine.

SHAFT – An interior space enclosed by walls or construction extending through one or more stories or basements that connects openings in successive floors, or floors and roof, to accommodate elevators, dumbwaiters, mechanical equipment, or similar devices to transmit light or ventilation air.

SHAFT ENCLOSURE – The walls or construction forming the boundaries of a shaft.

SHALL – Indicates a mandatory requirement.

SHOULD – Indicates a recommendation or that which is advised but not required.

SMOKE DETECTOR – An approved device that senses visible or invisible particles of combustion.

STANDARD – A document, the main text of which contains only mandatory provisions using the word "shall" to indicate requirements and that is in a form generally suitable for mandatory reference by another standard or code or for adoption into law. Nonmandatory provisions shall be located in an appendix, footnote, or fine-print note and are not to be considered a part of the requirements of a standard.

STATIONARY FUEL CELL POWER PLANT – A self-contained package or factory-matched packages that constitute an automatically operated assembly of integrated systems for generating useful electrical energy and recoverable energy that is permanently connected and fixed in place.

STRENGTH, ULTIMATE – The highest stress level that the component can tolerate without rupture.

SYSTEM OUTDOOR AIRFLOW (*Vot*) – The rate of outdoor airflow required at the ventilation system outdoor air intake.

222.0 –T–

TRAINED – A person who has become proficient in performing a skill reliably and safely through instruction and practice/field experience acceptable to the Authority Having Jurisdiction. [NFPA 96:3.3.48]

223.0 –U–

→ **UNIT HEATER** – A heating appliance designed for nonresidential space heating and equipped with an integral means for circulation of air.

UNUSUALLY TIGHT CONSTRUCTION is construction where:

(1) Walls and ceilings exposed to the outside atmosphere have a continuous water vapor retarder with a rating of one (1) perm or less with any openings gasketed or sealed;

(2) Weatherstripping is on openable windows and doors; and

(3) Caulking or sealants are applied to areas such as joints around window and door frames, between sole plates and floors, between wall-ceiling joints, between wall panels, and at penetrations for plumbing, electrical, and gas lines and at other openings.

224.0 –V–

VALVE, PRESSURE-RELIEF – A pressure-actuated valve held closed by a spring or other means and designed to automatically relieve pressure in excess of its setting; also called a safety valve.

VALVE, STOP – A device in a piping system to shut off the flow of the fluid.

VALVE, THREE-WAY-TYPE STOP – A manually operated valve with one inlet that alternately can stop flow to either of two outlets.

VALVES, COMPANION OR BLOCK – Pairs of mating stop valves valving off sections of refrigeration systems and arranged so that these sections may be joined before opening these valves or separated after closing them.

VENT – A listed factory-made vent pipe and vent fittings for conveying flue gases to the outside atmosphere.

Type B Gas Vent – A factory-made gas vent listed by a nationally recognized testing agency for venting listed or approved appliances equipped to burn only gas.

Type BW Gas Vent – A factory-made gas vent listed by a nationally recognized testing agency for venting listed or approved gas-fired vented wall furnaces.

Type L – A venting system consisting of listed vent piping and fittings for use with oil-burning appliances listed for use with Type L or with listed gas appliances.

VENT CONNECTOR, GAS – That portion of a gas-venting system that connects a listed gas appliance to a gas vent and is installed within the space or area in which the appliance is located.

VENTED APPLIANCE CATEGORIES

Category I – An appliance that operates with a nonpositive vent static pressure and with a vent gas temperature that avoids excessive condensate production in the vent.

Category II – An appliance that operates with a non-positive vent static pressure and with a vent gas temperature that may cause excessive condensate production in the vent.

Category III – An appliance that operates with a positive vent static pressure and with a vent gas temperature that avoids excessive condensate production in the vent.

Category IV – An appliance that operates with a positive vent static pressure and with a vent gas temperature that may cause excessive condensate production in the vent.

VENTED DECORATIVE APPLIANCE – A vented appliance whose only function is providing an aesthetic effect of flames.

VENTED WALL FURNACE – A vented environmental heating appliance designed for incorporation in, or permanent attachment to, a wall, floor, ceiling, or partition and arranged to furnish heated air by gravity or by a fan. This definition shall not include floor furnaces, unit heaters, and room heaters.

VENTILATING CEILING – A suspended ceiling containing many small apertures through which air, at low pressure, is forced downward from an overhead plenum dimensioned by the concealed space between the suspended ceiling and the floor or roof above.

VENTILATION SYSTEM – All of that equipment intended or installed for the purpose of supplying air to, or removing air from, any room or space by mechanical means, other than equipment that is a portion of an environmental heating, cooling, absorption, or evaporative cooling system.

VENTING COLLAR – The outlet opening of an appliance provided for connection of the vent system.

VENTING SYSTEM – The vent or chimney and its connectors, assembled to form a continuous open passageway from an appliance to the outside atmosphere for the purpose of removing products of combustion. This definition also shall include a venting assembly that is an integral part of an appliance.

VENTING SYSTEM—GRAVITY-TYPE – A system that depends entirely on the heat from the fuel being used to provide the energy required to vent an appliance.

VENTING SYSTEM—POWER-TYPE – A system that depends on a mechanical device to provide a positive draft within the venting system.

VOLUME, INTERNAL GROSS – The volume as determined from internal dimensions of the container, with no allowance for the volume of the internal parts.

225.0 –W–

WALL HEATER – See **Vented Wall Furnace.**

WARM AIR FURNACE – An environmental heating appliance designed or arranged to discharge heated air through any duct or ducts. This definition shall not include a unit heater.

WATER HEATER or HOT-WATER-HEATING BOILER – An appliance designed primarily to supply hot water and equipped with automatic controls limiting water temperature to a maximum of 210°F (99°C).

226.0 –X–

No definitions.

227.0 –Y–

No definitions.

228.0 –Z–

ZEOTROPE – A blend comprising multiple components of different volatiles that, when used in refrigeration cycles, change volumetric composition and saturation temperatures as they evaporate or condense at constant pressure.

CHAPTER 3

GENERAL REQUIREMENTS

301.0 Scope.

This chapter covers general requirements for heating, ventilating, air-conditioning, refrigeration, miscellaneous heat-producing, and energy-utilizing equipment. Such equipment shall conform to the requirements of this code.

Equipment shall not be installed or altered in violation of this code, nor shall the fuel input rate to equipment be increased in excess of the approved Btu/h (kW) rating at the altitude where it is being used.

Defective material or parts shall be replaced in such a manner as not to invalidate any approval.

302.0 Approval.

302.1 How Obtained. Each appliance shall be approved by the Authority Having Jurisdiction for safe use or comply with applicable nationally recognized standards as evidenced by the listing and label of an approved agency. A list of accepted standards is included in Chapter 17.

302.2 Complying Design of Equipment. Installers shall furnish satisfactory evidence that the appliance is constructed in conformity with the requirements of this code. The permanently attached label of an approved agency may be accepted as such evidence.

302.3 Oil-Burning Appliances. The tank, piping, and valves for appliances burning oil shall be installed in accordance with the requirements of NFPA 31, *Standard for the Installation of Oil-Burning Equipment*.

303.0 Installation.

303.1 Listed Appliances. Except as otherwise provided in the code, the installation of appliances regulated by this code shall conform to the conditions of listing. The appliance installer shall leave the manufacturer's installation and operating instructions attached to the appliance. Clearances of listed appliances from combustible materials shall be as specified in the listing or on the rating plate.

303.2 Room Large in Comparison to Size of Equipment. Central-heating furnaces not listed for closet or alcove installation shall be installed in a room or space having a volume at least twelve (12) times the total volume of the furnace; central-heating boilers not listed for closet or alcove installation shall be installed in a room or space having a volume sixteen (16) times the volume of the boiler.

Exception: The installation clearances for furnaces and boilers in rooms not large in comparison with the size of the equipment shall be as specified in the appliance listing regardless of whether the enclosure is of combustible or noncombustible materials and shall not be reduced by the protection methods described in Table 3-3 or any other method.

If the ceiling height of the room or space is greater than eight (8) feet (2,438 mm), the volume shall be calculated on the basis of an eight (8) foot (2,438 mm) height.

303.3 Unlisted Appliances. Unlisted appliances shall be installed with the standard clearances from combustible construction specified in Table 3-2. Unlisted appliances may have the standard clearances of Table 3-2 reduced by employing the forms of protection specified in Table 3-3. Forms of protection specified in Table 3-3 may be utilized to reduce clearances to combustible construction for all applicable appliances.

303.4 Anchorage of Appliances. Appliances designed to be fixed in position shall be securely fastened in place. Supports for appliances shall be designed and constructed to sustain vertical and horizontal loads within the stress limitations specified in the Building Code.

303.5 Movement of appliances with casters shall be limited by a restraining device installed in accordance with the connector and appliance manufacturer's installation instructions.

303.6 Identification of Equipment. When more than one heating, cooling, ventilating, or refrigerating system is installed on the roof of a building or within a building, it shall be permanently identified as to the area or space served by the equipment.

303.7 Liquefied Petroleum Gas Facilities. Containers, container valves regulating equipment, and appurtenances for the storage and supply of liquefied petroleum gas shall be installed in accordance with NFPA 58.

303.8 Equipment on Roofs.

303.8.1 General.

303.8.1.1 Enclosures. Equipment on roof shall be designed or enclosed so as to withstand climatic conditions in the areas in which it is installed. Where enclosures are provided, each enclosure shall be of reasonable height, and shall have at least a thirty (30) inch (762 mm) clearance between the entire service access panel(s) of the equipment and the wall of the enclosure.

303.8.1.2 Roof Support. Roofs on which equipment is to be installed shall be capable of supporting the additional load or shall be reinforced to support the additional load.

303.8.1.3 Corrosion Resistance. All access locks, screws, and bolts shall be of corrosion-resistant material.

303.8.1.4 Roof Drainage. Equipment shall be installed on a well-drained surface of the roof.

303.8.1.5 Guards or Rails. At least six (6) ft. (1,829 mm) between any part of the equipment and the edge of a roof or similar hazard, or rigidly fixed rail, guards, parapets, or other building structures at least forty-two (42) inches (1,067 mm) in height shall be provided on the exposed side.

303.8.1.6 Electrical Power. All equipment requiring an external source of electrical power for its operation shall be provided with:

(1) Readily accessible electrical disconnecting means within sight of the equipment.

(2) A 120 VAC grounding-type receptacle outlet on the roof adjacent to the equipment. The receptacle outlet shall be on the supply side of the disconnect switch.

304.0 Service and Access to Equipment and Appliances.

Equipment and appliances shall be accessible for inspection, service, repair, and replacement without removing permanent construction. Sufficient clearance shall be maintained to permit cleaning of heating surfaces; the replacement of filters, blowers, motors, burners, controls, and vent connections; the lubrication of moving parts where necessary; the adjustment and cleaning of burners and pilots; and the proper functioning of explosion vents, if provided. [NFPA 54-09:9.2.1]

Unless otherwise specified, not less than thirty (30) inches (762 mm) in depth, width, and height of working space shall be provided.

Exception: Unit heaters and room heaters may be installed with an eighteen (18) inch (457 mm) minimum depth working space. A platform shall not be required for unit heaters or room heaters. The operating instructions shall be attached to the appliance where they can be read easily.

304.1 Access to Equipment on Roofs.

304.1.1 General. Equipment and appliances located on roofs or other elevated locations shall be accessible. [NFPA 54:9.4.3.1]

304.1.2 Access from Inside. Buildings of more than fifteen (15) feet (4,572 mm) in height shall have an inside means of access to the roof, unless other means acceptable to the roof, unless other means acceptable to the Authority Having Jurisdiction are used. [NFPA 54:9.4.3.2]

304.1.2.1 Door or Scuttle. The inside means of access shall be a permanent or foldaway inside stairway or ladder, terminating in an enclosure, scuttle, or trap door. Such scuttles or trap doors shall be at least twenty-two (22) inches x twenty-four (24) inches (560 mm x 610 mm) in size, shall open easily and safely under all conditions, especially snow, and shall be constructed so as to permit access on the inside.

304.1.2.2 Guards or Rails. At least six (6) feet (1,829 mm) of clearance shall be available between the access opening and the edge of the roof or similar hazard, or rigidly fixed rails or guards a minimum of forty-two (42) inches (1,067 mm) in height shall be provided on the exposed side. Where parapets or other building structures are utilized in lieu of guards or rails, they shall be a minimum of forty-two (42) inches (1,067 mm) in height. [NFPA 54:9.4.3.3]

304.1.3 Permanent Lighting. Permanent lighting shall be provided at the roof access. The switch for such lighting shall be located inside the building near the access means leading to the roof. [NFPA 54:9.4.3.4]

304.1.4 Standing Water. Where water stands on the roof at the equipment, in the passageways to the equipment, where the roof is of a design having a water seal, a suitable platform, walkway, or both shall be provided above the waterline. Such platform(s) or walkway(s) shall be located adjacent to the equipment and control panels so that the equipment can be safely serviced where water stands on the roof.

305.0 Automatic Control Devices.

Heating appliances shall be equipped with a listed device or devices that will shut off the fuel supply to the main burner or burners in the event of pilot or ignition failure. In addition, liquefied petroleum gas-air-burning heating appliances shall be equipped with a listed automatic device or devices that will shut off the flow of gas to the pilot in the event of ignition failure.

Exception: The listed shutoff devices shall not be required on range or cooking tops, log lighters, lights, or other open-burner manually operated appliances, or listed appliances not requiring such devices and specific industrial appliances as approved by the Authority Having Jurisdiction.

Heating appliances whose manual fuel controls are not readily accessible from the main portion of the building being heated shall be equipped with remote controls.

Forced-air and gravity-type warm-air furnaces shall be equipped with a listed air outlet temperature limit control that cannot be set for temperatures higher than 250°F (121°C). Such controls shall be located in the bonnet or plenum, within two (2) feet (610 mm) of the discharge side of the heating element of gravity furnaces or in accordance with the conditions of listing.

Electric duct heaters shall be equipped with an approved automatic reset air outlet temperature limit control that will limit the outlet air temperature to not more than 200°F (93°C). The electric elements of the heater shall be equipped with fusible links or a manual reset temperature limit control that will prevent outlet air temperature in excess of 250°F (121°C).

306.0 Labeling.

306.1 Fuel-Burning Appliances. Fuel-burning heating appliances shall bear a permanent and legible factory-applied nameplate on which shall appear:

306.1.1 The manufacturer's name.

306.1.2 The approved fuel input rating of the appliance, expressed in Btu/h (W).

306.1.3 The model and serial number.

306.1.4 Instructions for the lighting, operation, and shutdown of the appliance.

306.1.5 The type of fuel approved for use with the appliance.

306.1.6 The symbol of an approved agency certifying compliance of the equipment with recognized standards.

306.1.7 Required clearances from combustible surfaces on which or adjacent to which it may be mounted.

306.2 Electric Heating Appliances. Electric heating appliances shall bear a permanent and legible factory-applied nameplate on which shall appear:

306.2.1 The name or trademark of the manufacturer.

306.2.2 The catalog (model) number or equivalent.

306.2.3 The electrical rating in volts, amperes (or watts), and; for other than single phase, the number of phases.

306.2.4 The output rating in Btu/h or kW.

306.2.5 The electrical rating in volts, amperes, or watts of each field-replaceable electrical component.

306.2.6 The symbol of an approved agency certifying compliance of equipment with recognized standards.

306.2.7 Required clearances from combustible surfaces on which or adjacent to which it may be mounted.

Also, the appliance shall be accompanied by clear and complete installation instructions, including required clearances from combustibles other than mounting or adjacent surfaces, and temperature rating of field-installed wiring connections of over 140°F (60°C).

306.3 Heat Pump and Electric Cooling Appliances. Heat pumps and electric cooling appliances shall bear a permanent and legible factory-applied nameplate on which shall appear:

306.3.1 The name or trademark of the manufacturer.

306.3.2 The catalog model nomenclature.

306.3.3 The amount and type of refrigerant.

306.3.4 The factory test pressures or pressures applied.

306.3.5 The electrical rating in volts, amperes, and, for other than single phase, the number of phases.

306.3.6 The output rating in Btu/h or W.

306.3.7 The electrical rating in volts, amperes, or watts of each field replaceable electrical component.

306.3.8 The symbol of an approved agency certifying compliance of the equipment with recognized standards.

306.3.9 Required clearances from combustible surfaces on which or adjacent to which it is permitted to be mounted.

Also, the appliance shall be accompanied by clear and complete installation instructions, including required clearances from combustible other than mounting or adjacent surfaces, and temperature rating of field-installed wiring connections over 140°F (60°C).

307.0 Location.

307.1 Protection Against Damage. Appliances installed in garages, warehouses, or other areas subject to mechanical damage shall be guarded against such damage by being installed behind protective barriers or by being elevated or located out of the normal path of vehicles.

Heating and cooling equipment located in a garage and that generates a glow, spark, or flame capable of igniting flammable vapors shall be installed with the pilots and burners or heating elements and switches at least eighteen (18) inches (457 mm) above the floor level.

Where such appliances installed within a garage are enclosed in a separate, approved compartment having access only from outside of the garage, such appliances may be installed at floor level, provided the required combustion air is taken from and discharged to the exterior of the garage.

Heating equipment located in rooms where cellulose nitrate plastic is stored or processed shall comply with the Fire Code.

307.2 Protection Against Flood Damage. For buildings located in flood hazard areas, heating, ventilating, air-conditioning, refrigeration, miscellaneous heat-producing, and energy-utilizing equipment and appliances shall be elevated at or above the design flood elevation.

Exception: Equipment and appliances are permitted to be located below the design flood elevation provided that they are designed and installed to prevent water from entering or accumulating within the components and to resist hydrostatic and hydrodynamic loads and stresses, including the effects of buoyancy, during the occurrence of flooding to the design flood elevation in compliance with the flood-resistant construction requirements of the Building Code.

307.2.1 Walls Below Buildings in Flood Hazard Areas Subject to High Velocity Wave Action. In flood hazard areas subject to high velocity wave action, equipment and appliances, including piping, shall not be mounted on or penetrate walls intended to break away under flood loads.

307.2.2 Air Exhaust and Intake Openings. Outside air exhaust openings and air intake openings shall be located at or above the design flood elevation.

308.0 Electrical Connections.

Equipment regulated by this code requiring electrical connections of more than fifty (50) volts shall have a positive means of disconnect adjacent to and in sight from the equipment served. A 120 volt receptacle shall be located within twenty-five (25) feet (7,620 mm) of the equipment for service and maintenance purposes. The receptacle need not be located on the same level as the equipment. Low-voltage wiring of fifty (50) volts or less within a structure shall be installed in a manner to prevent physical damage.

309.0 Condensate Wastes and Control.

309.1 Condensate Disposal. Condensate from air washers, air-cooling coils, fuel-burning condensing appliances, and the overflow from evaporative coolers and similar water-supplied equipment or similar air-conditioning equipment shall be collected and discharged to an approved plumbing fixture or disposal area. If discharged into the drainage system, equipment shall drain by means of an indirect waste pipe. The waste pipe shall have a slope of not less than 1/8 inch per foot (10.5 mm/m) or 1 percent slope and shall be of approved corrosion-resistant material not smaller than the outlet size as required in either Section 309.3 or 309.4 for air-cooling coils or condensing fuel-burning appliances, respectively. Condensate or wastewater shall not drain over a public way.

309.2 Condensate Control. When a cooling coil or cooling unit is located in an attic or furred space where damage may result from condensate overflow, an additional watertight pan of corrosion-resistant metal shall be installed beneath the cooling coil or unit top to catch the overflow condensate due to a clogged primary condensate drain, or one pan with a standing overflow and a separate secondary drain may be provided in lieu of the secondary drain pan. The additional pan or the standing overflow shall be provided with a drain pipe, minimum 3/4 inch (19.1 mm) nominal pipe size, discharging at a point that can be readily observed.

This requirement is in addition to the requirements in Sections 309.3 and 309.4.

309.3 Condensate Waste Sizing. Condensate waste pipes from air-cooling coils shall be sized in accordance with equipment capacity as follows:

TABLE 3-1
Minimum Condensate Pipe Size

Equipment Capacity in		Minimum Condensate Pipe Diameter	
Tons of Refrigeration	(kW)	Inches	(mm)
Up to 20	(Up to 70.34)	3/4	(20)
21 – 40	(73.85 – 140.67)	1	(25)
41 – 90	(144.19 – 316.6)	1-1/4	(32)
91 – 125	(320.03 – 439.6)	1-1/2	(40)
126 – 250	(443.12 – 879.2)	2	(50)

The size of condensate waste pipes may be for one unit or a combination of units, or as recommended by the manufacturer. The capacity of waste pipes assumes a one-eights (1/8) inch per foot (10.5 mm/m) or 1 percent slope, with the pipe running three-quarters (3/4) full at the following conditions:

Outside Air – 20%		Room Air – 80%	
DB	WB	DB	WB
90°F	73°F	75°F	62.5°F
(32°C)	(23°C)	(24°C)	(17°C)

Condensate drain sizing for other slopes or other conditions shall be approved by the Authority Having Jurisdiction.

309.4 Fuel-Burning Appliance Condensate Drains. Condensate drain lines from individual fuel-burning condensing appliances shall be sized according to the manufacturer's recommendations. Condensate drain lines serving more than one appliance shall be approved by the Authority Having Jurisdiction prior to installation.

309.5 Plastic Fittings. Female PVC screwed fittings shall be used with plastic male fittings and plastic male threads only.

310.0 Personnel Protection.

A suitable and substantial metal guard shall be provided around exposed flywheels, fans, pulleys, belts, and moving machinery that are portions of a heating, ventilating, or refrigerating system.

311.0 Heating or Cooling Air System.

311.1 Source. A heating or cooling air system shall be provided with return air, outside air, or both. A heating or cooling air system regulated by this code and designed to replace required ventilation shall be arranged to discharge into a conditioned space not less than the amount of outside air specified in Chapter 4.

311.2 Air Filters. Air filters shall be installed in a heating, cooling or makeup air system. Such filters shall comply with the standard, Air Filter Units, Test Performance of, that is referenced in Chapter 17, as Class I or II filters.

Exception: Systems serving single guest rooms or dwelling units shall not require a listed filter.

311.3 Prohibited Source. Outside or return air for a heating or cooling air system shall not be taken from the following locations:

(1) Closer than ten (10) feet (3,048 mm) from an appliance vent outlet, a vent opening of a plumbing drainage system, or the discharge outlet of an exhaust fan, unless the outlet is three (3) feet (914 mm) above the outside-air inlet.

(2) Where it is less than ten (10) feet above the surface of any abutting public way, driveway, sidewalk, street, alley or driveway.

(3) A hazardous or insanitary location or a refrigeration machinery room as defined in this code.

(4) From an area, the volume of which is less than 25 percent of the entire volume served by such system, unless there is a permanent opening to an area the volume of which is equal to 25 percent of the entire volume served.

Exception: Such openings when used for a heating or cooling air system in a dwelling unit shall be permitted to be reduced to not less than 50 percent of the required area, provided the balance of the required return air is taken from a room or hall having not less than three (3) doors leading to other rooms served by the furnace.

(5) A closet, bathroom, toilet room or kitchen.

(6) From rooms or spaces containing a fuel-burning appliance therein. Where such room or space serves as source of return air.

Exceptions:

(1) This shall not apply to fireplaces, fireplace appliance, residential cooking appliance, direct vent appliance, enclosed furnaces and domestic-type clothes dryers installed within the room or space.

(2) This shall not apply to a gravity-type or listed vented wall heating or cooling air system.

(3) This shall not apply to a blower-type heating or cooling air system complying with the following requirements:

(a) Where the return air is taken from a room or space having a volume exceeding one (1) cubic foot for each ten (10) Btu/h (2.93 W) fuel input rating of all fuel-burning appliances therein.

(b) Not less than 75 percent of the supply air is discharged back into the same room or space.

(c) Return-air inlets shall not be located within ten (10) feet (3,048 mm) from any appliance firebox or draft diverter in the same enclosed room or confined space.

311.4 Return-Air Limitations. Return air from one dwelling unit shall not discharge into another dwelling unit through the heating or cooling air system.

311.5 Outside Air Inlet Protection. Required outside-air inlets shall be covered with a screen having not less than one-fourth (1/4) inch (6.4 mm) openings.

TIA **Exception:** An outside-air inlet serving a nonres-
TIA idential portion of a building shall be permitted
TIA to be covered with a screen having openings
TIA larger than 1/4 inch (6.4 mm), but in no case
TIA exceeding one (1) inch (25.4 mm).
TIA
TIA

312.0 Water Supply.

Water supplies and backflow protection shall be as required by the Uniform Plumbing Code.

TABLE 3-2
Standard Installation Clearances in Inches for Unlisted Heat-Producing Appliances
[NFPA 54: Table 10.3.3]
See Section 303.0.

RESIDENTIAL-TYPE APPLIANCES	FUEL	APPLIANCE: ABOVE TOP OF CASING OR APPLIANCE	FROM TOP AND SIDES OF WARM-AIR BONNET OR PLENUM	FROM FRONT[1]	FROM BACK	FROM SIDES	
BOILERS AND WATER HEATERS[11]							
Steam Boilers—15 psi (103.4 kPa)	Automatic Oil or Comb. Gas-Oil	6		24	6	6	
Water Boilers—250°F (121°C)	Automatic Gas	6		18	6	6	
Water Heaters—200°F (93°C) All Water Walled or Jacketed	Solid	6		48	6	6	
FURNACES—CENTRAL; OR HEATERS[11]— ELECTRIC CENTRAL WARM-AIR FURNACES Gravity, Upflow, Downflow, Horizontal and Duct Warm-Air—250°F (121°C) max.	Automatic Oil or Comb. Gas-Oil	6[2]	6[2]	24	6	6	
	Automatic Gas	6[2]	6[2]	18	6	6	
	Solid	18[3]	18[3]	48	18	18	
	Electric	6[2]	6[2]	18	6	6	
FURNACES—FLOOR For Mounting in Combustible Floors	Automatic Oil or Comb. Gas-Oil	36		12	12	12	
	Automatic Gas	36		12	12	12	
HEAT EXCHANGERS Steam—15 psi (103.4 kPa) max. Hot Water—250°F (121°C) max.		1	1	1	1	1	
ROOM HEATERS[4] Circulating Type	Oil or Solid	36		24	12	12	
	Gas	36		24	12	12	
Radiant or Other Type	Oil or Solid	36		36	36	36	
	Gas	36		36	18	18	
	Gas with double metal or ceramic back	36		36	12	18	
Fireplace Stove	Solid	48[5]		54	48[5]	48[5]	
RADIATORS Steam or Hot Water[6]		36		6	6	6	
RANGES—COOKING STOVES						Firing Side	Opp. Side
	Oil	30[7]			9	24	18
	Gas	30[7]			6	6	6
	Solid Clay-Lined Firepot	30[7]			24	24	18
	Solid Unlined Firepot	30[7]			36	36	18
	Electric	30[7]			6	6	
INCINERATORS Domestic Types		36[8]		48	36	36	

COMMERCIAL INDUSTRIAL-TYPE APPLIANCES ANY AND ALL PHYSICAL SIZES EXCEPT AS NOTED[11]	FUEL	APPLIANCE: ABOVE TOP OF CASING OR APPLIANCE	FROM TOP AND SIDES OF WARM-AIR BONNET OR PLENUM	FROM FRONT[1]	FROM BACK[9]	FROM SIDES[9]
BOILERS AND WATER HEATERS 100 cu. ft. (2.832 m³) or less Steam, any pressure	All Fuels	18		48	18	18
50 psi (345 kPa) or less Any size	All Fuels	18		48	18	18
UNIT HEATERS Floor Mounted or Suspended—Any Size	Steam or Hot Water	1			1	1
Suspended—100 cu. ft. (2.832 m³) or less	Oil or Comb. Gas-Oil	6		24	18	18
Suspended—100 cu. ft. (2.832 m³) or less	Gas	6		18	18	18
Suspended—Over 100 cu. ft. (2.832 m³)	All Fuels	18		48	18	18
Floor Mounted—Any Size	All Fuels	18		48	18	18
RANGES—RESTAURANT-TYPE Floor Mounted	All Fuels	48		48	18	18
OTHER LOW-HEAT INDUSTRIAL APPLIANCES Floor Mounted or Suspended	All Fuels	18	18	48	18	18

TABLE 3-2 (continued)

Standard Installation Clearances in Inches for Unlisted Heat-Producing Appliances [NFPA 54: Table10.3.3]

COMMERCIAL INDUSTRIAL-TYPE APPLIANCES ANY AND ALL PHYSICAL SIZES EXCEPT AS NOTED[11]	FUEL	APPLIANCE: ABOVE TOP OF CASING OR APPLIANCE	FROM TOP AND SIDES OF WARM-AIR BONNET OR PLENUM	FROM FRONT[1]	FROM BACK[10]	FROM SIDES[10]
BOILERS AND WATER HEATERS Over 50 psi (345 kPa) Over 100 cu. ft. (2832 m^3)	All Fuels	48		96	36	36
OTHER MEDIUM-HEAT INDUSTRIAL APPLIANCES All Sizes	All Fuels	48	36	96	36	36
INCINERATORS All Sizes		48		96	36	36
INDUSTRIAL-TYPE HIGH-HEAT APPLIANCES HIGH-HEAT INDUSTRIAL APPLIANCES All Sizes	All Fuels	180		360	120	120

Notes:

1 The minimum dimension shall be that necessary for servicing the appliance, including access for cleaning and normal care, tube removal, etc.

2 For a listed oil, combination gas-oil, gas, or electric furnace, this dimension may be two (2) inches (51 mm) if the furnace limit control cannot be set higher than 250°F (121°C), or this dimension may be one (1) inch (25.4 mm) if the limit control cannot be set higher than 200°F (93°C), or the appliance shall be marked to indicate that the outlet air temperature cannot exceed 200°F (93°C).

3 The dimension may be six (6) inches (152 mm) for an automatically stoker-fired forced-warm-air furnace equipped with 250°F (121°C) limit control and with barometric draft control operated by draft intensity and permanently set to limit draft to a maximum intensity of 0.13 inch (3.3 mm) water gauge.

4 Unlisted appliances shall be installed on noncombustible floors and may be installed on protected combustible floors. Heating appliances approved for installation on protected combustible flooring shall be so constructed that flame and hot gases do not come in contact with the appliance base. Protection for combustible floors shall consist of four (4) inch (102 mm) hollow masonry covered with sheet metal at least 0.021 inch (0.53 mm) thick (No. 24 manufacturer's standard gauge). Masonry shall be permanently fastened in place in an approved manner with the ends unsealed and joints matched so as to provide free circulation of air through the masonry. Floor protection shall extend twelve (12) inches (305 mm) at the sides and rear of the appliance, except that at least eighteen (18) inches (457 mm) shall be required on the appliance-opening side or sides measured horizontally from the edges of the opening.

5 The forty-eight (48) inch (1,219 mm) clearance may be reduced to 36 inches (914 mm) when protection equivalent to that provided by (a)–(g) of Table 3-2 is applied to the combustible construction.

6 Steam pipes and hot water heating pipes shall be installed with a clearance of at least one (1) inch (25 mm) to all combustible construction or material, except that at the points where pipes carrying steam at not over fifteen (15) pounds gauge pressure (103.4 kPa) or hot water that emerge from a floor, wall, or ceiling, the clearance at the opening through the finish floorboards or wall-ceiling boards may be reduced to not less than one-half (1/2) inch (12.7 mm). Each such opening shall be covered with a plate of noncombustible material.

Such pipes passing through stock shelving shall be covered with not less than one (1) inch (25.4 mm) of approved insulation.

Wood boxes or casings enclosing uninsulated steam or hot water heating pipes or wooden covers to recesses in walls in which such uninsulated pipes are placed shall be lined with metal or insulating millboard.

Where the temperature of the boiler piping does not exceed 160°F (71°C), the provisions of this table shall not apply.

Coverings or insulation used on steam or hot water pipes shall be of material suitable for the operating temperature of the system. The insulation or jackets shall be of noncombustible materials, or the insulation or jackets and lap-seal adhesives shall be tested as a composite product. Such composite product shall have a flame-spread rating of not more than twenty-five (25) and a smoke-developed rating not to exceed fifty (50) when tested in accordance with UBC Standard No. 42-1.

7 To combustible material or metal cabinets. If the underside of such combustible material or metal cabinet is protected with insulating millboard at least one-quarter (1/4) inch (6.4 mm) thick covered with sheet metal of not less than 0.013 inch (0.33 mm) (No. 28 gauge), the distance may be reduced to twenty-four (24) inches (610 mm).

8 Clearance above charging door shall be at least forty-eight (48) inches (1,219 mm).

9 If the appliance is encased in brick, the eighteen (18) inch (457 mm) clearance above and at the sides and rear may be reduced to twelve (12) inches (305 mm).

10 If the appliance is encased in brick, the clearance above may be reduced to thirty-six (36) inches (914 mm) and at the sides and rear may be reduced to eighteen (18) inches (457 mm).

11 A central heating boiler or furnace shall be installed in accordance with the manufacturer's instructions and shall be installed on a floor of noncombustible construction with noncombustible flooring and surface finish and with no combustible material against the underside thereof, or on fire-resistive slabs or arches having no combustible material against the underside thereof.

Exception No. 1: Appliances listed for installation on a combustible floor.

Exception No. 2: Installation on a floor protected in an approved manner.

1 in. = 25.4 mm

TABLE 3-3
Clearances, in Inches, with Specified Forms of Protection[1, 2]

TYPE OF PROTECTION[4] Applied to the Combustible Material Unless Otherwise Specified and Covering All Surfaces within the Distance Specified as the Required Clearance with No Protection (Thicknesses Are Minimum)	Where the Standard Clearance in Table 3-1 with No Protection is:											
	36 inches			18 inches			12 inches		9 inches	6 inches		
	Above	Sides and Rear	Chimney or Vent Con-nector	Above	Sides and Rear	Chimney or Vent Con-nector	Above	Sides and Rear	Chimney or Vent Con-nector	Above	Sides and Rear	Chimney or Vent Con-nector
(a) 1/4" insulating millboard spaced out 1"[3]	30	18	30	15	9	12	9	6	6	3	2	3
(b) 0.013" (No. 28 manufacturer's standard gauge) steel sheet on 1/4" insulating millboard	24	18	24	12	9	12	9	6	4	3	2	2
(c) 0.013" (No. 28 manufacturer's standard gauge) steel sheet spaced out 1"[3]	18	12	18	9	6	9	6	4	4	2	2	2
(d) 0.013" (No. 28 manufacturer's standard gauge) steel sheet on 1/8" insulating millboard spaced out 1"[3]	18	12	18	9	6	9	6	4	4	2	2	2
(e) 1-1/2" insulating cement covering on heating appliance	18	12	36	9	6	18	6	4	9	2	1	6
(f) 1/4" insulating millboard on 1" mineral fiber batts reinforced with wire mesh or equivalent	18	12	18	6	6	6	4	4	4	2	2	2
(g) 0.027" (No. 22 manufacturer's standard gauge) steel sheet on 1" mineral fiber batts reinforced with wire or equivalent	18	12	12	4	3	3	2	2	2	2	2	2
(h) 1/4" insulating millboard	36	36	36	18	18	18	12	12	9	4	4	4

Notes:

[1] For appliances complying with Sections 303.2 and 303.3

[2] Except for the protection described in (e), all clearances shall be measured from the outer surface of the appliance to the combustible material, disregarding any intervening protection applied to the combustible material.

[3] Spacers shall be of noncombustible material.

[4] Insulating millboard is a factory-made product formed of noncombustible materials, normally fibers, and having a thermal conductivity of 1 Btu-inch per square foot per degree F [1.73 W/(m•K)] or less.

1 in = 25.4 mm

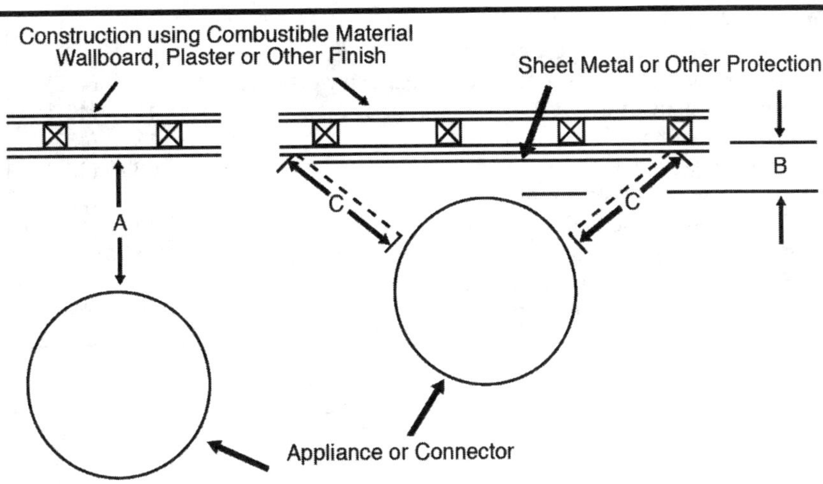

Notes:

1. Dimension shall not be less than the required clearance with no protection set forth in Tables 3-2 and 3-4 and in sections applying to various types of appliances.
2. Dimension shall be not less than the reduced clearance set forth in Table 3-3
3. Dimension shall not be less that the clearance required for dimension A.

FIGURE 3-1 Extent of Protection Required to Reduce Clearances from Appliance, Chimney, or Vent Connectors.

TABLE 3-4
Chimney Connector and Vent Connector Clearances from Combustible Materials

DESCRIPTION OF APPLIANCE	Minimum Clearance,[1] Inches	(mm)
RESIDENTIAL-TYPE APPLIANCES		
Single-Wall Metal Pipe Connectors[2]		
Gas appliances without draft hoods	18	(457)
Electric, gas, and oil incinerators	18	(457)
Oil and solid-fuel appliances	18	(457)
Unlisted gas appliances with draft hoods	9	(229)
Boilers and furnaces equipped with listed gas burners and with draft hoods	9[3]	(229)
Oil appliances listed as suitable for use with Type L venting systems (but only when connected to chimneys)	9	(229)
Listed gas appliances with draft hoods	6	(152)
Type L Venting System Piping Connectors		
Gas appliances without draft hoods	9	(229)
Electric, gas, and oil incinerators	9	(229)
Oil and solid-fuel appliances	9	(229)
Unlisted gas appliances with draft hoods	6	(152)
Boilers and furnaces equipped with listed gas burners and with draft hoods	6	(152)
Oil appliances listed as suitable for use with Type L venting systems[4]		
Listed gas appliances with draft hoods[5]		
Type B Gas Vent Piping Connectors		
Listed gas appliances with draft hoods[5]		
COMMERCIAL-INDUSTRIAL TYPE APPLIANCES		
Low-Heat Appliances		
Single-Wall Metal Pipe Connectors[2]		
Gas, oil, and solid-fuel boilers, furnaces, and water heaters	18	(457)
Ranges, restaurant-type	18	(457)
Oil unit heaters	18	(457)
Unlisted gas unit heaters	18	(457)
Listed gas unit heaters with draft hoods	6	(152)
Other low-heat industrial appliances	18	(457)
Medium-Heat Appliances		
Single-Wall Metal Pipe Connectors[2]		
All gas, oil, and solid-fuel appliances	36	(914)

Notes:

1 These clearances apply except if the listing of an appliance specifies different clearance, in which case the listed clearance takes precedence.

2 The clearances from connectors to combustible materials may be reduced if the combustible material is protected in accordance with Table 3-3.

3 The dimension may be six (6) inches (152 mm), provided the maximum flue temperatures entering the draft hood do not exceed 550°F (288°C).

4 If listed Type L venting system piping is used, the clearance may be in accordance with the venting system listing.

5 If listed Type B or Type L venting system piping is used, the clearance may be in accordance with the venting system listing.

TABLE 3-5
Capacities of Listed Metal Appliance Connectors[1]
For use with gas pressures 8-inch or more water column (1.99-kPa)

Semirigid Connector O.D.[2] Inches (mm)	Flexible Connector Nominal I.D.[3] Inches (mm)	Capacities for Various Lengths in Thousands Btu/h (based on a pressure drop of 0.2 inch (10.2 mm) water column) Natural Gas of 1,100 Btu/cu. ft. (11.4 W/L)[4]							
		1'	1-1/2'	2'	2-1/2'	3'	4'	5'	6'
		All Gas Appliances					Ranges and Clothes Dryers		
3/8 (10)	1/4 (8)	40	33	29	27	25			
1/2 (15)	3/8 (10)	93	76	66	62	58			
5/8 (18)	1/2 (15)	189	155	134	125	116	101	90	80
—	3/4 (20)	404	330	287	266	244			
—	1 (25)	803	661	573	534	500			

Notes:

1 Gas connectors are certified by the testing agency as complete assemblies, including the fittings and valves. Capacities shown are based on the use of fittings and valves supplied with the connector.

2 Outside diameter.

3 Internal diameter.

4 For liquefied petroleum gas, use 1.6 times the natural gas capacities shown.

1 ft. = 0.305 m, 1 Btu/h = 0.29 Watt

TABLE 3-6
Capacities of Listed Metal Appliance Connectors[1]
For use with gas pressures less than 8-inch water column (1.99-kPa)

Semirigid Connector O.D.[2] Inches (mm)	Flexible Connector Nominal I.D.[3] Inches (mm)	Capacities for Various Lengths in Thousands Btu/h (based on a pressure drop of 0.2 inch (10.2 mm) water column) Natural Gas of 1,100 Btu/cu. ft. (11.4 W/L)[4]							
		1'	1-1/2'	2'	2-1/2'	3'	4'	5'	6'
		All Gas Appliances					Ranges and Clothes Dryers		
3/8 (10)	1/4 (8)	28	23	20	19	17			
1/2 (15)	3/8 (10)	66	54	47	44	41			
5/8 (18)	1/2 (15)	134	110	95	88	82	72	63	57
—	3/4 (20)	285	233	202	188	174			
—	1 (25)	567	467	405	378	353			

Notes:

1 Gas connectors are certified by the testing agency as complete assemblies, including the fittings and valves. Capacities shown are based on the use of fittings and valves supplied with the connector.

2 Semirigid connector listings are based on the outside diameter.

3 Flexible connector listings are based on nominal diameter.

4 For liquefied petroleum gas, use 1.6 times the natural gas capacities shown.

1 ft. = 0.305 m, 1 Btu/h = 0.29 Watt

CHAPTER 4

VENTILATION AIR SUPPLY

401.0 General.

This chapter contains requirements for ventilation air supply and exhaust, evaporative cooling systems and makeup-air requirements for direct-gas-fired heaters, industrial air heaters, and miscellaneous heaters.

402.0 Ventilation Air.

402.1 General Requirements. All rooms and occupied spaces listed in Table 4-1 shall be designed to have ventilation (outdoor) air for occupants in accordance with this chapter.

402.1.1 Construction Documents. The outdoor air ventilation rate and air distribution assumptions made in the design of the ventilation system shall be clearly identified on the construction documents.

402.2 Natural Ventilation. Use of natural ventilation systems designed in accordance with this section shall be permitted in lieu of or in conjunction with mechanical ventilation systems. [ASHRAE 62.1:5.1]

Exception: An engineered natural ventilation system when approved by the Authority Having Jurisdiction need not meet the requirements of 402.2.1 and 402.2.2. [ASHRAE 62.1:5.1]

402.2.1 Location and Size of Openings. Naturally ventilated spaces shall be permanently open to and within twenty-five (25) feet (7,620 mm) of operable wall or roof openings to the outdoors, the openable area of which is a minimum of 4 percent of the net occupiable floor area. Where the openings are covered with louvers or otherwise obstructed, openable area shall be based on the free unobstructed area through the opening. Where interior spaces without direct openings to the outdoors are ventilated through adjoining rooms, the opening between rooms shall be permanently unobstructed and have a free area of not less than 8 percent of the area of the interior room nor less than 25 ft^2 (2.3 m^2). [ASHRAE 62.1:5.1.1]

402.2.2 Control and Accessibility. The means to open required operable openings shall be readily accessible to building occupants whenever the space is occupied. [ASHRAE 62.1:5.1.2]

402.3 Mechanical Ventilation. Where natural ventilation is not permitted by this section or the Building Code, mechanical ventilation systems shall be designed, constructed, and installed to provide a method of supply air and exhaust air. The system shall operate so that all rooms and spaces are continuously provided with the required ventilation rate while occupied.

403.0 Ventilation Rates.

The design outdoor air intake flow rate for a ventilation system shall be determined in accordance with Sections 403.1 through 403.6.

403.1 Zone Calculations. Zone parameters shall be determined in accordance with Sections 403.1.1 through 403.1.3. [ASHRAE 62.1:6.2.2]

403.1.1 Breathing Zone Outdoor Airflow. The design outdoor airflow required in the breathing zone of the occupiable space or spaces in a zone, i.e., the *breathing zone outdoor airflow* (V_{bz}), shall be determined in accordance with Equation 4-1. [ASHRAE 62.1:6.2.2.1]

$$V_{bz} = R_p P_z + R_a A_z \qquad \text{(Equation 4-1)}$$

WHERE:

A_z = *zone floor area*: the net occupiable floor area of the zone ft.^{2} (m^2).

P_z = *zone population*: The largest number of people expected to occupy the zone during typical usage. If the number of people expected to occupy the zone fluctuates, P_z shall be permitted to be estimated based on averaging approaches described in Section 403.5.2. If P_z cannot be accurately predicted during design, it shall be estimated based on the zone floor area and the default occupant density listed in Table 4-1.

R_p = outdoor airflow rate required per person as determined from Table 4-1.

R_a = outdoor airflow rate required per unit area as determined from Table 4-1. [ASHRAE 62.1:6.2.2.1]

403.1.2 Zone Air Distribution Effectiveness. The *zone air distribution effectiveness* (E_z) shall be determined using Table 4-2. [ASHRAE 62.1:6.2.2.2]

403.1.3 Zone Outdoor Airflow. The design *zone outdoor airflow* (V_{oz}), i.e., the outdoor airflow that must be provided to the zone by the supply air distribution system, shall be determined in accordance with Equation 4-2. [ASHRAE 62.1:6.2.2.3]

$$V_{oz} = V_{bz}/E_z \qquad \text{(Equation 4-2)}$$

403.2 Single-Zone Systems. When one air handler supplies a mixture of outdoor air and recirculated air to only one zone, the *outdoor air intake flow* (V_{ot}) shall be determined in accordance with Equation 4-3. [ASHRAE 62.1:6.2.3]

$$V_{ot} = V_{oz} \qquad \text{(Equation 4-3)}$$

403.3 100% Outdoor Air Systems. When one air handler supplies only outdoor air to one or more zones, the *outdoor air intake flow* (V_{ot}) shall be determined in accordance with Equation 4-4. [ASHRAE 62.1:6.2.4]

$$V_{ot} = \Sigma_{all\ zones}\ V_{oz} \qquad \text{(Equation 4-4)}$$

403.4 Multiple-Zone Recirculating Systems. When one air handler supplies a mixture of outdoor air and recirculated return air to more than one zone, the *outdoor air intake flow* (V_{ot}) shall be determined in accordance with Sections 403.4.1 through 403.4.4. [ASHRAE 62.1:6.2.5]

403.4.1 Primary Outdoor Air Fraction. When Table 4-3 is used to determine system ventilation efficiency, the *zone primary outdoor air fraction* (Z_p) shall be determined in accordance with Equation 4-5. [ASHRAE 62.1:6.2.5.1]

$$Z_p = V_{oz}/V_{pz} \qquad \text{(Equation 4-5)}$$

V_{pz} is the primary airflow to the zone from the air handler, including outdoor air and recirculated return air. [ASHRAE 62.1:6.2.5.1]

403.4.2 System Ventilation Efficiency. The *system ventilation efficiency* (E_v) shall be determined using Table 4-3 or Appendix A of ASHRAE 62.1. [ASHRAE 62.1:6.2.5.2]

403.4.3 Uncorrected Outdoor Air Intake. The design *uncorrected outdoor air intake* (V_{ou}) shall be determined in accordance with Equation 4-6.

$$V_{ou} = D\ \Sigma_{all\ zones}\ R_pP_z + \Sigma_{all\ zones}\ R_aA_z$$
(Equation 4-6)

The *occupant diversity, D,* shall be permitted to be used to account for variations in occupancy within the zones served by the system.

The *occupancy diversity* is defined as:

$$D = P_s/\Sigma_{all\ zones}\ P_z \qquad \text{(Equation 4-7)}$$

Where the *system population* (P_s) is the total population in the area served by the system. Alternative methods shall be permitted to be used to account for population diversity when calculating V_{ou}, provided that the resulting value is no less than that determined by Equation 4-6. [ASHRAE 62.1:6.2.5.3]

403.4.4 Outdoor Air Intake. The design *outdoor air intake flow* (V_{ot}) shall be determined in accordance with Equation 4-8. [ASHRAE 62.1:6.2.5.4]

$$V_{ot} = V_{ou}/E_v \qquad \text{(Equation 4-8)}$$

403.5 Design for Varying Operating Conditions.

403.5.1 Variable Load Conditions. Ventilation systems shall be designed to be capable of providing the required ventilation rates in the breathing zone whenever the zones served by the system are occupied, including all full and part-load conditions. [ASHRAE 62.1:6.2.6.1]

403.5.2 Short-Term Conditions. If it is known that peak occupancy will be of short duration or the ventilation rate will be varied or interrupted for a short period of time, the design shall be permitted to be based on the average conditions over a time period T determined by Equation 4-9. [ASHRAE 62.1:6.2.6.2]:

$$T = 3v\ /\ V_{bz}\ \text{(4-9) IP} \qquad \text{(Equation 4-9)}$$

WHERE:

T = averaging time period, (min)

v = the volume of the zone for which aver aging is being applied, ft.^{3} (m^3).

V_{bz} = *the breathing zone outdoor airflow* calculated using Equation 4-1 and the design value of the *zone population* P_z, cfm (L/s).

403.6 Dynamic Reset. The system shall be permitted to be designed to vary the design *outdoor air intake flow* (V_{ot}), or the space or zone airflow as operating conditions change.

403.7 Exhaust Ventilation. Exhaust airflow shall be provided in accordance with the requirements in Table 4-4. Exhaust makeup air shall be permitted to be any combination of outdoor air, recirculated air, and transfer air.

404.0 Makeup Air.

Makeup-air requirements for direct gas-fired heaters, industrial air heaters, and miscellaneous heaters are found in Chapters 5 and 9.

405.0 Evaporative Cooling Systems.

Evaporative cooling systems shall comply with this chapter.

Evaporative cooling systems shall be provided with outside air as specified for cooling systems in this code.

Air ducts and fire dampers that are a portion of an evaporative cooling system shall comply with this code.

405.1 Location. Evaporative cooling systems shall be installed so as to minimize the probability of damage from an external source.

405.2 Access, Inspection, and Repair. Evaporative coolers shall be accessible for inspection, service, and replacement without removing permanent construction.

405.3 Installation. An evaporative cooler supported by the building structure shall be installed on a substantial level base and shall be secured directly or indirectly to the building structure by suitable means to prevent displacement of the cooler.

Modifications made to the supporting framework of buildings as a result of the installation shall be in accordance with the requirements of the Building Code. Openings in exterior walls shall be flashed in an approved manner in accordance with the requirements of the Building Code.

An evaporative cooler supported directly by the ground shall be isolated from the ground by a level concrete slab extending not less than three (3) inches (76 mm) above the adjoining ground level.

An evaporative cooler supported on an above-ground platform shall be elevated at least six (6) inches (152 mm) above adjoining ground level.

Example: Determine the outdoor air rate required for a single zone AC unit serving an interior 2,000 ft.^{2} meeting/conference room with a design occupancy of 100 people. The system supplies and returns air from the ceiling.

Per Table 4-2, the zone air distribution effectiveness is 1.0 since the system supplies cooling only from the ceiling. Using the rates from Table 4-1 for a meeting/conference room, the minimum system outdoor air rate is calculated to be:

$$V_{ot} = \frac{R_p P_z + R_a A_z}{E_z}$$

$$= \frac{5 \times 100 + 0.06 \times 2{,}000}{1.0}$$

$$= 620 \text{ cfm}$$

TABLE 4-1
MINIMUM VENTILATION RATES IN BREATHING ZONE[1, 2]
[ASHRAE 62.1: Table 6-1]

	Occupancy Category[4]	People Outdoor Air Rate R_P (cfm/person)	Area Outdoor Air Rate R_A (cfm/ft.^{2})	Default Occupant Density[3] (people/1,000 ft.^{2})
	Correctional Facilities			
	Cell	5	0.12	25
	Day room	5	0.06	30
	Guard stations	5	0.06	15
	Booking/waiting	7.5	0.06	50
	Educational Facilities			
	Day care (through age 4)	10	0.18	25
ASHRAE ASHRAE	Day care sickroom	10	0.18	25
	Classrooms (ages 5-8)	10	0.12	25
	Classrooms (age 9 plus)	10	0.12	35
	Lecture classroom	7.5	0.06	65
	Lecture hall (fixed seats)	7.5	0.06	150
	Art classroom	10	0.18	20
	Science laboratories[E]	10	0.18	25
ASHRAE ASHRAE	University/college laboratories	10	0.18	25
	Wood/metal shop	10	0.18	20
	Computer lab	10	0.12	25
	Media center[A]	10	0.12	25
	Music/theater/dance	10	0.06	35
	Multi-use assembly	7.5	0.06	100
	Food and Beverage Service			
	Restaurant dining rooms	7.5	0.18	70
	Cafeteria/fast food dining	7.5	0.18	100
	Bars, cocktail lounges	7.5	0.18	100
	General			
ASHRAE ASHRAE	Break rooms	5	0.06	25
ASHRAE ASHRAE	Coffee stations	5	0.06	20
	Conference/meeting	5	0.06	50
	Corridors	-	0.06	-
	Storage rooms[B]	-	0.12	-
	Hotels, Motels, Resorts, Dormitories			
	Bedroom/living room	5	0.06	10
	Barracks sleeping areas	5	0.06	20
ASHRAE ASHRAE	Laundry rooms, central	5	0.12	10
ASHRAE ASHRAE	Laundry rooms within dwelling units	5	0.12	10
	Lobbies/pre-function	7.5	0.06	30
	Multipurpose assembly	5	0.06	120
	Office Buildings			
	Office space	5	0.06	5
	Reception areas	5	0.06	30
	Telephone/data entry	5	0.06	60
	Main entry lobbies	5	0.06	10
	Miscellaneous spaces			
	Bank vaults/safe deposit	5	0.06	5
	Computer (not printing)	5	0.06	4
ASHRAE ASHRAE	Electrical equipment rooms	-	0.06	-
ASHRAE ASHRAE	Elevator machine rooms	-	0.12	-
	Pharmacy (prep. area)	5	0.18	10

TABLE 4-1
MINIMUM VENTILATION RATES IN BREATHING ZONE[1, 2] (continued)
[ASHRAE 62.1: Table 6-1]

Photo studios	5	0.12	10
Shipping/receiving[B]	-	0.12	-
Transportation waiting	7.5	0.06	100
Warehouses[B]	-	0.06	-
Public Assembly Spaces			
Auditorium seating area	5	0.06	150
Places of religious worship	5	0.06	120
Courtrooms	5	0.06	70
Legislative chambers	5	0.06	50
Libraries	5	0.12	10
Lobbies	5	0.06	150
Museums (children's)	7.5	0.12	40
Museums/galleries	7.5	0.06	40
Residential			
Dwelling Unit[F, G]	5	0.06	F
Common corridors	-	0.06	-
Retail			
Sales (except as below)	7.5	0.12	15
Mall common areas	7.5	0.06	40
Barber shop	7.5	0.06	25
Beauty and nail salons	20	0.12	25
Pet shops (animal areas)	7.5	0.18	10
Supermarket	7.5	0.06	8
Coin-operated laundries	7.5	0.06	20
Sports and Entertainment			
Sports arena (play area)	-	0.30	-
Gym, stadium (play area)	-	0.30	30
Spectator areas	7.5	0.06	150
Swimming (pool & deck)[C]	-	0.48	-
Disco/dance floors	20	0.06	100
Health club/aerobics room	20	0.06	40
Health club/weight rooms	20	0.06	10
Bowling alley (seating)	10	0.12	40
Gambling casinos	7.5	0.18	120
Game arcades	7.5	0.18	20
Stages, studios[D]	10	0.06	70

Notes:

1 **Smoking:** This table applies to no-smoking areas. Rates for smoking-permitted spaces must be determined using other methods.

2 **Air Density:** Volumetric airflow rates are based on an air density of 1.2 kgda/m³ (0.075 lbda/ft³), which corresponds to dry air at a barometric pressure of 101.3 kPa (1 atm) and an air temperature of 21 °C (70 °F). Rates shall be permitted to be adjusted for actual density but such adjustment is not required for compliance with this chapter.

3 **Default Occupant Density:** The default occupant density shall be used when actual occupant density is not known.

4 **Unlisted Occupancies:** If the occupancy category for a proposed space or zone is not listed, the requirements for the listed occupancy category that is most similar in terms of occupant density, activities, and building construction shall be used.

1 ft. = 0.305 m

ITEM-SPECIFIC NOTES FOR TABLE 4-1

A For high school and college libraries, use values shown for Public Spaces – Library.

B Rate may not be sufficient when stored materials include those having potentially harmful emissions.

C Rate does not allow for humidity control. Additional ventilation or dehumidification may be required to remove moisture.

D Rate does not include special exhaust for stage effects, e.g., dry ice vapors, smoke.

E No class of air has been established for this occupancy category.

F Default occupancy for dwelling units shall be two persons for studio and one-bedroom units, with one additional person for each additional bedroom.

G Air from one residential dwelling shall not be recirculated or transferred to any other space outside of that dwelling.

TABLE 4-2
Zone Air Distribution Effectiveness
[ASHRAE 62.1: Table 6-2]

Air Distribution Configuration	Ez
Ceiling supply of cool air	1.0
Ceiling supply of warm air and floor return	1.0
Ceiling supply of warm air at least 15°F (8°C) above space temperature and ceiling return.	0.8
Ceiling supply of warm air less than 15°F (8°C) above space temperature and ceiling return provided that the 150 fpm (0.8 m/s) supply air jet reaches to within 4.5 ft. (1.4 m) of floor level.	1.0
Floor supply of cool air and ceiling return provided that the 150 fpm (0.8 m/s) supply jet reaches at least 4.5 ft. (1.4 m) above the floor.	1.0
Floor supply of cool air and ceiling return, provided low velocity displacement ventilation achieves unidirectional flow and thermal stratification.	1.2
Floor supply of warm air and floor return	1.0
Floor supply of warm air and ceiling return	0.7
Makeup supply drawn in on the opposite side of the room from the exhaust or return.	0.8
Makeup supply drawn in near to the exhaust or return location	0.5

Notes:
1. "Cool air" is air cooler than space temperature.
2. "Warm air" is air warmer than space temperature.
3. "Ceiling" includes any point above the breathing zone.
4. "Floor" includes any point below the breathing zone.
5. As an alternative to using the above values, determine E_z in accordance with ASHRAE Standard 129 for all air distribution configurations except unidirectional flow.

TABLE 4-3
System Ventilation Efficiency
[ASHRAE 62.1: Table 6-3]

Max (Z_p)	E_v
< 0.15	1.0
< 0.25	0.9
< 0.35	0.8
< 0.45	0.7
< 0.55	0.6
> 0.55	Use ASHRAE 62.1, Appendix A

Notes:
1. "Max Z_p" refers to the largest value of Z_p, calculated using Equation 4-5, among all the zones served by the system.
2. Interpolating between table values is permitted.

TABLE 4-4
Minimum Exhaust Rates
[ASHRAE 62.1: Table 6-4]

Occupancy Category[6]	Exhaust Rate cfm/unit	Exhaust Rate cfm/ft^2	Exhaust Rate L/s-unit	Exhaust Rate $L/s\text{-}m^2$
Art classrooms	-	0.70	-	3.5
Auto repair rooms [1]	-	1.50	-	7.5
Barber shop	-	0.50	-	2.5
Beauty and nail salons	-	0.60	-	3.0
Cell with toilet	-	1.00	-	5.0
Darkrooms	-	1.00	-	5.0
Arena [2]	-	0.50	-	2.5
Kitchen – commercial	-	0.70	-	3.5
Kitchenettes	-	0.30	-	1.5
Locker rooms	-	0.50	-	2.5
Locker/dressing rooms	-	0.25	-	1.25
Parking garages [3,7]	-	0.75	-	3.7
Janitor, trash, recycle	-	1.00	-	5.0
Pet shops (animal areas)	-	0.90	-	4.5
Copy, printing rooms	-	0.50	-	2.5
Science lab classrooms	-	1.00	-	5.0
Toilets – public [4]	50/70	-	25/35	-
Toilet – private [5]	25/50	-	12.5/25	-
Woodwork shop/classroom	-	0.50	-	2.5

Notes:

1 Stands where engines are run shall have exhaust systems that directly connect to the engine exhaust and prevent escape of fumes.

2 The rates do not include exhaust from vehicles or equipment with internal combustion engines.

3 Exhaust rate is not required for open parking garages as defined in accordance with the Building Code. TIA

4 Rate is per water closet or urinal. Provide the higher rate where periods of heavy use are expected to occur (e.g., toilets in theatres, schools, and sports facilities).

5 Rate is for a toilet room intended to be occupied by one person at a time. For continuous system operation during normal hours of use, the lower rate may be used. Otherwise use the higher rate.

6 For unlisted occupancies for a proposed space not listed in the table, the requirements for the listed occupancy that is most similar in terms of occupant density and occupancy type shall be used. TIA TIA

7 Exhaust rate is not required for enclosed parking garages having a floor area of one-thousand (1,000) square feet or less and used for the storage of five (5) or less motorized vehicles. TIA TIA TIA

CHAPTER 5

EXHAUST SYSTEMS

501.0 Scope.

This chapter includes requirements for environmental air ducts, product-conveying systems, and commercial hoods and kitchen ventilation.

Part I – Environmental Air Ducts and Product-Conveying Systems.

502.0 Definitions.

For the purposes of this chapter, the following definitions apply:

ACCESS PANEL – A closure device used to cover an opening into a duct, an enclosure, equipment, or an appurtenance. [NFPA 96:3.3.1]

AIR INTAKES – An opening in a building's envelope whose purpose is to allow outside air to be drawn into the structure to replace inside air that is removed by exhaust systems or to improve the quality of the inside air by providing a source of air having a lower concentration of odors, suspended particles, or heating content. [NFPA 96:3.3.2]

AIR POLLUTION CONTROL DEVICES – Equipment and devices used for the purpose of cleaning air passing through them or by them in such a manner as to reduce or remove the impurities contained therein. [NFPA 96:3.3.3]

APPURTENANCE – An accessory or a subordinate part that enables the primary device to perform or improve its intended function. [NFPA 96:3.3.5]

BAFFLE PLATE – An object placed in or near an appliance to change the direction or retard the flow of air, air-fuel mixtures, or flue gases. [NFPA 96:3.3.7]

CERTIFIED PERSON – A person trained and certified by the equipment manufacturer, or by a recognized organization through a formal certification program for the system to be serviced or cleaned, that is acceptable to the Authority Having Jurisdiction.

CLASSIFIED – See **Listed**.

CLEAN(ING) – For kitchen exhaust systems and cooking equipment, the act of removing grease, oil deposits, and other residue. [NFPA 96:3.3.11]

CLEARLY IDENTIFIED – Capable of being recognized by a person of normal vision without causing uncertainty and indecisiveness about the location or operating process of the identified item. [NFPA 96:3.3.12]

CLOSED COMBUSTIBLE CONSTRUCTION – Combustible building construction, including walls, structural framing, roofs, roof ceilings, floors, and floor-ceiling assemblies continuously enclosing a grease duct on four sides where one or more sides require protection per Section 507.2. [NFPA 96:3.3.13.1]

COMBUSTIBLE MATERIAL – Material subject to an increase in combustibility or flame-spread rating beyond the limits established in the definition of Limited-Combustible Material.

COMMERCIAL FOOD HEAT-PROCESSING EQUIPMENT – Equipment used in a food establishment for heat-processing food or utensils and that produces grease vapors, steam, fumes, smoke, or odors that are required to be removed through a local exhaust ventilation system.

COMPENSATING HOOD – A hood that has an outside-air supply with air delivered below or within the hood. When makeup air is diffused directly into the exhaust within the hood cavity, it becomes a short-circuit hood.

CONCEALED SPACES – That portion(s) of a building behind walls, over suspended ceilings, in pipe chases, attics, and elsewhere whose size might normally range from 1-3/4 inch (44.45 mm) stud spaces to eight (8) foot (2.44 m) interstitial truss spaces and that might contain combustible materials such as building structural members, thermal and/or electrical insulation, and ducting. Such spaces have sometimes been used as HVAC plenum chambers. [NFPA 96:3.3.45.1]

CONTINUOUS ENCLOSURE – A recognized architectural or mechanical component of a building having a fire resistance rating as required for the structure and whose purpose is to enclose the vapor removal duct for its full length to its termination point outside the structure without any portion of the enclosure having a fire resistance rating less than the required value. [NFPA 96:3.3.22.1]

CONTINUOUS WELD – A metal-joining method that produces a product without visible interruption or variation in quality. For the purpose of the definition, it specifically includes the exhaust compartment of hoods and welded joints of exhaust ducts, yet specifically does not include filter support frames or appendages inside hoods. [NFPA 96:3.3.14]

COOKING APPLIANCE FLUE OUTLET – The opening or openings in a cooking device where vapors, combustion gases, or both leave the cooking device. There might or might not be ductwork attached to this opening.

DAMPER – A valve or plate within a duct or its terminal components for controlling draft or the flow of gases, including air. [NFPA 96:3.3.15]

DETECTION DEVICES – Electrical, pneumatic, thermal, mechanical, or optical sensing instruments, or subcomponents of such instruments, whose purpose is to cause an automatic action upon the occurrence of some preselected event. In the context of this document, the event in question could be excessive temperature or flame, and the action could be the operation of a fire-extinguishing system.

DIPS – Depression or cup like places in horizontal duct runs in which liquids could accumulate. [NFPA 96:3.3.17]

DISCHARGE – The final portion of a duct or pipe where the product being conveyed is emptied or released from confinement; the termination point of the pipe or duct. [NFPA 96:3.3.18]

EASILY ACCESSIBLE – See **Accessible, Readily** in Section 203.0, Chapter 2.

ENVIRONMENTAL AIR DUCT – Ducting used for conveying air at temperatures not exceeding 250°F (121°C) to or from occupied areas of any occupancy through other than heating or air-conditioning systems, such as ventilation for human usage, domestic kitchen range exhaust, bathroom exhaust ducts, and domestic-type clothes dryer exhaust ducts.

FACTORY-BUILT GREASE ENCLOSURES – A listed factory-built grease duct enclosure system evaluated for reduced clearances to combustibles, and as an alternative to a duct with its fire-rated enclosure. [NFPA 96:3.3.22.2.1]

FIELD-APPLIED GREASE DUCT ENCLOSURES – A listed system evaluated for reduced clearances to combustibles, and as an alternative to a duct with its fire-rated enclosure. [NFPA 96:3.3.22.2.2]

FLAMMABLE VAPOR OR FUMES – The concentration of flammable constituents in air that exceeds twenty-five percent (25%) of its Lower Flammability Limit (LFL).

FIRE BARRIER WALL – A wall assembly complying with the requirements of NFPA 221, *Standard for Fire Walls and Fire Barrier Walls*, having a fire resistance rating of four (4) hours.

FIRE PARTITION – An interior wall or partition of a building that separates two areas and serves to restrict the spread of fire but does not qualify as a fire wall.

FIRE RESISTANCE RATING – A relative value in minutes or hours assigned to materials or assemblies that have withstood a fire exposure as established in accordance with NFPA 251, *Standard Methods of Tests of Fire Endurance of Building Construction and Materials.*

FUME INCINERATORS – Devices utilizing intense heat or fire to break down and/or oxidize vapors and odors contained in gases or air being exhausted into the atmosphere. [NFPA 96:3.3.27]

FUSIBLE LINK – A form of fixed-temperature heat-detecting device sometimes employed to restrain the operation of an electrical or mechanical control until its designed temperature is reached. Such devices are to be replaced following each operation.

GREASE – Rendered animal fat, vegetable shortening, and other such oily matter used for the purposes of and resulting from cooking and/or preparing foods. Grease might be liberated and entrained with exhaust air or might be visible as a liquid or solid.

GREASE DUCTS – A containment system for the transportation of air and grease vapors that is designed and installed to reduce the possibility of the accumulation of combustible condensation and the occurrence of damage if a fire occurs within the system. [NFPA 96:3.3.20.2]

GREASE FILTER – A removable component of the grease removal system designed to capture grease and direct it to a safe collection point before it enters a duct system. Filters are expected to minimize the projection of flames downstream when attacked by flame on the upstream side and are expected to maintain their strength, shape, and integrity when exposed to the anticipated rough handling, cleaning, and service found in the field.

GREASE FILTER, MESH–TYPE – A general-purpose air filter designed to collect and retain lint and grease from the air passing through it. This type of filter is not tested, listed, or acceptable for commercial cooking operations due to the increased fire hazard.

GREASE REMOVAL DEVICES – A system of components designed for and intended to process vapors, gases, and/or air as it is drawn through such devices by collecting the airborne grease particles and concentrating them for further action at some future time, leaving the exiting air with a lower amount of combustible matter. [NFPA 96:3.3.30]

GREASE-TIGHT – Constructed and performing in such a manner as not to permit the passage of any grease under normal cooking conditions. [NFPA 96:3.3.31]

HIGH BROILER – See **Upright Broiler**.[NFPA 96:3.3.8.1]

HIGH LIMIT CONTROL – An operating device installed in and serving as an integral component of a deep-fat fryer. Its purpose is the secondary limitation of temperature allowed by the cooking operation and, if that temperature is exceeded, the automatic interruption of the thermal energy input. [NFPA 96:3.3.32]

HOOD – An air-intake device connected to a mechanical exhaust system for collecting and removing grease, vapors, fumes, smoke, steam, heat, or odors from commercial food heat-processing equipment.

Fixed Baffle – A listed unitary exhaust hood design where the grease removal device is a nonremovable assembly that contains an integral fire-activated water-wash fire-extinguishing system listed for this purpose. [NFPA 96:3.3.33.1]

Type I – is a kitchen hood for collecting and removing grease and smoke.

Type II – is a general kitchen hood for collecting and removing steam, vapor, heat, or odors.

INTERCONNECTED – Mutually assembled to another component in such a manner that the operation of one directly affects the other or that the contents of one specific duct system are allowed to encounter or contact the products being moved by another duct system. [NFPA 96:3.3.34]

LIMITED-COMBUSTIBLE MATERIAL – A building construction material that does not comply with the definition of noncombustible material, that, in the form in which it is used, has a potential heat value not exceeding 3,500 Btu/lb. (8141 kJ/kg) (see NFPA 259, *Standard Test Method for Potential Heat of Building Materials*), and that complies with either of the following (A) or (B). Materials subject to an increase in combustibility or flame-spread rating beyond the limits herein established through the effects of age, moisture, or other atmospheric conditions shall be considered combustible.

(A) Materials having a structural base of noncombustible material, with a surfacing not exceeding a thickness of 1/8 inch (3.2 mm), and with a flame-spread rating not greater than fifty (50).

(B) Materials in the form and thickness used, other than as described in (A), having neither a flame-spread rating greater than twenty-five (25) nor evidence of continued progressive combustion and of such composition that surfaces that would be exposed by cutting through the material on any plane would have neither a flame-spread rating greater than twenty-five (25) nor evidence of continued progressive combustion.

LIQUID-TIGHT – Constructed and performing in such a manner as not to permit the passage of any liquid at any temperature. [NFPA 96:3.3.35]

NONCOMBUSTIBLE MATERIAL – A material that, in the form in which it is used and under the conditions anticipated, will not ignite, burn, support combustion, or release flammable vapors when subjected to fire or heat. Materials that are reported as passing ASTM E 136, *Standard Test Method for Behavior of Materials in a Vertical Tube Furnace* at 750°C, shall be considered noncombustible materials. [NFPA 220:3.3.4]

OPEN COMBUSTIBLE CONSTRUCTION – Combustible building constructions including wall, structural framing, roof, roof ceiling, floor, and floor ceiling assemblies adjacent to a grease duct on three or fewer sides where one or more sides require protection per Section 507.2. [NFPA 96:3.3.13.2]

PITCHED – To be fixed or set at a desired angle or inclination. [NFPA 96:3.3.37]

PRODUCT-CONVEYING DUCT – Ducting used for conveying solid particulates, such as refuse, dust, fumes, and smoke; liquid particulate matter, such as spray residue, mists, and fogs; vapors, such as vapors from flammable or corrosive liquids; noxious and toxic gases; and air at temperatures exceeding 250°F (121°C).

RECIRCULATING SYSTEMS – Systems for control of smoke or grease-laden vapors from commercial cooking equipment that do not exhaust to the outside. [NFPA 96:3.3.39]

REMOVABLE – Capable of being transferred to another location with a limited application of effort and tools. [NFPA 96:3.3.40]

REPLACEMENT AIR – See **Air, Makeup** in Section 203.0, Chapter 2.

SALAMANDER BROILER – See **Upright Broiler**. [NFPA 96:3.3.8.2]

SECONDARY FILTRATION – Fume incinerators, thermal recovery units, air pollution control devices or other filtration media installed in ducts or hoods located in the path of travel of exhaust products after the initial filtration.

SHALL – Indicates a mandatory requirement.

SINGLE HAZARD AREA – As considered in the applicable extinguishing system standard (see Section 513.2) or as determined by the Authority Having Jurisdiction.

SOLID COOKING FUEL – Any solid, organic, consumable fuel such as briquette, mesquite, hardwood, or charcoal. [NFPA 96:3.3.43]

SOLID-FUEL COOKING EQUIPMENT – Cooking equipment that is fired with solid cooking fuel. This equipment includes ovens, tandoori charcoal pots, grills, broilers, rotisseries, barbecue pits, or any other type of cooking equipment that derives all or part of its heat source from the burning of solid cooking fuel.

SOLVENT – A substance (usually liquid) capable of dissolving or dispersing another substance; a chemical compound designed and used to convert solidified grease into a liquid or semi-liquid state in order to facilitate a cleaning operation. [NFPA 96:3.3.44]

SPARK ARRESTER – A device or method that minimizes the passage of airborne sparks and embers into a plenum, duct, and flue. [NFPA 96:3.3.46]

→ **TERMINATION** – The final or intended end portion of a duct system that is designed and functions to fulfill the obligations of the system in a satisfactory manner.

THERMAL RECOVERY UNIT – A device or series of devices whose purpose is to reclaim only the heat content of air, vapors, gases, or fluids that are being expelled through the exhaust system and to transfer the thermal energy so reclaimed to a location whereby a useful purpose can be served. [NFPA 96:3.3.47]

→ **TRAP** – A cup like or U-shaped configuration located on the inside of a duct system component where liquids can accumulate. [NFPA 96:3.3.49]

UPRIGHT BROILER – An appliance used in the preparation of food whereby foods are exposed to intense radiant heat, and perhaps to convective heat, with the food and the radiant source not limited to a horizontal mode. [NFPA 96:3.3.8.3]

502.1 Product-Conveying Ducts – Classification. Product-conveying ducts shall be classified according to their use, as follows:

Class 1. Ducts conveying nonabrasives, such as smoke, spray, mists, fogs, noncorrosive fumes and gases, light fine dusts, or powders.

Class 2. Ducts conveying moderately abrasive particulate in light concentrations, such as sawdust and grain dust, and buffing and polishing dust.

Class 3. Ducts conveying Class 2 materials in high concentrations and highly abrasive materials in low concentrations, such as manganese, steel chips, and coke.

Class 4. Ducts conveying highly abrasive material in high concentrations.

Class 5. Ducts conveying corrosives, such as acid vapors.

503.0 Motors, Fans, and Filters.

503.1 General. Motors and fans shall be sized to provide the required air movement. Motors in areas that contain flammable vapors or dusts shall be of a type approved for such environments. A manually operated remote control installed at an approved location shall be provided to shut off fans or blowers in flammable vapor or dust systems. Electrical equipment used in operations that generate explosive or flammable vapors, fumes, or dusts shall be interlocked with the ventilation system so that the equipment cannot be operated unless the ventilation fans are in operation. Motors for fans used to convey flammable vapors or dusts shall be located outside the duct or shall be protected with approved shields and dustproofing. Motors and fans shall be accessible for servicing and maintenance.

503.2 Fans. Parts of fans in contact with explosive or flammable vapors, fumes, or dusts shall be of nonferrous or nonsparking materials or their casing shall be lined or constructed of such material. When the size and hardness of materials passing through a fan could produce a spark, both the fan and the casing shall be of nonsparking materials. When fans are required to be spark-resistant, their bearings shall not be within the airstream, and all parts of the fan shall be grounded. Fans in systems handling materials that are likely to clog the blades, and fans in buffing or woodworking exhaust systems, shall be of the radial-blade or tube-axial type.

Equipment used to exhaust explosive or flammable vapors, fumes, or dusts shall bear an identification plate stating the ventilation rate for which the system was designed.

Fans located in systems conveying corrosives shall be of materials that are resistant to the corrosive or shall be coated with corrosion-resistant materials.

503.3 Air filters shall be listed units. Liquid adhesive coatings used on filters shall have a flash point of 350°F (177°C) or higher, as determined by the Fire Code standards.

504.0 Environmental Air Ducts.

504.1 Makeup and Exhaust-Air Ducts. Environmental air ducts not regulated by other provisions of this code shall comply with this section. Ducts shall be substantially airtight and shall comply with the provisions of Chapter 6. Exhaust ducts under positive pressure shall not extend into or through ducts or plenums. Exhaust ducts shall terminate outside the building and shall be equipped with back-draft dampers. Environmental air ducts that have an alternate function as a part of an approved smoke-control system do not require design as Class 1 product-conveying ducts.

504.2 Domestic Range Vents. Ducts used for domestic kitchen range ventilation shall be of metal and shall have smooth interior surfaces. Ducts for domestic range hoods shall only serve cooking appliances.

Exception: Ducts for domestic kitchen downdraft grill-range ventilation installed under a concrete slab floor may be of approved Schedule 40 PVC provided:

(1) The under-floor trench in which the duct is installed shall be completely backfilled with sand or gravel.

(2) Not more than one (1) inch (25 mm) of six (6) inch diameter (152 mm) PVC coupling may protrude above the concrete floor surface.

(3) PVC pipe joints shall be solvent cemented to provided an air and grease-tight duct.

(4) The duct shall terminate above grade outside the building and shall be equipped with a back-draft damper.

504.3 Clothes Dryers.

504.3.1 Moisture Exhaust Ducts. Moisture exhaust ducts shall terminate on the outside of the building and shall be equipped with a back-draft damper. Screens shall not be installed at the duct termination. Ducts for exhausting clothes dryers shall not be connected or installed with sheet metal screws or other fasteners that will obstruct the flow. Clothes dryer moisture exhaust ducts shall not be connected to a gas vent connector, gas vent, or chimney, and shall only serve clothes dryers. Clothes dryer moisture exhaust ducts under positive pressure shall not extend into or through ducts or plenums.

504.3.2 Domestic Clothes Dryers. When a compartment or space for a domestic clothes dryer is provided, a minimum four (4) inch diameter (102 mm) moisture exhaust duct of approved material shall be installed in accordance with this section and Section 504.0.

When a closet is designed for the installation of a clothes dryer, a minimum opening of 100 square inches (64,516 mm^2) for makeup air shall be provided in the door or by other approved means.

504.3.2.1 Domestic Dryer Vents. Domestic clothes dryer moisture exhaust ducts shall be of metal and shall have smooth interior surfaces.

Exception: Listed clothes dryer transition ducts not more than six (6) feet (1,829 mm) in length may be used in connection with domestic dryer exhausts.

Flexible clothes dryer transition ducts shall not be concealed within construction.

504.3.2.2 Length Limitation. Unless otherwise permitted or required by the dryer manufacturer's installation instructions and approved by the Authority Having Jurisdiction, domestic dryer moisture exhaust ducts shall not exceed a total combined horizontal and vertical length of fourteen (14) feet (4,263 mm), including two (2) 90 degree (1.57 rad) elbows. Two (2) feet (610 mm) shall be deducted for each 90 degree (1.57 rad) elbow in excess of two.

504.3.3 Commercial Clothes Dryers. Commercial dryer exhaust ducts shall be installed in accordance with their listings. The installation of commercial clothes dryer exhaust ducts shall comply with the appliance manufacturer's installation instructions.

504.4 Heat (Energy) Recovery Ventilators.

504.4.1 Heat (energy) recovery ventilators shall be installed in accordance with their listings and comply with the appliance manufacturer's installation instructions. Heat (energy) recovery ventilator ducts shall comply with the provisions of Chapter 6.

504.5 Termination of Environmental Air Ducts. Environmental air duct exhaust shall terminate a minimum of three (3) feet (914 mm) from property line and three (3) feet (914 mm) from openings into the building.

504.6 Gypsum Wallboard Ducts. Bathroom and laundry room exhaust ducts may be of gypsum wallboard subject to the limitations of Section 602.1.

505.0 Design of Product-Conveying Ventilation Systems.

505.1 General. A mechanical ventilation or exhaust system shall be installed to control, capture, and remove emissions generated from product use or handling when required by the Building Code or Fire Code and when such emissions result in a hazard to life or property. The design of the system shall be such that the emissions are confined to the area in which they are generated by air currents, hoods, or enclosures and shall be exhausted by a duct system to a safe location or treated by removing contaminant's. Ducts conveying explosives or flammable vapors, fumes, or dusts shall extend directly to the exterior of the building without entering other spaces and shall not extend into or through ducts and plenums.

Exception: Ducts conveying vapor or fumes having flammable constituents less than 25 percent of their Lower Flammability Limit (LFL) may pass through other spaces.

505.1.1 Incompatible materials shall not be conveyed in the same system.

505.1.2 In systems conveying flammable vapors, gases, or mists, the concentration shall not exceed 25 percent of the lower flammability limit (LFL).

Exception: Higher concentrations shall be permitted if the exhaust system is designed and protected in accordance with the Standard on Explosion Prevention Systems in Chapter 17, using one or more of the following techniques:

(a) Combustible concentration reduction

(b) Oxidant concentration reduction

(c) Deflagration suppression

(d) Deflagration pressure containment

Separate and distinct systems shall be provided for incompatible materials.

Contaminated air shall not be recirculated to occupied areas unless contaminant's have been removed. Air contaminated with explosive or flammable vapors, fumes, or dusts; flammable or toxic gases; or radioactive material shall not be recirculated.

505.2 Minimum Velocities and Circulation. The velocity and circulation of air in work areas shall be such that contaminant's are captured by an airstream at the area where the emissions are generated and conveyed into a product-conveying duct system. Mixtures within work areas where contaminant's are generated shall be diluted below twenty-five (25) percent of their lower explosive limit or lower flammability limit with air that does not contain other contaminant's. The velocity of air within the duct shall be not less than set forth in Table 5-1.

Systems for removal of vapors, gases, and smoke shall be designed by the constant velocity or equal friction methods. Systems conveying particulate matter shall be designed employing the constant velocity method. Systems conveying explosive or radioactive materials shall be prebalanced through duct sizing. Other systems may be designed with balancing devices such as dampers. Dampers provided to balance airflow shall be provided with securely fixed minimum-position blocking devices to prevent restricting flow below the required volume or velocity.

505.3 Makeup Air. Makeup air shall be provided to replenish air exhausted by the ventilation system. Makeup-air intakes shall be located so as to avoid recirculation of contaminated air within enclosures.

505.4 Hoods and Enclosures. Hoods and enclosures shall be used when contaminant's originate in a concentrated area. The design of the hood or enclosure shall be such that air currents created by the exhaust systems will capture the contaminant's and transport them directly to the exhaust duct. The volume of air shall be sufficient to dilute explosive or flammable vapors, fumes, or dusts as set forth in Section 505.2. Hoods of steel shall have a base metal thickness not less than 0.027 inch (0.69 mm) (No. 22 gauge) for Class 1 and Class 5 metal duct systems; 0.033 inch (0.84 mm) (No. 20 gauge) for hoods serving a Class 2 duct system; 0.044 inch (1.12 mm) (No. 18 gauge) for hoods serving a Class 3 duct system; and 0.068 inch (1.73 mm) (No. 14 gauge) for hoods serving a Class 4 duct system.

Approved nonmetallic hoods and duct systems may be used for Class 5 corrosive systems when the corrosive mixture is nonflammable. Metal hoods used with Class 5 duct systems shall be protected with suitable corrosion-resistant material. Edges of

TABLE 5-1
Minimum Conveying Velocities

	Feet per Minute	(m/s)
Vapors, gases, smoke, fumes	Any	
Fine light dusts, such as cotton, lint, and wood flour (100 mesh and under)	2,000	(10.2)
Dry dusts; powders, such as fine rubber molding power, soap dust	2,500[1]	(12.7)[1]
Industrial dusts		
Average dusts, such as sawdust, grinding dust, coal dust	3,500	(17.8)
Heavy dusts, such as metal turnings, lead dusts	4,000	(20.3)
Moist dusts and chips, such as lead dust with chips, sticky buffing lint, quick-lime dust	4,500	(22.9)

Note:
[1] The velocity for aluminum and magnesium powder shall be not less than 4,000 feet per minute (20.3 m/s).

hoods shall be rounded. The minimum clearance between hoods and combustible construction shall be the clearance required by the duct system.

506.0 Product-Conveying Ducts.

506.1 Materials. Materials used in product-conveying duct systems shall be suitable for the intended use and shall be of metal.

Exceptions:

(1) Asbestos-cement, concrete, clay, or ceramic materials may be used when it is shown that these materials will be equivalent to metal ducts installed in accordance with this chapter.

(2) Ducts serving a Class 5 system may be constructed of approved nonmetallic material when the corrosive characteristics of the material being conveyed make a metal system unsuitable and when the mixture being conveyed is nonflammable.

Approved nonmetallic material shall be either a listed product having a flame-spread index of twenty-five (25) or less and a smoke-developed rating of fifty (50) or less on both inside and outside surfaces without evidence of continued progressive combustion, or shall have a flame-spread index of twenty-five (25) or less and shall be installed with an automatic fire-sprinkler protection system inside the duct.

(3) Ducts used in central vacuum cleaning systems within a dwelling unit shall be constructed of materials in compliance with the applicable standards referenced in Chapter 17. Penetrations of fire walls or floor-ceiling or roof-ceiling assemblies shall comply with the Building Code.

Copper or ferrous pipes or conduits extending from within the separation between a garage and dwelling unit to the central vacuuming unit may be used.

Aluminum ducts shall not be used in systems conveying flammable vapors, fumes, or explosive dusts, nor in Class 2, 3, or 4 systems. Galvanized steel and aluminum ducts shall not be used when the temperature of the material being conveyed exceeds 400°F (205°C).

Metal ducts used in Class 5 systems that are not resistant to the corrosiveness of the product shall be protected with appropriate corrosion-resistant material.

506.2 Construction. Ducts used for conveying products shall be of substantial airtight construction and shall not have openings other than those required for operation and maintenance of the system. Ducts constructed of steel shall comply with Table 5-5 or 5-6.

Exceptions:

(1) Class 1 product-conveying ducts that operate at less than four (4) inches (102 mm) water column (995.6 Pa) negative pressure and convey noncorrosive, nonflammable, and nonexplosive materials at temperatures not exceeding 250°F (121°C) may be constructed in accordance with Tables 6-1, 6-2, 6-3, 6-4, 6-5, 6-7, 6-8, or, with prior approval, UMC Standard No. 6-2.

(2) Ducts used in central vacuuming systems within a dwelling unit shall be constructed of materials in compliance with the applicable standards referenced in Chapter 17. Penetrations of fire-resistive walls, or floor-ceiling or roof-ceiling assemblies shall comply with the Building Code. Copper or ferrous pipes or conduit extending from within the separation between a garage and dwelling unit to the central vacuum unit may be used.

The use of rectangular ducts conveying particulates shall be subject to approval of the building official. The design of rectangular ducts shall consider the adhesiveness and buildup of products being conveyed within the duct.

Aluminum construction may be used in Class 1 duct systems only. The thickness of aluminum ducts shall be at least two Brown and Sharpe gauges thicker than the gauges required for steel ducts set forth in Tables 5-5 and 5-6.

506.3 Fittings. Fittings in Class 2, 3, and 4 systems shall be not less than two gauges thicker than the thickness required for straight runs. Flexible metallic duct may be used for connecting ductwork to vibrating equipment. Duct systems subject to wide temperature fluctuations shall be provided with expansion joints.

Branches shall connect to main ducts at the large end of transitions at an angle not exceeding forty-five (45) degrees (0.79 rad).

Except for ducts used to convey noncorrosive vapors with no particulate, accessible cleanouts shall be provided at ten (10) foot (3,048 mm) intervals and at changes in direction. Access openings shall also be provided for access to sprinklers and other equipment within the duct that require servicing.

506.4 Explosion Venting. Ducts conveying explosive dusts shall have explosion vents, openings protected by antiflashback swing valves, or rupture diaphragms. Openings to relieve explosive forces shall be located outside the building. When relief devices cannot provide sufficient pressure relief, ductwork shall be designed to withstand an internal pressure of not less than 100 pounds per square inch (689 kPa).

If a room or building contains a dust explosion hazard that is external to protected equipment, as defined in 2.2.3.1 of NFPA 654, such areas shall be provided with deflagration venting to a safe outside location.

506.5 Supports. Spacing of supports for ducts shall not exceed twelve (12) feet (3,658 mm) for eight (8) inch (203 mm) ducts nor twenty (20) feet (6,096 mm) for larger ducts, unless justified by the design.

506.5.1 Duct supports shall be designed to carry the weight of the duct half filled with material. Where sprinkler protection is provided in the duct, the hanger's design shall include the weight of the duct half filled with water or with the material being conveyed, whichever has the higher density. Loads shall not be placed on connecting equipment.

Exception: Where adequate drainage is provided, the weight of the water shall not require consideration.

506.5.2 Hangers and supports exposed to corrosive atmospheres shall be Type 316 SS or equivalent.

506.5.3 To avoid vibration and stress on the duct, hangers and supports shall be securely fastened to the building or structure.

506.5.4 Hangers and supports shall be designed to allow for expansion and contraction. [NFPA 91:4.5.6]

506.6 Fire Protection. Sprinklers or other fire-protection devices shall be installed within ducts having a cross-sectional dimension exceeding ten (10) inches (254 mm) when the duct conveys flammable vapors or fumes. Sprinklers shall be installed at twelve (12) foot (3,658 mm) intervals in horizontal ducts and at changes in direction. In vertical runs, sprinklers shall be installed at the top and at alternate floor levels.

506.7 Duct Clearances.

506.7.1 All ductwork and system components handling combustible material and operating at less than 140°F (60°C) shall have a clearance of not less than eighteen (18) inches (457 mm) from combustible construction or any combustible material.

Exception No. 1: When the ductwork system is equipped with an approved automatic extinguishing system designed for the specific hazard, the clearance shall be permitted to be reduced to six (6) inches (152 mm) from combustible materials and 1/2 inch (13 mm) from combustible construction.

Exception No. 2: When the combustible material and construction is protected by the use of materials or products listed for protection purposes or in accordance with Table 5-2.

506.7.1.1 Spacers and ties for protection materials shall be of noncombustible material and shall not be used directly behind the duct.

506.7.1.2 With all clearance reduction systems using a ventilated airspace, air circulation shall be provided as described in Table 5-2. There shall be at least one (1) inch (25 mm) between the wall protector and combustible walls and ceilings for clearance, reduction systems using a ventilated space.

506.7.1.3 Mineral wool batts (blanket or board) shall have a minimum density of eight (8) lb./ft.^{3} (3.6 kq/m^3) and have a minimum melting point of 1,500°F (816°C).

506.7.1.4 Insulation board used as a part of a clearance-reduction system shall have a thermal conductivity of 1 Btu in./ft.^{2} hr°F (0.14 W/m^2 hr°C) or less. Insulation board shall be formed of noncombustible material.

506.7.1.5 There shall be at least one (1) inch (25 mm) between the duct and the wall protector. In no case shall the clearance between the duct and the wall surface be reduced below that shown in Table 5-2.

506.7.2 Duct systems operating at elevated temperatures above 140°F (60°C) shall have clearances from combustible building construction or any combustible material of not less than eighteen (18) inch (457 mm).

506.7.3 Where clearance is reduced by using an airspace between the combustible wall and the wall protector, air circulation shall be provided by one of the following methods.

506.7.3.1 Air circulation shall be permitted to be provided by leaving all edges of the wall protector open with at least a one (1) inch (25 mm) airgap.

506.7.3.2 If the wall protector is mounted on a single flat wall away from corners, air circulation shall be permitted to be provided by one of the following:

(A) Leaving top and bottom edges open to circulation by maintaining the one (1) inch (25 mm) airgap.

TABLE 5-2
Reduction of Duct Clearance with Specified Forms of Protection

Clearance reduction applied to and covering all combustible surfaces with the distance specified as required clearance with no protection in 506.7.2	Minimum Allowable Reduction in Clearance (%)	
Form of Protection	**As Wall Protector**	**As Ceiling Protector**
(a) 3-1/2 in. (90 mm) thick masonry wall without ventilated airspace	33	—
(b) 1/2 in. (13 mm) thick noncombustible insulation board over 1 in. (25.4 mm) glass fiber or mineral wool batts without ventilated airspace	50	33
(c) 0.024 in. (0.61 mm) (No. 24 gauge) sheet metal over 1 in. (25.4 mm) glass fiber or mineral wool batts reinforced with wire, or equivalent, on rear face with at least a 1 in. (25.4 mm) air gap	66	66
(d) 3-1/2 in (90 mm) thick masonry wall with at least a 1 in (25.4 mm) air gap	66	—
(e) 0.024 in. (0.61 mm) (No. 24 gauge) sheet metal with at least a 1 in. (25.4 mm) air gap	66	50
(f) 1/2 in. (13 mm) thick noncombustible insulation board with at least a 1 in. (25.4 mm) air gap	66	50
(g) 0.024 in. (0.61mm) (No. 24 gauge) sheet metal with ventilated airspace over 0.024 in. (0.61 mm) (No. 24 gauge) sheet metal with at least a 1 in. (25.4 mm) air gap	66	50
(h) 1 in. (25.4 mm) glass fiber or mineral wool batts sandwiched between two sheets 0.024 in. (0.61 mm) (No. 24 gauge) sheet metal with at least a 1 in. (25.4 mm) air gap	66	50

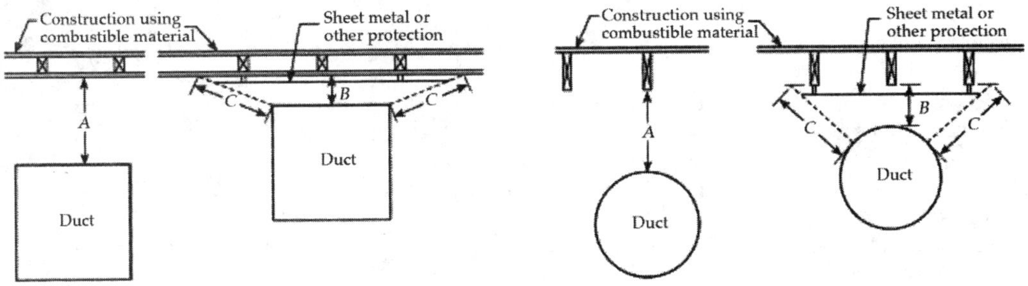

Notes:
1. A = required clearance with no protection.
2. B = reduced clearance permitted.
3. The protection applied to the construction using combustible material shall extend far enough in each direction make C = A

Extent of protection required to reduce clearances from ducts.

(B) Leaving top and both side edges open to circulation by maintaining the (1) inch (25 mm) air gap.

506.7.3.3 Wall protectors that cover two walls in a corner shall be permitted to be open at the top and bottom edges with at least a one (1) inch (25 mm) air gap. [NFPA 91:4.6.3.3]

506.8 Protection from Physical Damage. Ducts installed in locations where they are subject to physical damage shall be protected by suitable guards.

506.9 Exhaust Outlets. The termination point for exhaust ducts discharging to the atmosphere shall be not less than the following:

506.9.1 Ducts conveying explosive or flammable vapors, fumes, or dusts: thirty (30) feet (9,144 mm) from property line; ten (10) feet (3,048 mm) from openings into the building, six (6) feet (1,829 mm) from exterior walls or roofs; thirty (30) feet (9,144 mm) from combustible walls or openings into the building that are in the direction of the exhaust discharge; ten (10) feet (3,048 mm) above adjoining grade.

506.9.2 Other product-conveying outlets: ten (10) feet (3,048 mm) from property line; three (3) feet (914 mm) from exterior wall or roof; ten (10) feet (3,048 mm) from openings into the building; ten (10) feet (3,048 mm) above adjoining grade.

TABLE 5-3
Reduction of Clearances with Specified Forms of Protection
[NFPA 54: Table 10.2.3(b)]

Type of protection applied to and covering surfaces of combustible material within the distance specified as the required clearance with no protection [See Figures 5-4 through 5-6.]	Where the required clearance with no protection from appliance, vent connector, or single-wall metal pipe is: 36 in.		18 in.		12 in.		9 in.		6 in.	
	Allowable Clearances with Specified Protection (in.) — Use Col. 1 for clearances above appliance or horizontal connector. Use Col. 2 for clearances from appliances, vertical connector, and single-wall metal pipe.									
	Above Col. 1	Sides and Rear Col. 2	Above Col. 1	Sides and Rear Col. 2	Above Col. 1	Sides and Rear Col. 2	Above Col. 1	Sides and Rear Col. 2	Above Col. 1	Sides and Rear Col. 2
(1) 3-1/2 in. thick masonry wall without ventilated air space	--	24	--	12	--	9	--	6	--	5
(2) 1/2 in. insulation board over 1 in. glass fiber or mineral wool batts	24	18	12	9	9	6	6	5	4	3
(3) 0.024 sheet metal over 1 in. glass fiber or mineral wool batts reinforced with wire on rear face with ventilated air space	18	12	9	6	6	4	5	3	3	3
(4) 3-1/2 in. thick masonry wall with ventilated air space	--	12	--	6	--	6	--	6	--	6
(5) 0.024 sheet metal with ventilated air space	18	12	9	6	6	4	5	3	3	2
(6) 1/2 in. thick insulation board with ventilated air space	18	12	9	6	6	4	5	3	3	3
(7) 0.024 sheet metal with ventilated air space over 0.024 sheet metal with ventilated air space	18	12	9	6	6	4	5	3	3	3
(8) 1 in. glass fiber or mineral wool batts sandwiched between two sheets 0.024 sheet metal with ventilated air space	18	12	9	6	6	4	5	3	3	3

Notes:

1 Reduction of clearances from combustible materials shall not interfere with combustion air, draft hood clearance and relief, and accessibility of servicing.

2 Clearances shall be measured from the outer surface of the combustible material to the nearest point on the surface of the appliance, disregarding any intervening protection applied to the combustible material.

3 Spacers and ties shall be of noncombustible material. No spacer or tie shall be used directly opposite the appliance or connector.

4 Where clearance reduction systems use a ventilated air space, adequate provision for air circulation shall be provided as described. [See Figure 5-5 and Figure 5-6.]

5 There shall be not less than one (1) in. (25 mm) between clearance reduction systems and combustible walls and ceilings for reduction systems using a ventilated air space.

6 Where a wall protector is mounted on a single flat wall away from corners, it shall have a minimum one (1) inch (25 mm) air gap. To provide adequate air circulation, the bottom and top edges, or only the side and top edges, or edges shall be left open.

7 Mineral wool batts (blanket or board) shall have a minimum density of eight (8) lb/ft.^{3} (128 kg/m^3) and a minimum melting point of 1,500°F (816°C).

8 Insulation material used as part of a clearance reduction system shall have a thermal conductivity of 1.0 Btu in./ft^2/h-°F (0.144 W/m-K) or less.

9 There shall be not less than (1) inch (25 mm) between the appliance and the protector. In no case shall the clearance between the appliance and the combustible surface be reduced below that allowed in this table.

10 Clearances and thicknesses are minimum; larger clearances and thicknesses are acceptable.

11 Listed single-wall connectors shall be installed in accordance with the terms of their listing and the manufacturer's instructions.

1 in. = 25.4 mm

Part II – Commercial Hoods and Kitchen Ventilation.

507.0 General Requirements.

507.1 Cooking equipment used in processes producing smoke or grease-laden vapors shall be equipped with an exhaust system that complies with all the equipment and performance requirements of this standard, and all such equipment and performance shall be maintained per this standard during all periods of operation of the cooking equipment. Specifically, the following equipment shall be kept in good working condition:

(A) Cooking equipment

(B) Hoods

(C) Ducts (if applicable)

(D) Fans

(E) Fire suppression systems

(F) Special effluent or energy control equipment

All airflows shall be maintained. Maintenance and repairs shall be performed on all components at intervals necessary to maintain these conditions.

507.1.1 Cooking equipment that has been listed in accordance with UL 197 or an equivalent standard for reduced emissions shall not be required to be provided with an exhaust system. [NFPA 96:4.1.1.1*]

507.1.2 The listing evaluation of cooking equipment covered by Section 507.1.1 shall demonstrate that the grease discharge at the exhaust duct of a test hood placed over the appliance shall not exceed 5 mg/m^3 when operated with a total airflow of 0.236 cubic meters per second (500 cfm). [NFPA 96:4.1.1.2]

507.1.3 The responsibility for inspection, maintenance, and cleanliness of the ventilation control and fire protection of the commercial cooking operations shall be the ultimate responsibility of the owner of the system provided that this responsibility has not been transferred in written form to a management company or other party. [NFPA 96:4.1.5]

507.1.4 All solid-fuel cooking equipment shall comply with the requirements of Section 517.0.

507.1.5 Multiple-tenancy applications shall require the concerted cooperation of design, installation, operation, and maintenance responsibilities by tenants and by the building owner.

507.1.6 All interior surfaces of the exhaust system shall be accessible for cleaning and inspection purposes.

507.1.7 Cooking equipment used in fixed, mobile, or temporary concessions, such as trucks, buses, trailers, pavilions, tents, or any form of roofed enclosure, shall comply with this standard unless all or part of the installation is exempted by the Authority Having Jurisdiction.

507.2 Clearance.

507.2.1 Where enclosures are not required, hoods, grease removal devices, exhaust fans, and ducts shall have a clearance of at least eighteen (18) inches (457 mm) to combustible material, three (3) inches (76 mm) to limited-combustible material, and zero (0) inches (0 mm) to noncombustible material. [NFPA 96:4.2.1]

507.2.2 Where a hood, duct, or grease removal device is listed for clearances less than those required in 507.2.1, the listing requirements shall be permitted. [NFPA 96:4.2.2]

507.2.3 Clearance Reduction.

507.2.3.1 Where a clearance reduction system consisting of 0.013 in. (0.33 mm) (28 gauge) sheet metal spaced out one (1) inch (25 mm) on noncombustible spacers is provided, there shall be a minimum of nine (9) inches (229 mm) clearance to combustible material. [NFPA 96:4.2.3.1]

507.2.3.2 Where a clearance reduction system consisting of 0.027 in. (0.69 mm) (22 gauge) sheet metal on one (1) inch (25 mm) mineral wool batts or ceramic fiber blanket reinforced with wire mesh or equivalent spaced out one (1) inch (25 mm) on noncombustible spacers is provided, there shall be a minimum of three (3) inches (76 mm) clearance to combustible material. [NFPA 96:4.2.3.2]

507.2.3.3 Zero clearance to limited-combustible materials shall be permitted where protected by metal lath and plaster, ceramic tile, quarry tile, other noncombustible materials or assembly of noncombustible materials, or materials and products that are listed for the purpose of reducing clearance. [NFPA 96:4.2.3.3]

507.2.4 Clearance Integrity.

507.2.4.1 In the event of damage, the material or product shall be repaired and restored to meet its intended listing or clearance requirements and shall be acceptable to the authority having jurisdiction. [NFPA 96:4.2.4.1]

507.2.4.2 In the event of a fire within a kitchen exhaust system, the duct and its enclosure (rated shaft, factory-built grease duct enclosure, or field-applied grease duct enclosure) shall be inspected by qualified personnel to determine whether the duct and protection method are structurally sound, capable of maintaining their fire protection function, and in compliance with this standard for continued operation. [NFPA 96:4.2.4.2]

507.2.4.3 Protection shall be provided on the wall from the bottom of the hood to the floor, or to the top of the noncombustible material extending to the floor, to the same level as required in 507.2.1. [NFPA 96:4.2.4.3]

507.2.4.4 The protection methods for ducts to reduce clearance shall be applied to the combustible or limited-combustible construction, not to the duct itself. [NFPA 96:4.2.4.4]

507.2.5 Factory-built grease duct enclosures shall be protected with a through-penetration firestop system classified in accordance with ASTM E814 having an "F" and "T" rating equal to the fire resistance rating of the assembly being penetrated from the point at which the duct penetrates a ceiling, wall or floor to the outlet terminal, and the factory-built grease duct protection system shall be listed in accordance with UL 2221, *Standard for Tests of Fire Resistive Grease Duct Enclosure Assemblies* and installed in accordance with the manufacturer's instructions and the listing requirements. [NFPA 96:4.3.3, 4.3.3.1, 4.3.3.2]

507.2.6 Field-applied grease duct enclosures shall be protected with a through penetration firestop system classified in accordance with ASTM E814 having an "F" and "T" rating equal to the fire resistance rating of the assembly being penetrated. The surface of the field fabricated grease duct shall be continuously covered on all sides from the point at which the duct enclosure penetrates a ceiling, wall or floor to the outlet terminal, and listed in accordance with ASTM E 2336 *Standard Test Methods for Fire resistive Grease Duct Enclosure Systems*, and installed in accordance with the manufacturer's instructions and the listing requirements. [NFPA 96:4.3.1, 4.3.1.1, 4.3.1.2]

507.2.7 Field-applied grease duct enclosures and factory-built grease duct enclosures shall demonstrate that they provide sufficient mechanical and structural integrity, resiliency, and stability when subjected to expected building environmental conditions, duct movement under general operating conditions, and duct movement due to fire conditions. [NFPA 96:4.3.4]

507.2.7.1 Measures shall be taken to prevent physical damage to any material or product used for the purpose of reducing clearances.

Exception: When the duct is protected with a field-applied grease duct enclosure or factory-built grease duct enclosure.

507.2.7.2 The specifications of material, gauge, and construction of the duct used in the testing and listing of field-applied grease duct enclosures and factory-built grease duct enclosures shall be included as minimum requirements in their listing and installation documentation. [NFPA 96:4.3.5]

507.2.7.3 The following clearance options for which field-applied grease duct enclosures and factory-built grease duct enclosures have been successfully evaluated shall be clearly identified in their listing and installation documentation and on their label: [NFPA 96:4.3.6]

(1) Open combustible clearance at manufacturer's requested dimensions. [NFPA 96:4.3.6(1)]

(2) Closed combustible clearance at manufacturer's requested dimensions, with or without specified ventilation. [NFPA 96:4.3.6(2)]

(3) Rated shaft clearance at manufacturer's requested dimensions, with or without specified ventilation. [NFPA 96:4.3.6(3)]

507.2.8 A duct shall be permitted to contact noncombustible floors, interior walls, and other noncombustible structures or supports, but it shall not be in contact for more than fifty (50) percent of its surface area per each lineal foot of contact length. [NFPA 96:4.4.1]

507.2.8.1 Where duct contact must exceed the requirements of 507.2.8, the duct shall be protected from corrosion. [NFPA 96:4.4.2]

507.2.8.2 Where the duct is listed for zero clearance to combustibles or otherwise protected with a material or product listed for the purpose of reducing clearance to zero, the duct shall be permitted to exceed the contact limits of 507.2.8 without additional corrosion protection. [NFPA 96:4.4.3]

507.2.9 Clearances between the duct and interior surfaces of enclosures shall meet the requirements of Section 507.2.

507.3 A drawing(s) of the exhaust system installation along with a copy of operating instructions for subassemblies and components used in the exhaust system, including electrical schematics, shall be available on the premises.

507.4 If required by the Authority Having Jurisdiction, notification in writing shall be given of any alteration, replacement, or relocation of any exhaust or extinguishing system or part thereof or cooking equipment. Satisfaction shall be provided to the Authority Having Jurisdiction that the complete exhaust system as addressed in this standard is installed and operable in accordance with the approved design and the manufacturer's instructions.

508.0 Hoods.

508.1 Where Required. Hoods shall be installed at or above all commercial-type deep fat fryers, broilers, fry grills, steam-jacketed kettles, hot-top ranges, ovens, barbecues, rotisseries, dishwashing machines, and similar equipment that produces comparable amounts of steam, smoke, grease, or heat in a food-processing establishment. For the purpose of this section, a food-processing establishment shall include any building or portion thereof used for the processing of food, but shall not include a dwelling unit.

508.1.1 Construction. The hood or that portion of a primary collection means designed for collecting cooking vapors and residues shall be constructed of and be supported by steel not less than 0.043 inch (1.09 mm) (No. 18 MSG) in thickness, stainless steel not less than 0.037 inch (0.94 mm) (No. 20 MSG) in thickness, or other approved material of equivalent strength and fire and corrosion resistance.

Exceptions:

(1) Listed exhaust hoods with or without exhaust dampers.

(2) Type II hoods shall be constructed of at least 0.024 inch (0.61 mm) (No. 24 gauge steel. Hoods constructed of copper shall be of copper sheets weighing at least twenty-four (24) ounces per square foot (7.32 kg/m^2). Joints and seams shall be substantially tight. Solder shall not be used except for sealing a joint or seam.

All hoods shall be secured in place by noncombustible supports.

508.1.2 Wall mounted exhaust hood assemblies shall be tight fitting against the back wall as to not permit passage of grease vapor behind the hood, or between the back wall and the hood assembly. [NFPA 96:5.1.13]

508.2 All seams, joints, and penetrations of the hood enclosure that direct and capture grease-laden vapors and exhaust gases shall have a liquid-tight continuous external weld to the hood's lower outermost perimeter. [NFPA 96:5.1.2]

508.2.1 Seams, joints, and penetrations of the hood shall be permitted to be internally welded, provided that the weld is formed smooth or ground smooth, so as to not trap grease, and is cleanable. [NFPA 96:5.1.3]

508.2.2 Internal hood joints, seams, filter support frames, and appurtenances attached inside the hood shall be sealed or otherwise made grease-tight. [NFPA 96:5.1.4]

508.2.3 Penetrations shall be permitted to be sealed by devices that are listed for such use and whose presence does not detract from the hood's or duct's structural integrity. [NFPA 96:5.1.5]

508.2.4 Listed exhaust hoods with or without exhaust dampers shall be permitted to be constructed of materials required by the listing. [NFPA 96:5.1.6]

508.2.5 Listed exhaust hoods with or without exhaust dampers shall be permitted to be assembled in accordance with the listing requirements. [NFPA 96:5.1.7]

508.2.6 Eyebrow-type hoods over gas or electric ovens shall be permitted to have a duct constructed as required in Section 510.0 from the oven flue(s) connected to the hood canopy upstream of the exhaust plenum as shown in Figure 5-1. [NFPA 96:5.1.8.1]

508.2.7 The duct connecting the oven flue(s) to the hood canopy shall be connected with a continuous weld or have a duct-to-duct connection. [See Figure 5-6(a) through Figure 5-6(d)] [NFPA 96:5.1.8.2]

508.2.8 Insulation materials other than electrical insulation shall have a flame spread rating of 25 or less when tested in accordance with UL 723. [NFPA 96:5.1.9]

508.2.9 Adhesives or cements used in the installation of insulating materials shall comply with the requirements of 508.2.8, when tested with the specific insulating material. [NFPA 96:5.1.10]

508.2.10 Penetrations shall be sealed with listed devices in accordance with the requirements of 508.2.11. [NFPA 96:5.1.11]

508.2.11 Devices that require penetration of the hood, such as pipe and conduit penetration fittings and fasteners, shall be listed in accordance with UL 1978. [NFPA 96:5.1.12]

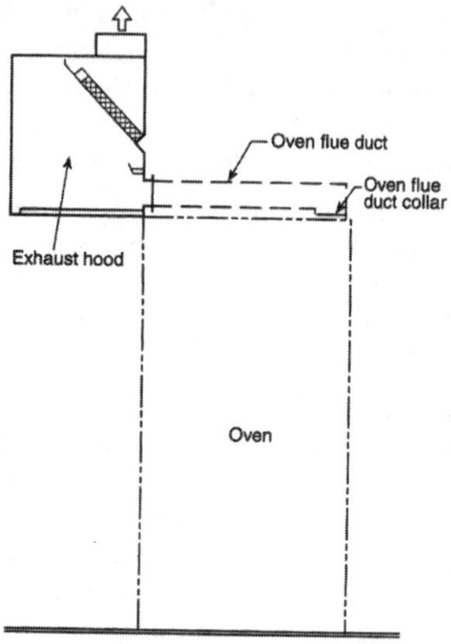

FIGURE 5-1 Typical Section of Eyebrow-Type Hood. [NFPA 96:5.1.8.1]

508.3 Insulation materials other than electrical insulation shall have a flame-spread rating of twenty-five (25) or less when tested in accordance with UL 723, *Standard for Test for Surface Burning Characteristics of Building Materials.* Adhesives or cements used in the installation of insulating materials shall comply with the preceding requirements when tested with the specific insulating material.

508.4 Hood Size. Hoods shall be sized according to the following calculations and configured to provide for the capture and removal of grease-laden vapors. (See Section 511.2.3)

508.4.1 Canopy Size and Location. For canopy-type commercial cooking hoods, the inside edge thereof shall overhang or extend a horizontal distance of not less than six (6) inches (152 mm) beyond the edge of the cooking surface on all open sides, and the vertical distance between the lip of the hood and the cooking surface shall not exceed four (4) feet (1,219 mm).

Exception: Listed exhaust hoods are to be installed in accordance with the terms of their listings and the manufacturer's installation instructions.

508.4.1.1 Capacity of Hoods. Canopy-type commercial cooking hoods shall exhaust through the hood a minimum quantity of air determined by application of the following formulas:

WHERE:

A = the horizontal surface area of the hood, in square feet (m^2).

P = that part of the perimeter of the hood that is open, in feet (mm).

D = distance in feet (mm) between the lower lip of the hood and the cooking surface.

Q = quantity of air, in cubic feet per minute (L/s).

When cooking equipment is installed back to back and is covered by a common island-type hood, the airflow required may be calculated using the formula for three sides exposed. Type II hood airflow requirements shall be in accordance with the requirements for low-temperature appliance hoods.

508.4.1.2 The minimum airflow for solid-fuel cooking equipment, grease-burning charbroilers, and undefined equipment shall be:

Number of Exposed Sides	Formula
4 (island or central hood)	$Q = 300A$
3 or less	$Q = 200A$
Alternate formula	$Q = 100PD$

508.4.1.3 Type I hoods when the cooking equipment includes high-temperature appliances such as deep-fat fryers:

Number of Exposed Sides	Formula
4 (island or central hood)	$Q = 150A$
3 or less	$Q = 100A$
Alternate formula	$Q = 100PD$

508.4.1.4 Type I hoods where the cooking equipment includes medium-temperature appliances such as rotisseries, grills, and ranges:

Number of Exposed Sides	Formula
4 (island or central hood)	$Q = 100A$
3 or less	$Q = 75A$
Alternate formula	$Q = 50PD$

508.4.1.5 Type I hoods where the cooking equipment includes low-temperature appliances such as medium-to-low-temperature ranges, roasters, roasting ovens, pastry ovens, and equipment approved for use under a Type II hood, such as pizza ovens:

Number of Exposed Sides	Formula
4 (island or central hood)	$Q = 75A$
3 or less	$Q = 50A$
Alternate formula	$Q = 50PD$

Exception: Listed exhaust hoods are to be installed in accordance with the terms of their listing and the manufacturer's installation instructions.

508.4.2 Capacity for Noncanopy Hoods. In addition to all other requirements for hoods specified in this section, the volume of air exhausting through a noncanopy-type hood to the duct system shall be not less than 300 cubic feet per minute per lineal foot (465 L/s per m) of cooking equipment. Listed noncanopy exhaust hoods and filters shall be sized and installed in accordance with the terms of their listing and the manufacturer's installation instructions.

Exception: Listed hood assemblies designed and installed specifically for the intended use.

508.5 Exhaust Hood Assemblies with Integrated Supply-Air Plenums.

508.5.1 The construction and size of exhaust hood assemblies with integrated supply air plenums shall comply with the requirements of Sections 508.1 and 508.4. [NFPA 96:5.3.1]

508.5.2 The construction of the outer shell or the inner exhaust shell shall comply with Section 508.1. [NFPA 96:5.3.2]

508.5.3 Where the outer shell is welded, the inner shell shall be of grease-tight construction. [NFPA 96:5.3.3]

508.5.4 Fire Dampers.

508.5.4.1 A fire-actuated damper shall be installed in the supply air plenum at each point where a supply air duct inlet or a supply air outlet penetrates the continuously welded shell of the assembly. [NFPA 96:5.3.4.1]

508.5.4.2 The fire damper shall be listed for such use or be part of a listed exhaust hood with or without exhaust damper. [NFPA 96:5.3.4.2]

508.5.4.3 The actuation device shall have a maximum temperature rating of 141°C (286°F). [NFPA 96:5.3.4.3]

508.5.5 Supply air plenums that discharge air from the face rather than from the bottom or into the exhaust hood and that are isolated from the exhaust hood by the continuously welded shell extending to the lower outermost perimeter of the entire hood assembly shall not require a fire-actuated damper. [NFPA 96:5.3.4.4]

508.6 Listed Hood Assemblies. Listed hood assemblies shall be installed in accordance with the terms of their listing and the manufacturer's instructions. [NFPA 96:5.4.1] Listed hood assemblies shall be tested in accordance with UL 710 or equivalent. [NFPA 96:5.4.2]

508.7 Solid-Fuel Hood Assemblies. Where solid-fuel cooking equipment is to be used, the solid-fuel hood assembly shall comply with Section 517.0.

508.8 Listed Ultra-Violet Hoods. Listed ultra-violet hoods shall be installed and maintained in accordance with the terms of their listing and the manufacturer's instructions. [NFPA 96:5.5]

508.9 Exhaust Outlets. An exhaust outlet within an unlisted hood shall be located so as to optimize the capture of particulate matter. Each outlet shall serve not more than a twelve (12) foot (3,658 mm) section of an unlisted hood.

509.0 Grease Removal Devices in Hoods.

509.1 Grease Removal Devices. Listed grease filters, baffles, or other approved grease removal devices for use with commercial cooking equipment shall be provided. Listed grease filters shall be tested in accordance with UL 1046, Grease Filters for Exhaust Ducts. Mesh filters shall not be used.

509.2 Installation.

509.2.1 The distance between the grease removal device and the cooking surface shall be as great as possible but not less than eighteen (18) inches (457 mm). [NFPA 96:6.2.1.1]

509.2.2 Where grease removal devices are used in conjunction with charcoal or charcoal-type broilers, including gas or electrically heated charbroilers, a minimum vertical distance of four (4) feet (1,219 mm) shall be maintained between the lower edge of the grease removal device and the cooking surface. [NFPA 96:6.2.1.2]

Exception No. 1: Grease removal devices supplied as part of listed hood assemblies shall be installed in accordance with the terms of the listing and the manufacturer's instructions. [NFPA 96:6.2.1.5]

Exception No. 2: For cooking equipment without exposed flame and where flue gases bypass grease removal devices, the minimum vertical distance shall be permitted to be reduced to not less than six (6) inches (152 mm). [NFPA 96:6.2.1.3]

Exception No. 3: Where a grease removal device is listed for separation distances less than those required in 509.2.1 and 509.2.2, the listing requirements shall be permitted. [NFPA 96:6.2.1.4]

509.2.3 Grease Removal Device Protection. Grease removal devices shall be protected from combustion gas outlets and from direct flame impingement occurring during normal operation of cooking appliances producing high flue gas temperatures, such as deep-fat fryers or upright or high broilers (salamander broilers), where the distance between the grease removal device and the appliance flue outlet (heat source) is less than eighteen (18) inches (457 mm). [NFPA 96:6.2.2.1]

509.2.3.1 This protection shall be permitted to be accomplished by the installation of a steel or stainless steel baffle plate between the heat source and the grease removal device. [NFPA 96:6.2.2.2]

509.2.3.2 The baffle plate shall be sized and located so that flames or combustion gases shall travel a distance not less than eighteen (18) inches (457 mm) from the heat source to the grease removal device. [NFPA 96:6.2.2.3]

509.2.3.3 The baffle shall be located not less than six (6) inches (152 mm) from the grease removal devices. [NFPA 96:6.2.2.4]

509.2.4 Grease Filters. Grease filters shall be listed and constructed of steel or listed equivalent material and shall be of rigid construction that will not distort or crush under normal operation, handling, and cleaning conditions. [NFPA 96:6.2.3.1 and 6.2.3.2]

509.2.4.1 Filters shall be tight fitting and firmly held in place and grease filters shall be arranged so that all exhaust air passes through the grease filters. [NFPA 96:6.2.3.3]

509.2.4.2 Filters shall be easily accessible and removable for cleaning. [NFPA 96:6.2.3.4]

509.2.4.3 Filters shall be installed at an angle not less than forty-five (45) degrees from the horizontal. [NFPA 96:6.2.3.5]

509.2.5 Grease Drip Trays. Filters shall be equipped with a drip tray beneath their lower edges. [NFPA 96:6.2.4.1]

509.2.5.1 The tray shall be kept to the minimum size needed to collect grease and shall be pitched to drain into an enclosed metal container having a capacity not exceeding one (1) gallon (3.785 L) [NFPA 96:6.2.4.2, and 6.2.4.3].

509.2.6 Grease Filter Orientation. Grease filters that require a specific orientation to drain grease shall be clearly so designated, or the hood shall be constructed so that filters cannot be installed in the wrong orientation. [NFPA 96:6.2.5]

509.3 Solid-Fuel Grease Removal Devices. Where solid-fuel cooking equipment is provided with grease removal devices, these devices shall comply with Section 517.0.

510.0 Exhaust Duct Systems.

510.1 General.

510.1.1 Ducts shall not pass through fire walls or fire partitions.

510.1.2 All ducts shall lead as directly as is practicable to the exterior of the building, so as not to unduly increase any fire hazard.

510.1.3 Duct systems shall not be interconnected with any other building ventilation or exhaust system.

510.1.4 All ducts shall be installed without forming dips or traps that might collect residues. In manifold (common duct) systems, the lowest end of the main duct shall be connected flush on the bottom with the branch duct. Duct systems serving a Type I hood shall be so constructed and installed that grease cannot become pocketed in any portion thereof, and the system shall slope not less than 1/4 inch per lineal foot (21 mm/m) toward the hood or toward an approved grease reservoir. Where horizontal ducts exceed seventy-five (75) feet (22,860 mm) in length, the slope shall be not less than one (1) inch per lineal foot (83 mm/m).

510.1.5 Openings required for accessibility shall comply with Section 510.3.

510.1.6 A sign shall be placed on all access panels stating the following:

ACCESS PANEL—DO NOT OBSTRUCT

510.1.7 Duct bracing and supports shall be of noncombustible material, securely attached to the structure and designed to carry gravity and lateral loads within the stress limitations of the Building Code. Bolts, screws, rivets, and other mechanical fasteners shall not penetrate duct walls.

510.1.8 Ducts, Non-Grease. Ducts and plenums serving Type II hoods shall be constructed of rigid metallic materials as set forth in Chapter 6. Duct bracing and supports shall comply with Chapter 6. Ducts subject to positive pressure shall be adequately sealed.

510.2 Clearance.

510.2.1 Clearance between ducts and combustible materials shall be provided in accordance with the requirements of Section 507.2.

510.2.2 Listed grease ducts shall be installed in accordance with the terms of their listings and manufacturers' instructions.

510.3 Openings.

510.3.1 Openings shall be provided at the sides or at the top of the duct, whichever is more accessible, and at changes of direction. Openings shall be protected by approved access panels that comply with Section 510.3.4.5.

Exception: Openings shall not be required in portions of the duct that are accessible from the duct entry or discharge.

510.3.2 For hoods with dampers in the exhaust or supply collar, an access panel for cleaning and inspection shall be provided in the duct or the hood collar. This access panel shall be as close to the hood as possible but shall not exceed eighteen (18) inches (457 mm).

Exception: Dampers that are accessible from under the hood.

510.3.3 Exhaust fans with ductwork connected to both sides shall have access for cleaning and inspection within three (3) feet (914 mm) of each side of the fan. [NFPA 96:7.3.7]

510.3.4 Openings in Ducts. Openings shall conform with Sections 510.3.4.1 through 510.3.4.5. [NFPA 96:7.4]

510.3.4.1 Horizontal Ducts. On horizontal ducts, at least one 20 inch by 20 inch (508 mm x 508 mm) opening shall be provided for personnel entry. [NFPA 96:7.4.1.1]

510.3.4.1.1 Horizontal ducting shall be secured sufficiently to allow for the weight of personnel entry into the duct. Where an opening of this size is not possible, openings large enough to permit thorough cleaning shall be provided at twelve (12) feet (3,658 mm) intervals. [NFPA 96:7.4.1.2]

510.3.4.1.2 Openings on horizontal grease duct systems shall be provided with safe access and a work platform when not easily accessible from a 10 feet (3 m) stepladder. [NFPA 96:7.4.1.3]

510.3.4.1.3 Support systems for horizontal grease duct systems twenty-four (24) inches (609 mm) and larger in any cross-sectional dimension shall be designed for the weight of the ductwork plus 363 kg (800 lb) at any point in the duct systems. [NFPA 96:7.4.1.4]

510.3.4.2 On nonlisted ductwork, the edge of the opening shall be not less than 1-1/2 inch (38.1 mm) from all outside edges of the duct or welded seams. [NFPA 96:7.4.1.5]

510.3.4.3 Vertical Ducts. On vertical ductwork where personnel entry is possible, access shall be provided at the top of the vertical riser to accommodate descent. [NFPA 96:7.4.2.1]

510.3.4.3.1 Where personnel entry is not possible, adequate access for cleaning shall be provided on each floor. [NFPA 96:7.4.2.2]

510.3.4.4 Access Panels. Access panels shall be of the same material and thickness as the duct (Section 510.5.1). Access panels shall have a gasket or sealant that is rated for 1,500°F (815.6°C) and shall be grease-tight. Fasteners, such as bolts, weld studs, latches, or wing nuts, used to secure the access panels shall be carbon steel or stainless steel and shall not penetrate duct walls.

Exception: Listed grease duct access door assemblies (access panels) shall be installed in accordance with their terms of the listings and the manufacturer's instructions.

510.3.4.5 Openings for installation, servicing, and inspection of listed fire protection system devices and for duct cleaning shall be provided in ducts and enclosures and shall conform to the requirements of 510.3. Enclosure openings required to reach access panels in the ductwork shall be large enough for the removal of the access panel.

510.4 Listed Grease Ducts. Listed grease ducts shall be installed in accordance with the terms of the listing and the manufacturer's instructions.

510.5 Other Grease Ducts. Other grease ducts shall comply with the requirements of this section.

510.5.1 Materials. Ducts shall be constructed of and supported by carbon steel not less than 0.054 inch (1.37 mm) (No. 16 MSG) in thickness or stainless steel not less than 0.043 inch (1.09 mm) (No. 18 MSG) in thickness.

510.5.2 Installation.

510.5.2.1 All seams, joints, penetrations, and duct-to-hood collar connections shall have a liquid-tight continuous external weld.

Exception No. 1: Duct-to-hood collar connections as shown in Figure 5-2 shall be permitted.

Exception No. 2: Penetrations shall be permitted to be sealed by other listed devices that are tested to be grease-tight and are evaluated under the same conditions of fire severity as the hood or enclosure of listed grease extractors and whose presence does not detract from the hood's or duct's structural integrity.

Exception No. 3: Internal welding shall be permitted, provided the joint is formed or ground smooth and is readily accessible for inspection.

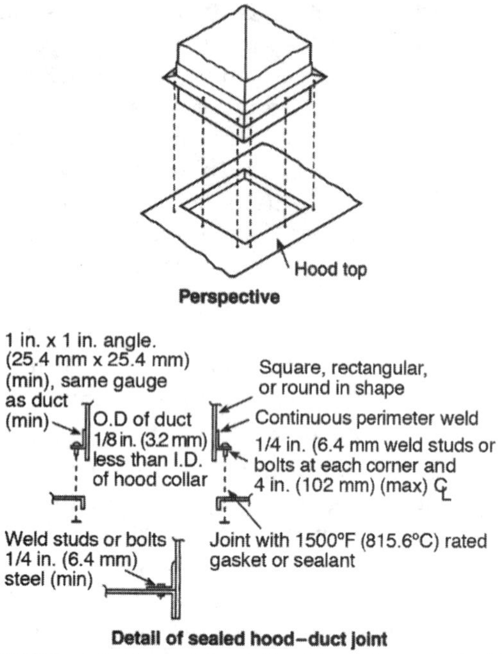

FIGURE 5-2 Permitted Duct-to-Hood Collar Connection. [NFPA 96:7.5.2.2]

510.5.2.2 Overlapping duct connections of either the telescoping or the bell type shall be used for welded field joints, not butt-weld connections. The inside duct section shall always be uphill of the outside duct section. The difference between inside dimensions of overlapping sections shall not exceed 1/4 inch (6.4 mm). The overlap shall not exceed two (2) inches (50.8 mm). (See Figure 5-3.)

510.6 Exterior Installations.

510.6.1 The exterior portion of the ductwork shall be vertical wherever possible and shall be installed and adequately supported on the exterior of a building. Bolts, screws, rivets, and other mechanical fasteners shall not penetrate duct walls. Clearance of any ducts shall comply with Section 507.2.

510.6.2 All ducts shall be protected on the exterior by paint or other suitable weather-protective coating or shall be constructed of noncorrosive stainless steel. Ductwork subject to corrosion shall have minimal contact with the building surface.

510.7 Interior Installations.

510.7.1 In all buildings more than one story in height, and in one-story buildings where the roof-ceiling assembly is required to have a fire resistance rating, the ducts shall be enclosed in a continuous enclosure extending from the lowest fire-rated ceiling or floor above the hood, through any concealed spaces, to or through the roof so as to maintain the integrity of the fire separations

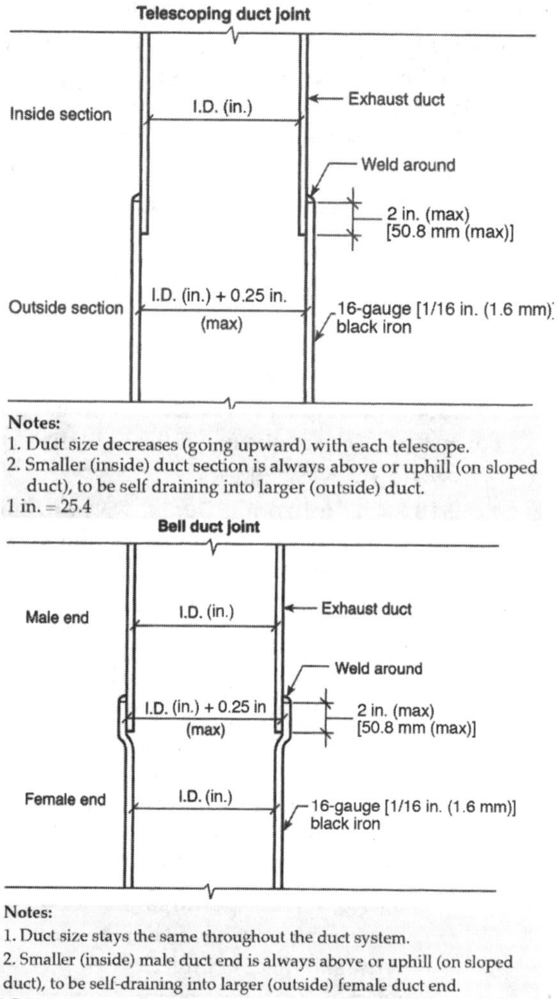

FIGURE 5-3 Telescoping and Bell-Type Duct Connections. [NFPA 96:7.5.5.1(a)(b)]

required by the applicable Building Code provisions. The enclosure shall be sealed around the duct at the point of penetration of the lowest fire-rated ceiling or floor above the hood in order to maintain the fire resistance rating of the enclosure and shall be vented to the exterior of the building through weather-protected openings.

Exception: The continuous enclosure provisions shall not be required where a field-applied grease duct enclosure or a factory-built grease duct enclosure (see Section 507.2.3) is protected with a listed duct-through-penetration protection system equivalent to the fire resistance rating of the assembly being penetrated, and the materials are installed in accordance with the conditions of their listings and the manufacturers' instructions and are acceptable to the Authority Having Jurisdiction.

510.7.2 The enclosure required in Section 510.7.1 shall conform to Sections 510.7.2.1 through 510.7.2.3.

510.7.2.1 If the building is less than four stories in height, the enclosure wall shall have a fire resistance rating of not less than one (1) hour.

510.7.2.2 If the building is four stories or more in height, the enclosure wall shall have a fire resistance rating of not less than two (2) hours.

510.7.2.3 Clearance from the duct or the exhaust fan to the interior surface of enclosures of combustible construction shall be not less than eighteen (18) inches (457 mm), and clearance from the duct to the interior surface of enclosures of noncombustible or limited-combustible construction shall be not less than six (6) inches (152 mm). Provisions for reducing clearances as described in Section 507.2 are not applicable to enclosures.

Exception: Clearance from the outer surfaces of field-applied grease duct enclosures and factory-built grease duct enclosures to the interior surfaces of construction installed around them shall be permitted to be reduced where the field-applied grease duct enclosure materials and the factory-built grease duct enclosures are installed in accordance with the conditions of their listings and the manufacturers' instructions and are acceptable to the Authority Having Jurisdiction.

510.7.3 For field-applied grease duct enclosures and factory-built grease duct enclosures, the materials and products shall conform to Sections 510.7.3.1 through 510.7.3.3.

510.7.3.1 Field-applied grease duct enclosures and factory-built grease duct enclosures shall demonstrate that they provide sufficient mechanical and structural integrity, resiliency, and stability when subjected to expected building environmental conditions, duct movement under general operating conditions, and duct movement due to interior and exterior fire conditions.

510.7.3.2 Measures shall be taken to prevent physical damage to any covering or enclosure material. Any damage to the covering or enclosure shall be repaired and the covering or enclosure restored to meet its intended listing and fire-resistive rating and to be acceptable to the Authority Having Jurisdiction.

510.7.3.3 In the event of a fire within a kitchen exhaust system, the duct, the enclosure, or the covering directly applied to the duct shall be inspected by qualified personnel to determine whether the duct, the enclosure, and the covering directly applied to the duct are structurally sound, capable of maintaining their fire protection functions, suitable for continued operation, and acceptable to the Authority Having Jurisdiction.

510.7.4 For listed grease ducts, see Section 510.4.

510.7.5 If openings in the enclosure walls are provided, they shall be protected by approved self-closing fire doors of proper rating. Fire doors shall be installed in accordance with NFPA 80, *Standard for Fire Doors and Fire Windows*. Openings on other listed materials or products shall be clearly identified and labeled according to the terms of the listing and the manufacturer's instructions and shall be acceptable to the Authority Having Jurisdiction. The panels shall be readily accessible.

510.7.6 Each duct system shall constitute an individual system serving only exhaust hoods in one fire zone on one floor. Multiple ducts shall not be permitted in a single enclosure unless acceptable to the Authority Having Jurisdiction.

510.8 Termination of Exhaust System.

510.8.1 The exhaust system shall terminate as follows:

(A) Outside the building with a fan or duct.

(B) Through the roof, or to the roof from outside, as in Section 510.8.2, or through a wall, as in Section 510.8.3.

510.8.2 Rooftop Terminations.

510.8.2.1 Rooftop terminations shall be arranged with or provided with the following:

(A) A minimum of ten (10) feet (3,048 mm) of clearance from the outlet to adjacent buildings, property lines, and air intakes. Where space limitations absolutely prevent a ten (10) foot (3,048 mm) horizontal separation from an air intake, a vertical separation shall be permitted, with the exhaust outlet being a minimum of three (3) feet (914 mm) above any air intake located within ten (10) feet (3,048 mm) horizontally.

(B) The exhaust flow directed up and away from the surface of the roof and a minimum of forty (40) inches (1,016 mm) above the roof surface.

(C) The ability to drain grease out of any traps or low points formed in the fan or duct near the termination of the system into a collection container that is noncombustible, closed, rainproof, structurally sound for the service to which it is applied, and will not sustain combustion. A grease collection device that is applied to exhaust systems shall not inhibit the performance of any fan.

Exception: Grease containers that are evaluated for equivalency with the preceding requirements and listed as such.

(D) A listed grease duct complying with Section 510.4, or with ductwork complying with Section 510.5.

(E) A hinged upblast fan supplied with flexible weatherproof electrical cable and service hold-open retainer to permit proper inspection and cleaning that is listed for commercial cooking equipment, provided the ductwork extends a minimum of eighteen (18) inches (457 mm) above the roof surface and the fan discharges a minimum of forty (40) inches (1,016 mm) above the roof surface. (See Section 511.1.1.)

(F) Other approved fan, provided (1) it meets the requirements of Sections 510.8.2(C) and 511.1.3, and (2) its discharge or its extended duct discharge meets the requirements of Section 510.8.2(B).

510.8.2.2 Listed flexible connectors may be used on exterior roof locations when required for proper equipment vibration isolation.

510.8.2.3 Fans shall be provided with safe access and a work surface for inspection and cleaning. [NFPA 96:7.8.2.2]

510.8.3 Wall Terminations.

510.8.3.1 Wall terminations shall be arranged with or provided with the following properties:

(A) Through a noncombustible wall with a minimum of ten (10) feet (3,048 mm) of clearance from the outlet to adjacent buildings, property lines, grade level, combustible construction, or electrical equipment or lines, and the closest point of any air intake or operable door or window at or below the plane of the exhaust termination. The closest point of any air intake or operable door or window above the plane of the exhaust termination shall be a minimum of ten (10) feet (3,048 mm) distant, plus 0.25 feet (0.076m) for each one (1) degree from horizontal, the angle of degree being measured from the center of the exhaust termination to the center of the air intake or operable door or window. (See Figure 5-4.)

Exception: A wall termination in a secured area can be at a lower height above grade if acceptable to the Authority Having Jurisdiction.

(B) The exhaust flow directed perpendicularly outward from the wall face or upward.

(C) All the ductwork pitched to drain the grease back into the hood(s), or with a drain provided to bring the grease back into a container within the building or into a remote grease trap.

(D) A listed grease duct complying with Section 510.4, or other ducts complying with Section 510.5.

(E) An approved fan, provided it meets the requirements of Sections 510.8.3.1(C) and 511.1.1 or 511.1.3.

510.9 Solid-Fuel Duct Systems. Where solid-fuel cooking equipment is to be vented, the duct system shall comply with Section 517.0.

511.0 Air Movement.

511.1 Exhaust Fans for Commercial Cooking Equipment.

Where solid-fuel cooking equipment is to be vented, the duct system shall comply with Section 517.0.

511.1.1 Approved upblast fans with motors surrounded by the airstream shall be hinged, supplied with flexible weatherproof electrical cable and service hold-open retainers, and listed for this use. (See Figure 5-5.) Installation shall conform to the requirements of Section 510.8.

511.1.2 In-Line Exhaust Fans.

511.1.2.1 In-line fans shall be of the type with the motor located outside the airstream and with belts and pulleys protected from the airstream by a grease-tight housing. They shall be connected to the exhaust duct by flanges securely bolted as shown in Figures 5-6(a) – 5-6(d), or by a system specifically listed for such use. Flexible connectors shall not be used. [NFPA 96:8.1.2.1 – 8.1.2.3]

511.1.2.2 If the design or positioning of the fan allows grease to be trapped, a drain directed to a readily accessible and visible grease receptacle, not exceeding one (1) gallon (3.8 L), shall be provided. In-line exhaust fans shall be located in an easily accessible area of

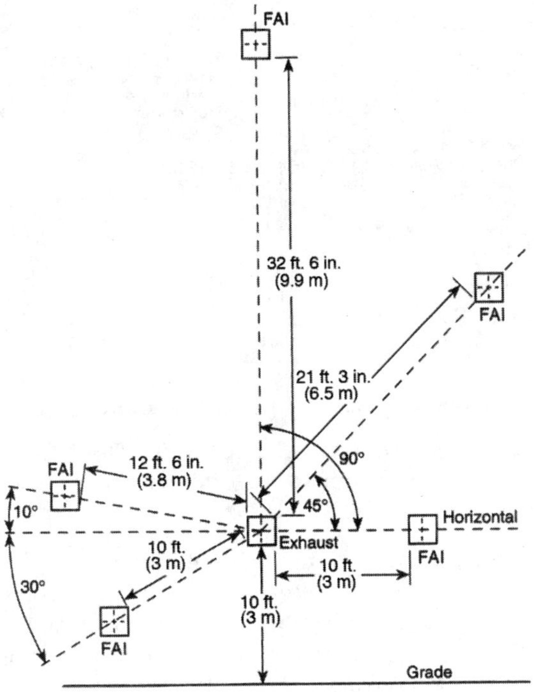

Notes:

1. Fresh air intake (FAI) applies to any air intake, including an operable door or window.
2. Example:
   FAI is same plane as exhaust or lower: 10 ft. [3 m (min.)] between closet edges.
   FAI above plane of exhaust: 10 ft. + 0.25 ft. (3 m + 0.076 m) degree between closet edge.

FIGURE 5-4 Exhaust Termination Distance from Fresh Air Intake (FAI) or Operable Door or Window. [NFPA 96:7.8.3]

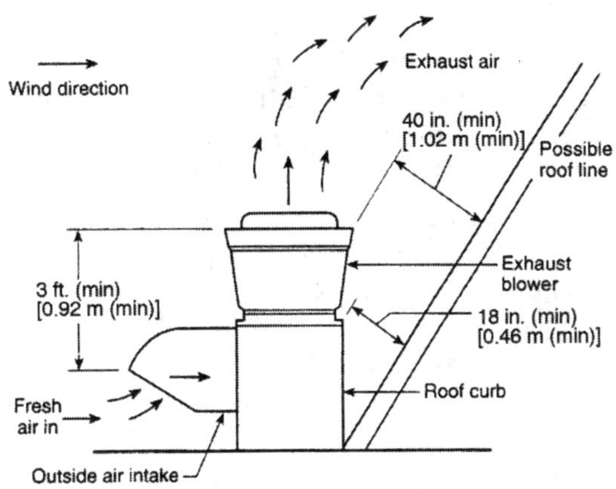

FIGURE 5-5 Upblast Fan Clearances. [NFPA 96:7.8.2.1]

adequate size to allow for service or removal. If the duct system connected to the fan is in an enclosure, the space or room in which the exhaust fan is located shall have the same fire resistance rating as the enclosure. [NFPA 96:8.1.2.4 – 8.1.2.6]

511.1.3 Utility set exhaust fans, if installed at the rooftop termination point, shall meet the requirements of Sections 510.8.2.1(A) and Section 510.8.2.3. If the fan is installed within the building, it shall be located in an easily accessible area of adequate size to allow for service or removal. If the duct system connected to the fan is in an enclosure, the space or room in which the exhaust fan is located shall have the same fire resistance rating as the enclosure. The fan shall be connected to the exhaust duct by flanges securely bolted as shown in Figures 5-6(a) through 5-6(d) or by a system specifically listed for such use. Flexible connectors shall not be used. Exhaust fans shall have a drain directed to a readily accessible and visible grease receptacle not to exceed one (1) gallon (3.8 L). [NFPA 96:8.1.3.1 - 8.1.3.6]

511.1.4 Exhaust fan housings shall be constructed of carbon steel not less than 0.054 inch (1.37 mm) (No. 16 MSG) in thickness or stainless steel not less than 0.043 inch (1.09 mm) (No. 18 MSG) in thickness or, if listed, constructed in accordance with the terms of the listing. [NFPA 96:8.1.4]

511.1.5 Openings for cleaning, servicing, and inspection shall conform to the requirements of Section 510.3.3. Clearances shall conform to the requirements of Sections 507.2, or 510.7.2.3 if installed within an enclosure.

511.1.6 All wiring and electrical equipment shall comply with NFPA 70, *National Electrical Code*.

511.2 Airflow.

511.2.1 The air velocity through any duct shall be not less than 500 feet per minute (152.4 meters per minute). [NFPA 96:8.2.1.1]

Exception: Transition duct sections shall be permitted to be connected to hoods and exhaust fans that do not meet this velocity, provided that they do not exceed three (3) feet (914 mm) in length and do not contain traps for grease.

511.2.2 Transition duct sections that do not exceed three (3) feet (914 mm) in length and do not contain grease traps shall be permitted to be connected to hoods and exhaust fans that do not meet this velocity. [NFPA 96:8.2.1.2]

511.2.3 Exhaust-air volumes for hoods shall be of sufficient level to provide for capture and removal of grease-laden cooking vapors. Test data, performance acceptable to the Authority Having Jurisdiction, or both, shall be provided, displayed, or both, upon request. [NFPA 96:8.2.2.1, 8.2.2.2]

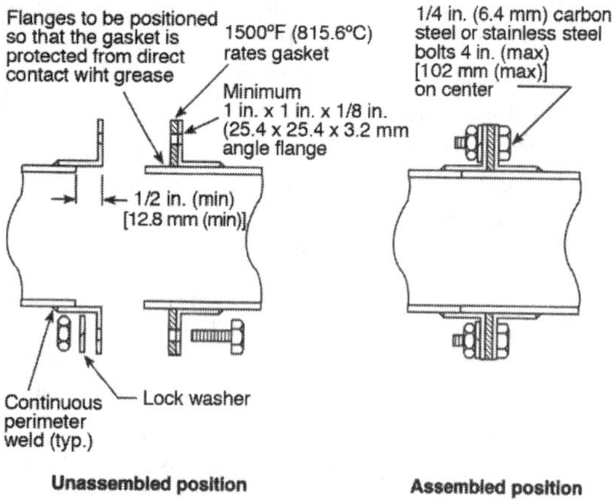

FIGURE 5-6(a) Typical Section of Duct-to-Fan Connection–Butt Joint Method. [NFPA 96:8.1.2.2(a)]

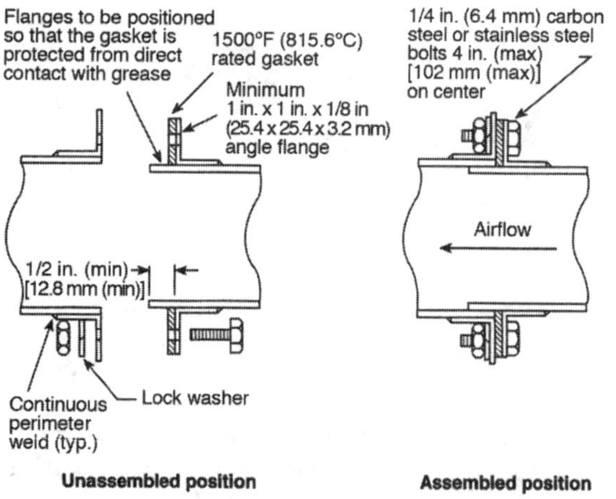

FIGURE 5-6(b) Typical Section of Duct-to-Fan Connection–Overlapping Method. [NFPA 96: 8.1.2.2(b)]

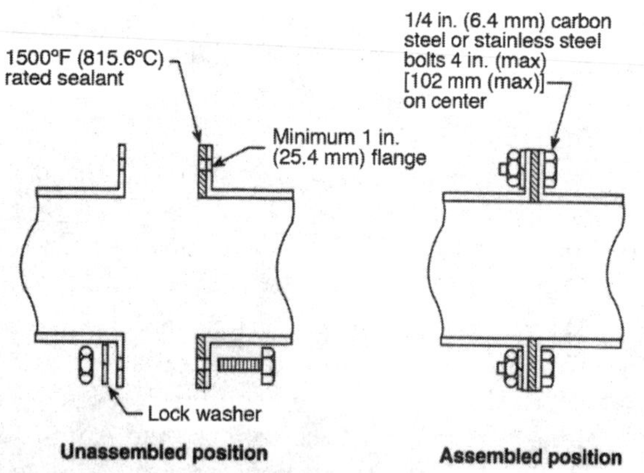

FIGURE 5-6(c) Typical Section of Duct-to-Fan Connection–Sealant Method. [NFPA 96: 8.1.2.2(c)]

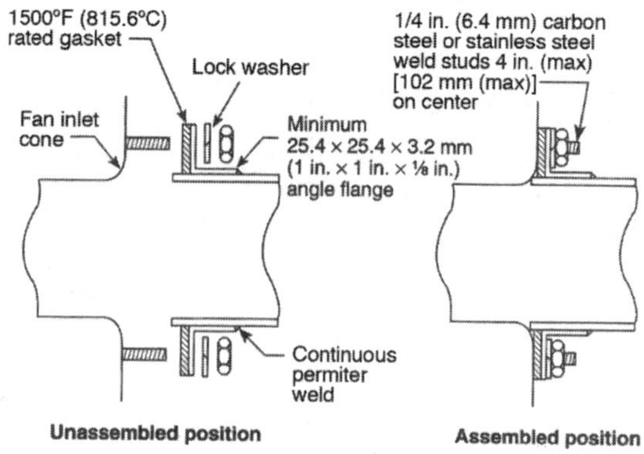

FIGURE 5-6(d) Typical Section of Duct-to-Fan Connection–Direct to Fan Inlet Cone Method. [NFPA 96:8.1.2.2(d)]

Exception: Lower exhaust-air volumes shall be permitted during no-load cooking conditions, provided they are sufficient to capture and remove flue gases and residual vapors from cooking equipment.

511.2.4 A hood exhaust fan(s) shall continue to operate after the extinguishing system has been activated, unless fan shutdown is required by a listed component of the ventilation system or by the design of the extinguishing system. It shall not be required to restart the hood exhaust fan when the extinguishing system is activated if the exhaust fan and all cooking equipment served by the fan had previously been shut down. The hood exhaust fan shall not be required to start upon activation of the extinguishing system if the exhaust fan and all cooking equipment served by the fan have previously been shut down. [NFPA 96:8.2.3.1 - 8.2.3.2]

511.3 Replacement Air. Replacement air quantity shall be adequate to prevent negative pressures in the commercial cooking area(s) from exceeding 0.02 inch water column (4.98 kPa). When its fire-extinguishing system discharges, makeup air supplied internally to a hood shall be shut off. [NFPA 96:8.3.1 - 8.3.2]

Exception: Compensating hoods shall meet the airflow requirements specified in Sections 508.4.1.3 through 508.4.1.5. Compensating hoods shall extract at least 20 percent of their required exhaust airflow from the kitchen area.

511.4 Common Duct (Manifold) Systems.

511.4.1 Master kitchen exhaust ducts that serve multiple tenants shall include provision to bleed air from outdoors or from adjacent spaces into the master exhaust duct where required to maintain the necessary minimum air velocity in the master exhaust duct. [NFPA 96:8.4.1]

511.4.2 The bleed-air ducts shall connect to the top or side of the master exhaust duct. [NFPA 96:8.4.2]

511.4.3 The bleed-air duct shall have a fire damper at least twelve (12) inches (305 mm) from the exhaust duct connection. [NFPA 96:8.4.3]

511.4.4 The bleed-air duct shall have the same construction and clearance requirements as the main exhaust duct from the connection to the exhaust duct to at least twelve (12) inches (305 mm) on both sides of the fire damper. [NFPA 96:8.4.4]

511.4.5 Each bleed air duct shall have a means of adjusting (e.g., using volume dampers) the bleed-air quantity. [NFPA 96:8.4.5]

511.4.6 Such means to adjust the bleed air quality shall be installed in between the fire damper and the source of bleed air. [NFPA 96:8.4.6]

511.4.7 A bleed air duct shall not be used for the exhaust of grease-laden vapors and shall be so labeled. [NFPA 96:8.4.7]

511.4.8 Unused tenant exhaust connections to the master exhaust duct that are not used as bleed air connections shall be disconnected and sealed at the main duct. [NFPA 96:8.4.8]

511.5 Solid-Fuel Air Movement Requirements. Where solid-fuel cooking equipment is used, exhaust and replacement air also shall comply with Section 517.0.

512.0 Auxiliary Equipment.

512.1 Dampers. Dampers shall not be installed in exhaust ducts or exhaust duct systems. [NFPA 96:9.1.1]

512.1.1 Where specifically listed for such use or where required as part of a listed device or system, dampers in exhaust ducts or exhaust duct systems shall be permitted. [NFPA 96:9.1.2]

512.2 Electrical Equipment.

512.2.1 Wiring systems of any type shall not be installed in ducts. [NFPA 96:9.2.1]

512.2.2 Only where specifically listed for such use shall motors, lights, and other electrical devices be permitted to be installed in ducts or hoods or located in the path of travel of exhaust products. [NFPA 96:9.2.2]

512.2.3 Lighting units on hoods shall not be located in concealed spaces.

512.2.4 Lighting units shall be permitted in concealed spaces where such units are part of a listed exhaust hood. [NFPA 96:9.2.3.3]

512.2.5 Listed lighting units specifically listed for such use and installed in accordance with the terms of the listing shall be permitted to be installed in concealed spaces. [NFPA 96:9.2.3.4]

512.2.6 All electrical equipment shall be installed in accordance with NFPA 70, *National Electrical Code*, with due regard to the effects of heat, vapor, and grease on the equipment.

512.3 Other Equipment.

512.3.1 Fume incinerators, thermal recovery units, air pollution control devices, or other devices shall be permitted to be installed in ducts or hoods or located in the path of travel of exhaust products where specifically approved for such use. Downgrading other parts of the exhaust system due to the installation of these approved devices, whether listed or not, shall not be allowed. [NFPA 96:9.3.1 - 9.3.2]

512.3.2 Any equipment, listed or otherwise, that provides secondary filtration or air pollution control and that is installed in the path of travel of exhaust products shall be provided with an approved automatic fire-extinguishing system for the protection of the component sections of the equipment and shall include protection of the ductwork downstream of the equipment, whether or not the equipment is provided with a damper. Filter media used in secondary filtration or air pollution control units and not complying with Section 509.2.4 shall have fire protection that is adequate for the filter media being used in accordance with the fire-extinguishing system manufacturer's instructions. If the equipment can be a source of ignition, it shall be provided with appropriate detection to operate the fire-extinguishing system. [NFPA 96:9.3.3, 9.3.3.1, 9.3.4]

512.3.3 Where a cooking exhaust system employs an air pollution control device that recirculates air into the building, the provisions of Section 516.0 shall apply. [NFPA 96:9.3.5]

512.4 Solid-Fuel Auxiliary Equipment. Where solid fuel cooking comprises any part of a cooking operation, additional provisions and equipment as described in Section 517.0 shall be used where required.

513.0 Fire-Extinguishing Equipment.

513.1 General Requirements.

513.1.1 Fire-extinguishing equipment for the protection of grease removal devices, hood exhaust plenums, and exhaust duct systems shall be provided. [NFPA 96:10.1.1]

513.1.2 Cooking equipment that produces grease-laden vapors (such as, but not limited to, deep-fat fryers, ranges, griddles, broilers, woks, tilting skillets, and braising pans) shall be protected by fire-extinguishing equipment. [NFPA 96:10.1.2]

513.2 Types of Equipment.

513.2.1 Fire-extinguishing equipment shall include both automatic fire-extinguishing systems as primary protection and portable fire extinguishers as secondary backup. [NFPA 96:10.2.1]

513.2.1.1 A placard identifying the use of the extinguisher as secondary backup means to the automatic fire suppression system shall be conspicuously placed near each portable fire extinguisher in the cooking area. [NFPA 96:10.2.2]

513.2.2 Automatic fire-extinguishing systems shall comply with standard UL 300, *Fire Testing of Fire Extinguishing Systems for Protection of Restaurant Cooking Areas*, or other equivalent standards and shall be installed in accordance with their listings. [NFPA 96:10.2.3]

Exception: Automatic fire-extinguishing equipment provided as part of listed recirculating systems complying with standard UL 710B, *Outline of Investigation for Recirculating Exhaust System.*

513.2.2.1 Automatic fire-extinguishing systems shall be installed in accordance with the terms of their listing, the manufacturer's instructions, and the following standards where applicable. [NFPA 96:10.2.6]

(1) NFPA 12, *Standard on Carbon Dioxide Extinguishing Systems*

(2) NFPA 13, *Standard for the Installation of Sprinkler Systems*

(3) NFPA 17, *Standard for Dry Chemical Extinguishing Systems*

(4) NFPA 17A, *Standard for Wet Chemical Extinguishing Systems*

513.2.3 In existing systems, when changes in the cooking media, positioning, or replacement of cooking equipment occur, the fire-extinguishing system shall be made to comply with 513.2.2. [NFPA 96:10.2.3.1]

513.2.4 Modification of Existing Hood Systems.

513.2.4.1 Any abandoned pipe or conduit from a previous installation shall be removed from within the hood, plenum, and exhaust duct. [NFPA 96:10.2.7.1]

513.2.4.2 Penetrations and holes resulting from the removal of conduit or piping shall be sealed with listed or equivalent liquid tight sealing devices. [NFPA 96:10.2.7.2]

513.2.4.3 The addition of obstructions to spray patterns from the cooking appliance nozzle(s) such as baffle plates, shelves, or any modification shall not be permitted. [NFPA 96:10.2.7.3]

513.2.4.4 Changes or modifications to the hazard after installation of the fire extinguishing systems shall result in reevaluation of the system design by a properly trained, qualified, and certified person(s). [NFPA 96:10.2.7.4]

513.2.5 Fixed Baffle Hoods with Water Wash. Grease removal devices, hood exhaust plenums, and exhaust ducts requiring protection in accordance with Section 513.1.1 shall be permitted to be protected by a listed fixed baffle hood containing a constant or fire-actuated water wash system that is listed and in compliance with UL 300 or other equivalent standards and shall be installed in accordance with the requirements of their listing. [NFPA 96:10.2.8.1]

513.2.5.1 The water for listed, fixed baffle hood assemblies shall be permitted to be supplied from the domestic water supply when the minimum water pressure and flow are provided in accordance with the terms of the listing. [NFPA 96:10.2.8.3]

513.2.5.2 The water supply shall be controlled by a supervised water supply control valve. [NFPA 96:10.2.8.4]

513.2.5.3 The water wash in the fixed baffle hood shall be activated by the cooking equipment extinguishing system. [NFPA 96:10.2.8.5]

513.2.5.4 A water-wash system approved to be used for protection of the grease removal device(s), hood exhaust plenum(s), exhaust duct(s), or combination thereof shall include instruction and appropriate electrical interface for simultaneous activation of the water-wash system from an automatic fire-extinguishing system, where the automatic fire-extinguishing system is used for cooking equipment protection only. [NFPA 96:10.2.8.6]

513.2.5.5 Where the fire-extinguishing system provides protection for the cooking equipment, hood, and duct, activation of the water wash shall not be required. [NFPA 96:10.2.8.7]

513.2.5.6 The water required for listed automatic fire-extinguishing systems shall be permitted to be supplied from the domestic water supply where the minimum water pressure and flow are provided in accordance with the terms of the listing. The water supply shall be controlled by a supervised water supply control valve. [NFPA 96:10.2.9]

513.2.6 Water Valve Supervision. Valves controlling the water supply to listed fixed baffle hood assemblies, automatic fire-extinguishing systems, or both shall be listed indicating type of valve and shall be supervised open by one of the following methods:

(1) Central station, proprietary, or remote station alarm service

(2) Local alarm service that will cause the sounding of an audible signal at a constantly attended point

(3) Locking valves open

(4) Sealing of valves and approved weekly recorded inspection. [NFPA 96:10.2.10]

513.3 Simultaneous Operation. Fixed-pipe extinguishing systems in a single hazard area shall be arranged for simultaneous automatic operation upon actuation of anyone of the systems. [NFPA 96:10.3.1]

513.3.1 Simultaneous operation shall not be required where the fixed pipe extinguishing system is an automatic sprinkler system. [NFPA 96:10.3.2]

513.3.2 Simultaneous operation shall not be required where dry or wet chemical system shall be permitted to be used to protect common exhaust ductwork by one of the methods specified in NFPA 17 or NFPA 17A. [NFPA 96:10.3.3]

513.4 Fuel Shutoff.

513.4.1 Upon activation of any fire-extinguishing system for a cooking operation, all sources of fuel and electric power that produce heat to all equipment requiring protection by that system shall automatically shut off. [NFPA 96:10.4.1]

Exception: Solid-fuel cooking operations.

513.4.2 Steam supplied from an external source shall not be required to automatically shut off. [NFPA 96:10.4.2]

513.4.3 Any gas appliance not requiring protection, but located under the same ventilating equipment, shall also automatically shut off upon activation of any extinguishing system. [NFPA 96:10.4.3]

513.4.4 Shutoff devices shall require manual reset. [NFPA 96:10.4.4]

513.5 Manual Activation.

513.5.1 A readily accessible means for manual activation shall be located between forty-two (42) inches and forty-eight (48) inches (1,067 mm and 1,219 mm) above the floor, located in a path of exit or egress, and clearly identify the hazard protected. A manual actuation device shall be located a minimum of ten (10) feet (3 m) when possible and a maximum of twenty (20) feet (6 m) from the protected kitchen appliance(s) within the path of egress. Manual actuation using a cable-operated pull station shall not require more than forty (40) pounds (178 N) of force, with a pull movement not to exceed fourteen (14) inches (356 mm) to actuate the fire suppression system. The automatic and manual means of system activation external to the control head or releasing device shall be separate and independent of each other so that failure of one will not impair the operation of the other. [NFPA 96:10.5.1 – 10.5.2]

513.5.2 The manual means of system activation shall be permitted to be common with the automatic means if the manual activation device is located between the control head or releasing device and the first fusible link. [NFPA 96:10.5.3]

513.5.3 An automatic sprinkler system shall not require a manual means of system activation. [NFPA 96:10.5.4]

513.5.4 The means for manual actuator(s) shall be mechanical or rely on electrical power for actuation. [NFPA 96:10.5.5]

513.5.5 Electrical power shall be permitted to be used for manual activation if a standby power supply is provided or if supervision is provided in accordance with Section 513.7. [NFPA 96:10.5.6]

513.6 System Annunciation.

513.6.1 Upon activation of an automatic fire extinguishing system, an audible alarm or visual indicator shall be provided to show that the system has activated. [NFPA 96:10.6.1]

513.6.2 Where a fire alarm signaling system is serving the occupancy where the extinguishing system is located, the activation of the automatic fire-extinguishing system shall activate the fire alarm signaling system. [NFPA 96:10.6.2]

513.7 System Supervision.

513.7.1 Where electrical power is required to operate the automatic fire-extinguishing system, it shall be monitored by a supervisory alarm, with a standby power supply provided. [NFPA 96:10.7.1]

513.7.2 System supervision shall not be required where an automatic fire-extinguishing system(s) includes automatic mechanical detection and actuation as a backup detection system. [NFPA 96:10.7.2]

513.7.3 System supervision shall not be required where a fire-extinguishing system(s) is interconnected or interlocked with the cooking equipment power source(s) so that if the fire extinguishing system becomes inoperable due to power failure, all sources of fuel or electric power that produce heat to all cooking equipment serviced by that hood shall automatically shut off. [NFPA 96:10.7.3]

513.7.4 System supervision shall not be required where an automatic fire-extinguishing system, including automatic mechanical detection and actuation, is electrically connected to a listed fire-actuated water-wash system for simultaneous operation of both systems. [NFPA 96:10.7.4]

513.8 Special Design and Application.

513.8.1 Hoods containing automatic fire extinguishing systems are protected areas; therefore, these hoods are not considered obstructions to overhead sprinkler systems and shall not require floor coverage underneath. [NFPA 96:10.8.1]

513.8.2 A single listed detection device, listed with the extinguishing system, shall be permitted for more than one (1) appliance when installed in accordance with the terms of the listing. [NFPA 96:10.8.2]

513.9 Review and Certification.

513.9.1 Where required, complete drawings of the system installation, including the hood(s), exhaust duct(s), and appliances, along with the interface of the fire-extinguishing system detectors, piping, nozzles, fuel shutoff devices, agent storage container(s), and manual actuation device(s), shall be submitted to the Authority Having Jurisdiction. [NFPA 96:10.9.1]

513.10 Installation Requirements.

513.10.1 Installation of systems shall be performed only by persons properly trained and qualified to install the specific system being provided. The installer shall provide certification that the installation is in agreement with the terms of the listing and the manufacturer's instructions and/or approved design. [NFPA 96:10.9.2]

513.11 Portable Fire Extinguishers.

513.11.1 Portable fire extinguishers shall be installed in kitchen cooking areas in accordance with NFPA 10, *Standard for Portable Fire extinguishers.* Such extinguishers shall use agents that specify upon contact with hot grease, such as sodium bicarbonate and potassium bicarbonate dry chemical and potassium carbonate solutions. Class B gas-type portables such as CO_2 and halon shall not be permitted in kitchen cooking areas. Manufacturer's recommendations shall be followed. [NFPA 96:10.10.1 – 10.10.4]

513.11.1.1 Portable fire extinguishers listed specifically for use in the kitchen cooking areas shall also be permitted.

513.11.2 Other fire extinguishers in the kitchen area shall be installed in accordance with NFPA 10, *Standard for Portable Fire Extinguishers.* [NFPA 96:10.10.5]

513.12 Solid-Fuel Fire-Extinguishing Equipment. Where solid-fuel cooking equipment is served by fire extinguishing equipment, the provisions of Section 517.0 shall apply.

514.0 Procedures for the Use and Maintenance of Equipment.

514.1 Operating Procedures.

514.1.1 Exhaust systems shall be operated whenever cooking equipment is turned on. [NFPA 96:11.1.1]

514.1.2 Filter-equipped exhaust systems shall not be operated with filters removed. [NFPA 96:11.1.2]

514.1.3 Openings provided for replacing air exhausted through ventilating equipment shall not be restricted by covers, dampers, or any other means that would reduce the operating efficiency of the exhaust system. [NFPA 96:11.1.3]

514.1.4 Instructions for manually operating the fire-extinguishing system shall be posted conspicuously in the kitchen and shall be reviewed periodically with employees by the management. [NFPA 96:11.1.4]

514.1.5 Listed exhaust hoods shall be operated in accordance with the terms of their listings and the manufacturer's instructions. [NFPA 96:11.1.5]

514.1.6 Cooking equipment shall not be operated while its fire-extinguishing system or exhaust system is nonoperational or otherwise impaired. [NFPA 96:11.1.6]

514.1.7 Secondary filtration and pollution control equipment shall be operated in accordance with the terms of its listing and the manufacturer's recommendations. [NFPA 96:11.1.7]

514.1.8 Inspection and maintenance of equipment allowed in Section 512.3.1 shall be conducted by properly trained and qualified persons at a frequency determined by the manufacturer's instructions or equipment listing. [NFPA 96:11.1.8]

514.2 Inspection. Maintenance of the fire-extinguishing systems and listed exhaust hoods containing a constant or fire-activated water system that is listed to extinguish a fire in the grease removal devices, hood exhaust plenums, and exhaust ducts shall be made by properly trained, qualified, and certified person(s) or company acceptable to the Authority Having Jurisdiction at least every six (6) months. [NFPA 96:11.2.1]

514.2.1 All actuation components, including remote manual pull stations, mechanical or electrical devices, detectors, actuators, and fire-actuated dampers, shall be checked for proper operation during the inspection in accordance with the manufacturer's listed procedures. In addition to these requirements, the specific inspection requirements of the applicable NFPA standard shall also be followed. [NFPA 96-04:11.2.2 - 11.2.3]

514.2.2 Fusible links (including fusible links on fire damper assemblies) and automatic sprinkler heads shall be replaced at least semiannually, or more frequently if necessary where required by the manufacturer. [NFPA 96-04:11.2.4]

514.2.3 The year of manufacture and the date of installation of the fusible links shall be marked on the system inspection tag. The tag shall be signed or initialed by the installer. [NFPA 96:11.2.5]

Other detection devices not including fusible links and automatic sprinklers shall be serviced or replaced in accordance with the manufacturer's recommendations. [NFPA 96-04:11.2.6]

514.2.4 Where automatic bulb-type sprinklers or spray nozzles are used and annual examination shows no buildup of grease or other material on the sprinkler or spray nozzles, annual replacement shall not be required. [NFPA 96:11.2.7]

514.2.5 If required, certificates of inspection and maintenance shall be forwarded to the Authority Having Jurisdiction. [NFPA 96:11.2.8]

514.3 Inspection for Grease Buildup.

514.3.1 The entire exhaust system shall be inspected for grease buildup by a properly trained, qualified, and certified company or person(s) acceptable to the Authority Having Jurisdiction in accordance with Table 5-4.

TABLE 5-4
Exhaust System Inspection Schedule
[NFPA 96: Table 11.4]

Type or Volume of Cooking Frequency	Frequency
Systems serving solid-fuel cooking operations	Monthly
Systems serving high-volume cooking operations such as 24-hour cooking, charbroiling, or wok cooking	Quarterly
Systems serving moderate-volume cooking operations	Semi-annually
Systems serving low-volume cooking operations, such as churches, day camps, seasonal businesses, or senior centers.	Annually

514.4 Cleaning of Exhaust Systems.

514.4.1 Upon inspection, if the exhaust system is found to be contaminated with deposits from grease-laden vapors, the contaminated portions of the exhaust system shall be cleaned by a properly trained, qualified, and certified company or person(s) acceptable to the Authority Having Jurisdiction. [NFPA 96:11.6.1]

514.4.2 Hoods, grease removal devices, fans, ducts, and other appurtenances shall be cleaned to remove combustible contaminant's prior to surfaces becoming heavily contaminated with grease or oily sludge. [NFPA 96:11.6.2]

514.4.3 At the start of the cleaning process, electrical switches that could be activated accidentally shall be locked out. [NFPA 96:11.6.3]

514.4.4 Components of the fire suppression system shall not be rendered inoperable during the cleaning process. [NFPA 96:11.6.4]

514.4.5 Fire-extinguishing systems shall be permitted to be rendered inoperable during the cleaning process where serviced by properly trained and qualified persons. [NFPA 96:11.6.5]

514.4.6 Flammable solvents or other flammable cleaning aids shall not be used. [NFPA 96:11.6.6]

514.4.7 Cleaning chemicals shall not be applied on fusible links or other detection devices of the automatic extinguishing system. [NFPA 96:11.6.7]

514.4.8 After the exhaust system is cleaned, it shall not be coated with powder or other substance. [NFPA 96:11.6.8]

514.4.9 When cleaning procedures are completed, all access panels (doors) and cover plates shall be restored to their normal operational condition. [NFPA 96:11.6.9]

514.4.10 When an access panel is removed, a service company label or tag preprinted with the name of the company and giving the date of inspection or cleaning shall be affixed near the affected access panels. [NFPA 96:11.6.10]

514.4.11 Dampers and diffusers shall be positioned for proper airflow. [NFPA 96:11.6.11]

514.4.12 When cleaning procedures are completed, all electrical switches and system components shall be returned to an operable state. [NFPA 96:11.6.12]

514.4.13 When an exhaust cleaning service is used, a certificate showing the name of the servicing company, the name of the person performing the work, and the date of inspection or cleaning shall be maintained on the premises. [NFPA 96:11.6.13]

514.4.14 After cleaning or inspection is completed, the exhaust cleaning company and the person performing the work at the location shall provide the owner of the system with a written report that also specifies areas that were inaccessible or not cleaned. [NFPA 96:11.6.14]

514.4.15 Where required, certificates of inspection and cleaning and reports of areas not cleaned shall be submitted to the Authority Having Jurisdiction. [NFPA 96:11.6.15]

515.0 Minimum Safety Requirements for Cooking Equipment.

515.1 Cooking Equipment.

515.1.1 Cooking equipment shall be approved based on one of the following criteria:

(A) Listings by a testing laboratory.

(B) Test data acceptable to the Authority Having Jurisdiction. [NFPA 96:12.1.1]

515.1.2 Installation.

515.1.2.1 All listed appliances shall be installed in accordance with the terms of their listings and the manufacturer's instructions. [NFPA 96:12.1.2.1]

515.1.2.2 Cooking appliances requiring protection shall not be moved, modified, or rearranged without prior reevaluation of the fire-extinguishing system by the system installer or servicing agent, unless otherwise allowed by the design of the fire-extinguishing system. [NFPA 96:12.1.2.2]

515.1.2.3 The fire-extinguishing system shall not require re-evaluation where the cooking appliances are moved to perform maintenance and cleaning provided the appliances are returned to approved design location prior to cooking operations, and any disconnected fire-extinguishing system nozzles attached to the appliances are reconnected in accordance with the manufacturer's listed design manual. [NFPA 96:12.1.2.3]

515.1.2.4 All deep-fat fryers shall be installed with at least a sixteen (16) inch (406 mm) space between the fryer and surface flames from adjacent cooking equipment. [NFPA 96:12.1.2.4]

515.1.2.5 Where a steel or tempered glass baffle plate is installed at a minimum eight (8) inches (203 mm) in height between the fryer and surface flames of the adjacent appliance, the requirement for a sixteen (16) inch (406 mm) space shall not apply. [NFPA 96:12.1.2.5]

515.1.2.6 If the fryer and the surface flames are at different horizontal planes, the minimum height of eight (8) inches (203 mm) shall be measured from the higher of the two. [NFPA 96:12.1.2.5.1]

515.2 Operating Controls. Deep-fat fryers shall be equipped with a separate high-limit control in addition to the adjustable operating control (thermostat) to shut off fuel or energy when the fat temperature reaches 475°F (246°C) at one (1) inch (25 mm) below the surface. [NFPA 96:12.2]

515.3 Cooking Equipment Maintenance.

515.3.1 An inspection and servicing of the cooking equipment shall be made at least annually by properly trained and qualified persons. [NFPA 96:11.5.1]

515.3.2 Cooking equipment that collects grease below the surface, behind the equipment, or in cooking equipment flue gas exhaust, such as griddles or charbroilers, shall be inspected and, if found with grease accumulation, cleaned by a properly trained, qualified, and certified person acceptable to the authority having jurisdiction. [NFPA 96:11.5.2]

516.0 Recirculating Systems.

516.1 General Requirements. Recirculating systems containing or for use with appliances used in processes producing smoke or grease-laden vapors shall be equipped with components complying with the following:

(A) The clearance requirements of Section 507.2.

(B) A hood complying with the requirements of Section 508.0.

(C) Grease removal devices complying with Section 509.0.

(D) The air movement requirements of Sections 511.2.1 and 511.2.2.

(E) Auxiliary equipment (such as particulate and odor removal devices) complying with Section 512.0.

(F) Fire-extinguishing equipment complying with the requirements of Section 513.0.

Exception: References to ducts in Sections 513.1.1 and 513.5.1.

(G) The use and maintenance requirements of 514.0.

(H) The minimum safety requirements of Section 515.0.

(I) All the requirements of this chapter. [NFPA 96:13.1]

516.2 Design Restrictions. All recirculating systems shall comply with Section 516.2. [NFPA 96:13.2]

516.2.1 Only gas- or electrically fueled cooking appliances shall be used. Gas-fueled appliances shall not have the combustion flue outlet(s) directly connected to the recirculating system. Gas-fueled appliances shall have a minimum eighteen (18) inch (457 mm) clearance from the flue outlet to the filter inlet in accordance with Section 509.2.3 and shall meet the installation requirements of NFPA 54, *National Fuel Gas Code,* or NFPA 58, *Liquefied Petroleum Gas Code.* [NFPA 96:13.2.1 – 13.2.3]

516.2.2 Recirculating systems shall be listed with a testing laboratory. The recirculating system listing shall include integral fire protection for recirculating hoods, including canopy type [NFPA 96:13.2.4.1]

516.2.3 Cooking appliances that require protection and that are under a recirculating hood shall be protected by either the integral fire protection system for the hood or a system in accordance with Section 513.0. [NFPA 96:13.2.4.2]

516.2.4 A recirculating system shall not use cooking equipment that exceeds that recirculating system's labeled maximum limits for that type of equipment, stated in maximum energy input, maximum cooking temperature, and maximum square area of cooking surface or cubic volume of cooking cavity. [NFPA 96:13.2.6]

516.2.5 The listing label shall show the type(s) of cooking equipment tested and the maximum limits specified in 516.2.4. [NFPA 96:13.2.7]

516.2.6 A fire-actuated damper shall be installed at the exhaust outlet of the system. The actuation device shall have a maximum temperature rating of 375°F (190°C). [NFPA 96:13.2.8, 13.2.10]

516.2.7 No electrical wiring shall be installed in the interior sections of the hood plenum that might become exposed to grease.

Exception: As permitted by NFPA 70, *National Electrical Code.*

516.2.8 The power supply of any electrostatic precipitator (ESP) shall be of the "cold spark," ferro-resonant type in that the voltage falls off as the current draw of a short increases. [NFPA 96:13.2.11]

516.2.9 Listing evaluation shall include the following:

(A) Capture and containment of vapors at published and labeled airflows.

(B) Grease discharge at the exhaust outlet of the system not to exceed an average of 5 mg/m^3 of exhausted air sampled from that equipment at maximum amount of product that is capable of being processed over a continuous eight (8) hour test per EPA Test Method 202, *Determination of Condensable Particulate Emissions for Stationary Sources,* with the system operating at its minimum listed airflow.

(C) Listing and labeling of clearance to combustibles from all sides, top, and bottom.

(D) Electrical connection in the field in accordance with NFPA 70, *National Electrical Code.*

(E) Interlocks on all removable components that lie in the path of airflow within the unit to ensure that they are in place during operation of the cooking appliance. [NFPA 96:13.2.12]

516.3 Interlocks. The recirculating system shall be provided with interlocks of all critical components and operations as indicated in Sections 516.3.1 through 516.3.4 such that, if any of these interlocks are interrupted, the cooking appliance shall not be able to operate. [NFPA 96:13.3.1]

516.3.1 All closure panels encompassing airflow sections shall have interlocks to ensure the panels are in place and fully sealed.[NFPA 96:13.3.2]

516.3.2 Each filter component (grease and odor) shall have an interlock to prove the component is in place. [NFPA 96:13.3.3]

516.3.3 ESP Interlocks.

516.3.3.1 Each ESP shall have a sensor to prove its performance is as designed, with no interruption of the power to exceed two (2) minutes. [NFPA 96:13.3.4.1]

516.3.3.2 The sensor shall be a manual reset device or circuit. [NFPA 96:13.3.4.2]

516.3.4 An airflow switch or transducer shall be provided after the last filter component to ensure that a minimum airflow is maintained. This device opens the interlock circuit when the airflow falls twenty-five (25) percent below the system's normal operating flow or ten (10) percent below its listed minimum rating, whichever is lower. This switch or transducer shall be a manual reset device or circuit. [NFPA 96:13.3.5]

516.4 Location and Application Restrictions. The location of recirculating systems shall be approved by the Authority Having Jurisdiction. Items to be reviewed in the fire risk assessment shall include, but not be limited to, life safety, combustibility of surroundings, proximity to air vents, and total fuel load. [NFPA 96:13.4]

516.5 Additional Fire Safety Requirements.

516.5.1 In addition to the appliance nozzle(s), a recirculating system shall be listed with the appropriate fire protection for grease filters, grease filtration, odor filtration units, and ductwork where applicable. [NFPA 96:13.5.1]

516.5.2 In addition to any other fire-extinguishing system actuation device, there shall be a fire-extinguishing system actuation device installed downstream of any ESP.[NFPA 96:13.5.2]

516.5.3 The requirements of Section 513.6 shall also apply to recirculating system locations. [NFPA 96:13.5.3]

516.6 Use and Maintenance.

516.6.1 Automatic or manual covers on cooking appliances, especially fryers, shall not interfere with the application of the fire suppression system.[NFPA 96:13.6.1]

516.6.2 All filters shall be cleaned or replaced in accordance with the manufacturer's instructions. [NFPA 96:13.6.2]

516.6.3 All ESPs shall be cleaned a minimum of once per week following manufacturer's cleaning instructions. [NFPA 96:13.6.3]

516.6.4 The entire hood plenum and the blower section shall be cleaned a minimum of once every three (3) months. [NFPA 96:13.6.4]

516.6.5 Inspection and testing of the total operation and all safety interlocks in accordance with the manufacturer's instructions shall be performed by qualified service personnel a minimum of once every six (6) months, or more frequently if required. [NFPA 96:13.6.5]

516.6.6 Fire-extinguishing equipment shall be inspected in accordance with Section 514.2. [NFPA 96:13.6.6]

516.6.7 A signed and dated log of maintenance as performed in accordance with Sections 516.6.4 and 516.6.5 shall be available on the premises for use by the Authority Having Jurisdiction. [NFPA 96:13.6.7]

517.0 Solid-Fuel Cooking Operations.

517.1 Venting Application. Venting requirements of solid-fuel cooking operations shall be determined in accordance with Sections 517.1.1 through 517.1.6.

517.1.1 Where solid-fuel cooking equipment is required by the manufacturer to have a natural draft, the vent shall comply with Section 517.4.

517.1.2 Where the solid-fuel cooking equipment has a self-contained top, is the only appliance to be vented in an isolated space (except for a single water heater with its own separate vent), has a separate makeup-air system, and is provided with supply and return air (not supplied or returned from other spaces), the system shall comply with Sections 517.4 and 517.6.

517.1.3 Where the solid-fuel cooking equipment is located in a space with other vented equipment, all vented equipment shall have an exhaust system interlocked with a makeup-air system for the space as per Section 517.6.

517.1.4 Natural draft ventilation systems and power-exhausted ventilation systems shall comply with Sections 517.3, 517.4, and 517.6.

517.1.5 Where a solid-fuel cooking appliance allows effluent to escape from the appliance opening, this opening shall be covered by a hood and an exhaust system that meets the requirements of Sections 517.3, 517.4, and 517.6.

517.1.6 Solid-fuel cooking operations shall have spark arresters to minimize the passage of airborne sparks and embers into plenums and ducts. Where the solid-fuel cooking operation is not located under a hood, a spark arrester shall be provided to minimize the passage of sparks and embers into flues and chimneys.

517.2 Location of Appliances.

517.2.1 Every appliance shall be located with respect to building construction and other equipment so as to permit access to the appliance.

517.2.2 Solid-fuel cooking appliances shall not be installed in confined spaces.

Exception: Solid-fuel cooking appliances listed for installation in confined spaces shall be installed in accordance with the terms of the listing and the manufacturer's instructions.

517.2.3 Solid-fuel cooking appliances shall not be installed in any location where gasoline or any other flammable vapors or gases are likely to be present.

517.3 Hoods for Solid-Fuel Cooking.

517.3.1 Hoods shall be sized and located in a manner capable of capturing and containing all of the effluent discharging from the appliances.

The hood and its exhaust system shall comply with the provisions of the code.

517.3.2 All solid-fuel cooking equipment served by hood and duct systems shall be separate from all other exhaust systems.

Exception: Cooking equipment not requiring automatic fire-extinguishing equipment (as per the provisions of Section 510.0) shall be permitted to be installed under a common hood with solid-fuel cooking equipment that is served by a duct system separate from all other exhaust systems.

517.4 Exhaust Systems for Solid-Fuel Cooking. Where a hood is not required, in buildings where the duct system is three (3) stories or less in height, a duct complying with Section 510.0 shall be provided.

517.4.1 If a hood is used, the duct system shall conform with the Section 510.0.

517.4.2 A listed or approved grease duct system shall be provided for solid-fuel cooking exhaust systems that are four (4) stories in height or greater.

517.4.3 Where a hood is used, the duct system shall conform with the requirements of Section 510.0.

517.4.4 Wall terminations of solid-fuel exhaust systems shall be prohibited.

517.5 Grease Removal Devices for Solid-Fuel Cooking. Grease removal devices shall be constructed of steel or stainless steel or be approved for solid-fuel cooking.

517.5.1 If airborne sparks and embers can be generated by the solid-fuel cooking operation, spark arrester devices shall be used prior to the grease removal device to minimize the entrance of these sparks and embers into the grease removal device and into the hood and duct system.

517.5.2 Filters shall be a minimum of four (4) feet (1.2 m) above the appliance cooking surface.

517.6 Air Movement for Solid-Fuel Cooking. Exhaust system requirements shall comply with the provisions of the code for hooded operation or shall be installed in accordance with the manufacturer's recommendation for unhooded applications.

517.6.1 A replacement or makeup-air system shall be provided to ensure a positive supply of replacement air at all times during cooking operations.

517.6.2 Makeup-air systems serving solid-fuel cooking operations shall be interlocked with the exhaust-air system and powered, if necessary, to prevent the space from attaining a negative pressure while the solid-fuel appliance is in operation.

517.7 Fire-Extinguishing Equipment for Solid-Fuel Cooking.

517.7.1 Approved fire-extinguishing equipment shall be provided to protect all solid-fuel-burning cooking appliances that might be a source of ignition of grease in the hood, grease removal device, or duct.

Exception: If acceptable to the Authority Having Jurisdiction, solid-fuel-burning cooking appliances constructed of solid masonry or reinforced portland or refractory cement concrete and vented in accordance with NFPA 211, *Standard for Chimneys, Fireplaces, Vents, and Solid Fuel–Burning Appliances*, Chapters 3 and 4, shall not require fixed automatic fire-extinguishing equipment.

517.7.2 Approved fire-extinguishing equipment shall be provided for the protection of grease removal devices, hoods, and duct systems.

Exception: If acceptable to the Authority Having Jurisdiction, solid-fuel-burning cooking appliances constructed of solid masonry or reinforced portland or refractory cement concrete and vented in accordance with NFPA 211, *Standard for Chimneys, Fireplaces, Vents, and Solid Fuel–Burning Appliances*, Chapters 3 and 4, shall not require automatic fire-extinguishing equipment.

517.7.3 Approved fire-extinguishing equipment for solid-fuel-burning cooking appliances, where required, shall comply with NFPA 211 and shall be comprised of water-based agents.

517.7.4 Fire-extinguishing equipment shall be rated and designed to extinguish solid-fuel cooking fires, in accordance with the manufacturer's recommendations. The fire-extinguishing equipment shall be of sufficient size to totally extinguish fire in the entire hazard area and prevent reignition of the fuel.

517.7.5 All solid-fuel appliances (whether or not under a hood) with fireboxes of five (5) $ft.^3$ (0.14 m^3) volume or less shall at least have a listed 2-A rated water-type fire extinguisher or a 1.6 gal (6 L) wet chemical fire extinguisher listed for Class K fires in accordance with NFPA 10, in the immediate vicinity of the appliance. [NFPA 96:14.7.8]

517.7.6 Solid-fuel appliances with fireboxes exceeding five (5) $ft.^3$ (0.14 m^3) shall be provided with a fixed-water-pipe system with a hose in

TABLE 5-5 (continued)
Minimum Sheet Metal Thickness for Round Ducts for Product-Conveying System Ducts

Negative Pressure	Reinf. Spacing (Inches)	Class 3								
		Up to 7"	8" to 11"	12" to 15"	16" to 19"	20" to 23"	24" to 35"	36" to 47"	48" to 59"	60"
12" to 15"	0	.033 (20 ga.)	.033 (20 ga.)	.055 (16 ga.)	.068 (14 ga.)	.097 (12 ga.)	.097 (12 ga.)			
	96	.033 (20 ga.)	.033 (20 ga.)	.044 (18 ga.)	.044 (18 ga.)	.055 (16 ga.)	.055 (16 ga.)	.097 (12 ga.)	.112 (11 ga.)	.112 (11 ga.)
	48	.033 (20 ga.)	.033 (20 ga.)	.044 (18 ga.)	.044 (18 ga.)	.055 (16 ga.)	.055 (16 ga.)	.068 (14 ga.)	.068 (14 ga.)	.068 (14 ga.)
	24	.033 (20 ga.)	.033 (20 ga.)	.044 (18 ga.)	.044 (18 ga.)	.055 (16 ga.)	.055 (16 ga.)	.068 (14 ga.)	.068 (14 ga.)	.068 (14 ga.)
16" to 20"	0	.033 (20 ga.)	.044 (18 ga.)	.068 (14 ga.)	.097 (12 ga.)	.112 (11 ga.)				
	96	.033 (20 ga.)	.033 (20 ga.)	.055 (16 ga.)	.055 (16 ga.)	.068 (14 ga.)	.068 (14 ga.)	.112 (11 ga.)	.112 (11 ga.)	.112 (11 ga.)
	48	.033 (20 ga.)	.033 (20 ga.)	.044 (18 ga.)	.044 (18 ga.)	.055 (16 ga.)	.055 (16 ga.)	.068 (14 ga.)	.068 (14 ga.)	.097 (12 ga.)
	24	.033 (20 ga.)	.033 (20 ga.)	.044 (18 ga.)	.044 (18 ga.)	.055 (16 ga.)	.055 (16 ga.)	.068 (14 ga.)	.068 (14 ga.)	.068 (14 ga.)
		Class 4								
To 7"	0	.055 (16 ga.)	.055 (16 ga.)	.055 (16 ga.)	.055 (16 ga.)	.068 (14 ga.)	.068 (14 ga.)	.127 (10 ga.)		
	96	.055 (16 ga.)	.055 (16 ga.)	.055 (16 ga.)	.055 (16 ga.)	.068 (14 ga.)	.068 (14 ga.)	.097 (12 ga.)	.097 (12 ga.)	.097 (12 ga.)
	48	.055 (16 ga.)	.055 (16 ga.)	.055 (16 ga.)	.055 (16 ga.)	.068 (14 ga.)	.068 (14 ga.)	.097 (12 ga.)	.097 (12 ga.)	.097 (12 ga.)
	24	.055 (16 ga.)	.055 (16 ga.)	.055 (16 ga.)	.055 (16 ga.)	.068 (14 ga.)	.068 (14 ga.)	.097 (12 ga.)	.097 (12 ga.)	.097 (12 ga.)
8" to 11"	0	.055 (16 ga.)	.055 (16 ga.)	.055 (16 ga.)	.055 (16 ga.)	.068 (14 ga.)	.097 (12 ga.)			
	96	.055 (16 ga.)	.055 (16 ga.)	.055 (16 ga.)	.055 (16 ga.)	.068 (14 ga.)	.068 (14 ga.)	.097 (12 ga.)	.097 (12 ga.)	.097 (12 ga.)
	48	.055 (16 ga.)	.055 (16 ga.)	.055 (16 ga.)	.055 (16 ga.)	.068 (14 ga.)	.068 (14 ga.)	.097 (12 ga.)	.097 (12 ga.)	.097 (12 ga.)
	24	.055 (16 ga.)	.055 (16 ga.)	.055 (16 ga.)	.055 (16 ga.)	.068 (14 ga.)	.068 (14 ga.)	.097 (12 ga.)	.097 (12 ga.)	.097 (12 ga.)
12" to 15"	0	.055 (16 ga.)	.055 (16 ga.)	.055 (16 ga.)	.068 (14 ga.)	.097 (12 ga.)	.097 (12 ga.)			
	96	.055 (16 ga.)	.055 (16 ga.)	.055 (16 ga.)	.055 (16 ga.)	.068 (14 ga.)	.068 (14 ga.)	.097 (12 ga.)	.112 (11 ga.)	.112 (11 ga.)
	48	.055 (16 ga.)	.055 (16 ga.)	.055 (16 ga.)	.055 (16 ga.)	.068 (14 ga.)	.068 (14 ga.)	.097 (12 ga.)	.097 (12 ga.)	.097 (12 ga.)
	24	.055 (16 ga.)	.055 (16 ga.)	.055 (16 ga.)	.055 (16 ga.)	.068 (14 ga.)	.068 (14 ga.)	.097 (12 ga.)	.097 (12 ga.)	.097 (12 ga.)
16" to 20"	0	.055 (16 ga.)	.055 (16 ga.)	.068 (14 ga.)	.097 (12 ga.)	.112 (11 ga.)				
	96	.055 (16 ga.)	.055 (16 ga.)	.055 (16 ga.)	.055 (16 ga.)	.068 (14 ga.)	.068 (14 ga.)	.112 (11 ga.)	.112 (11 ga.)	.112 (11 ga.)
	48	.055 (16 ga.)	.055 (16 ga.)	.055 (16 ga.)	.055 (16 ga.)	.068 (14 ga.)	.068 (14 ga.)	.097 (12 ga.)	.097 (12 ga.)	.097 (12 ga.)
	24	.055 (16 ga.)	.055 (16 ga.)	.055 (16 ga.)	.055 (16 ga.)	.068 (14 ga.)	.068 (14 ga.)	.097 (12 ga.)	.097 (12 ga.)	.097 (12 ga.)

Note: 1 in. = 25.4 mm

TABLE 5-6
Minimum Sheet Metal Thickness for Rectangular Ducts

Negative Pressure	Reinf. Spacing	Longest Side of Duct									
		Class 1					Class 2				
		Up to 12"	13" to 24"	25" to 36"	37" to 48"	49" to 60"	Up to 12"	13" to 24"	25" to 36"	37" to 48"	49" to 60"
To 7"	48	.021 (24 ga.)	.033 (20 ga.)	.055 (16 ga.)	.068 (14 ga.)		.033 (20 ga.)	.044 (18 ga.)	.055 (16 ga.)	.068 (14 ga.)	
To 7"	24	.021 (24 ga.)	.027 (22 ga.)	.033 (20 ga.)	.033 (20 ga.)	.033 (20 ga.)	.033 (20 ga.)	.044 (18 ga.)	.055 (16 ga.)	.055 (16 ga.)	.055 (16 ga.)
To 7"	12	.021 (24 ga.)	.021 (24 ga.)	.021 (24 ga.)	.021 (24 ga.)	.021 (24 ga.)	.033 (20 ga.)	.044 (18 ga.)	.055 (16 ga.)	.055 (16 ga.)	.055 (16 ga.)
8" to 11"	48	.027 (22 ga.)	.068 (14 ga.)	.097 (12 ga.)	.097 (12 ga.)		.033 (20 ga.)	.068 (14 ga.)	.097 (12 ga.)	.097 (12 ga.)	
8" to 11"	24	.027 (22 ga.)	.055 (16 ga.)	.055 (16 ga.)	.068 (14 ga.)	.068 (14 ga.)	.033 (20 ga.)	.055 (16 ga.)	.055 (16 ga.)	.068 (14 ga.)	.068 (14 ga.)
8" to 11"	12	.021 (24 ga.)	.027 (22 ga.)	.027 (22 ga.)	.027 (22 ga.)	.027 (22 ga.)	.033 (20 ga.)	.044 (18 ga.)	.055 (16 ga.)	.055 (16 ga.)	.055 (16 ga.)
12" to 15"	48	.044 (18 ga.)	.097 (12 ga.)				.044 (18 ga.)	.044 (18 ga.)			
12" to 15"	24	.044 (18 ga.)	.055 (16 ga.)	.097 (12 ga.)	.097 (12 ga.)	.097 (12 ga.)	.044 (18 ga.)	.044 (18 ga.)	.097 (12 ga.)	.097 (12 ga.)	.097 (12 ga.)
12" to 15"	12	.027 (22 ga.)	.044 (18 ga.)	.044 (18 ga.)	.044 (18 ga.)	.044 (18 ga.)	.033 (20 ga.)	.044 (18 ga.)	.055 (16 ga.)	.055 (16 ga.)	.055 (16 ga.)
16" to 20"	48	.068 (14 ga.)	.112 (11 ga.)				.068 (14 ga.)	.112 (11 ga.)			
16" to 20"	24	.068 (14 ga.)	.068 (14 ga.)	.112 (11 ga.)	.112 (11 ga.)	.112 (11 ga.)	.068 (14 ga.)	.068 (14 ga.)	.112 (11 ga.)	.112 (11 ga.)	.112 (11 ga.)
16" to 20"	12	.033 (20 ga.)	.068 (14 ga.)	.068 (14 ga.)	.068 (14 ga.)	.068 (14 ga.)	.033 (20 ga.)	.044 (18 ga.)	.068 (14 ga.)	.068 (14 ga.)	.068 (14 ga.)
		Class 3					Class 4				
To 7"	48	.044 (18 ga.)	.055 (16 ga.)	.068 (14 ga.)	.068 (14 ga.)		.055 (16 ga.)	.068 (14 ga.)	.097 (12 ga.)	.097 (12 ga.)	
To 7"	24	.044 (18 ga.)	.055 (16 ga.)	.068 (14 ga.)	.068 (14 ga.)	.068 (14 ga.)	.055 (16 ga.)	.068 (14 ga.)	.097 (12 ga.)	.097 (12 ga.)	.097 (12 ga.)
To 7"	12	.044 (18 ga.)	.055 (16 ga.)	.068 (14 ga.)	.068 (14 ga.)	.068 (14 ga.)	.055 (16 ga.)	.068 (14 ga.)	.097 (12 ga.)	.097 (12 ga.)	.097 (12 ga.)
8" to 11"	48	.044 (18 ga.)	.068 (14 ga.)	.097 (12 ga.)	.097 (12 ga.)		.055 (16 ga.)	.068 (14 ga.)	.097 (12 ga.)	.097 (12 ga.)	
8" to 11"	24	.044 (18 ga.)	.055 (16 ga.)	.068 (14 ga.)	.068 (14 ga.)	.068 (14 ga.)	.055 (16 ga.)	.068 (14 ga.)	.097 (12 ga.)	.097 (12 ga.)	.097 (12 ga.)
8" to 11"	12	.044 (18 ga.)	.055 (16 ga.)	.068 (14 ga.)	.068 (14 ga.)	.068 (14 ga.)	.055 (16 ga.)	.068 (14 ga.)	.097 (12 ga.)	.097 (12 ga.)	.097 (12 ga.)
12" to 15"	48	.044 (18 ga.)	.097 (12 ga.)				.055 (16 ga.)	.097 (12 ga.)			
12" to 15"	24	.044 (18 ga.)	.055 (16 ga.)	.097 (12 ga.)	.097 (12 ga.)	.097 (12 ga.)	.055 (16 ga.)	.068 (14 ga.)	.097 (12 ga.)	.097 (12 ga.)	.097 (12 ga.)
12" to 15"	12	.044 (18 ga.)	.055 (16 ga.)	.068 (14 ga.)	.068 (14 ga.)	.068 (14 ga.)	.055 (16 ga.)	.068 (14 ga.)	.097 (12 ga.)	.097 (12 ga.)	.097 (12 ga.)
16" to 20"	48	.068 (14 ga.)	.112 (11 ga.)				.055 (16 ga.)	.112 (11 ga.)			
16" to 20"	24	.068 (14 ga.)	.068 (14 ga.)	.112 (11 ga.)	.112 (11 ga.)	.112 (11 ga.)	.055 (16 ga.)	.068 (14 ga.)	.112 (11 ga.)	.112 (11 ga.)	.112 (11 ga.)
16" to 20"	12	.044 (18 ga.)	.068 (14 ga.)	.068 (14 ga.)	.068 (14 ga.)	.068 (14 ga.)	.055 (16 ga.)	.068 (14 ga.)	.097 (12 ga.)	.097 (12 ga.)	.097 (12 ga.)

Note: 1 in. = 25.4 mm

CHAPTER 6

DUCT SYSTEMS

601.0 Scope.

601.1 Ducts and plenums that are portions of a heating, cooling, absorption or evaporative cooling, or product-conveying system shall comply with the requirements of this chapter.

601.2 Sizing Requirements. Duct systems used with blower-type equipment that are portions of a heating, cooling, absorption, evaporative cooling, or outdoor-air ventilation system shall be sized in accordance with Chapter 17, or by other approved methods.

602.0 Material.

602.1 General. Supply air, return air, and outside air for heating, cooling, or evaporative cooling systems shall be conducted through duct systems constructed of metal as set forth in Tables 6-1, 6-2, 6-3, 6-4, 6-7, 6-8, 6-9, and 6-10, or metal ducts complying with UMC Standard No. 6-2 or the referenced HVAC duct construction standard in Chapter 17. Rectangular ducts in excess of two (2) inches w.g. shall comply with UMC Standard No. 6-2 or the referenced HVAC duct construction standard in Chapter 17. Ducts, plenums, and fittings may be constructed of concrete, clay, or ceramics when installed in the ground or in a concrete slab, provided the joints are tightly sealed.

Corridors shall not be used to convey air to or from rooms if the corridor is required to be of fire-resistive construction per the Building Code.

Concealed building spaces or independent construction within buildings may be used as ducts or plenums.

When gypsum products are exposed in ducts or plenums, the air temperature shall be restricted to a range from 50°F (10°C) to 125°F (52°C), and moisture content shall be controlled so that the material is not adversely affected. For the purpose of this section, gypsum products shall not be exposed in ducts serving as supply from evaporative coolers, and in other air-handling systems regulated by this chapter when the temperature of the gypsum product will be below the dew point temperature.

See Chapter 8 for limitations on combustion products venting systems extending into or through ducts or plenums.

See Chapter 5 for limitations on environmental air systems exhaust ducts extending into or through ducts or plenums.

Exhaust ducts under positive pressure and venting systems shall not extend into or pass through ducts or plenums. For appliance vents and chimneys, see Chapter 8.

602.2 Combustibles Within Ducts or Plenums. Materials exposed within ducts or plenums shall be noncombustible or shall have a flame spread index not greater than twenty-five (25) and a smoke developed index not greater than fifty (50), when tested as a composite product in accordance with one of the following test methods: NFPA 255, *Method of Test of Surface Burning Characteristics of Building Materials*, ASTM E84, *Surface Burning Characteristics of Building Materials*, or UL 723, *Test for Surface Burning Characteristics of Building Materials*, except as indicated below.

Exceptions:

(1) Return-air and outside-air ducts, plenums, or concealed spaces that serve a dwelling unit shall be permitted to be of combustible construction.

(2) Air filters meeting the requirements of Sections 311.2 and 503.3.

(3) Water evaporation media in an evaporative cooler.

(4) Charcoal filters when protected with an approved fire suppression system.

(5) Electrical wiring in plenums shall comply with NFPA 70, *National Electrical Code*. Electrical wires and cables and optical fiber cables shall be listed and labeled as suitable for use in plenums and shall have a flame spread distance not greater than five (5) feet (1,524 mm), an average optical density not greater than 0.15, and a peak optical density not greater than 0.5, when tested in accordance with NFPA 262, *Flame Travel and Smoke of Wires and Cables for Use in Air-Handling Spaces*.

(6) Nonmetallic fire sprinkler piping in plenums shall be listed and labeled as suitable for use in plenums and shall have a flame spread distance not greater than five (5) feet (1,524 mm), an average optical density not greater than 0.15 and, a peak optical density not greater than 0.5, when tested in accordance with UL 1887, *Fire Test of Plastic Sprinkler Pipe for Visible Flame and Smoke Characteristics*.

(7) Nonmetallic pneumatic tubing in plenums shall be listed and labeled as suitable for use in plenums and shall have a flame spread distance not greater than five (5) feet (1,524 mm), an average optical density not greater than 0.15, and a peak optical density not greater than 0.5, when tested in accordance with UL 1820, *Fire Test of Pneumatic Tubing for Flame and Smoke Characteristics*.

(8) Loudspeakers and recessed lighting fixtures, including their assemblies and accessories, in plenums shall be listed and labeled as suitable for use in plenums and shall have a peak rate of heat release not greater than 100 kilowatts, an average optical density not greater than 0.15, and a peak optical density not greater than 0.5, when tested in accordance with UL 2043, *Fire Test for Heat and Visible Smoke Release for Discrete Products and their Accessories Installed in Air-Handling Spaces*.

(9) Smoke detectors.

(10) Duct insulation, coverings, and linings and other supplementary materials complying with 605.0.

(11) Materials in a Group H, Division 6, fabrication area including the areas above and below the fabrication area sharing a common air recirculation path with the fabrication area.

602.3 Factory-Made Air Ducts. Factory-made air ducts shall be approved for the use intended or shall conform to the requirements of the referenced standard for air ducts in Chapter 17. Each portion of a factory-made air duct system shall be identified by the manufacturer with a label or other suitable identification indicating compliance with the referenced standard for air ducts in Chapter 17 and its class designation. These ducts shall be listed and shall be installed in accordance with the terms of their listing and the requirements of UMC Standard No. 6-5. Flexible air connectors are not permitted.

602.4 Joints and Seams of Ducts. Joints of duct systems shall be made substantially air-tight by means of tapes, mastics, gasketing, or other means.

Crimp joints for round ducts shall have a contact lap of at least 1-1/2 inch (38 mm) and shall be mechanically fastened by means of at least three (3) sheet-metal screws equally spaced around the joint, or an equivalent fastening method.

Joints and seams for 0.016 inch (0.41 mm) (No. 28 gauge) and 0.013 inch (0.33 mm) (No. 30 gauge) residential rectangular ducts shall be as specified in Table 6-1 for 0.019 inch (0.48 mm) (No. 26 gauge) material.

Joints and seams for rectangular duct systems shall be as specified in Table 6-1.

Joints and seams for flat oval ducts and round ducts in other than single-dwelling units shall be as specified in Table 6-8.

Joints and seams and all reinforcements for factory-made air ducts and plenums shall meet with the conditions of prior approval in accordance with the installation instructions that shall accompany the product. Closure systems for rigid air ducts and plenums shall be listed in accordance with UL 181A, *Standard for Closure Systems for Use with Rigid Air Ducts and Air Connectors*. Closure systems for flexible air ducts shall be listed in accordance with UL 181B, *Standard for Closure Systems for Use with Flexible Air Ducts and Air Connectors*.

602.5 Metal. Every duct, plenum, or fitting of metal shall comply with Table 6-1 or 6-8.

Exceptions:

(1) Ducts, plenums, and fittings for systems serving single-dwelling units may comply with Table 6-9.

(2) Duct systems complying with UMC Standard No. 6-2 or the referenced HVAC duct construction standard in Chapter 17, with prior approval, or duct systems complying with UL 181, *Standard for Factory-Made Air Ducts and Air Connectors*.

(3) Duct systems complying with the UMC Standard No. 6-2 or the referenced HVAC duct construction standard in Chapter 17, with prior approval.

602.6 Tin. Existing tin ducts may be used when cooling coils are added to a heating system, provided the first ten (10) feet (3,048 mm) of the duct or plenum measured from the cooling coil discharge are constructed of metal of the gauge thickness set forth in Tables 6-1, 6-8, or 6-9 of this chapter or are of approved material and construction. Tin ducts completely enclosed in inaccessible concealed areas need not be replaced. All accessible ducts shall be insulated to comply with Table 6-6 of this chapter. For the purpose of this subsection, ducts shall be considered accessible where the access space is thirty (30) inches (762 mm) or greater in height.

602.7 Vibration Isolators. Vibration isolators installed between mechanical equipment and metal ducts (or casings) shall be made of an approved material and shall not exceed ten (10) inches (254 mm) in length.

603.0 Quality of Material.

Galvanized steel shall be of lock-forming quality with a minimum coating of 1.25 ounces of zinc per square foot (0.04 kg/m^3) conforming to the requirements of UMC Standard No. 2-2.

604.0 Installation of Ducts.

604.1 General Ducts Under Floor or Crawl Space. Air ducts installed under a floor in a crawl space shall be installed as follows:

(A) Shall not prevent access to any area of the crawl space.

(B) Where it is required to move under ducts for access to areas of the crawl space a minimum vertical clearance of eighteen (18) inches (457 mm) shall be provided.

604.2 Metal Ducts. Ducts shall be securely fastened in place at each change of direction and as set forth in Table 6-7. Vertical rectangular ducts and vertical round ducts shall be supported as set forth in Table 6-7, Part A. Riser ducts shall be held in place by means of metal straps or angles and channels to secure the riser to the structure.

Metal ducts shall be installed with at least four (4) inches (102 mm) separation from earth. Metal ducts when installed in or under a concrete slab shall be encased in at least two (2) inches (51 mm) of concrete.

Ducts shall be installed in a building with adequate clearance so as to permit retaining the full thickness of fireproofing on structural members.

Supports for rectangular ducts as set forth in Table 6-7, when suspended from above, shall be installed on two opposite sides of each duct and shall be riveted, bolted, or metal screwed to each side of the duct at not more than the intervals specified.

Horizontal round ducts forty (40) inches (1,016 mm) or less in diameter when suspended from above shall be supported at intervals not more than as set forth in Table 6-7 with one hanger installed to comply with the requirements listed below:

604.2.1 Ducts shall be equipped with tight-fitting circular bands extending around the entire perimeter of the duct at each specified support interval.

604.2.2 Circular bands shall be not less than one (1) inch (25 mm) wide nor less than equivalent to the gauge of the duct material it supports.

Exception: Ducts ten (10) inches (254 mm) and less in diameter may be supported by No. 18 gauge galvanized steel wire.

604.2.3 Each circular band shall be provided with a suitable means of connecting to the suspending support.

604.2.4 Ducts shall be braced and guyed to prevent lateral or horizontal swing.

604.3 Factory-Made Air Ducts. Listed Class 0 or Class 1 factory-made air ducts may be installed in any occupancy covered by this code.

Factory-made air ducts shall not be used for vertical risers in air-duct systems serving more than two stories. Such ducts shall not penetrate construction where fire dampers are required.

Factory-made air ducts shall be installed with at least four (4) inches (102 mm) of separation from earth, except when installed as a liner inside of concrete, tile, or metal pipe; they shall be protected from physical damage.

The temperature of the air to be conveyed in any of these classes of ducts shall not exceed 250°F (122°C).

604.4 Protection of Ducts. Ducts installed in locations where they are exposed to mechanical damage by vehicles or from other causes shall be protected by approved barriers.

604.5 Support of Ducts. Installers shall provide the manufacturer's field fabrication and installation instructions.

In the absence of specific supporting materials and spacing, approved factory-made air ducts may be installed as set forth in Table 6-10.

604.6 Protection Against Flood Damage. In flood hazard areas, ducts shall be located above the design flood elevation or shall be designed and constructed to prevent water from entering or accumulating within the ducts during floods up to the design flood elevation. If the ducts are located below the design flood elevation, the ducts shall be capable of resisting hydrostatic and hydrodynamic loads and stresses, including the effects of buoyancy, during the occurrence of flooding to the design flood elevation.

605.0 Insulation of Ducts.

Supply-, return-air ducts and plenums of a heating or cooling system shall be insulated to achieve the minimum thermal (R) value as set forth in Tables 6-6 A and B.

Exceptions:

(A) Factory-installed plenums, casings, or ductwork furnished as a part of HVAC equipment tested and rated in accordance with approved energy efficiency standards.

(B) Ducts or plenums located in conditioned spaces.

(C) For runouts less than ten (10) feet (3,048 mm) in length to air terminals or air outlets, the rated R value of insulation need not exceed R-3.5 (R-0.6).

(D) Backs of air outlets and outlet plenums exposed to unconditioned or indirectly conditioned spaces with face areas exceeding five (5) square feet (0.5 m^2) need not exceed

R-2 (R-0.4); those five (5) square feet (0.5 m^2) or smaller need not be insulated.

(E) Ducts and plenums used exclusively for evaporative cooling systems.

Approved materials shall be installed within ducts and plenums for insulating, sound deadening, or other purposes. Materials shall have a mold, humidity, and erosion-resistant surface that meets the requirements of the referenced standard for air ducts in Chapter 17. Duct liners in systems operating with air velocities exceeding 2,000 feet per minute (10.16 m/s) shall be fastened with both adhesive and mechanical fasteners, and exposed edges shall have adequate treatment to withstand the operating velocity.

Insulation applied to the surface of ducts, including duct coverings, linings, tapes, and adhesives, located in buildings shall have a flame-spread index not greater than twenty-five (25) and a smoke developed index not greater than fifty (50), when tested in accordance with NFPA 255, *Method of Test of Burning Characteristics of Building Materials*, or in accordance with ASTM E 84, *Surface Burning Characteristics of Building Materials*, or in accordance with the provisions of UL 723, *Test for Surface Burning Characteristics of Building Materials*. The specimen preparation and mounting procedures of ASTM E 2231, *Specimen Preparation and Mounting of Pipe and Duct Insulation Materials to Assess Surface Burning Characteristics*, shall be used. Air duct coverings and linings shall not flame, glow, smolder, or smoke when tested in accordance with ASTM C 411, *Hot-Surface Performance of High-Temperature Thermal Insulation*, at the temperature to which they are exposed in service. In no case shall the test temperature be below 250°F (121°C).

Factory-made air ducts and faced insulations intended for installation on the exterior of ducts shall be legibly printed with the name of the manufacturer, the thermal resistance (R) value at installed thickness, and the flame-spread index and smoke developed index of the composite material.

606.0 Smoke Dampers, Fire Dampers, and Ceiling Dampers.

606.1 Smoke Dampers. Smoke dampers shall comply with the standards for leakage-rated dampers referenced in Chapter 17, and shall be installed in accordance with approved manufacturer's installation instructions when required by the Building Code. Smoke dampers shall be labeled by an approved agency.

606.2 Fire Dampers. Fire dampers shall comply with the standard for fire dampers referenced in Chapter 17, and shall be installed in accordance with the approved manufacturer's installation instructions when required by the Building Code. Fire dampers shall have been tested for closure under airflow conditions and shall be labeled for both maximum airflow permitted and direction of flow. When more than one damper is installed at a point in a single air path, the entire airflow shall be assumed to be passing through the smallest damper area. Fire dampers shall be labeled by an approved agency.

Ductwork shall be connected to damper sleeves or assemblies in accordance with the fire damper manufacturer's installation instructions.

606.3 Ceiling Radiation Dampers. Ceiling radiation dampers shall comply with the standard for ceiling radiation dampers referenced in Chapter 17, and shall be installed in accordance with the manufacturer's approved installation instructions in the fire-resistive ceiling element of floor-ceiling and roof-ceiling assemblies when required by the Building Code. Fire dampers not meeting the temperature limitation of ceiling radiation dampers shall not be used as a substitute. Ceiling radiation dampers shall be labeled by an approved agency.

606.4 Multiple Arrangements. When size requires the use of multiple dampers, the installation shall be framed in an approved manner to ensure that the dampers remain in place.

606.5 Access and Identification. Dampers shall be provided with an approved means of access large enough to permit inspection and maintenance of the damper and its operating parts. The access shall not impair fire-resistive construction. Access shall not require the use of tools, keys, or special knowledge. Access points shall be permanently identified on the exterior by a label with letters not less than 1/2 inch (12.7 mm) in height reading: SMOKE DAMPER or FIRE DAMPER. Access doors in ducts shall be tight fitting and suitable for the required duct construction.

606.6 Freedom from Interference. Dampers shall be installed in a manner to ensure positive closing or opening as required by function. Interior liners or insulation shall be held back from portions of a damper, its sleeve, or adjoining duct that would interfere with the damper's proper operation. Exterior materials shall be installed so as to avoid interference with the operation or maintenance of external operating devices needed for proper function.

606.7 Temperature Classification of Operating Elements. Fusible links, thermal sensors, and pneumatic or electric operators shall have a temperature rating or classification as required by the Building Code.

607.0 Ventilating Ceilings.

607.1 General. Perforated ceilings may be used for air supply within the limitations of this section. Exit corridors, when required to be of fire-resistive construction by the Building Code, shall not have ventilating ceilings.

607.2 Requirements. Ventilating ceilings shall comply with the following provisions:

607.2.1 Suspended ventilating ceiling material shall have a Class I flame-spread classification on both sides, determined in accordance with the Building Code. Suspended ventilating ceiling supports shall be of noncombustible materials.

607.2.2 Lighting fixtures recessed into ventilating ceilings shall be of a type approved for that purpose.

608.0 Use of Under-Floor Space as Supply Plenum for Dwelling Units.

An under-floor space may be used as a supply plenum, provided:

608.1 The use of under-floor space shall be limited to dwelling units not more than two stories in height. Except for the floor immediately above the under-floor plenum, supply ducts shall be provided extending from the plenum to registers on other floor levels.

Exception: In flood hazard areas, under-floor spaces shall not be used as supply plenums unless the flood opening requirements in the Building Code are satisfied.

608.2 Such spaces shall be cleaned of all loose combustible scrap material and shall be tightly and substantially enclosed.

608.3 The enclosing material of the under-floor space, including the sidewall insulation, shall be not more flammable than one (1) inch (25 mm) (nominal) wood boards (flame-spread index of 200). Installation of foam plastics is regulated by the Building Code.

608.4 Access shall be through an opening in the floor and shall be not less than twenty-four (24) inches by twenty-four (24) inches (610 mm x 610 mm).

608.5 A furnace supplying warm air to under-floor space shall be equipped with an automatic control that will start the air-circulating fan when the air in the furnace bonnet reaches a temperature not higher than 150°F (66°C). Such control shall be one that cannot be set higher than 150°F (66°C).

608.6 A furnace supplying warm air to such space shall be equipped with an approved temperature limit control that will limit outlet air temperature to 200°F (93°C).

608.7 A noncombustible receptacle shall be placed below each floor opening into the air chamber, and such receptacle shall conform to the following:

608.7.1 The receptacle shall be securely suspended from the floor members and shall be not more than eighteen (18) inches (457 mm) below the floor opening.

608.7.2 The area of the receptacle shall extend three (3) inches (76 mm) beyond the opening on all sides.

608.7.3 The perimeter of the receptacle shall have a vertical lip at least one (1) inch (25.4 mm) high at the open sides if it is at the level of the bottom of the joists, or three (3) inches (76 mm) high if the receptacle is suspended.

608.8 Floor registers shall be designed for easy removal in order to give access for cleaning the receptacles.

608.9 Exterior walls and interior stud partitions shall be fire-stopped at the floor.

608.10 Each wall register shall be connected to the air chamber by a register box or boot.

608.11 A duct conforming with Section 602.0 shall extend from the furnace supply outlet at least six (6) inches (152 mm) below combustible framing.

608.12 The entire ground surface of the under-floor space shall be covered with a vapor barrier having a minimum thickness of four (4) mils (0.1016 mm) and a flame-spread index of 200 or less.

608.13 Fuel-gas lines and plumbing waste cleanouts are not located within the space.

609.0 Automatic Shutoffs.

Air-moving systems supplying air in excess of 2,000 cubic feet per minute (944 L/s) to enclosed spaces within buildings shall be equipped with an automatic shutoff. Automatic shutoff shall be accomplished by interrupting the power source of the air-moving equipment upon detection of smoke in the main supply-air duct served by such equipment. Smoke detectors shall be labeled by an approved agency for air duct installation and shall be installed in accordance with the manufacturer's approved instructions. Such devices shall be compatible with the operating velocities, pressures, temperatures, and humidities of the system. Where fire-detection or alarm systems are provided for the building, the smoke detectors required by this section shall be supervised by such systems.

Exceptions:

(1) When the space supplied by the air-moving equipment is served by a total coverage smoke-detection system complying with the

Fire Code, interconnection to such system may be used to accomplish the required shutoff.

(2) Automatic shutoff is not required when all occupied rooms served by the air-handling equipment have direct exit to the exterior and the travel distance does not exceed 100 feet (30,480 mm).

(3) Automatic shutoff is not required for Group R, Division 3 and Group U Occupancies.

(4) Automatic shutoff is not required for approved smoke-control systems or where analysis demonstrates shutoff would create a greater hazard, such as may be encountered in air-moving equipment supplying specialized portions of Group H Occupancies. Such equipment shall be required to have smoke detection with remote indication and manual shutoff capability at an approved location.

(5) Smoke detectors that are factory installed in listed air-moving equipment may be used in lieu of smoke detectors installed in the main supply-air duct served by such equipment.

TABLE 6-1 Duct Construction for 4 Feet Duct Lengths [SMACNA HVAC Duct Construction Standards; Second Edition - 1995]

2" w.g STATIC POS. OR NEG.				TRANSVERSE JOINT REINFORCEMENT							INTERMEDIATE REINFORCEMENT	
(1)	(2)	(3)	(4)	(5)		(6)	(7)	(8)	(9)	(10)	(11)	(12)
DUCT DIM.	DUCT GA. (MIN.)	REINF. SPACING (MAX.)	REINF. CODE GRADE	DRIVE SLIP T-3 / HEMMED "S" SLIP T-6: SLIP GAGE	BACKUP	STANDING S T-10	STANDING S T-12	STANDING S (BAR REINFORCED) T-13 / STANDING S (ANGLE REINFORCED) T-14	T-24 / T-25a / T-25b, H = 1-3/8" (WITH GASKET)	T-24a (WITH GASKET)	ANGLE	ZEE
						H x T	H x T	H x T + HR	H = 1 ⅜"	H x T	H x T (MIN)	H x B x T (MIN)
10" dn	26	None		24	None							
11"-20"	26	4'	C	24	*	1x 22 ga.	1x 24 ga.	NOT USED	26 ga.	1 x 22 ga.	1 x ⅛	1 x ¾ x 18 ga.
21"-28"	26	4'	D	24	*	1 ⅛ x 20 ga.	1 x 20 ga. or 1½ x 22 ga.				1 ¼ x 12 ga. or 1 x ⅛ HR	
29"-30"	26	4'	E	24	*	1 ⅛ x 18 ga.	1 x 18 ga. or 1 ½ x 20 ga.		24 ga.	1 ½ x 20 ga.		1 ½ x ¾ x 16 ga.
31"-42"	24	4'	F	24	*	1 ⅝ x 18 ga.	1 ½ x 18 ga.	1 ⅝ x 22 ga. 1 ½ x ⅛ Bar	22 ga			
43"-48"	22	4'	G	24	*	1 ⅝ x 18 ga.	1 ½ x 18 ga.	1 ⅝ x 20 ga. 1 ½ x ⅛ Bar	20 ga.	1 ½ x 18 ga.	1 ½ x ⅛	
49"-60"	24	2'	F	24	*	1 ⅝ x 18 ga.	1 ½ x 18 ga.	1 ⅝ x 22 ga. 1 ½ x ⅛ Bar	22 ga.	1 ½ x 20 ga.		
61"-72"	24	2'	H	24	*	NOT GIVEN	NOT GIVEN	1 ⅝ x 18 ga. 1 ½ x ⅛ Bar	20 ga. + rods	NOT GIVEN	2 x ⅛	1 ½ x ¾ x ⅛
73"-84"	22	2'	I	22	*			2 ⅛ x 20 ga. 2 x ⅛ Angle			2 x 3/16	2 x 1 ⅛ x 12 ga.
85"-96"	22	2'	I	22	*							

Notes:

1. *Means use back-up member from columns 11 or 12. Exception: The drive only requires back-up over 18" length.
2. Spacing in column 3 refers to joint-to-joint, joint-to-intermediate, or intermediate-to intermediate.
3. The same sheet thickness must be used on all side of duct. Each duct dimension, width, or depth, controls the minimum reinforcement requirements for that particular side.
4. Duct sides 19" wide and larger which have more than ten square feet of unbraced panel shall be beaded or crossbroken unless the ducts will have external insulation or internal liner. The applies to duct of 20 ga. or less.
5. Contents of the narrow scope tables and these notes do not constitute all requirements for the construction's compliance. Complete details are provided in the manual entitled HVAC Duct Construction standards, Metal and Flexible. The manual is available from local SMACNA chapter offices of the National Association office; inquire for terms.
6. Construction conforming to the standard does not prevent some traditionally acceptable rumble (noise) under sudden pressure change conditions such as start up and shut down of systems.

1 in. = 25.4 mm

Comment:
Construction tables are prepared in "narrow scope" for 2" water gage and 1" water gage pressure classification. These tables have been produced for the convenience of those contractors or engineers who wish to have fewer options.

Sheet Metal and Air Conditioning
Contractors' National Association, Chantilly, VA

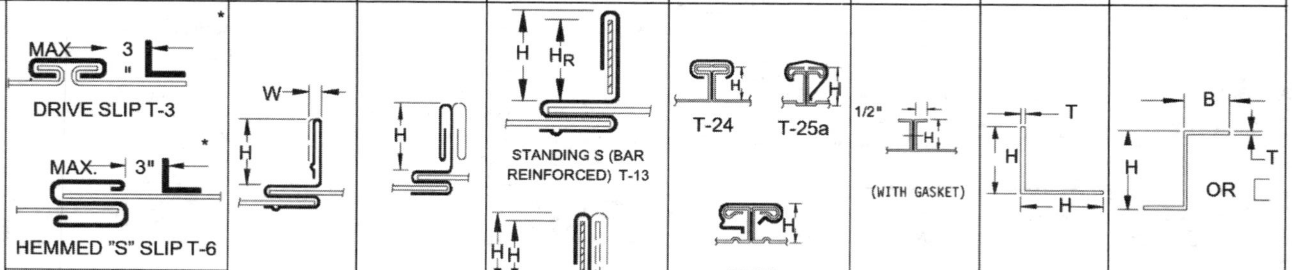

2" w.g STATIC POS. OR NEG.				TABLE 6-2 Duct Construction for 5 Feet Duct Lengths [SMACNA HVAC Duct Construction Standards; Second Edition - 1995] TRANSVERSE JOINT REINFORCEMENT							INTERMEDIATE REINFORCEMENT	
(1)	(2)	(3)	(4)	(5)		(6)	(7)	(8)	(9)	(10)	(11)	(12)
DUCT DIM.	DUCT GA. (MIN.)	REINF. SPACING (MAX.)	REINF. CODE GRADE	SLIP GAGE	BACKUP	STANDING S T-10	STANDING S T-12	STANDING S (ANGLE REINFORCED) T-14	H = 1-3/8" (WITH GASKET)	T-24a	ANGLE	ZEE
10" dn	26	None		24	None	H x T	H x T	H x T + HR	H x 1 ⅝"	H x T	H x T (MIN)	H x B x T (MIN)
11"-20"	26	5'	C	24	*	1x 22 ga.	1 x 22 ga.	NOT USED	26 ga.	1 x 22 ga.	1 x ⅛	1 x ¾ x 18 ga.
21"-26"	26	5'	D	24	*	1 ⅛ x 20 ga.	1 x 20 ga.				1 ¼ x 12 ga. or 1 x ⅛	
27"-30"	24	5'	E	24	*	1 ⅛ x 18 ga.	1 x 18 ga.		24 ga.	1 ½ x 20 ga.		1 ½ x ¾ x 16 ga.
31"-36"	22	5'	F	24	*	1 ⅝ x 18 ga.	1 ½ x 18 ga.	1 ⅝ x 22 ga. 1 ½ x ⅛ Bar	22 ga			
37"-54"	24	2 ½'	F	24	*	1 ⅝ x 18 ga.	1 ½ x 18 ga.	1 ⅝ x 20 ga. 1 ½ x ⅛ Bar	22 ga.		1 ½ x ⅛	
55"-60"	24	2 ½'	G	24	*	1 ⅝ x 18 ga.	1 ½ x 18 ga.	1 ⅝ x 22 ga. 1 ½ x ⅛ Bar	22 ga. + rods	1 ½ x 18 ga.		
61"-72"	22	2 ½'	H	24	*	NOT GIVEN	NOT GIVEN	1 ⅝ x 18 ga. 1 ½ x ⅛ Angle	20 ga. + rods	NOT GIVEN	2 x ⅛	1 ½ x ¾ x ⅛
73"-84"	22	2 ½'	I	22	*			2 ⅛ x 20 ga. 2 x ⅛ Angle			2 x 3/16	2 x 1 ⅛ x 12 ga.
85"-96"	20	2 ½'	I	22	*							

Notes:

1.*Means use back-up member from columns 11 or 12. Exception: The drive only requires back-up over 18" length.
2. Spacing in column 3 refers to joint-to-joint, joint-to-intermediate, or intermediate-to intermediate.
3. The same sheet thickness must be used on all side of duct. Each duct dimension, width, or depth, controls the minimum reinforcement requirements for that particular side.
4. Duct sides 19" wide and larger which have more than ten square feet of unbraced panel shall be beaded or crossbroken unless the ducts will have external insulation or internal liner. The applies to duct of 20 ga. or less.
5. Contents of the narrow scope tables and these notes do not constitute all requirements for the construction's compliance. Complete details are provided in the manual entitled HVAC Duct Construction standards, Metal and Flexible. The manual is available from local SMACNA chapter offices of the National Association office; inquire for terms.
6. Construction conforming to the standard does not prevent some traditionally acceptable rumble (noise) under sudden pressure change conditions such as start up and shut down of systems.

1 in. = 25.4 mm

Comment:

Construction tables are prepared in "narrow scope" for 2" water gage and 1" water gage pressure classification. These tables have been produced for the convenience of those contractors or engineers who wish to have fewer options.

Sheet Metal and Air Conditioning
Contractors' National Association, Chantilly, VA

RECTANGULAR ALUMINUM DUCT
ADAPTED FROM 3" W.G. (750 Pa) or LOWER

TABLE 6-3
THICKNESS ADJUSTMENTS

Galv. Steel Ga (mm) nominal	28 (0.48)	26 (0.55)	24 (0.70)	22 (0.78)	20 (1.00)	18 (1.31)	16 (1.61)
Min. Alum. equivalent[1] (mm)	0.023 (0.58)	0.027 (0.69)	0.034 (0.86)	0.043 (1.09)	0.052 (1.32)	0.067 (1.70)	0.083 (2.11)
Commercial size (mm)	0.025 (0.60)	0.032 (0.80)	0.04 (1.00)	0.05 (1.27)	0.063 (1.60)	0.071 (1.80)	0.09 (2.29)
Lbs wt/Sf. Alum.	Consult Appendix-5 for Weights						

Note:
[1]Alloy 3003-H-14.

TABLE 6-4
DIMENSION ADJUSTMENTS

Galv. Rigidity Class	A	B	C	D	E	F	G	H	I	J	K	L
Alum. dim. per Galv. Class	C	E	E	F	H	I	I	K	[1]	[1]	[1]	[1]

Note:
[1]Calculate an effective I_x= 3x that used for steel.

TABLE 6-5
REINFORCEMENTS

Steel Angle Size, Inches (mm)	Cod	Equivalent Alumn,[1] Angle Size, Inches	Steel Bar	Aluminum Bar[1]
1 x 1 x16 Ga. (25 x 25 x 1.61)	C	1-1/4 x 1-1/4 x 1/8 (31.8 x 31.8 x 3.2)	1 x 1/8 (25 x 3.2)	1-1/2 x 1/8 or 1-1/4 x 3/16 (38.1 x 3.2 or 31.8 x 4.8)
1 x 1 x 1/8 (25 x 25 x 3.2)	D	1-1/2 x 1-1/2 x 1/8 (38.1 x 38.1 x 3.2)	1-1/2 x 1/8 (38.1 x 3.2)	1-1/2 x 1/8 x or 1-1/4 x 3/16 (38.1 x 3.2 or 38.1 x 4.8)
1-1/4 x 1-1/4 x 1/8 (31.8 x 31.8 x 3.2)	F	1-3/4 x 1-3/4 x 1/8 (44.5 x 44.5 x 3.2)		
1-1/2 x 1-1/2 x 1/8 (38.1 x 38.1 x 3.2)	G	2-1/2 x 2-1/2 x 1/8 (63.5 x 63.5 x 3.2)		
2 x 2 x 1/8 (51 x 51 x 3.2)	H	2-1/2 x 2-1/2 x 3/16 (63.5 x 63.5 x 4.8)		
2 x 2 x 3/16 (51 x 51 x 4.8)	I	2-1/2 x 2-1/2 x 3/8 or 3 x 3 x 1/4 (63.5 x 63.5 x 9.5 or 76.2 x 76.2 x 6.4)		
2 x 2 x 1/4 (51 x 51 x 6.4)	J	2-1/2 x 2-1/2 x 3/8 or 3 x 3 x 1/4 (63.5 x 63.5 x 7.9 or 76.2 x 76.2 x 6.4)		
2-1/2 x 2-1/2 x 3/16 (63.5 x 63.5 x 4.8)	K	3 x 3 x 3/8 or 3-1/2 x 3-1/2 x 1/4 (76.2 x 76.2 x 9.5 or 88.9 x 88.9 x 6.4)		

Notes:
[1]Alloy 6061-T6 Strength normally.
Any aluminum shape substitute must have a moment of inertia three times that of steel and have 30,000 psi minimum yield strength.

TABLE 6-6 A (I-P Units)
Minimum Duct Insulation R Value[a] Cooling and Heating Only Supply Ducts and Return Ducts[d, e]

Climate Zone			Duct Location						
Envelope Criteria Table	HDD65	CDD50	Exterior	Ventilated Attic	Unvented Attic w/ Backloaded Ceiling	Unvented Attic w/ Roof Insulation[a]	Uncon-ditioned Space[b]	Indirectly Con-ditioned Space[c]	Buried
Heating-Only Ducts									
5-1 to 5-7	0–1,800	all	none	none	none	none	none	none	none
5-8 to 5-12	1,801–3,600	all	R-3.5	none	none	none	none	none	none
5-13 to 5-15	3,601–5,400	all	R-3.5	none	none	none	none	none	none
5-16 to 5-18	5,401–7,200	all	R-6	R-3.5	none	none	none	none	R-3.5
5-19 to 5-20	7,201–9,000	all	R-6	R-6	R-3.5	none	none	none	R-3.5
5-21 to 5-22	9,001–10,800	all	R-8	R-6	R-6	none	R-3.5	none	R-3.5
5-23	10,801–12,600	all	R-8	R-6	R-6	none	R-6	none	R-6
5-24	12,601–16,200	all	R-8	R-8	R-6	none	R-6	none	R-6
5-25	16,201–19,800	all	R-10	R-8	R-8	none	R-6	none	R-6
5-26	19,801+	all	R-10	R-10	R-8	none	R-8	none	R-6
Cooling-Only Ducts									
5-15, 18, 20, 22 to 26	all	0–1,800	R-1.9	R-1.9	R-1.9	R-1.9	R-1.9	none	none
5-12, 14, 17, 19, 21	all	1,801–3,600	R-3.5	R-1.9	R-3.5	R-1.9	R-1.9	none	none
5-7, 9, 11, 13, 16	all	3,601–5,400	R-3.5	R-3.5	R-6	R-1.9	R-1.9	none	none
5-4, 6, 8, 10	all	5,401–7,200	R-6	R-6	R-6	R-3.5	R-1.9	none	none
5-3, 5-5	all	7,201–9,000	R-6	R-6	R-6	R-3.5	R-3.5	none	R-3.5
5-2	all	9,001–10,800	R-6	R-6	R-8	R-3.5	R-3.5	none	R-3.5
5-1	all	10,801+	R-8	R-8	R-8	R-3.5	R-3.5	none	R-3.5
Return Ducts									
5-15 to 5-26	all climates	R-3.5	R-3.5	R-3.5	none	none	none	none	none

Notes:

[a] Insulation R-values, measured in (h·ft^2·°F)/Btu, are for the insulation as installed and do not include film resistance. The required minimum thicknesses do not consider water vapor transmission and possible surface condensation. Where exterior walls are used as plenum walls, wall insulation shall be as required by the most restrictive condition. Insulation resistance measured on a horizontal plane in accordance with ASTM C518 at a mean temperature of 75°F (24°C) at the installed thickness.

[b] Includes crawl spaces, both ventilated and nonventilated.

[c] Includes return-air plenums with or without exposed roofs above.

[d] Vapor retarders shall be installed on supply ducts in spaces vented to the outside in geographic areas where the average summer dewpoint temperature based on the 2-1/2 percent column of dry-bulb and mean coincident wet-bulb temperature exceeds 60°F (16°C).

[e] Insulation shall have an approved weatherproof barrier.

TABLE 6-6 B (I-P Units)
Minimum Duct Insulation R Value[a] Combined Heating and Cooling Ducts[d, e]

Climate Zone			Duct Location						
Envelope Criteria Table	HDD65	CDD50	Exterior	Ventilated Attic	Unvented Attic w/ Backloaded Ceiling	Unvented Attic w/ Roof Insulation[a]	Unconditioned Space[b]	Indirectly Conditioned Space[c]	Buried
5-1	0–900	10,801+	R-8	R-6	R-8	R-3.5	R-3.5	none	R-3.5
5-2	0–900	9,001–10,800	R-6	R-6	R-8	R-3.5	R-3.5	none	R-3.5
5-3	0–900	7,201–9,000	R-6	R-6	R-6	R-3.5	R-3.5	none	R-3.5
5-4	0–900	0–7,200	R-6	R-3.5	R-6	R-3.5	R-1.9	none	R-3.5
5-5	901–1,800	7,201+	R-6	R-6	R-6	R-3.5	R-3.5	none	R-3.5
5-6	901–1,800	5,401–7,200	R-6	R-6	R-6	R-3.5	R-3.5	none	R-3.5
5-7	901–1,800	0–5,400	R-3.5	R-3.5	R-6	R-1.9	R-1.9	none	R-1.9
5-8	1,801–2,700	5,401+	R-6	R-6	R-6	R-3.5	R-3.5	none	R-3.5
5-9	1,801–2,700	0–5,400	R-6	R-3.5	R-6	R-1.9	R-1.9	none	R-1.9
5-10	2,701–3,600	5,401+	R-6	R-6	R-6	R-3.5	R-3.5	none	R-3.5
5-11	2,701–3,600	3,601–5,400	R-6	R-6	R-6	R-3.5	R-3.5	none	R-1.9
5-12	2,701–3,600	0–3,600	R-3.5	R-3.5	R-3.5	R-1.9	R-1.9	none	R-1.9
5-13	3,601–5,400	3,601+	R-6	R-6	R-6	R-3.5	R-3.5	none	R-3.5
5-14	3,601–5,400	1,801–3,600	R-6	R-3.5	R-6	R-1.9	R-3.5	none	R-1.9
5-15	3,601–5,400	0-1,800	R-3.5	R-3.5	R-3.5	R-1.9	R-1.9	none	R-1.9
5-16	5,401–7,200	3,601+	R-6	R-6	R-6	R-3.5	R-3.5	none	R-3.5
5-17	5,401–7,200	1,801–3,600	R-6	R-6	R-6	R-1.9	R-3.5	none	R-3.5
5-18	5,401–7,200	0–1,800	R-6	R-3.5	R-3.5	R-1.9	R-3.5	none	R-3.5
5-19	7,201–9,000	1,801+	R-8	R-6	R-6	R-1.9	R-3.5	none	R-3.5
5-20	7,201–9,000	0–1,800	R-6	R-6	R-6	R-1.9	R-3.5	none	R-3.5
5-21	9,001–10,800	1,801+	R-8	R-6	R-6	R-1.9	R-6	none	R-3.5
5-22	9,001–10,800	0–1,800	R-8	R-6	R-6	R-1.9	R-3.5	none	R-3.5
5-23	10,801–12,600	all	R-8	R-6	R-6	R-1.9	R-6	none	R-6
5-24	12,601–16,200	all	R-8	R-8	R-8	R-1.9	R-6	none	R-6
5-25	16,201–19,800	all	R-10	R-8	R-8	R-3.5	R-6	none	R-6
5-26	19,801+	all	R-10	R-10	R-8	R-3.5	R-8	R-3.5	R-6

Notes:

[a] Insulation R-values, measured in (h·ft^2·°F)/Btu, are for the insulation as installed and do not include film resistance. The required minimum thicknesses do not consider water vapor transmission and possible surface condensation. Where exterior walls are used as plenum walls, wall insulation shall be as required by the most restrictive condition. Insulation resistance measured on a horizontal plane in accordance with ASTM C518 at a mean temperature of 75°F (24°C) at the installed thickness.

[b] Includes crawl spaces, both ventilated and nonventilated.

[c] Includes return air plenums with or without exposed roofs above.

[d] Vapor retarders shall be installed on cooling supply ducts in spaces vented to the outside in geographic areas where the average summer dewpoint temperature based on the 2-1/2 percent column of dry-bulb and mean coincident wet-bulb temperature exceeds 60°F (16°C).

[e] Insulation shall have an approved weatherproof barrier.

TABLE 6-7
Duct Support

A. Vertical Ducts

Maximum Side of Rectangular Duct	Metal Strap or Angle Bracket	Maximum Diameter of Round Ducts	Strap
24"	1" x 1/8" strap[1]	10"	0.047" (No. 18 gauge) galvanized steel 2" wide[1]
36"	1" x 1" x 1/8" angle[1]	20"	0.058" (No. 16 gauge) galvanized steel 2" wide[1]
48"	1-1/8" x 1-1/8" x 1/8" angle[1]	40"	1/8" steel x 1-1/2"[1]
60"	1-1/2" x 1-1/2" x 1/8" angle[1]	60"	1/8" steel x 2"[1]
Over 60"	2" x 2" x 1/8" angle[1]	Over 60"	3/16" steel x 2"[1]

B. Horizontal Ducts

18"	1" x 18 gauge[2]	10"	Same gauge as galvanized steel duct, 1" wide or (No. 18 gauge galvanized steel wire) on 10′ centers
30"	1" x 18 gauge[2]		
48"	1" x 1/8"[2]	20"	Same gauge as galvanized steel duct, 1" wide or (No. 8 gauge galvanized steel wire) tied to 1" galvanized steel band around duct on 10′ centers
60"	1" x 1/8"[2]		
80"	1" x 1/8"[2]	40"	
		60"	Same gauge as galvanized steel duct, 1-1/2" wide on 6' centers
		Over 60"	Same gauge as galvanized steel duct, 1-1/2" wide on 4' centers

C. Horizontal Ducts — Trapeze-Type Supports

Maximum Diameter of Round Duct or Side of Rectangular Duct	Horizontal Support Angle[3]	Hanger
36"	1-1/2" x 1-1/2" x 1/8"	1/4" round rod or 1" x 1" x 1/8" angle
48"	2" x 2" x 1/8"	1/4" round rod or 1" x 1" x 1/8" angle
60"	2" x 2" x 1/8"	5/16" round rod or 1" x 1" x 1/8" angle
84"	2" x 2" x 1/8"	3/8" round rod or 1" x 1" x 1/8" angle

Notes:
[1] Spaced vertically not more than 12 feet on centers.
[2] Spaced horizontally not more than 10 feet on centers.
[3] Spaced not more than 8 feet on centers.
1 in. = 25.4 mm, 1 ft. = 0.305 m

TABLE 6-8
Construction Details for Round and Flat-Oval Ducts

Duct Diameter Max. Width (In Inches)	Aluminum B. & S. Gage	STEEL — THICKNESS IN INCHES (STEEL — GALVANIZED SHEET GAUGE)					GIRTH JOINTS[1]	
	Pressure < 2" W.C.	Pressure < 2" W.C.[2]		Pressure > 2" < 10" W.C.				Minimum Girth Reinforcing, Maximum Spacing and Angle Size
	Round	Round	Flat-Oval	Spiral Seam	Longitudinal Seam	Welded Fittings	Pressure > 2" < 10" W.C.	
Up to 9	24	0.019 (26)	0.024 (24)	0.019 (26)	0.024 (24)	0.030 (22)	2" Slip	None
Over 9 Up to 14	24	0.019 (26)	0.024 (24)	0.024 (24)	0.030 (22)	0.036 (20)	4" Slip	None
Over 14 Up to 23	22	0.024 (24)	0.030 (22)	0.024 (24)	0.030 (22)	0.036 (20)	4" Slip	None
Over 23 Up to 37	20	0.030 (22)	0.036 (20)	0.030 (22)	0.036 (20)	0.036 (20)	4" Slip	None
Over 37 Up to 51	18	0.036 (20)	0.047 (18)	0.036 (20)	0.036 (20)	0.047 (18)	1-1/4" x 1-1/4" x 1/8" Flange	1-1/4" x 1-1/4" x 1/8" on 72"
Over 51 Up to 61	16	0.047 (18)	0.058 (16)	X (18)	0.047 (18)	0.047 Flange	1-1/4" x 1-1/4" x 1/8" on 72"	1-1/4" x 1-1/4" x 1/8"
Over 61 Up to 84	14	0.058 (16)	0.070 (14)	X	0.058 (16)	0.058 (16)	1-1/2" x 1-1/2" x 1/8" Flange	1-1/2" x 1-1/2" x 1/8" on 48"

Notes:

[1] For pressure < 2 inches W.C., any of the following joints are acceptable: butt slip; pipe slip; pipe lock; roll slip; snap slip; plenum lock; and companion flange.

[2] Acceptable longitudinal seams for pressure < 2 inches W.C.: Acme (grooved); snap lock; standing; and spiral.

1 in. = 25.4 mm, 1 in. W.C. = 0.249 kPa

TABLE 6-9
Thickness of Metal Ducts and Plenums Used for Heating or Cooling for a Single-Dwelling Unit

	GALVANIZED STEEL		
	Minimum Thickness, Inches (mm)	Equivalent Galvanized Sheet Gauge No.	Approximate Aluminum B.&S. Gauge
Round ducts and enclosed rectangular ducts			
14" (356 mm) or less	0.013 (0.33)	30	26
Over 14" (356 mm)	0.016 (0.41)	28	24
Exposed rectangular ducts			
14" (356 mm) or less	0.016 (0.41)	28	24
Over 14" (356 mm)	0.019 (0.48)	26	22

TABLE 6-10
Alternate Supports for Factory-Made Air Ducts

Class 1 Duct Material	Minimum Support Material	Maximum Support Spacing, Feet
Rigid round fibrous	No. 26 gauge x 5/8" wide galvanized iron strap	6
Corrugated metal	No. 18 gauge galvanized iron wire	6
Flexible	No. 26 gauge x 1-1/2" wide galvanized iron strap	4

Note: 1 in. = 25.4 mm, 1 ft. = 0.305 m

CHAPTER 7

COMBUSTION AIR

701.0 Air for Combustion and Ventilation.

701.1 General.

701.1.1 Air for combustion, ventilation, and dilution of flue gases for gas utilization equipment installed in buildings shall be obtained by application of one of the methods covered in Sections 701.2 through 701.8.3. Gas utilization equipment of other than natural draft and Category I vented appliances shall be provided with combustion, ventilation, and dilution air in accordance with the equipment manufacturer's instructions. Where infiltration does not provide the necessary air, outdoor air shall be introduced in accordance with methods covered in Sections 701.4 through 701.8.3.

Exceptions:

(1) This provision shall not apply to direct-vent appliances.

(2) Type 1 clothes dryers that are provided with makeup air in accordance with Section 504.3.2.

701.1.2 Gas appliances of other than natural draft design and other than Category I vented appliances shall be provided with combustion, ventilation, and dilution air in accordance with the appliance manufacturer's instructions. [NFPA 54:9.3.1.2]

701.1.3 Where used, a draft hood or a barometric draft regulator shall be installed in the same room or enclosure as the equipment served so as to prevent any difference in pressure between the hood or regulator and the combustion-air supply. [NFPA 54:9.3.1.4]

701.1.4 Where exhaust fans, clothes dryers, and kitchen ventilation systems interfere with the operation of appliances, makeup air shall be provided. [NFPA 54-09:9.3.1.5]

701.2 Indoor Combustion Air. The required volume of indoor air shall be determined in accordance with the method in section 701.2.1 or 701.2.2, except that where the air infiltration rate is known to be less than 0.40 ACH, the method in Section 701.2.2 shall be used. The total required volume shall be the sum of the required volume calculated for all appliances located within the space. Rooms communicating directly with the space in which the appliances are installed through openings not furnished with doors, and through combustion air openings sized and located in accordance with Section 701.3.1, are considered a part of the required volume. [NFPA 54:9.3.2]

701.2.1 Standard Method. The minimum required volume shall be fifty (50) cubic feet per 1,000 Btu/hour (4.8 m^3/kW) [NFPA 54:9.3.2.1].

701.2.1.1 Known Air Infiltration Rate Method Equations.

Equation 701.2.1.1(a):

$$\text{Required Volume}_{other} \geq (21 \text{ ft.}^3/\text{ACH}) \times (I_{other}/1{,}000 \text{ Btu/h})$$

Equation 701.2.1.1(b):

$$\text{Required Volume}_{fan} \geq (15 \text{ ft.}^3/\text{ACH}) \times (I_{fan}/1{,}000 \text{ Btu/h})$$

WHERE:

I_{other} = all appliances other than fan-assisted input in Btu per hour

I_{fan} = fan-assisted appliance input in Btu per hour

ACH = air change per hour (percent of volume of space exchanged per hour, expressed as a decimal)

701.2.2 Known Air Infiltration Rate Method. Where the air infiltration rate of a structure is known, the minimum required volume shall be determined as follows [NFPA 54:9.3.2.2]:

(1) For appliances having other than fan-assisted combustion systems: calculate using equation 701.2.1.1(a) but no smaller than thirty-five (35) cubic feet per 1,000 Btu/hour (3.4 m^3/kW). [NFPA 54:9.3.2.2(1)]

(2) For fan-assisted combustion system appliances: calculate using equation 701.2.1.1(b) but no smaller than twenty-five (25) cubic feet per 1,000 Btu/hour (2.4 m^3/kW). [NFPA 54: 9.3.2.2(2)]

(3) For purposes of this calculation, an infiltration rate greater than 0.60 ACH shall not be used in the equations. [NFPA 54: 9.3.2.2(3)]

701.3 Indoor Opening Size and Location.

701.3.1 Openings used to connect indoor spaces shall be sized and located in accordance with the following [NFPA 54:9.3.2.3]:

(1) Combining spaces on the same story. Each opening shall have a minimum free area of 1 in.^{2}/1,000 Btu/h. (220 mm^3/kW) of the total input rating of all gas utilization equipment in the space, but not less than 100 in.^{2} (0.06 m^2). One opening shall commence within twelve

(12) inches (300 mm) of the top, and one opening shall commence within twelve (12) in. (300 mm) of the bottom, of the enclosure. (See Figure 7-1.) The minimum dimension of air openings shall be not less than three (3) inches (80 mm). [NFPA 54:9.3.2.3(1)]

(2) Combining spaces in different stories. The volumes of spaces in different stories shall be considered as communicating spaces where such spaces are connected by one or more openings in doors or floors having a total minimum free area of 2 $in.^2$/1,000 Btu/h (4,400 mm^2/kw) of total input rating of all gas utilization equipment. [NFPA 54:9.3.2.3(2)]

701.4 Outdoor Combustion Air. Outdoor combustion air shall be provided through opening(s) to the outdoors in accordance with the methods in Section 701.4.1 or 701.4.2. The minimum dimension of air openings shall not be less than three (3) inches (80 mm). [NFPA 54:9.3.3]

701.4.1 Two Permanent Openings Method. Two permanent openings, one commencing within twelve (12) inches (300 mm) of the top and one commencing within twelve (12) inches (300 mm) of the bottom of the enclosure, shall be provided. The openings shall communicate directly, or by ducts, with the outdoors or spaces that freely communicate with the outdoors, as follows [NFPA 54:9.3.3.1]:

(1) Where directly communicating with the outdoors or where communicating to the outdoors through vertical ducts, each opening shall have a minimum free area of 1 $in.^2$/4,000 Btu/h (550 mm^2/kW) of total input rating of all appliances in the enclosure. (See Figures 7-2 and 7-3.) [NFPA 54:9.3.3.1(1)]

(2) Where communicating with the outdoors through horizontal ducts, each opening shall have a minimum free area of 1 $in.^2$/2,000 Btu/h (1,100 mm^2/kW) of total input rating of all appliances in the enclosure. (See Figure 7-4.) [NFPA 54:9.3.3.1(2)]

701.4.2 One Permanent Opening Method. One permanent opening, commencing within twelve (12) inches (300 mm) of the top of the enclosure, shall be provided. The equipment shall have clearances of at least one (1) inch (25 mm) from the sides and back and six (6) inches (160 mm) from the front of the appliance. The opening shall directly communicate with the outdoors or shall communicate through a vertical or horizontal duct to the outdoors or spaces that freely communicate with the outdoors (see Figure 7-5) and shall have a minimum free area of [NFPA 54:9.3.3.2]:

(1) 1 $in.^2$/3,000 Btu/h (700 mm^2/kW) of the total input rating of all equipment located in the enclosure, and [NFPA 54:9.3.3.2(1)]

(2) Not less than the sum of the areas of all vent connectors in the space. [NFPA 54:9.3.3.2 (2)]

701.5 Combination Indoor and Outdoor Combustion Air. The use of a combination of indoor and outdoor combustion air shall be in accordance with Sections 701.5.1 through 701.5.3. [NFPA 54:9.3.4]

701.5.1 Indoor Openings. Where used, openings connecting the interior spaces shall comply with Section 701.3.1. [NFPA 54:9.3.4(1)]

701.5.2 Outdoor opening(s) shall be located in accordance with Section 701.4. [NFPA 54: 9.3.4(2)]

701.5.3 Outdoor Opening(s) Size. The outdoor opening(s) size shall be calculated in accordance with the following [NFPA 54:9.3.4(3)]:

(1) The ratio of interior spaces shall be the available volume of all communicating spaces divided by the required volume. [NFPA 54:9.3.4(3)(a)]

(2) The outdoor size reduction factor shall be one (1) minus the ratio of interior spaces. [NFPA 54:9.3.4(3)(b)]

(3) The minimum size of outdoor opening(s) shall be the full size of outdoor opening(s) calculated in accordance with Section 701.4, multiplied by the reduction factor. The minimum dimension of air openings shall not be less than three (3) inches (80 mm). [NFPA 54:9.3.4(3)(c)]

701.6 Engineered Installations. Engineered combustion air installations shall provide adequate supply of combustion, ventilation, and dilution air and shall be approved by the Authority Having Jurisdiction. [NFPA 54:9.3.5]

701.7 Mechanical Combustion Air Supply. Where all combustion air is provided by a mechanical air supply system, the combustion air shall be supplied from outdoors at the minimum rate of 0.35 $feet^3$/min per 1,000 Btu/h (0.034 m^3/min per kW) for all appliances located within the space. [NFPA 54:9.3.6]

701.8 Mechanical Combustion Air Requirements.

701.8.1 Where exhaust fans are installed, additional air shall be provided to replace the exhausted air. [NFPA 54:9.3.6.1]

701.8.2 Each of the appliances served shall be interlocked to the mechanical air supply system to prevent main burner operation where the mechanical air supply system is not in operation. [NFPA 54:9.3.6.2]

701.8.3 Where combustion air is provided by the building's mechanical ventilation system, the system shall provide the specified combustion air rate in addition to the required ventilation air. [NFPA 54:9.3.6.3]

701.9 Louvers, Grilles and Screens.

(A) Louvers and Grilles. The required size of openings for combustion, ventilation, and dilution air shall be based on the net free area of each opening. Where the free area through a design of louver, grille or screen is known, it shall be used in calculating the size opening required to provide the free area specified. Where the louver and grille design and free area are not known, it shall be assumed that wood louvers will have a 25 percent free area and metal louvers and grilles will have a 75 percent free area. Nonmotorized louvers and grilles shall be fixed in the open position. [NFPA 54:9.3.7.1]

(B) Minimum Screen Mesh Size. Screens shall not be smaller than 1/4 inch (6.4 mm) mesh. [NFPA 54:9.3.7.2]

(C) Motorized Louvers. Motorized louvers shall be interlocked with the equipment so they are proven in the full open position prior to main burner ignition and during main burner operation. Means shall be provided to prevent the main burner from igniting should the louver fail to open during burner startup and to shut down the main burner if the louvers close during burner operation. [NFPA 54:9.3.7.3]

701.10 Combustion-Air Ducts. Combustion-air ducts shall comply with the following: {NFPA 54:9.3.8]

(1) Ducts shall be of galvanized steel or a material having equivalent corrosion resistance, strength, and rigidity.

Exception: Within dwellings units, unobstructed stud and joist spaces shall not be prohibited from conveying combustion air, provided that not more than one fireblock is removed. [NFPA 54:9.3.8.1]

(2) Ducts shall terminate in an unobstructed space, allowing free movement of combustion air to the appliances. [NFPA 54:9.3.8.2]

(3) Ducts shall serve a single space. [NFPA 54:9.3.8.3]

(4) Ducts shall not service both upper and lower combustion air openings where both such openings are used.

The separation between ducts serving upper and lower combustion-air openings shall be maintained to the source of combustion air. [NFPA 54:9.3.8.4]

(5) Ducts terminating in attics shall not be screened. [NFPA 54:9.3.8.5]

(6) Intakes for combustion-air ducts located exterior to the building shall have the lowest side of the combustion air intake openings located at least 12 inches vertically from the adjoining grade level. [NFPA 54:9.3.8.8]

(7) Horizontal upper combustion-air ducts shall not slope downward toward the source of combustion air. [NFPA 54:9.3.8.6]

(8) The remaining space surrounding a chimney liner, gas vent, special gas vent, or plastic piping installed within a masonry, metal, or factory-built chimney shall not be used to supply combustion air, unless it is listed and shown in the manufacturer's installation instructions. [NFPA 54:9.3.8.7]

701.11 Dampers Prohibited. Combustion-air ducts or plenums shall not be installed so as to require openings in or penetrations through construction where fire dampers are required. Manually operated dampers shall not be installed in combustion-air openings. With prior approval, power-actuated movable louvers admitting combustion air may be used and, if installed, shall be electrically interlocked with the main burner fuel-supply valve so as to prevent fuel delivery unless the louvers are in the fully open position.

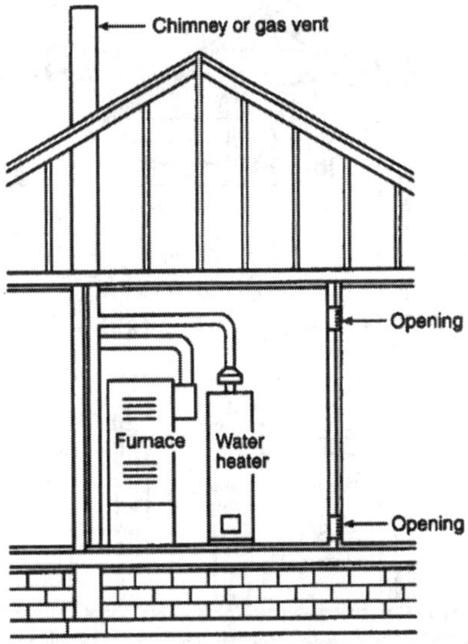

FIGURE 7-1 All Combustion Air from Indoor Spaces Through Indoor Combustion Air Openings. [NFPA 54:A.9.3.2.3(1)]

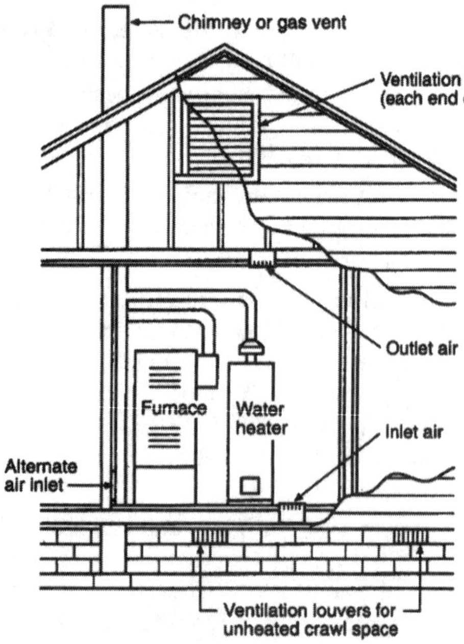

FIGURE 7-2 All Combustion Air from Outdoors—Inlet Air from Ventilated Crawl Space and Outlet Air to Ventilated Attic. [NFPA 54:A.9.3.3.1(1)(A)]

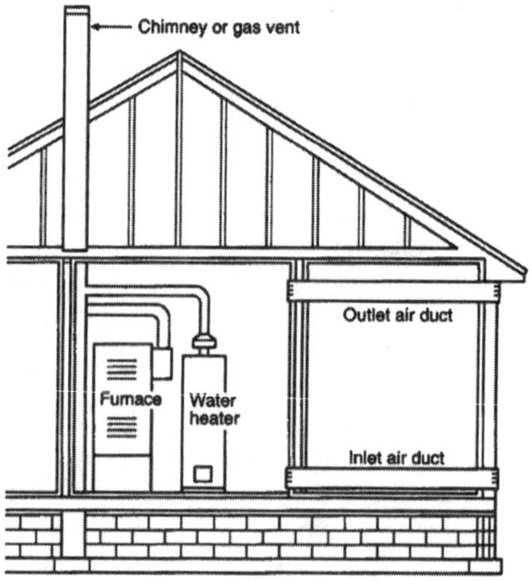

FIGURE 7-4 All Combustion Air from Outdoors Through Horizontal Ducts. [NFPA 54:A.9.3.3.1(2)]

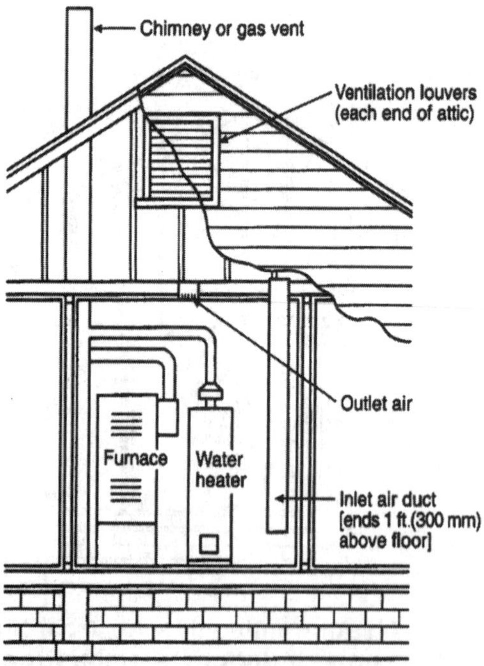

FIGURE 7-3 All Combustion Air from Outdoors Through Ventilated Attic. [NFPA 54:A.9.3.3.1(1)(B)]

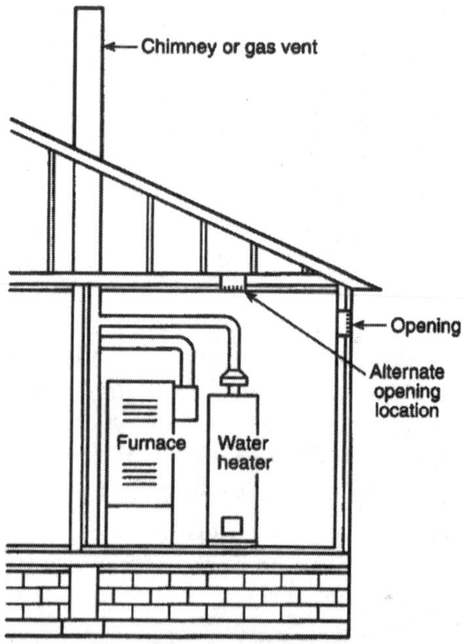

FIGURE 7-5 All Combustion Air from Outdoors Through Single Combustion Air Opening. [NFPA 54:A.9.3.3.2]

CHAPTER 8

CHIMNEYS AND VENTS

801.0 Scope.

The requirements of this chapter shall govern the venting of fuel-burning appliances.

801.1 Venting of Gas Appliances. Low-heat gas appliances shall be vented in accordance with this chapter. Other gas appliances shall be vented in accordance with NFPA 211 or other applicable standards.

801.2 Appliances Fueled by Other Fuels. Appliances fueled by fuels other than gas shall be vented in accordance with NFPA 211 and the appliance manufacturer's instructions.

802.0 Venting of Appliances.

802.1 General. This section recognizes that the choice of venting materials and the methods of installation of venting systems are dependent on the operating characteristics of the appliance.

802.1.1 Categories. The operating characteristics of vented appliances can be categorized with respect to positive or negative pressure within the venting system and whether or not the appliance generates flue or vent gases that can condense in the venting system. See Section 224.0 for the definition of these vented appliance categories. [NFPA 54:12.2.2]

802.1.2 Installation. Listed vents shall be installed in accordance with Chapter 8 and the manufacturer's installation instructions. [NFPA 54:12.2.3]

802.1.3 Appliance vents shall not discharge into any space enclosed by screens having openings smaller than one-quarter (1/4) inch (6.4 mm) mesh.

802.2 Specification for Venting.

802.2.1 Connection to Venting Systems. Except as permitted in Section 802.2.2 through 802.2.6, all appliances shall be connected to venting systems. [NFPA 54:12.3.1]

802.2.2 Appliances Not Required to Be Vented. The following appliances shall not be required to be vented [NFPA 54:12.3.2]:

802.2.2.1 Listed ranges. [NFPA 54:12.3.2(1)]

802.2.2.2 Built-in domestic cooking units listed and marked for optional venting. [NFPA 54:12.3.2(2)]

802.2.2.3 Listed hot plates and listed laundry stoves. [NFPA 54:12.3.2(3)]

802.2.2.4 Listed Type 1 clothes dryers exhausted in accordance with Section 504.3.2. [NFPA 54:12.3.2(4)]

802.2.2.5 A single listed booster-type (automatic instantaneous) water heater, when designed and used solely for the sanitizing rinse requirements of a dishwashing machine, provided that the appliance is installed with the draft hood in place and unaltered, if a draft hood is required, in a commercial kitchen having a mechanical exhaust system; where installed in this manner, the draft hood outlet shall not be less than thirty-six (36) inches (910 mm) vertically and six (6) inches (150 mm) horizontally from any surface other than the appliance. [NFPA 54:12.3.2(5)]

802.2.2.6 Listed refrigerators. [NFPA 54:12.3.2(6)]

802.2.2.7 Counter appliances. [NFPA 54:12.3.2(7)]

802.2.2.8 Room heaters listed for unvented use (see Sections 924.1 and 924.2). [NFPA 54:12.3.2(8)]

802.2.2.9 Direct gas-fired makeup air heaters. [NFPA 54:12.3.2(9)]

802.2.2.10 Other appliances listed for unvented use and not provided with flue collars. [NFPA 54:12.3.2(10)]

802.2.2.11 Specialized appliances of limited input such as laboratory burners or gas lights. [NFPA 54:12.3.2(11)]

Where any or all of the appliances in Sections 802.2.2.5 through 802.2.2.11 are installed so the aggregate input rating exceeds twenty (20) Btu/h/ft.^{3} (207 W/m^3) of room or space in which it is installed, one or more shall be provided with venting systems or other approved means for removing the vent gases to the outside atmosphere so the aggregate input rating of the remaining unvented appliances do not exceed twenty (20) Btu/h/ft.^{3} (207 W/m^3). Where the calculation includes the volume of an adjacent room or space, the room or

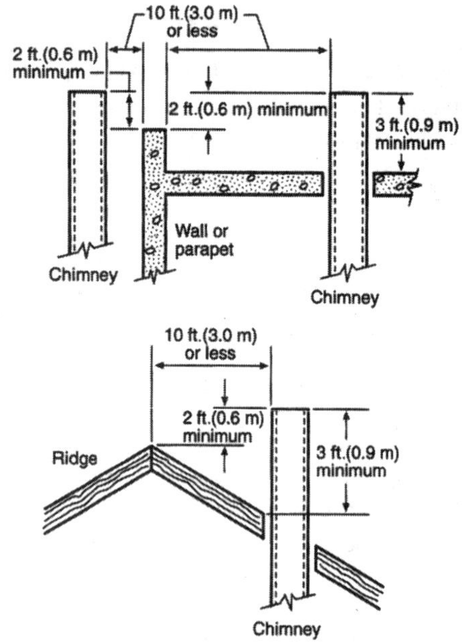

(a) Termination 10 ft.(3.0 m) or Less from Ridge, Wall, or Parapet

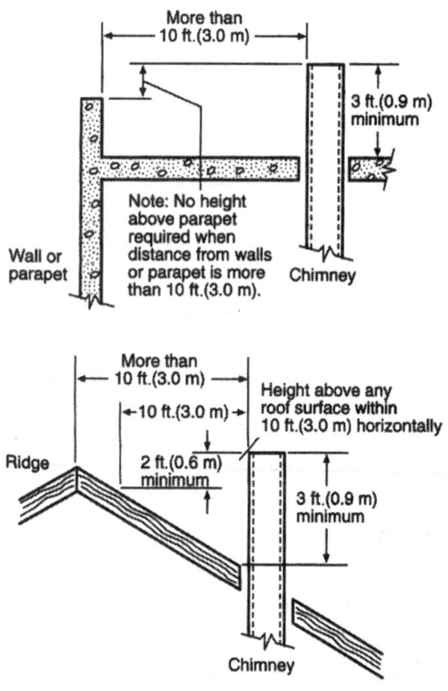

(b) Termination More Than 10 ft.(3.0 m) from Ridge, Wall, or Parapet

FIGURE 8-1 Typical Termination Locations for Chimneys and Single-Wall Metal Pipes Serving Residential-Type and Low-Heat Appliance. [NFPA 54: Figure 12.6.2.1]

factory-built chimney system and are installed in accordance with manufacturer' s installation instructions. [NFPA 54:12.6.2.4]

802.5.3 Size of Chimneys. The effective area of a chimney venting system serving listed gas appliances with draft hoods, Category I appliances, and other appliances listed for use with Type B vents shall be in accordance with one of the following methods [NFPA 54:12.6.3.1]:

(1) Section 803.0. [NFPA 54:12.6.3.1(1)]

(2) For sizing an individual chimney venting system for a single appliance with a draft hood, the effective areas of the vent connector and chimney flue shall be not less than the area of the appliance flue collar or draft hood outlet or greater than seven times the draft hood outlet area. [NFPA 54:12.6.3.1(2)]

(3) For sizing a chimney venting system connected to two (2) appliances with draft hoods, the effective area of the chimney flue shall be not less than the area of the larger draft hood outlet plus 50 percent of the area of the smaller draft hood outlet, or greater than seven (7) times the smallest draft hood outlet area. [NFPA 54:12.6.3.1(3)]

(4) Other approved engineering methods. [NFPA 54:12.6.3.1(5)]

(5) Chimney venting systems using mechanical draft shall be sized in accordance with approved engineering methods. [NFPA 54:12.6.3.1(4)]

802.5.4 Inspection of Chimneys.

802.5.4.1 Before replacing an existing appliance or connecting a vent connector to a chimney, the chimney passageway shall be examined to ascertain that it is clear and free of obstructions and shall be cleaned if previously used for venting solid or liquid fuel-burning appliances or fireplaces. [NFPA 54:12.6.4.1]

802.5.4.2 Chimneys shall be lined in accordance with NFPA 211, *Standard for Chimneys, Fireplaces, Vents, and Solid Fuel-Burning Appliances*. [NFPA 54:12.6.4.2]

Exception: Existing chimneys shall be permitted to have their use continued when an appliance is replaced by an appliance or similar type, input rating, and efficiency, where the chimney complies with 802.5.4, and the sizing of the chimney is in accordance with 802.5.3.

802.5.4.3 Cleanouts shall be examined to determine that they will remain tightly closed when not in use. [NFPA 54:12.6.4.3]

802.5.4.4 When inspection reveals that an existing chimney is not safe for the intended application, it shall be repaired, rebuilt, lined, relined, or replaced with a vent or chimney to conform to NFPA 211, *Standard for Chimneys, Fireplaces, Vents, and Solid-Fuel-Burning Appliances*, and shall be suitable for the appliance to be attached. [NFPA 54:12.6.4.4]

802.5.5 Chimney Serving Appliances Burning Other Fuels.

802.5.5.1 Appliances shall not be connected to a chimney flue serving a separate appliance designed to burn solid fuel. [NFPA 54:12.6.5.1]

802.5.5.2 Where one (1) chimney serves gas appliances and an appliance burning liquid fuel, the appliances shall be connected through separate openings or shall be connected through a single opening where joined by a suitable fitting located as close as practical to the chimney. Where two (2) or more openings are provided into one (1) chimney flue, they shall be at different levels. Where the appliance is automatically controlled, it shall be equipped with a safety shutoff device. [NFPA 54:12.6.5.2]

802.5.5.3 A listed combination gas and solid fuel-burning appliance connected to a single chimney flue shall be equipped with a manual reset device to shut off gas to the main burner in the event of sustained back-draft or flue gas spillage. The chimney flue shall be sized to properly vent the appliance. [NFPA 54: 12.6.5.3]

802.5.5.4 A single chimney flue serving a listed combination gas- and oil-burning appliance shall be sized to properly vent the appliance. [NFPA 54:12.6.5.4]

802.5.6 Support of Chimneys. All portions of chimneys shall be supported for the design and weight of the materials employed. Listed factory-built chimneys shall be supported and spaced in accordance with their listings and the manufacturers' instructions. [NFPA 54:12.6.6]

802.5.7 Cleanouts. Where a chimney that formerly carried flue products from liquid or solid fuel-burning appliances is used with an appliance using fuel gas, an accessible cleanout shall be provided. The cleanout shall have a tight-fitting cover and be installed so its upper edge is at least six (6) inches (150 mm) below the lower edge of the lowest chimney inlet opening. [NFPA 54:12.6.7]

802.5.8 Space Surrounding Lining or Vent. The remaining space surrounding a chimney liner, gas vent, special gas vent, or plastic piping installed within a masonry chimney flue shall not be used to vent another appliance. [NFPA 54:12.6.8.1]

Exception: The insertion of another liner or vent within the chimney as provided in this code and the liner or vent manufacturer's instructions.

The remaining space surrounding a chimney liner, gas vent, special gas vent, or plastic piping installed within a masonry, metal or factory-built chimney flue shall not be used to supply combustion air. [NFPA 54:12.6.8.2]

Exception: Direct vent gas-fired appliances designed for installation in a solid-fuel-burning fireplace where installed in accordance with the listing and the manufacturer's instruction.

802.6 Gas Vents.

802.6.1 A gas vent passing through a roof shall extend through the entire roof flashing, roof jack, or roof thimble and be terminated with a listed termination cap. [NFPA 54:12.7.2(6)]

802.6.1.1 Type B or Type L vents shall extend in a generally vertical direction with offsets not exceeding 45 degrees, except that a vent system having not more than one 60 degree offset shall be permitted. Any angle greater than 45 degrees from the vertical is considered horizontal. The total horizontal distance of a vent plus the horizontal vent connector serving draft hood equipped appliances shall not be greater than 75 percent of the vertical height of the vent. [NFPA 54:12.7.3.2]

Exception: Systems designed and sized as provided in Section 803.0 or in accordance with other approved engineering methods.

Vents serving Category I fan-assisted appliances shall be installed in accordance with the appliance manufacturer's instructions and Section 803.0 or other approved engineering methods. [NFPA 54:12.7.1(3)]

802.6.2 A gas vent shall terminate in accordance with one of the following:

(1) Above the roof surface with a listed cap or listed roof assembly. Gas vents twelve (12) inches (300 mm) in size or smaller with listed caps shall be permitted to be termi-

nated in accordance with Figure 8-2, provided they are at least eight (8) feet (2,438 mm) from a vertical wall or similar obstruction. All other gas vents shall terminate not less than two (2) feet (610 mm) above the highest point where they pass through the roof and at least two (2) feet (610 mm) higher than any portion of a building within ten (10) feet (3,048 mm). [NFPA 54:12.7.2(1)(a) and (b)]

(2) Industrial appliance provided in Section 802.2.4. [NFPA 54:12.7.2(1)(c)]

(3) Direct-vent systems as provided in Section 802.2.5. [NFPA 54:12.7.2(1)(d)]

(4) Appliance with integral vents as provided in Section 802.2.6. [NFPA 54:12.7.2(1)(e)]

(5) Mechanical draft systems as provided in Section 802.3.4. [NFPA 54:12.7.2(1)(f)]

(6) Ventilating hoods and exhaust systems as provided in Section 802.3.5. [NFPA 54:12.7.2(1)(g)]

802.6.2.1 A Type B or a Type L gas vent shall terminate at least five (5) feet (1,524 mm) in vertical height above the highest connected appliances draft hood or flue collar. [NFPA 54:12.7.2(2)]

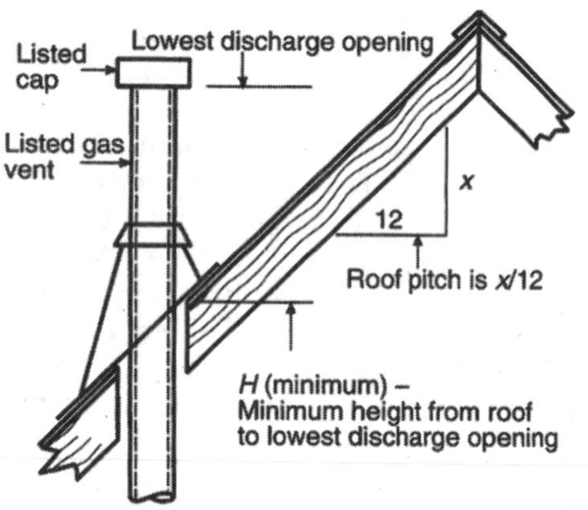

Roof pitch	H (minimum) ft	m
Flat to 6/12	1.0	0.30
6/12 to 7/12	1.25	0.38
Over 7/12 to 8/12	1.5	0.46
Over 8/12 to 9/12	2.0	0.61
Over 9/12 to 10/12	2.5	0.76
Over 10/12 to 11/12	3.25	0.99
Over 11/12 to 12/12	4.0	1.22
Over 12/12 to 14/12	5.0	1.52
Over 14/12 to 16/12	6.0	1.83
Over 16/12 to 18/12	7.0	2.13
Over 18/12 to 20/12	7.5	2.27
Over 20/12 to 21/12	8.0	2.44

FIGURE 8-2 Gas Vent Termination Locations for Listed Caps 12 in. (300 mm) or Less in Size at Least 8 ft. (2.4 m) from a Vertical Wall. [NFPA 54: Figure 12.7.2]

802.6.2.2 A Type B-W gas vent shall terminate at least twelve (12) feet (3,658 mm) in vertical height above the bottom of the wall furnace. [NFPA 54:12.7.2(3)]

802.6.2.3 A gas vent extending through an exterior wall shall not terminate adjacent to the wall or below eaves or parapets, except as provided in Sections 802.2.5 and 802.3.4. [NFPA 54:12.7.2(4)]

802.6.2.4 Decorative shrouds shall not be installed at the termination of gas vents except where such shrouds are listed for use with the specific gas venting system and are installed in accordance with manufacturers' installation instructions. [NFPA 54:12.7.2(5)]

802.6.2.5 All gas vents shall extend through the roof flashing, roof jack, or roof thimble and terminate with a listed cap or listed roof assembly. [NFPA 54:12.7.2(6)]

802.6.2.6 A gas vent shall terminate at least three (3) feet (914 mm) above a forced air inlet located within ten (10) feet (3,048 mm). [NFPA 54:12.7.2(7)]

802.6.3 Size of Gas Vents. Venting systems shall be sized and constructed in accordance with Section 803.0 or other approved engineering methods and the gas vent and appliance manufacturers' instructions. [NFPA 54:12.7.3]

802.6.3.1 Category I Appliances. The sizing of natural draft venting systems serving one (1) or more listed appliances equipped with a draft hood or appliances listed for use with Type B gas vent, installed in a single story of a building, shall be in accordance with one of the following methods. [NFPA 54:12.7.3.1]

(1) The provisions of Section 803.0. [NFPA 54:12.7.3.1(1)]

(2) Vents serving fan-assisted combustion system appliances, or combinations of fan-assisted combustion system and draft hood-equipped appliances shall be sized in accordance with Section 803.0 or other approved engineering methods. [NFPA 54:12.7.3.1(2)]

(3) For sizing an individual gas vent for a single, draft hood-equipped appliance, the effective area of the vent connector and the gas vent shall be not less than the area of the appliance draft hood outlet or greater than seven (7) times the draft hood outlet area. [NFPA 54:12.7.3.1(3)]

(4) For sizing a gas vent connected to two (2) appliances, with draft hoods, the

effective area of the vent shall be not less than the area of the larger draft hood outlet plus 50 percent of the area of the smaller draft hood outlet or greater than seven (7) times the smaller draft hood outlet area. [NFPA 54:12.7.3.1(4)]

(5) Approved engineering practices. [NFPA 54:12.7.3.1(5)]

802.6.3.2 Category II, Category III, and Category IV Appliances. The sizing of gas vents for Category II, Category III, and Category IV appliances shall be in accordance with the appliance manufacturer's instructions. [NFPA 54:12.7.3.3]

802.6.3.3 Sizing. Chimney venting systems using mechanical draft shall be sized in accordance with approved engineering methods. [NFPA 54:12.7.3.4]

802.6.4 Gas Vents Serving Appliances on More Than One Floor. A common gas vent shall be permitted in multistory installations to vent Category I appliances located on more than one floor level, provided the venting system is designed and installed in accordance with approved engineering methods. For the purpose of this section, crawl spaces, basements, and attics shall be considered as floor levels. [NFPA 54:12.7.4.1]

All appliances connected to the common vent shall be located in rooms separated from a habitable space. Each of these rooms shall have provisions for an adequate supply of combustion, ventilation, and dilution air that is not supplied from a habitable space. (See Figure 8-3.) [NFPA 54:12.7.4.2]

The size of the connectors and common segments of multistory venting systems for appliances listed for use with Type B double-wall gas vent shall be in accordance with Table 8-10, provided [NFPA 54:12.7.4.3]:

(1) The available total height (H) for each segment of a multistory venting system is the vertical distance between the level of the highest draft hood outlet or flue collar on that floor and the centerline of the next highest interconnection tee. (See Figure 8-4.) [NFPA 54:12.7.4.3(1)]

(2) The size of the connector for a segment is determined from its appliance heat input and available connector rise, and shall not be smaller than the draft hood outlet or flue collar size. [NFPA 54:12.7.4.3(2)]

(3) The size of the common vertical vent segment, and of the interconnection tee at the base of that segment, shall be based on the total appliance heat input entering that segment and its available total height. [NFPA 54:12.7.4.3(3)]

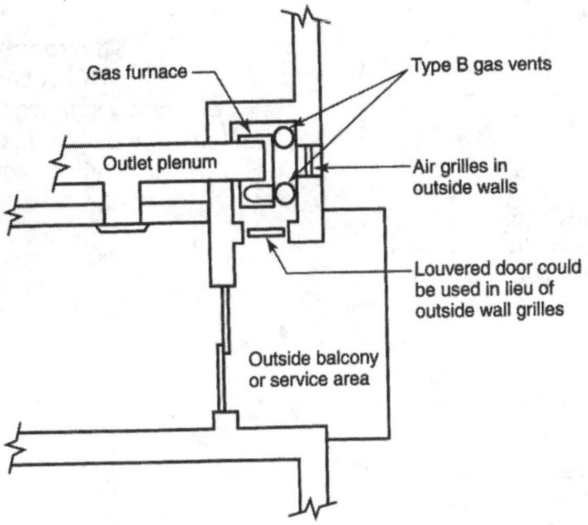

FIGURE 8-3 Plan View of Practical Separation Method for Multistory Gas Venting. [NFPA 54: Figure 12.7.4.2]

802.6.5 Support of Gas Vents. Gas vents shall be supported and spaced in accordance with their listings and the manufacturers' instructions. [NFPA 54:12.7.5]

802.6.6 Marking. In those localities where solid and liquid fuels are used extensively, gas vents shall be permanently identified by a label attached to the wall or ceiling at a point where the vent connector enters the gas vent. The label shall read: "This gas vent is for appliances that burn gas. Do not connect to solid or liquid-fuel-burning appliances or incinerators." The Authority Having Jurisdiction shall determine whether its area constitutes such a locality. [NFPA 54:12.7.6]

802.6.7 Screws, rivets, and other fasteners shall not penetrate the inner wall of double wall gas vents, except at the transition from the appliance draft hood outlet, flue collar or single wall metal connector to a double wall vent.

802.7 Single-Wall Metal Pipe.

802.7.1 Construction. Single-wall metal pipe shall be constructed of galvanized sheet steel not less than 0.0304 inch (0.7 mm) thick or of other approved, noncombustible, corrosion-resistant material. [NFPA 54:12.8.1]

802.7.2 Cold Climate. Uninsulated single-wall metal pipe shall not be used outdoors in cold climates for venting appliances in regions where the 99 percent winter design temperature is less than 32°F (0°C). [NFPA 54:12.8.2]

TABLE 8-2
Clearance for Connectors
[NFPA 54: Table 12.8.4.4]

	Minimum Distance from Combustible Material			
Appliance	**Listed Type B Gas Vent Material**	**Listed Type L Vent Material**	**Single-Wall Metal Pipe**	**Factory-Built Chimney Sections**
Listed appliance with draft hoods and -appliance listed for use with Type B gas vents	As listed	As listed	6 in.	As listed
Residential boilers and furnaces with listed gas conversion burner and with draft hood	6 in.	6 in.	9 in.	As listed
Residential appliances listed for use with Type L vents	Not permitted	As listed	9 in.	As listed
Residential incinerators	Not permitted	9 in.	18 in.	As listed
Listed gas-fired toilets	Not permitted	As listed	As listed	As listed
Unlisted residential appliances with draft hood	Not permitted	6 in.	9 in.	As listed
Residential and low-heat appliance other than those above	Not permitted	9 in.	18 in.	As listed
Medium-heat appliance	Not permitted	Not permitted	36 in.	As listed

Notes:
These clearances shall apply unless the listing of an appliance or connector specifies clearances, in which case the listed clearances shall apply.
1 in.= 25.4 mm

802.7.3 Termination. The termination of single-wall metal pipe shall comply with the following requirements: [NFPA 54:12.8.3]

802.7.3.1 Single-wall metal pipe shall terminate at least five (5) feet (1,524 mm) in vertical height above the highest connected appliance draft hood outlet or flue collar. [NFPA 54:12.8.3(1)]

802.7.3.2 Single-wall metal pipe shall extend at least two (2) feet (610 mm) above the highest point where it passes through a roof of a building and at least two (2) feet (610 mm) higher than any portion of a building within a horizontal distance of ten (10) feet (3,048 mm). (See Figure 8-1.) [NFPA 54:12.8.3(2)]

802.7.3.3 An approved cap or roof assembly shall be attached to the terminus of a single-wall metal pipe. (Also see Section 802.7.4.4) [NFPA 54:12.8.3(3)]

802.7.4 Installation with Appliances Permitted by Section 802.4.1.

802.7.4.1 Prohibited Use. Single-wall metal pipe shall not be used as a vent in dwellings and residential occupancies. [NFPA 54-09:12.8.4.1]

802.7.4.2 Single-wall metal pipe shall be used only for runs directly from the space in which the appliance is located through the roof or exterior wall to the outer air. A pipe passing through a roof shall extend without interruption through the roof flashing, roof jacket, or roof thimble. [NFPA 54:12.8.4.1]

802.7.4.3 Single-wall metal pipe shall not originate in any unoccupied attic or concealed space and shall not pass through any attic, inside wall, concealed space, or floor. For the installation of a single-wall metal pipe through an exterior combustible wall see Section 802.10.14.2. [NFPA 54:12.8.4.2]

802.7.4.4 Single-wall metal pipe used for venting an incinerator shall be exposed and readily examinable for its full length and shall have suitable clearances maintained. [NFPA 54:12.8.4.3]

802.7.4.5 Minimum clearances from single-wall metal pipe to combustible material shall be in accordance with Table 8-2. Reduced clearances from single-wall metal pipe to combustible material shall be as specified for vent connectors in Table 5-3. [NFPA 54:12.8.4.4]

802.7.4.6 Where a single-wall metal pipe passes through a roof constructed of combustible material, a noncombustible, nonventilating thimble shall be used at the point of passage. The thimble shall extend at least eighteen (18) inches (460 mm) above and six (6) inches (150 mm) below the roof, with the annular space open at the bottom and closed only at the top. The thimble shall be sized in accordance with Section 802.10.14.2. [NFPA 54:12.8.4.5]

802.7.5 Size of Single-Wall Metal Pipe. Single-wall metal piping shall comply with the following requirements: [NFPA 54:12.8.5]

802.7.5.1 A venting system of a single-wall metal pipe shall be sized in accordance with one of the following methods and the appliance manufacturer's instructions: [NFPA 54:12.8.5(1)]

(1) For a draft hood–equipped appliance, in accordance with Section 803.0. [NFPA 54:12.8.5(1)(a)]

(2) For a venting system for a single appliance with a draft hood, the areas of the connector and the pipe each shall not be less than the area of the appliance flue collar or draft hood outlet, whichever is smaller. The vent area shall not be greater than seven times the draft hood outlet area. [NFPA 54:12.8.5(1)(b)]

(3) Other approved engineering methods. [NFPA 54:12.8.5(1)(c)]

802.7.5.2 Where a single-wall metal pipe is used and has a shape other than round, it shall have an equivalent effective area equal to the effective area of the round pipe for which it is substituted, and the minimum internal dimension of the pipe shall be two (2) inches (50 mm). [NFPA 54:12.8.5(2)]

802.7.5.3 The vent cap or a roof assembly shall have a venting capacity not less than that of the pipe to which it is attached. [NFPA 54:12.8.5(3)]

802.7.6 Support of Single-Wall Metal Pipe. All portions of single-wall metal pipe shall be supported for the design and weight of the material employed. [NFPA 54:12.8.6]

802.7.7 Marking. Single-wall metal pipe shall comply with the marking provisions of Section 802.6.6. [NFPA 54:12.8.7]

802.8 Through-the-Wall Vent Termination. Through the wall vent termination's shall be in accordance with the following:

802.8.1 A mechanical draft venting system shall terminate at least three (3) feet (914 mm) above any forced air inlet located within ten (10) feet (3,048 mm). [NFPA 54:12.9.1]

Exception 1: This provision shall not apply to the combustion-air intake of a direct-vent appliance.

Exception 2: This provision shall not apply to the separation of the integral outdoor-air inlet and flue gas discharge of listed outdoor appliances.

802.8.2 A mechanical draft venting system of other than direct-vent type shall terminate at least four (4) feet (1,219 mm) below, four (4) feet (1,219 mm) horizontally from, or one (1) foot (300 mm) above any door, operable window, or gravity air inlet into any building. The bottom of the vent terminal shall be located at least twelve (12) inches (300 mm) above grade. [NFPA 54:12.9.2]

802.8.3 The vent terminal of a direct-vent appliance with an input of 10,000 Btu/h (3 kW) or less shall be located at least six (6) inches (150 mm) from any air opening into a building, and such an appliance with an input over 10,000 Btu/h (3 kW) but not over 50,000 Btu/h (14.7 kW) shall be installed with a nine (9) inches (230 mm) vent termination clearance, and an appliance with an input over 50,000 Btu/h (14.7 kW) shall have at least a twelve (12) inches (300 mm) vent termination clearance. The bottom of the vent terminal and the air intake shall be located at least twelve (12) inches (300 mm) above grade. [NFPA 54:12.9.3]

802.8.4 Through-the-wall vents for Category II and Category IV appliances and noncategorized condensing appliances shall not terminate over public walkways or over an area where condensate or vapor could create a nuisance or hazard or could be detrimental to the operation of regulators, relief valves, or other equipment. Where local experience indicates that condensate is a problem with Category I and Category III appliances, this provision shall also apply. Drains for condensate shall be installed in accordance with the manufacturer's installation instructions. [NFPA 54-09:12.9.4]

802.8.5 Where vents, including those for direct-vent appliances, penetrate outside walls of buildings, the annular spaces around such penetrations shall be permanently sealed using approved materials to prevent entry of combustion products into the building. [NFPA 54-09:12.9.6]

802.9 Condensation Drain.

802.9.1 Provision shall be made to collect and dispose of condensate from venting systems serving Category II and Category IV appliances and noncategorized condensing appliances in accordance with Section 802.8.4. [NFPA 54:12.10.1]

802.9.2 Where local experience indicates that condensation is a problem, provision shall be made to drain off and dispose of condensate from venting systems serving Category I and Category III appliances in accordance with Section 802.8.4. [NFPA 54:12.10.2]

802.10 Vent Connectors for Category I Appliances.

802.10.1 Where Required. A vent connector shall be used to connect appliances to a gas vent, chimney, or single-wall metal pipe, except where the gas vent, chimney, or single-wall metal pipe is directly connected to the appliance. [NFPA 54:12.11.1]

802.10.2 Materials.

802.10.2.1 A vent connector shall be made of noncombustible, corrosion-resistant material capable of withstanding the vent gas temperature produced by the appliance and of sufficient thickness to withstand physical damage. [NFPA 54:12.11.2.1]

802.10.2.2 Where the vent connector used for appliances having a draft hood or a Category I appliance is located in or passes through an unconditioned area, that portion of the vent connector shall be listed Type B, Type L, or listed vent material having equivalent insulation qualities. [NFPA 54:12.11.2.2]

Exception: Single-wall metal pipe located within the exterior walls of the building and located in areas having a local 99 percent winter design temperature of 5°F (-15°C) or higher.

802.10.2.3 Where the vent connector used for appliances having a draft hood or a Category I appliance is located in or passes through attics and crawl spaces, that portion of the vent connector shall be listed Type B, Type L, or listed vent material having equivalent insulation qualities. [NFPA 54:12.11.2.3]

802.10.2.4 Vent connectors for residential-type appliances shall comply with the following: [NFPA 54:12.11.2.4]

(1) Vent connectors for listed gas appliances having draft hoods and for appliances having draft hoods and equipped with listed conversion burners that are not installed in attics, crawl spaces, or other unconditioned areas shall be one of the following: [NFPA 54:12.11.2.4(1)]

(a) Type B or Type L vent material.

(b) Galvanized sheet steel not less than 0.018 in. (0.46 mm) thick.

(c) Aluminum (1100 or 3003 alloy or equivalent) sheet not less than 0.027 in. (0.69 mm) thick.

(d) Stainless steel sheet not less than 0.012 in. (0.31 mm) thick.

(e) Smooth-interior-wall metal pipe having resistance to heat and corrosion equal to or greater than that of b, c, or d above.

(f) A listed vent connector.

(2) Vent connectors shall not be covered with insulation. [NFPA 54:12.11.2.4(2)]

Exception: Listed insulated vent connectors shall be installed according to the terms of their listing in accordance with the manufacturer's installation instructions.

802.10.2.5 A vent connector for non-residential low-heat appliance shall be a factory-built chimney section or steel pipe having resistance to heat and corrosion equivalent to that for the appropriate galvanized pipe as specified in Table 8-3. Factory-built chimney sections shall be joined together in accordance with the chimney manufacturer's instructions. [NFPA 54:12.11.2.5]

802.10.2.6 Vent connectors for medium-heat appliances, commercial and industrial incinerators shall be constructed of factory-built, medium-heat chimney sections or steel of a thickness not less than that speci-

TABLE 8-3
Minimum Thickness for Galvanized Steel Vent Connectors for Low-Heat Appliances
[NFPA 54: Table 12.11.2.5]

Diameter of Connector (in.)	Minimum Thickness (in.)
Less than 6	0.019
6 to less than 10	0.023
10 to 12 inclusive	0.029
14 to 16 inclusive	0.034
Over 16	0.056

Notes: 1 in. = 25.4 mm, 1 in.^{2} = 645 mm^2

fied in Table 8-4, and shall comply with the following: [NFPA 54:12.11.2.6]

(1) A steel vent connector for appliances with a vent gas temperature in excess of 1,000°F (538°C) measured at the entrance to the connector shall be lined with medium-duty fire brick (ASTM C 64, *Specification for Refractories for Incinerators and Boilers, Type F*) or the equivalent.

(2) The lining shall be at least two and one-half (2-1/2) inches (64 mm) thick for a vent connector having a diameter or greatest cross-sectional dimension of eighteen (18) inches (460 mm) or less.

(3) The lining shall be at least four and one-half (4-1/2) inches (110 mm) thick laid on the four and one-half (4-1/2) inches (110 mm) bed for a vent connector having diameter or greatest cross-sectional dimension greater than eighteen (18) inches (460 mm).

(4) Factory-built chimney sections, if employed, shall be joined together in accordance with the chimney manufacturer's instructions.

802.10.3 Size of Vent Connector.

802.10.3.1 A vent connector for gas utilization appliances with a single draft hood or for a Category I fan-assisted combustion system appliance shall be sized and installed in accordance with Section 803.0 or other approved engineering methods. [NFPA 54:12.11.3.1]

802.10.3.2 For a single appliance having more than one (1) draft hood outlet or flue collar, the manifold shall be constructed according to the instructions of the appliance manufacturer. Where there are no instructions, the manifold shall be designed and constructed in accordance with approved engineering practices. As an alternate method, the effective area of the manifold shall equal the combined area of the flue collars or draft hood outlets and the vent connectors shall have a minimum one (1) foot (305 mm) rise. [NFPA 54:12.11.3.2]

802.10.3.3 Where two or more gas appliances are connected to a common vent or chimney, each vent connector shall be sized in accordance with Section 803.0 or other approved engineering methods. As an alternative method applicable only when all of the appliances are draft hood equipped, each vent connector shall have an effective area not less than the area of the draft hood outlet of the appliance to which it is connected. [NFPA 54:12.11.3.3, 12.11.3.4]

802.10.3.4 Where two (2) or more gas appliances are vented through a common vent connector or vent manifold, the common vent connector or vent manifold shall be located at the highest level consistent with available head-room and clearance to combustible material and shall be sized in accordance with Section 803.0 or other approved engineering methods. [NFPA 54:12.11.3.5]

As an alternate method applicable only where there are two (2) draft hood–equipped appliances, the effective area of the common vent connector or vent manifold and all junction fittings shall be not less than the area of the larger vent connector plus 50 percent of the areas of smaller flue collar outlets. [NFPA 54:12.11.3.6]

802.10.3.5 Where the size of a vent connector is increased to overcome installation limitations and obtain connector capacity equal to the appliance input, the size increase shall be made at the appliance draft hood outlet. [NFPA 54:12.11.3.7]

802.10.4 Two or More Appliances Connected to a Single Vent.

802.10.4.1 Where two (2) or more openings are provided into one (1) chimney flue or vent, the openings shall either be at different levels, or the connectors shall be attached to the vertical portion of the chimney or vent at an angle of 45 degrees or less relative to the vertical. [NFPA 54-09:12.11.4.1]

802.10.4.2 Where two (2) or more vent connectors enter a common gas vent, chimney flue, or single-wall metal pipe, the smaller connector shall enter at the highest level consistent with the available headroom or clearance to combustible material. [NFPA 54:12.11.4.1]

TABLE 8-4
Minimum Thickness for Steel Vent Connectors for Medium-Heat Appliances and Commercial and Industrial Incinerators
[NFPA 54: Table 12.11.2.6]

Vent Connector Size		Minimum
Diameter (in.)	Area (in.^{2})	Thickness (in.)
Up to 14	Up to 154	0.053
Over 14 to 16	154 to 201	0.067
Over 16 to 18	201 to 254	0.093
Over 18	Larger than 254	0.123

Notes: 1 in. = 25.4 mm, 1 in.^{2} = 645 mm^2

802.10.4.3 Vent connectors serving Category I appliances shall not be connected to any portion of a mechanical draft system operating under positive static pressure, such as those serving Category III or Category IV appliances. [NFPA 54:12.11.4.2]

802.10.5 Clearance. Minimum clearances from vent connectors to combustible material shall be in accordance with Table 8-2. [NFPA 54:12.11.5]

Exception: The clearance between a vent connector and combustible material shall be permitted to be reduced where the combustible material is protected as specified for vent connectors in Table 5-3.

802.10.6 Avoid Unnecessary Bends. A vent connector shall be installed so as to avoid turns or other construction features that create excessive resistance to flow of vent gases. [NFPA 54:12.11.6]

802.10.7 Joints. Joints between sections of connector piping and connections to flue collars or draft hood outlets shall be fastened in accordance with one of the following methods: [NFPA 54:12.11.7]

(1) By sheet metal screws.

(2) Vent connectors of listed vent material shall be assembled and connected to flue collars or draft hood outlets in accordance with the manufacturers' instructions.

(3) Other approved means.

802.10.8 Slope. A vent connector shall be installed without any dips or sags and shall slope upward toward the vent or chimney at least one-quarter (1/4) in./ft. (20 mm/m). [NFPA 54:12.11.8]

Exception: Vent connectors attached to a mechanical draft system installed in accordance with the draft system manufacturers' instructions.

802.10.9 Length of Vent Connector.

802.10.9.1 A vent connector shall be as short as practical, and the appliance located as close as practical to the chimney or vent. [NFPA 54:12.11.9.1]

802.10.9.2 The maximum horizontal length of a single-wall connector shall be 75 percent of the height of the chimney or vent except for engineered systems. [NFPA 54:12.11.9.2]

802.10.9.3 The maximum horizontal length of a Type B double-wall connector shall be 100 percent of the height of the chimney or vent, except for engineered systems. The maximum length of an individual connector for a chimney or vent system serving multiple appliances, from the appliance outlet to the junction with the common vent or another connector, shall be 100 percent of the height of the chimney or vent. [NFPA 54: 12.11.9.3]

802.10.10 Support. A vent connector shall be supported for the design and weight of the material employed to maintain clearances and prevent physical damage and separation of joints. [NFPA 54:12.11.10]

802.10.11 Chimney Connection. Where entering a flue in a masonry or metal chimney, the vent connector shall be installed above the extreme bottom to avoid stoppage. Where a thimble or slip joint is used to facilitate removal of the connector, the connector shall be firmly attached to or inserted into the thimble or slip joint to prevent the connector from falling out. Means shall be employed to prevent the connector from entering so far as to restrict the space between its end and the opposite wall of the chimney flue. [NFPA 54:12.11.11]

802.10.12 Inspection. The entire length of a vent connector shall be readily accessible for inspection, cleaning, and replacement. [NFPA 54:12.11.12]

802.10.13 Fireplaces. A vent connector shall not be connected to a chimney flue serving a fireplace unless the fireplace flue opening is permanently sealed. [NFPA 54:12.11.13]

802.10.14 Passage Through Ceilings, Floors, or Walls.

802.10.14.1 A vent connector shall not pass through any ceiling, floor, or fire-resistance rated wall. A single-wall metal pipe connector shall not pass through any interior wall.

Exception: Vent connectors made of listed Type B or Type L vent material and serving listed appliances with draft hoods and other appliances listed for use with Type B gas vents that pass through walls or partitions constructed of combustible material shall be installed with not less than the listed clearance to combustible material.

802.10.14.2 A vent connector made of a single-wall metal pipe shall not pass through a combustible exterior wall unless guarded at the point of passage by a ventilated metal thimble not smaller than the following: [NFPA 54:12.11.14.2]

(1) For listed appliances equipped with draft hoods and appliances listed for

use with Type B gas vents, the thimble shall be a minimum of four (4) inches (100 mm) larger in diameter than the vent connector. Where there is a run of not less than six (6) feet (305 mm) of vent connector in the opening between the draft hood outlet and the thimble, the thimble shall be a minimum of two (2) inches (50 mm) larger in diameter than the vent connector.

(2) For unlisted appliances having draft hoods, the thimble shall be a minimum of six (6) inches (150 mm) larger in diameter than the vent connector.

(3) For residential and low-heat appliances, the thimble shall be a minimum of twelve (12) inches (300 mm) larger in diameter than the vent connector.

Exception: In lieu of thimble protection, all combustible material in the wall shall be removed from the vent connector a sufficient distance to provide the specified clearance from such vent connector to combustible material. Any material used to close up such opening shall be noncombustible.

802.10.14.3 Vent connectors for medium-heat appliances shall not pass through walls or partitions constructed of combustible material. [NFPA 54:12.11.14.3]

802.11 Vent Connectors for Category II, Category III, and Category IV Appliances. (See Section 802.4.) [NFPA 54:12.12]

802.12 Draft Hoods and Draft Controls.

802.12.1 Appliances Requiring Draft Hoods. Vented appliances shall be installed with draft hoods. [NFPA 54:12.13.1]

Exception: Dual oven–type combination ranges; incinerators; direct-vent appliances; fan-assisted combustion system appliances; appliances requiring chimney draft for operation; single-firebox boilers equipped with conversion burners with inputs greater than 400,000 Btu/h (117 kW); appliances equipped with blast, power, or pressure burners that are not listed for use with draft hoods; and appliances designed for forced venting.

802.12.2 Installation. A draft hood supplied with or forming a part of listed vented appliance shall be installed without alteration, exactly as furnished and specified by the appliance manufacturer. If a draft hood is not supplied by the appliance manufacturer where one (1) is required, a draft hood shall be installed, be of a listed or approved type, and, in the absence of other instructions, be of the same size as the appliance flue collar. Where a draft hood is required with a conversion burner, it shall be of a listed or approved type. Where it is determined that a draft hood of special design is needed or preferable for a particular installation, the installation shall be in accordance with the recommendations of the appliance manufacturer and shall be with the approval of the Authority Having Jurisdiction. [NFPA 54:12.13.2, 12.13.2.1, 12.13.2.2]

802.12.3 Draft Control Devices. Where a draft control device is part of the appliance or is supplied by the appliance manufacturer, it shall be installed in accordance with the manufacturer's instructions. In the absence of manufacturer's instructions, the device shall be attached to the flue collar of the appliance or as near to the appliance as practical. [NFPA 54:12.13.3]

802.12.4 Additional Devices. Appliances (except incinerators) requiring controlled chimney draft shall be permitted to be equipped with a listed double-acting barometric draft regulator installed and adjusted in accordance with the manufacturers' instructions. [NFPA 54:12.13.4]

802.12.5 Location. Draft hoods and barometric draft regulators shall be installed in the same room or enclosure as the appliance in such a manner as to prevent any difference in pressure between the hood or regulator and the combustion air supply. [NFPA 54:12.13.5]

802.12.6 Positioning. Draft hoods and draft regulators shall be installed in the position for that they were designed with reference to the horizontal and vertical planes and shall be located so that the relief opening is not obstructed by any part of the appliance or adjacent construction. The appliance and its draft hood shall be located so that the relief opening is accessible for checking vent operation. [NFPA 54:12.13.6]

802.12.7 Clearance. A draft hood shall be located so that its relief opening is not less than six (6) inches (150 mm) from any surface except that of the appliance it serves and the venting system to which the draft hood is connected. Where a greater or lesser clearance is indicated on the appliance label, the clearance shall not be less than that specified on the label. Such clearances shall not be reduced. [NFPA 54:12.13.7]

802.13 Manually Operated Dampers. A manually operated damper shall not be placed in any appliance vent connector. Fixed baffles shall not be classified as manually operated dampers. [NFPA 54:12.14]

802.14 Automatically Operated Vent Dampers. An automatically operated vent damper shall be of a listed type. [NFPA 54:12.15]

802.15 Obstructions. Devices that retard the flow of vent gases shall not be installed in a vent connector, chimney, or vent. The following shall not be considered as obstructions:

(1) Draft regulators and safety controls specifically listed for installation in venting systems and installed in accordance with the terms of their listing.

(2) Approved draft regulators and safety controls designed and installed in accordance with approved engineering methods.

(3) Listed heat reclaimers and automatically operated vent dampers installed in accordance with the terms of their listing.

(4) Vent dampers serving listed appliances installed in accordance with Sections 803.1 and 803.2.1 or other approved engineering methods.

(5) Approved economizers, heat reclaimers, and recuperators installed in venting systems of appliances not required to be equipped with draft hoods, provided the appliance manufacturer's instructions cover the installation of such a device in the venting system and performance in accordance with Sections 802.3.1 and 802.3.2 is obtained. [NFPA 54:12.16]

803.0 Sizing of Category I Venting Systems.

803.1 These venting tables shall not be used where obstructions (see Section 802.15) are installed in the venting system. The installation of vents serving listed appliances with vent dampers shall be in accordance with the appliance manufacturer's instructions or in accordance with the following: [NFPA 54:13.1.1]

(1) The maximum capacity of the vent system shall be determined using the NAT Max column.

(2) The minimum capacity shall be determined as if the appliance were a fan-assisted appliance, using the FAN Min column to determine the minimum capacity of the vent system. Where the corresponding FAN Min is NA, the vent configuration shall not be permitted and an alternative venting configuration shall be utilized.

803.1.1 Where the vent size determined from the tables is smaller than the appliance draft hood outlet or flue collar, the use of the smaller size shall be permitted provided that the installation complies with all of the following requirements: [NFPA 54:13.1.2]

(1) The total vent height (*H*) is at least ten (10) feet (3,048 mm).

(2) Vents for appliance draft hood outlets or flue collars twelve (12) inches (300 mm) in diameter or smaller are not reduced more than one table size.

(3) Vents for appliance draft hood outlets or flue collars larger than twelve (12) inches (300 mm) in diameter are not reduced more than two table sizes.

(4) The maximum capacity listed in the tables for a fan-assisted appliance is reduced by ten (10) percent (0.90 maximum table capacity).

(5) The draft hood outlet is greater than four (4) inches (100 mm) in diameter. Do not connect a three (3) inches (80 mm) diameter vent to a four (4) inches (100 mm) diameter draft hood outlet. This provision shall not apply to fan-assisted appliances.

803.1.2 Single-appliance venting configurations with zero (0) lateral lengths in Tables 8-5, 8-6, and 8-9 shall not have elbows in the venting system. For vent configurations with lateral lengths, the venting tables include allowance for two (2) 90 degree turns. For each additional elbow up to and including 45 degrees, the maximum capacity listed in the venting tables shall be reduced by 5 percent. For each elbow greater than 45 degrees up to and including 90 degrees, the maximum capacity listed in the venting tables shall be reduced by 10 percent. Where multiple offsets occur in a vent, the total lateral length of all offsets combined shall not exceed that specified in Tables 8-5 through 8-9. [NFPA 54-09:13.1.3]

803.1.3 Zero (0) lateral (*L*) shall apply only to a straight vertical vent attached to a top-outlet draft hood or flue collar. [NFPA 54:13.1.4]

803.1.4 Sea level input ratings shall be used when determining maximum capacity for high-altitude installation. Actual input (derated for altitude) shall be used for determining minimum capacity for high-altitude installation. [NFPA 54:13.1.5]

803.1.5 For appliances with more than one (1) input rate, the minimum vent capacity (FAN Min) determined from the tables shall be less than the lowest appliance input rating, and the maximum vent capacity (FAN Max/NAT Max) determined from the tables shall be greater than the highest appliance rating input. [NFPA 54:13.1.6]

803.1.6 Listed corrugated metallic chimney liner systems in masonry chimneys shall be sized by using Table 8-5 or 8-6 for Type B vents, with the maximum capacity reduced by 20 percent (0.80 maximum capacity) and the minimum capacity as shown in Table 8-5 or 8-6. Corrugated metallic liner systems installed with bends or offsets shall have their maximum capacity further reduced in accordance with Section 803.1.2. The 20 percent reduction for corrugated metallic chimney liner systems includes an allowance for one (1) long radius 90 degree turn at the bottom of the liner. [NFPA 54:13.1.7]

803.1.7 Where the vertical vent has a larger diameter than the vent connector, the vertical vent diameter shall be used to determine the minimum vent capacity, and the connector diameter shall be used to determine the maximum vent capacity. The flow area of the vertical vent shall not exceed seven (7) times the flow area of the listed appliance categorized vent area, flue collar area, or draft hood outlet area unless designed in accordance with approved engineering methods. [NFPA 54:13.1.9]

803.1.8 Connection to Chimney Liners. Connections between chimney liners and listed double-wall connectors shall be made with listed adapters designed for such purpose. [NFPA 54:13.1.8]

803.1.8.1 Tables 8-5 through 8-9 shall be used for chimneys and vents not exposed to the outdoors below the roof line. A Type B vent or listed chimney lining system passing through an unused masonry chimney flue shall not be considered to be exposed to the outdoors. A Type B vent passing through an unventilated enclosure or chase insulated to a value of not less than R8 shall not be considered to be exposed to the outdoors. Table 8-7 in combination with Table 8-15 shall be used for clay-tile-lined exterior masonry chimneys, provided all of the following are met:

(1) The vent connector is Type B double wall.

(2) The vent connector length is limited to one and one-half (1-1/2) feet for each inch (18 mm/mm) of vent connector diameter.

(3) The appliance is draft hood equipped.

(4) The input rating is less than the maximum capacity given in Table 8-7.

(5) For a water heater, the outdoor design temperature shall not be less than 5°F (-15°C).

(6) For a space-heating appliance, the input rating is greater than the minimum capacity given by Table 8-15. [NFPA 54:13.1.11]

803.1.9 Corrugated vent connectors shall not be smaller than the listed appliance categorized vent diameter, flue collar diameter, or draft hood outlet diameter. [NFPA 54:13.1.12]

803.1.10 Vent connectors shall not be upsized more than two (2) sizes greater than the listed appliance categorized vent diameter, flue collar diameter, or draft hood outlet diameter. [NFPA 54:13.1.13]

803.1.11 In a single run of vent or vent connector, more than one diameter and type shall be permitted to be used, provided that all the sizes and types are permitted by the tables. [NFPA 54:13.1.14]

803.1.12 Interpolation shall be permitted in calculating capacities for vent dimensions that fall between table entries. (See Example 8-1 in Section 803.2.23.) [NFPA 54:13.1.15]

803.1.13 Extrapolation beyond the table entries shall not be permitted. [NFPA 54:13.1.16]

803.1.14 For vent heights lower than six (6) feet (1,829 mm) and higher than shown in the tables, engineering methods shall be used to calculate vent capacities. [NFPA 54:13.1.17]

803.1.15 Draft Hood Conversion Accessories. Draft hood conversion accessories for use with masonry chimneys venting listed Category I fan-assisted appliances shall be listed and installed in accordance with the listed accessory manufacturer's installation instructions. [NFPA 54:13.1.10]

803.2 Additional Requirements to Multiple Appliance Vent Tables 8-10 through 8-14 and Tables 8-16 through 8-19.

803.2.1 These venting tables shall not be used where obstructions (see Section 802.15) are installed in the venting system. The installation of vents serving listed appliances with vent dampers shall be in accordance with the appliance manufacturer's instructions or in accordance with the following:

(1) The maximum capacity of the vent connector shall be determined using the NAT Max column.

(2) The maximum capacity of the vertical vent or chimney shall be determined using the FAN+NAT column when the second appliance is a fan-assisted appliance, or the

NAT+NAT column when the second appliance is equipped with a draft hood.

(3) The minimum capacity shall be determined as if the appliance were a fan-assisted appliance.

(a) The minimum capacity of the vent connector shall be determined using the FAN Min column.

(b) The FAN+FAN column shall be used when the second appliance is a fan-assisted appliance, and the FAN+NAT column shall be used when the second appliance is equipped with a draft hood, to determine whether the vertical vent or chimney configuration is not permitted (NA). Where the vent configuration is NA, the vent configuration shall not be permitted, and an alternative venting configuration shall be utilized. [NFPA 54:13.2.1]

803.2.2 The maximum vent connector horizontal length shall be 18 in./in. (18 mm/mm) of connector diameter (see the Vent Connector Maximum Length Table). [NFPA 54:13.2.2]

Vent Connector Maximum Length [NFPA 54: Table 13.2.2]

Connector Diameter Maximum (in.)	Connector Horizontal Length (ft.)
3	4-1/2
4	6
5	7-1/2
6	9
7	10-1/2
8	12
9	13-1/2
10	15
12	18
14	21
16	24
18	27
20	30
22	33
24	36

Notes: 1 in. = 25.4 mm, 1 ft. = 0.305 m

803.2.3 The vent connector shall be routed to the vent utilizing the shortest possible route. Connectors with longer horizontal lengths than those listed in Section 803.2.2 are permitted under the following conditions: [NFPA 54:13.2.3]

(A) The maximum capacity (FAN Max or NAT Max) of the vent connector shall be reduced 10 percent for each additional multiple of the length listed in 803.2.2. For example, the maximum length listed for a four (4) inches (100 mm) connector is six (6) feet (1,829 mm). With a connector length greater than six (6) feet (1,829 mm) but not exceeding twelve (12) feet (3,658 mm), the maximum capacity must be reduced by 10 percent (0.90 maximum vent connector capacity). With connector length greater than twelve (12) feet (3,658 mm) but not exceeding eighteen (18) feet (5,486 mm), the maximum capacity must be reduced by 20 percent (0.80 maximum vent capacity). [NFPA 54:13.2.3(1)]

(B) For a connector serving a fan-assisted appliance, the minimum capacity (FAN Min) of the connector shall be determined by referring to the corresponding single-appliance table. For Type B double-wall connectors, Table 8-5 shall be used. For single-wall connectors, Table 8-6 shall be used. The height (*H*) and lateral (*L*) shall be measured according to the procedures for a single-appliance vent, as if the other appliances were not present. [NFPA 54:13.2.3(2)]

803.2.4 Where the vent connectors are combined prior to entering the vertical portion of the common vent to form a common vent manifold, the size of the common vent manifold and the common vent shall be determined by applying a 10 percent reduction (0.90 x maximum common vent capacity) to the Common Vent Capacity part of the common vent tables. The length of the common vent connector manifold (*LM*) shall not exceed 18 in./in. (18 mm/mm) of common vent connector manifold diameter (*D*). (See Figure 8-4.) [NFPA 54:13.2.4]

803.2.5 Where the common vertical vent is offset, the maximum capacity of the common vent shall be reduced in accordance with Section 803.2.6, and the horizontal length of the common vent offset shall not exceed 18 in./in. (18 mm/mm) of common vent diameter. Where multiple offsets occur in a common vent, the total horizontal length of all offsets combined shall not exceed 18 in./in. (18 mm/mm) of common vent diameter. [NFPA 54-09:13.2.5]

803.2.6 For each elbow up to and including 45 degrees in the common vent, the maximum common vent capacity listed in the venting tables shall be reduced by 5 percent. For each

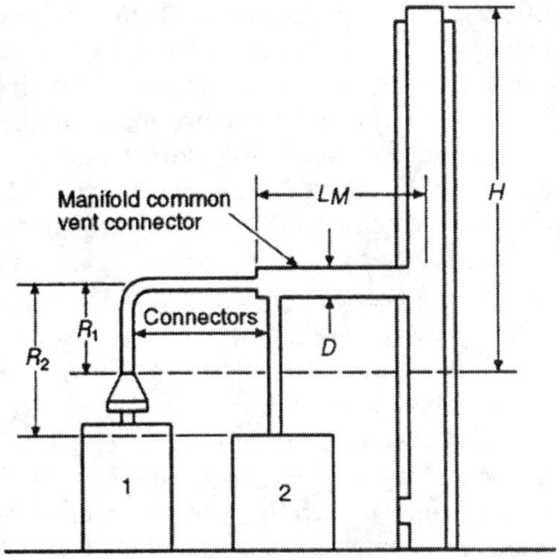

FIGURE 8-4 Use of Manifolded Common Vent Connector. [NFPA 54: Figure G.1(k)]

elbow exceeding 45 degrees up to and including 90 degrees, the maximum common vent capacity listed in the venting tables shall be reduced by 10 percent. [NFPA 54:13.2.6]

→ **803.2.7 Common Vent Minimum Size.** The cross-sectional area of the common vent shall be equal to or greater than the cross-sectional area of the largest connector. [NFPA 54:13.2.8]

803.2.8 Tee and Wye Fittings. Tee and wye fittings connected to a common gas vent shall be considered as part of the common gas vent and constructed of materials consistent with that of the common gas vent. [NFPA 54-09:13.2.9]

803.2.8.1 At the point where tee or wye fittings connect to a common vent, the opening size of the fitting shall be equal to the size of the common vent. Such fittings shall not be prohibited from having reduced size openings at the point of connection of appliance vent connectors. [NFPA 54:13.2.10]

803.2.9 Sea level input ratings shall be used when determining maximum capacity for high-altitude installation. Actual input (derated for altitude) shall be used for determining minimum capacity for high-altitude installation. [NFPA 54:13.2.11]

803.2.10 The connector rise (*R*) for each appliance connector shall be measured from the draft hood outlet or flue collar to the centerline where the vent gas streams come together. [NFPA 54:13.2.12]

803.2.11 For multiple units of appliances all located on one (1) floor, available total height (*H*) shall be measured from the highest draft hood outlet or flue collar up to the level of the outlet of the common vent. [NFPA 54:13.2.13]

803.2.12 For multistory installations, available total height (*H*) for each segment of the system shall be the vertical distance between the highest draft hood outlet or flue collar entering that segment and the centerline of the next higher interconnection tee. (See Figure 8-5.) [NFPA 54:13.2.14]

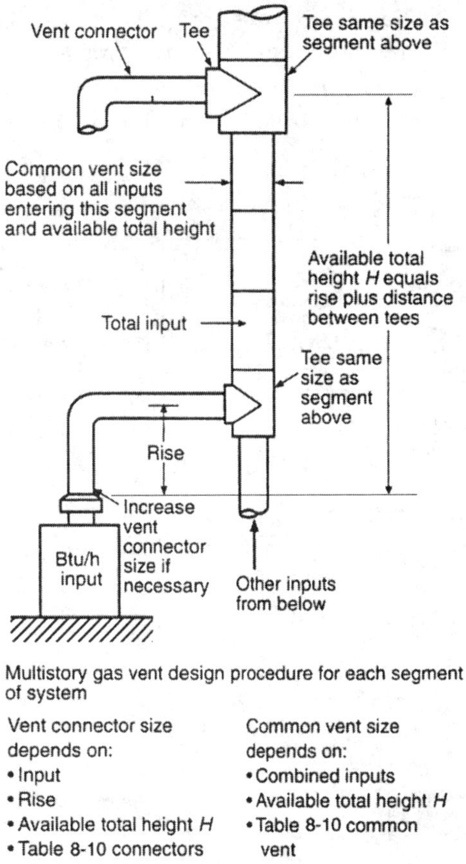

FIGURE 8-5 Multistory Gas Vent Design Procedure for Each Segment of System. [NFPA 54: Figure G.1(m)]

803.2.13 The size of the lowest connector and of the vertical vent leading to the lowest interconnection of a multistory system shall be in accordance with Table 8-5 or 8-6 for available total height (*H*) up to the lowest interconnection. (See Figure 8-6.) [NFPA 54:13.2.15]

803.2.14 Where used in multistory systems, vertical common vents shall be Type B double-wall and shall be installed with a listed vent cap. [NFPA 54:13.2.16]

803.2.15 Offsets in multistory common vent systems shall be limited to a single offset in each system, and systems with an offset shall comply with all of the following:

(1) The offset angle shall not exceed 45 degrees from vertical.

(2) The horizontal length of the offset shall not exceed eighteen (18) inches for each inch (180 mm/mm) of common vent diameter of the segment in that the offset is located.

(3) For the segment of the common vertical vent containing the offset, the common vent capacity listed in the common venting tables shall be reduced by 20 percent (0.80 x maximum common vent capacity).

(4) A multistory common vent shall not be reduced in size above the offset. [NFPA 54:13.2.17]

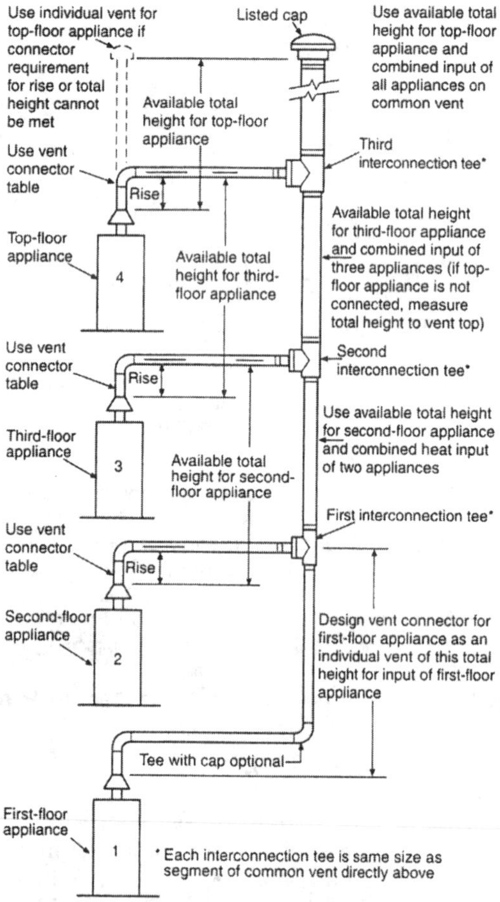

FIGURE 8-6 Principle of Design of Multistory Vents Using Vent Connector and Common Vent Design Tables. [NFPA 54: Figure G.1(n)]

803.2.16 Where two or more appliances are connected to a vertical vent or chimney, the flow area of the largest section of vertical vent or chimney shall not exceed seven times the smallest listed appliance categorized vent areas, flue collar area, or draft hood outlet area, unless designed in accordance with approved engineering methods. [NFPA 54:13.2.18]

803.2.17 For appliances with more than one input rate, the minimum vent connector capacity (FAN Min) determined from the tables shall be less than the lowest appliance input rating, and the maximum vent connector capacity (FAN Max or NAT Max) determined from the tables shall be greater than the highest appliance input rating. [NFPA 54:13.2.19]

803.2.18 Listed, corrugated metallic chimney liner systems in masonry chimneys shall be sized by using Table 8-10 or 8-11 for Type B vents, with the maximum capacity reduced by 20 percent (0.80 maximum capacity) and the minimum capacity as shown in Table 8-10 or 8-11. Corrugated metallic liner systems installed with bends or offsets shall have their maximum capacity further reduced in accordance with Sections 803.2.5 and 803.2.6. The 20 percent reduction for corrugated metallic chimney liner systems includes an allowance for one long radius 90 degree turn at the bottom of the liner. [NFPA 54:13.2.20]

803.2.19 Tables 8-10 and 8-11 shall be used for chimneys and vents not exposed to the outdoors below the roof line. A Type B vent passing through an unventilated enclosure or chase insulated to a value of not less than R 8 shall not be considered to be exposed to the outdoors. Tables 8-16 through 8-19 shall be used for clay-tile-lined exterior masonry chimneys, provided all of the following conditions are met:

(1) Vent connector is Type B double-wall.

(2) At least one appliance is draft hood-equipped.

(3) The combined appliance input rating is less than the maximum capacity given by Table 8-16 (for NAT+NAT) or Table 8-18 (for FAN+NAT).

(4) The input rating of each space-heating appliance is greater than the minimum input rating given by Table 8-17 (for NAT+NAT) or Table 8-19 (for FAN+NAT).

(5) The vent connector sizing is in accordance with Table 8-12. [NFPA 54:13.2.22]

803.2.20 Vent connectors shall not be increased more than two sizes greater than the listed appliance categorized vent diameter, flue collar diameter, or draft hood outlet diameter. Vent connectors for draft hood-equipped appliances shall not be smaller than the draft hood outlet diameter. Where a vent connector size(s) determined from the tables for a fan-assisted appliance(s) is smaller than the flue collar diameter, the use of the smaller size(s) shall be permitted provided that the installation complies with all of the following:

(1) Vent connectors for fan-assisted appliance flue collars twelve (12) inches (300 mm) in diameter or smaller are not reduced by more than one table size [e.g., 12 in. to 10 in. (300 mm to 250 mm) is a one-size reduction] and those larger than twelve (12) inches (300 mm) in diameter are not reduced more than two table sizes [e.g., 24 in. to 20 in. (610 mm to 510 mm) is a two-size reduction].

(2) The fan-assisted appliance(s) is common vented with a draft hood equipped appliance(s).

(3) The vent connector has a smooth interior wall. [NFPA 54:13.2.24]

803.2.21 All combinations of pipe sizes and single-wall and double-wall metal pipe shall be allowed within any connector run(s) or within the common vent provided ALL of the appropriate tables permit ALL of the desired sizes and types of pipe, as if they were used for the entire length of the subject connector or vent. Where single-wall and Type B double-wall metal pipes are used for vent connectors within the same venting system, the common vent must be sized using Table 8-11 or 8-13, as appropriate. [NFPA 54:13.2.25]

803.2.22 Where a table permits more than one (1) diameter of pipe to be used for a connector or vent, all the permitted sizes shall be permitted to be used. [NFPA 54:13.2.26]

803.2.23 Interpolation shall be permitted in calculating capacities for vent dimensions that fall between table entries. (See Example 8-1.) [NFPA 54:13.2.27]

Example 8-1: Interpolating between Table Values. An installer has an 80,000 Btu/h. input appliance with a four (4) inch diameter draft hood outlet that needs to be vented into a twelve (12) foot high Type B vent. The vent connector has a five (5) foot lateral length and is also Type B. Can this appliance be vented using a four (4) inch diameter vent?

Solution

Table 8-5 is used in the case of an all Type B vent system. However, since there is no entry in Table 8-5 for a height of 12 ft., interpolation must be used. Read down the 4 in. diameter NAT Max column to the row associated with 10 ft. height and 5 ft. lateral to find the capacity value of 77,000 Btu/h. Read further down to the 15 ft. height, 5 ft. lateral row to find the capacity value of 87,000 Btu/h. The difference between the 15 ft. height capacity value and the 10 ft. height capacity value is 10,000 Btu/h. The capacity for a vent system with a 12 ft. height is equal to the capacity for a 10 ft. height plus 2/5 of the difference between the 10 ft. and 15 ft. height values, or 77,000 + 2/5 x 10,000 = 81,000 Btu/h. Therefore, a 4 in. diameter vent can be used in the installation.

803.2.24 Extrapolation beyond the table entries shall not be permitted. [NFPA 54:13.2.28]

803.2.25 For vent heights lower than six (6) feet and higher than shown in the tables, engineering methods shall be used to calculate vent capacities. [NFPA 54:13.2.29]

Table 8-5

Type B Double-Wall Gas Vent [NFPA 54: Table 13.1(a)]

						Number of Appliances:	Single									
						Appliance Type:	Category I									
						Appliance Vent Connection:	Connected Directly to Vent									
		Vent Diameter – *D* (in.)														
		3			4			5			6			7		
		Appliance Input Rating in Thousands of Btu per Hour														
Height *H*	Lateral *L*	FAN		NAT	FAN		NAT	FAN		NAT	FAN		NAT	FAN		NAT
(ft)	(ft)	Min	Max	Max	Min	Max	Max	Min	Max	Max	Min	Max	Max	Min	Max	Max
6	0	0	78	46	0	152	86	0	251	141	0	375	205	0	524	285
	2	13	51	36	18	97	67	27	157	105	32	232	157	44	321	217
	4	21	49	34	30	94	64	39	153	103	50	227	153	66	316	211
	6	25	46	52	36	91	61	47	149	100	59	223	149	78	310	205
8	0	0	84	50	0	165	94	0	276	155	0	415	235	0	583	320
	2	12	57	40	16	109	75	25	178	120	28	263	180	42	365	247
	5	23	53	38	32	103	71	42	171	115	53	255	173	70	356	237
	8	28	49	35	39	98	66	51	164	109	64	247	165	84	347	227
10	0	0	88	53	0	175	100	0	295	166	0	447	255	0	631	345
	2	12	61	42	17	118	81	23	194	129	26	289	195	40	402	273
	5	23	57	40	32	113	77	41	187	124	52	280	188	68	392	263
	10	30	51	36	41	104	70	54	176	115	67	267	175	88	376	245
15	0	0	94	58	0	191	112	0	327	187	0	502	285	0	716	390
	2	11	69	48	15	136	93	20	226	150	22	339	225	38	475	316
	5	22	65	45	30	130	87	39	219	142	49	330	217	64	463	300
	10	29	59	41	40	121	82	51	206	135	64	315	208	84	445	288
	15	35	53	37	48	112	76	61	195	128	76	301	198	98	429	275
20	0	0	97	61	0	202	119	0	349	202	0	540	307	0	776	430
	2	10	75	51	14	149	100	18	250	166	20	377	249	33	531	346
	5	21	71	48	29	143	96	38	242	160	47	367	241	62	519	337
	10	28	64	44	38	133	89	50	229	150	62	351	228	81	499	321
	15	34	58	40	46	124	84	59	217	142	73	337	217	94	481	308
	20	48	52	35	55	116	78	69	206	134	84	322	206	107	464	295
30	0	0	100	64	0	213	128	0	374	220	0	587	336	0	853	475
	2	9	81	56	13	166	112	14	283	185	18	432	280	27	613	394
	5	21	77	54	28	160	108	36	275	176	45	421	273	58	600	385
	10	27	70	50	37	150	102	48	262	171	59	405	261	77	580	371
	15	33	64	NA	44	141	96	57	249	163	70	389	249	90	560	357
	20	56	58	NA	53	132	90	66	237	154	80	374	237	102	542	343
	30	NA	NA	NA	73	113	NA	88	214	NA	104	346	219	131	507	321
50	0	0	101	67	0	216	134	0	397	232	0	633	363	0	932	518
	2	8	86	61	11	183	122	14	320	206	15	497	314	22	715	445
	5	20	82	NA	27	177	119	35	312	200	43	487	308	55	702	438
	10	26	76	NA	35	168	114	45	299	190	56	471	298	73	681	426
	15	59	70	NA	42	158	NA	54	287	180	66	455	288	85	662	413
	20	NA	NA	NA	50	149	NA	63	275	169	76	440	278	97	642	401
	30	NA	NA	NA	69	131	NA	84	250	NA	99	410	259	123	605	376
100	0	NA	NA	NA	0	218	NA	0	407	NA	0	665	400	0	997	560
	2	NA	NA	NA	10	194	NA	12	354	NA	13	566	375	18	831	510
	5	NA	NA	NA	26	189	NA	33	347	NA	40	557	369	52	820	504
	10	NA	NA	NA	33	182	NA	43	335	NA	53	542	361	68	801	493
	15	NA	NA	NA	40	174	NA	50	321	NA	62	528	353	80	782	482
	20	NA	NA	NA	47	166	NA	59	311	NA	71	513	344	90	763	471
	30	NA	NA	NA	NA	NA	NA	78	290	NA	92	483	NA	115	726	449
	50	NA	NA	NA	NA	NA	NA	NA	NA	NA	147	428	NA	180	651	405

Table 8-5

Type B Double-Wall Gas Vent [NFPA 54: Table 13.1(a)] (continued)

								Number of Appliances:	Single							
								Appliance Type:	Category I							
								Appliance Vent Connection:	Connected Directly to Vent							
		Vent Diameter – *D* (in.)														
		8			9			10			12			14		
		Appliance Input Rating in Thousands of Btu per Hour														
Height *H* (ft)	Lateral *L* (ft)	FAN Min	FAN Max	NAT Max	FAN Min	FAN Max	NAT Max	FAN Min	FAN Max	NAT Max	FAN Min	FAN Max	NAT Max	FAN Min	FAN Max	NAT Max
6	0	0	698	370	0	897	470	0	1121	570	0	1645	850	0	2267	1170
	2	53	425	285	63	543	370	75	675	455	103	982	650	138	1346	890
	4	79	419	279	93	536	862	110	668	445	147	975	640	191	1338	880
	6	93	413	273	110	530	354	128	661	435	171	967	630	219	1330	870
8	0	0	780	415	0	1006	537	0	1261	660	0	1858	970	0	2571	1320
	2	50	483	322	60	619	418	71	770	515	98	1124	745	130	1543	1020
	5	83	473	313	99	607	407	115	758	503	154	1110	733	199	1528	1010
	8	99	463	303	117	596	396	137	746	490	180	1097	720	231	1514	1000
10	0	0	847	450	0	1096	585	0	1377	720	0	2036	1060	0	2825	1450
	2	48	533	355	57	684	457	68	852	560	93	1244	850	124	1713	1130
	5	81	522	346	95	671	446	112	839	547	149	1229	829	192	1696	1105
	10	104	504	330	122	651	427	142	817	525	187	1204	795	238	1669	1080
15	0	0	970	525	0	1263	682	0	1596	840	0	2380	1240	0	3323	1720
	2	45	633	414	53	815	544	63	1019	675	86	1495	985	114	2062	1350
	5	76	620	403	90	800	529	105	1003	660	140	1476	967	182	2041	1327
	10	99	600	386	116	777	507	135	977	635	177	1446	936	227	2009	1289
	15	115	580	373	134	755	491	155	953	610	202	1418	905	257	1976	1250
20	0	0	1057	575	0	1384	752	0	1756	930	0	2637	1350	0	3701	1900
	2	41	711	470	50	917	612	59	1150	755	81	1694	1100	107	2343	1520
	5	73	697	460	86	902	599	101	1133	738	135	1674	1079	174	2320	1498
	10	95	675	443	112	877	576	130	1105	710	172	1641	1045	220	2282	1460
	15	111	654	427	129	853	557	150	1078	688	195	1609	1018	248	2245	1425
	20	125	634	410	145	830	537	167	1052	665	217	1578	990	273	2210	1390
30	0	0	1173	650	0	1548	855	0	1977	1060	0	3004	1550	0	4252	2170
	2	33	826	535	42	1072	700	54	1351	865	74	2004	1310	98	2786	1800
	5	69	811	524	82	1055	688	96	1332	851	127	1981	1289	164	2759	1775
	10	91	788	507	107	1028	668	125	1301	829	164	1944	1254	209	2716	1733
	15	105	765	490	124	1002	648	143	1272	807	187	1908	1220	237	2674	1692
	20	119	743	473	139	977	628	160	1243	784	207	1873	1185	260	2633	1650
	30	149	702	444	171	929	594	195	1189	745	246	1807	1130	305	2555	1585
50	0	0	1297	708	0	1730	952	0	2231	1195	0	3441	1825	0	4034	2550
	2	26	975	615	33	1276	813	41	1620	1010	66	2431	1513	86	3409	2125
	5	65	960	605	77	1259	798	90	1600	996	118	2406	1495	151	3380	2102
	10	86	935	589	101	1230	773	118	1567	972	154	2366	1466	196	3332	2064
	15	100	911	572	117	1203	747	136	1536	948	177	2327	1437	222	3285	2026
	20	113	888	556	131	1176	722	151	1505	924	195	2288	1408	244	3239	1987
	30	141	844	522	161	1125	670	183	1446	876	232	2214	1349	287	3150	1910
100	0	0	1411	770	0	1908	1040	0	2491	1310	0	3925	2050	0	5729	2950
	2	21	1155	700	25	1536	935	30	1975	1170	44	3027	1820	72	4313	2550
	5	60	1141	692	71	1519	926	82	1955	1159	107	3002	1803	136	4282	2531
	10	80	1118	679	94	1492	910	108	1923	1142	142	2961	1775	180	4231	2500
	15	93	1095	666	109	1465	895	126	1892	1124	163	2920	1747	206	4182	2469
	20	105	1073	653	122	1438	880	141	1861	1107	181	2880	1719	226	4133	2438
	30	131	1029	627	149	1387	849	170	1802	1071	215	2803	1663	265	4037	2375
	50	197	944	575	217	1288	787	241	1688	1000	292	2657	1550	350	3856	2250

Table 8-5

Type B Double-Wall Gas Vent [NFPA 54: Table 13.1(a)] (continued)

		Number of Appliances:			Single											
		Appliance Type:			Category I											
		Appliance Vent Connection:			Connected Directly to Vent											
		Vent Diameter – *D* (in.)														
		16			18			20			22			24		
		Appliance Input Rating in Thousands of Btu per Hour														
Height *H* (ft)	Lateral *L* (ft)	FAN		NAT	FAN		NAT	FAN		NAT	FAN		NAT	FAN		NAT
		Min	Max	Max	Min	Max	Max	Min	Max	Max	Min	Max	Max	Min	Max	Max
6	0	0	2983	1530	0	3802	1960	0	4721	2430	0	5737	2950	0	6853	3520
	2	178	1769	1170	225	2250	1480	296	2782	1850	360	3377	2220	426	4030	2670
	4	242	1761	1160	300	2242	1475	390	2774	1835	469	3370	2215	555	4023	2660
	6	276	1753	1150	341	2235	1470	437	2767	1820	523	3363	2210	618	4017	2650
8	0	0	3399	1740	0	4333	2220	0	5387	2750	0	6555	3360	0	7838	4010
	2	168	2030	1340	212	2584	1700	278	3196	2110	336	3882	2560	401	4634	3050
	5	251	2013	1330	311	2563	1685	398	3180	2090	476	3863	2545	562	4612	3040
	8	289	2000	1320	354	2552	1670	450	3163	2070	537	3850	2530	630	4602	3030
10	0	0	3742	1925	0	4782	2450	0	5955	3050	0	7254	3710	0	8682	4450
	2	161	2256	1480	202	2868	1890	264	3556	2340	319	4322	2840	378	5153	3390
	5	243	2238	1461	300	2849	1871	382	3536	2318	458	4301	2818	540	5132	3371
	10	298	2209	1430	364	2818	1840	459	3504	2280	546	4268	2780	641	5099	3340
15	0	0	4423	2270	0	5678	2900	0	7099	3620	0	8665	4410	0	10,393	5300
	2	147	2719	1770	186	3467	2260	239	4304	2800	290	5232	3410	346	6251	4080
	5	229	2696	1748	283	3442	2235	355	4278	2777	426	5204	3385	501	6222	4057
	10	283	2659	1712	346	3402	2193	432	4234	2739	510	5159	3343	599	6175	4019
	15	318	2623	1675	385	3363	2150	479	4192	2700	564	5115	3300	665	6129	3980
20	0	0	4948	2520	0	6376	3250	0	7988	4060	0	9785	4980	0	11,753	6000
	2	139	3097	2000	175	3955	2570	220	4916	3200	269	5983	3910	321	7154	4700
	5	219	3071	1978	270	3926	2544	337	4885	3174	403	5950	3880	475	7119	4662
	10	273	3029	1940	334	3880	2500	413	4835	3130	489	5896	3830	573	7063	4600
	15	306	2988	1910	372	3835	2465	459	4786	3090	541	5844	3795	631	7007	4575
	20	335	2948	1880	404	3791	2430	495	4737	3050	585	5792	3760	689	6953	4550
30	0	0	5725	2920	0	7420	3770	0	9341	4750	0	11,483	5850	0	13,848	7060
	2	127	3696	2380	159	4734	3050	199	5900	3810	241	7194	4650	285	8617	5600
	5	206	3666	2350	252	4701	3020	312	5863	3783	373	7155	4622	439	8574	5552
	10	259	3617	2300	316	4647	2970	386	5803	3739	456	7090	4574	535	8505	5471
	15	292	3570	2250	354	4594	2920	431	5744	3695	507	7026	4527	590	8437	5391
	20	319	3523	2200	384	4542	2870	467	5686	3650	548	6964	4480	639	8370	5310
	30	369	3433	2130	440	4442	2785	540	5574	3565	635	6842	4375	739	8239	5225
50	0	0	6711	3440	0	8774	4460	0	11,129	5635	0	13,767	6940	0	16,694	8430
	2	113	4554	2840	141	5864	3670	171	7339	4630	209	8980	5695	251	10,788	6860
	5	191	4520	2813	234	5826	3639	283	7295	4597	336	8933	5654	394	10,737	6818
	10	243	4464	2767	295	5763	3585	355	7224	4542	419	8855	5585	491	10,652	6749
	15	274	4409	2721	330	5701	3534	396	7155	4511	465	8779	5546	542	10,570	6710
	20	300	4356	2675	361	5641	3481	433	7086	4479	506	8704	5506	586	10,488	6670
	30	347	4253	2631	412	5523	3431	494	6953	4421	577	8557	5444	672	10,328	6603
100	0	0	7914	4050	0	10,485	5300	0	13,454	6700	0	16,817	8600	0	20,578	10,300
	2	95	5834	3500	120	7591	4600	138	9577	5800	169	11,803	7200	204	14,264	8800
	5	172	5797	3475	208	7548	4566	245	9528	5769	293	11,748	7162	341	14,204	8756
	10	223	5737	3434	268	7478	4509	318	9447	5717	374	11,658	7100	436	14,105	8683
	15	252	5678	3392	304	7409	4451	358	9367	5665	418	11,569	7037	487	14,007	8610
	20	277	5619	3351	330	7341	4394	387	9289	5613	452	11,482	6975	523	13,910	8537
	30	319	5505	3267	378	7209	4279	446	9136	5509	514	11,310	6850	592	13,720	8391
	50	415	5289	3100	486	6956	4050	572	8841	5300	659	10,979	6600	752	13,354	8100

For SI units, 1 in. = 25.4 mm, 1 ft. = 0.305 m, 1 in.^{2} = 645 mm^2, 1,000 Btu/hr = 0.293 kW

Table 8-6

Type B Double-Wall Gas Vent [NFPA 54: Table 13.1(b)]

							Number of Appliances:	Single								
							Appliance Type:	Category I								
							Appliance Vent Connection:	Single-Wall Metal Connector								
		Vent Diameter – *D* (in.)														
		3			4			5			6			7		
		Appliance Input Rating in Thousands of Btu per Hour														
Height *H* (ft)	Lateral *L* (ft)	FAN		NAT	FAN		NAT	FAN		NAT	FAN		NAT	FAN		NAT
		Min	Max	Max	Min	Max	Max	Min	Max	Max	Min	Max	Max	Min	Max	Max
6	0	38	77	45	59	151	85	85	249	140	126	373	204	165	522	284
	2	39	51	36	60	96	66	85	156	104	123	231	156	159	320	213
	4	NA	NA	33	74	92	63	102	152	102	146	225	152	187	313	208
	6	NA	NA	31	83	89	60	114	147	99	163	220	148	207	307	203
8	0	37	83	50	58	164	93	83	273	154	123	412	234	161	580	319
	2	39	56	39	59	108	75	83	176	119	121	261	179	155	363	246
	5	NA	NA	37	77	102	69	107	168	114	151	252	171	193	352	235
	8	NA	NA	33	90	95	64	122	161	107	175	243	163	223	342	225
10	0	37	87	53	57	174	99	82	293	165	120	444	254	158	628	344
	2	39	61	41	59	117	80	82	193	128	119	287	194	153	400	272
	5	52	56	39	76	111	76	105	185	122	148	277	186	190	388	261
	10	NA	NA	34	97	100	68	132	171	112	188	261	171	237	369	241
15	0	36	93	57	56	190	111	80	325	186	116	499	283	153	713	388
	2	38	69	47	57	136	93	80	225	149	115	337	224	148	473	314
	5	51	63	44	75	128	86	102	216	140	144	326	217	182	459	298
	10	NA	NA	39	95	116	79	128	201	131	182	308	203	228	438	284
	15	NA	NA	NA	NA	NA	72	158	186	124	220	290	192	272	418	269
20	0	35	96	60	54	200	118	78	346	201	114	537	306	149	772	428
	2	37	74	50	56	148	99	78	248	165	113	375	248	144	528	344
	5	50	68	47	73	140	94	100	239	158	141	363	239	178	514	334
	10	NA	NA	41	93	129	86	125	223	146	177	344	224	222	491	316
	15	NA	NA	NA	NA	NA	80	155	208	136	216	325	210	264	469	301
	20	NA	NA	NA	NA	NA	NA	186	192	126	254	306	196	309	448	285
30	0	34	99	63	53	211	127	76	372	219	110	584	334	144	849	472
	2	37	80	56	55	164	111	76	281	183	109	429	279	139	610	392
	5	49	74	52	72	157	106	98	271	173	136	417	271	171	595	382
	10	NA	NA	NA	91	144	98	122	255	168	171	397	257	213	570	367
	15	NA	NA	NA	115	131	NA	151	239	157	208	377	242	255	547	349
	20	NA	NA	NA	NA	NA	NA	181	223	NA	246	357	228	298	524	333
	30	NA	NA	NA	NA	NA	NA	NA	NA	NA	NA	NA	NA	389	477	305
50	0	33	99	66	51	213	133	73	394	230	105	629	361	138	928	515
	2	36	84	61	53	181	121	73	318	205	104	495	312	133	712	443
	5	48	80	NA	70	174	117	94	308	198	131	482	305	164	696	435
	10	NA	NA	NA	89	160	NA	118	292	186	136	461	292	203	671	420
	15	NA	NA	NA	112	148	NA	145	275	174	199	441	280	244	646	405
	20	NA	NA	NA	NA	NA	NA	176	257	NA	236	420	267	285	622	389
	30	NA	NA	NA	NA	NA	NA	NA	NA	NA	315	376	NA	373	573	NA
100	0	NA	NA	NA	49	214	NA	69	403	NA	100	659	395	131	991	555
	2	NA	NA	NA	51	192	NA	70	351	NA	98	563	373	125	828	508
	5	NA	NA	NA	67	186	NA	90	342	NA	125	551	366	156	813	501
	10	NA	NA	NA	85	175	NA	113	324	NA	153	532	354	191	789	486
	15	NA	NA	NA	132	162	NA	138	310	NA	188	511	343	230	764	473
	20	NA	NA	NA	NA	NA	NA	168	295	NA	224	487	NA	270	739	458
	30	NA	NA	NA	NA	NA	NA	231	264	NA	301	448	NA	355	685	NA
	50	NA	NA	NA	NA	NA	NA	NA	NA	NA	NA	NA	NA	540	584	NA

Table 8-6

Type B Double-Wall Gas Vent [NFPA 54: Table 13.1(b)] (continued)

		Number of Appliances:					Single						
		Appliance Type:					Category I						
		Appliance Vent Connection:					Single-Wall Metal Connector						
		Vent Diameter – *D* (in.)											
		8			9			10			12		
		Appliance Input Rating in Thousands of Btu per Hour											
Height *H* (ft)	Lateral *L* (ft)	FAN		NAT	FAN		NAT	FAN		NAT	FAN		NAT
		Min	Max	Max	Min	Max	Max	Min	Max	Max	Min	Max	Max
6	0	211	695	369	267	894	469	371	1118	569	537	1639	849
	2	201	423	284	251	541	368	347	673	453	498	979	648
	4	237	416	277	295	533	360	409	664	443	584	971	638
	6	263	409	271	327	526	352	449	656	433	638	962	627
8	0	206	777	414	258	1002	536	360	1257	658	521	1852	967
	2	197	482	321	246	617	417	339	768	513	486	1120	743
	5	245	470	311	305	604	404	418	754	500	598	1104	730
	8	280	458	300	344	591	392	470	740	486	665	1080	715
10	0	202	844	449	253	1093	584	351	1373	718	507	2031	1057
	2	193	531	354	242	681	456	332	849	559	475	1242	848
	5	241	518	344	299	667	443	409	834	544	584	1224	825
	10	296	497	325	363	643	423	492	808	520	688	1194	788
15	0	195	966	523	244	1259	681	336	1591	838	488	2374	1237
	2	187	631	413	232	812	543	319	1015	673	457	1491	983
	5	231	616	400	287	795	526	392	997	657	562	1469	963
	10	284	592	381	349	768	501	470	966	628	664	1433	928
	15	334	568	367	404	742	484	540	937	601	750	1399	894
20	0	190	1053	573	238	1379	750	326	1751	927	473	2631	1346
	2	182	708	468	227	914	611	309	1146	754	443	1689	1098
	5	224	692	457	279	896	596	381	1126	734	547	1665	1074
	10	277	666	437	339	866	570	457	1092	702	646	1626	1037
	15	325	640	419	393	838	549	526	1060	677	730	1587	1005
	20	374	616	400	448	810	526	592	1028	651	808	1550	973
30	0	184	1168	647	229	1542	852	312	1971	1056	454	2996	1545
	2	175	823	533	219	1069	698	296	1346	863	424	1999	1308
	5	215	806	521	269	1049	684	366	1324	846	524	1971	1283
	10	265	777	501	327	1017	662	440	1287	821	620	1927	1243
	15	312	750	481	379	985	638	507	1251	794	702	1884	1205
	20	360	723	461	433	955	615	570	1216	768	780	1841	1166
	30	461	670	426	541	895	574	704	1147	720	937	1759	1101
50	0	176	1292	704	220	1724	948	295	2223	1189	428	3432	1818
	2	168	971	613	209	1273	811	280	1615	1007	401	2426	1509
	5	204	953	602	257	1252	795	347	1591	991	496	2396	1490
	10	253	923	583	313	1217	765	418	1551	963	589	2347	1455
	15	299	894	562	363	1183	736	481	1512	934	668	2299	1421
	20	345	866	543	415	1150	708	544	1473	906	741	2251	1387
	30	442	809	502	521	1086	649	674	1399	848	892	2159	1318
100	0	166	1404	765	207	1900	1033	273	2479	1300	395	3912	2042
	2	158	1152	698	196	1532	933	259	1970	1168	371	3021	1817
	5	194	1134	688	240	1511	921	322	1945	1153	460	2990	1796
	10	238	1104	672	293	1477	902	389	1905	1133	547	2938	1763
	15	281	1075	656	342	1443	884	447	1865	1110	618	2888	1730
	20	325	1046	639	391	1410	864	507	1825	1087	690	2838	1696
	30	418	988	NA	491	1343	824	631	1747	1041	834	2739	1627
	50	617	866	NA	711	1205	NA	895	1591	NA	1138	2547	1489

For SI units, 1 in. = 25.4 mm, 1 ft. = 0.305 m, 1 in.^{2} = 645 mm^2, 1,000 Btu/hr = 0.293 kW

Table 8-7

Masonry Chimney [NFPA 54-09: Table 13.1(c)]

		Number of Appliances: Single														
		Appliance Type: Category I														
		Appliance Vent Connection: Type B Double-Wall Connector														
		Type B Double-Wall Connector Diameter – *D* (in.) To be used with chimney areas within the size limits at bottom														
		3			4			5			6			7		
		Appliance Input Rating in Thousands of Btu per Hour														
Height *H* (ft)	Lateral *L* (ft)	FAN		NAT	FAN		NAT	FAN		NAT	FAN		NAT	FAN		NAT
		Min	Max	Max	Min	Max	Max	Min	Max	Max	Min	Max	Max	Min	Max	Max
6	2	NA	NA	28	NA	NA	52	NA	NA	86	NA	NA	130	NA	NA	180
	5	NA	NA	25	NA	NA	49	NA	NA	82	NA	NA	117	NA	NA	165
8	2	NA	NA	29	NA	NA	55	NA	NA	93	NA	NA	145	NA	NA	198
	5	NA	NA	26	NA	NA	52	NA	NA	88	NA	NA	134	NA	NA	183
	8	NA	NA	24	NA	NA	48	NA	NA	83	NA	NA	127	NA	NA	175
10	2	NA	NA	31	NA	NA	61	NA	NA	103	NA	NA	162	NA	NA	221
	5	NA	NA	28	NA	NA	57	NA	NA	96	NA	NA	148	NA	NA	204
	10	NA	NA	25	NA	NA	30	NA	NA	87	NA	NA	139	NA	NA	191
15	2	NA	NA	35	NA	NA	67	NA	NA	114	NA	NA	179	53	475	250
	5	NA	NA	35	NA	NA	62	NA	NA	107	NA	NA	164	NA	NA	231
	10	NA	NA	28	NA	NA	55	NA	NA	97	NA	NA	153	NA	NA	216
	15	NA	NA	NA	NA	NA	48	NA	NA	89	NA	NA	141	NA	NA	201
20	2	NA	NA	38	NA	NA	74	NA	NA	124	NA	NA	201	51	522	274
	5	NA	NA	36	NA	NA	68	NA	NA	116	NA	NA	184	80	503	254
	10	NA	NA	NA	NA	NA	60	NA	NA	107	NA	NA	172	NA	NA	237
	15	NA	NA	NA	NA	NA	NA	NA	NA	97	NA	NA	159	NA	NA	220
	20	NA	NA	NA	NA	NA	NA	NA	NA	83	NA	NA	148	NA	NA	206
30	2	NA	NA	41	NA	NA	82	NA	NA	137	NA	NA	216	47	581	303
	5	NA	NA	NA	NA	NA	76	NA	NA	128	NA	NA	198	75	561	281
	10	NA	NA	NA	NA	NA	67	NA	NA	115	NA	NA	184	NA	NA	263
	15	NA	NA	NA	NA	NA	NA	NA	NA	107	NA	NA	171	NA	NA	243
	20	NA	NA	NA	NA	NA	NA	NA	NA	91	NA	NA	159	NA	NA	227
	30	NA	NA	NA	NA	NA	NA	NA	NA	NA	NA	NA	NA	NA	NA	188
50	2	NA	NA	NA	NA	NA	92	NA	NA	161	NA	NA	251	NA	NA	351
	5	NA	NA	NA	NA	NA	NA	NA	NA	131	NA	NA	230	NA	NA	323
	10	NA	NA	NA	NA	NA	NA	NA	NA	138	NA	NA	215	NA	NA	304
	15	NA	NA	NA	NA	NA	NA	NA	NA	127	NA	NA	199	NA	NA	282
	20	NA	NA	NA	NA	NA	NA	NA	NA	NA	NA	NA	185	NA	NA	264
	30	NA	NA	NA	NA	NA	NA	NA	NA	NA	NA	NA	NA	NA	NA	NA
Minimum internal area of chimney (in.^{2})		12			19			28			38			50		
Maximum internal area of chimney (in.^{2})		Seven times the listed appliance categorized vent area, flue collar area, or draft hood outlet area.														

Table 8-7

Masonry Chimney [NFPA 54-09: Table 13.1(c)] (continued)

		Number of Appliances:					Single						
		Appliance Type:					Category I						
		Appliance Vent Connection:					Type B Double-Wall Connector						
		Type B Double-Wall Connector Diameter – *D* (in.) To be used with chimney areas within the size limits at bottom											
		8			9			10			12		
		Appliance Input Rating in Thousands of Btu per Hour											
Height *H* (ft)	Lateral *L* (ft)	FAN		NAT	FAN		NAT	FAN		NAT	FAN		NAT
		Min	Max	Max	Min	Max	Max	Min	Max	Max	Min	Max	Max
6	2	NA	NA	247	NA	NA	320	NA	NA	401	NA	NA	581
	5	NA	NA	231	NA	NA	298	NA	NA	376	NA	NA	561
8	2	NA	NA	266	84	590	350	100	728	446	139	1024	651
	5	NA	NA	247	NA	NA	328	149	711	423	201	1007	640
	8	NA	NA	239	NA	NA	318	173	695	410	231	990	623
10	2	68	519	298	82	655	388	98	810	491	136	1144	724
	5	NA	NA	277	124	638	365	146	791	466	196	1124	712
	10	NA	NA	263	155	610	347	182	762	444	240	1093	668
15	2	64	613	336	77	779	441	92	968	562	127	1376	841
	5	99	594	313	118	759	416	139	946	533	186	1352	828
	10	126	565	296	148	727	394	173	912	567	229	1315	777
	15	NA	NA	281	171	698	375	198	880	485	259	1280	742
20	2	61	678	375	73	867	491	87	1083	627	121	1548	953
	5	95	658	350	113	845	463	133	1059	597	179	1523	933
	10	122	627	332	143	811	440	167	1022	566	221	1482	879
	15	NA	NA	314	165	780	418	191	987	541	251	1443	840
	20	NA	NA	296	186	750	397	214	955	513	277	1406	807
30	2	57	762	421	68	985	558	81	1240	717	111	1793	1112
	5	90	741	393	106	962	526	125	1216	683	169	1766	1094
	10	115	709	373	135	927	500	158	1176	648	210	1721	1025
	15	NA	NA	353	156	893	476	181	1139	621	239	1679	981
	20	NA	NA	332	176	860	450	203	1103	592	264	1638	940
	30	NA	NA	288	NA	NA	416	249	1035	555	318	1560	877
50	2	51	840	477	61	1106	633	72	1413	812	99	2080	1243
	5	83	819	445	98	1083	596	116	1387	774	155	2052	1225
	10	NA	NA	424	126	1047	567	147	1347	733	195	2006	1147
	15	NA	NA	400	146	1010	539	170	1307	702	222	1961	1099
	20	NA	NA	376	165	977	511	190	1269	669	246	1916	1050
	30	NA	NA	327	NA	NA	468	233	1196	623	295	1832	984
Minimum internal area of chimney (in.^{2})		63			78			95			132		
Maximum internal area of chimney (in.^{2})		Seven times the listed appliance categorized vent area, flue collar area, or draft hood outlet area.											

For SI units, 1 in. = 25.4 mm, 1 ft. = 0.305 m, 1 in.^{2} = 645 mm^2, 1,000 Btu/hr = 0.293 kW

Table 8-8

Masonry Chimney [NFPA 54-09: Table 13.1(d)]

					Number of Appliances:					**Single**						
					Appliance Type:					**Category I**						
					Appliance Vent Connection:					**Single-Wall Metal Connector**						
		Single-Wall Metal Connector Diameter – *D* (in.) To be used with chimney areas within the size limits at bottom														
		3			**4**			**5**			**6**			**7**		
		Appliance Input Rating in Thousands of Btu per Hour														
Height *H* (ft)	**Lateral *L* (ft)**	**FAN**		**NAT**	**FAN**		**NAT**	**FAN**		**NAT**	**FAN**		**NAT**	**FAN**		**NAT**
		Min	**Max**	**Max**	**Min**	**Max**	**Max**	**Min**	**Max**	**Max**	**Min**	**Max**	**Max**	**Min**	**Max**	**Max**
6	2	NA	NA	28	NA	NA	52	NA	NA	86	NA	NA	130	NA	NA	180
	5	NA	NA	25	NA	NA	48	NA	NA	81	NA	NA	116	NA	NA	164
8	2	NA	NA	29	NA	NA	55	NA	NA	93	NA	NA	145	NA	NA	197
	5	NA	NA	26	NA	NA	51	NA	NA	87	NA	NA	133	NA	NA	182
	8	NA	NA	23	NA	NA	47	NA	NA	82	NA	NA	126	NA	NA	174
10	2	NA	NA	31	NA	NA	61	NA	NA	102	NA	NA	161	NA	NA	220
	5	NA	NA	28	NA	NA	56	NA	NA	95	NA	NA	147	NA	NA	203
	10	NA	NA	24	NA	NA	49	NA	NA	86	NA	NA	137	NA	NA	189
15	2	NA	NA	35	NA	NA	67	NA	NA	113	NA	NA	178	166	473	249
	5	NA	NA	32	NA	NA	61	NA	NA	106	NA	NA	163	NA	NA	230
	10	NA	NA	27	NA	NA	54	NA	NA	96	NA	NA	151	NA	NA	214
	15	NA	NA	NA	NA	NA	46	NA	NA	87	NA	NA	138	NA	NA	198
20	2	NA	NA	38	NA	NA	73	NA	NA	123	NA	NA	200	163	520	273
	5	NA	NA	35	NA	NA	67	NA	NA	115	NA	NA	183	NA	NA	252
	10	NA	NA	NA	NA	NA	59	NA	NA	105	NA	NA	170	NA	NA	235
	15	NA	NA	NA	NA	NA	NA	NA	NA	95	NA	NA	156	NA	NA	217
	20	NA	NA	NA	NA	NA	NA	NA	NA	80	NA	NA	144	NA	NA	202
30	2	NA	NA	41	NA	NA	81	NA	NA	136	NA	NA	215	158	578	302
	5	NA	NA	NA	NA	NA	75	NA	NA	127	NA	NA	196	NA	NA	279
	10	NA	NA	NA	NA	NA	66	NA	NA	113	NA	NA	182	NA	NA	260
	15	NA	NA	NA	NA	NA	NA	NA	NA	105	NA	NA	168	NA	NA	240
	20	NA	NA	NA	NA	NA	NA	NA	NA	88	NA	NA	155	NA	NA	223
	30	NA	NA	NA	NA	NA	NA	NA	NA	NA	NA	NA	NA	NA	NA	182
50	2	NA	NA	NA	NA	NA	91	NA	NA	160	NA	NA	250	NA	NA	350
	5	NA	NA	NA	NA	NA	NA	NA	NA	149	NA	NA	228	NA	NA	321
	10	NA	NA	NA	NA	NA	NA	NA	NA	136	NA	NA	212	NA	NA	301
	15	NA	NA	NA	NA	NA	NA	NA	NA	124	NA	NA	195	NA	NA	278
	20	NA	NA	NA	NA	NA	NA	NA	NA	NA	NA	NA	180	NA	NA	258
	30	NA	NA	NA	NA	NA	NA	NA	NA	NA	NA	NA	NA	NA	NA	NA
Minimum internal area of chimney (in.^{2})		12			19			28			38			50		
Maximum internal area of chimney (in.^{2})		Seven times the listed appliance categorized vent area, flue collar area, or draft hood outlet area.														

Table 8-8

Masonry Chimney [NFPA 54-09: Table 13.1(d)] (continued)

		Number of Appliances: Single											
		Appliance Type: Category I											
		Appliance Vent Connection: Single-Wall Metal Connector											
		Single-Wall Metal Connector Diameter – *D* (in.) To be used with chimney areas within the size limits at bottom											
		8			9			10			12		
		Appliance Input Rating in Thousands of Btu per Hour											
Height *H* (ft)	Lateral *L* (ft)	FAN		NAT	FAN		NAT	FAN		NAT	FAN		NAT
		Min	Max	Max	Min	Max	Max	Min	Max	Max	Min	Max	Max
6	2	NA	NA	247	NA	NA	319	NA	NA	400	NA	NA	580
	5	NA	NA	230	NA	NA	297	NA	NA	375	NA	NA	560
8	2	NA	NA	265	NA	NA	349	382	725	445	549	1021	650
	5	NA	NA	246	NA	NA	327	NA	NA	422	673	1003	638
	8	NA	NA	237	NA	NA	317	NA	NA	408	747	985	621
10	2	216	518	297	271	654	387	373	808	490	536	1142	722
	5	NA	NA	276	334	635	364	459	789	465	657	1121	710
	10	NA	NA	261	NA	NA	345	547	758	441	771	1088	665
15	2	211	611	335	264	776	440	362	965	560	520	1373	840
	5	261	591	312	325	755	414	444	942	531	637	1348	825
	10	NA	NA	294	392	722	392	531	907	504	749	1309	774
	15	NA	NA	278	452	692	372	606	873	481	841	1272	738
20	2	206	675	374	258	864	490	252	1079	625	508	1544	950
	5	255	655	348	317	842	461	433	1055	594	623	1518	930
	10	312	622	330	382	806	437	517	1016	562	733	1475	875
	15	NA	NA	311	442	773	414	591	979	539	823	1434	835
	20	NA	NA	292	NA	NA	392	663	944	510	911	1394	800
30	2	200	759	420	249	982	556	340	1237	715	489	1789	1110
	5	245	737	391	306	958	524	417	1210	680	600	1760	1090
	10	300	703	370	370	920	496	500	1168	644	708	1713	1020
	15	NA	NA	349	428	884	471	572	1128	615	798	1668	975
	20	NA	NA	327	NA	NA	445	643	1089	585	883	1624	932
	30	NA	NA	281	NA	NA	408	NA	NA	544	1055	1539	865
50	2	191	837	475	238	1103	631	323	1408	810	463	2076	1240
	5	NA	NA	442	293	1078	593	398	1381	770	571	2044	1220
	10	NA	NA	420	355	1038	562	447	1337	728	674	1994	1140
	15	NA	NA	395	NA	NA	533	546	1294	695	761	1945	1090
	20	NA	NA	370	NA	NA	504	616	1251	660	844	1898	1040
	30	NA	NA	318	NA	NA	458	NA	NA	610	1009	1805	970
Minimum internal area of chimney ($in.^2$)		63			78			95			132		
Maximum internal area of chimney ($in.^2$)		Seven times the listed appliance categorized vent area, flue collar area, or draft hood outlet area.											

For SI units, 1 in. = 25.4 mm, 1 ft. = 0.305 m, 1 $in.^2$ = 645 mm^2, 1,000 Btu/hr = 0.293 kW

Table 8-9

Single-Wall Metal Pipe or Type B Asbestos Cement Vent [NFPA 54: Table 13.1(e)]

				Number of Appliances:			Single		
				Appliance Type:			Draft Hood-Equipped		
				Appliance Vent Connection:			Connected Directly to Pipe or Vent		
Height *H* (ft)	Lateral *L* (ft)	Diameter – *D* (in.) To be used with chimney areas within the size limits at bottom							
		3	4	5	6	7	8	10	12
		Appliance Input Rating in Thousands of Btu per Hour							
		Maximum Appliance Input Rating in Thousands of Btu per Hour							
6	0	39	70	116	170	232	312	500	750
	2	31	55	94	141	194	260	415	620
	5	28	51	88	128	177	242	390	600
8	0	42	76	126	185	252	340	542	815
	2	32	61	102	154	210	284	451	680
	5	29	56	95	141	194	264	430	648
	10	24	49	86	131	180	250	406	325
10	0	45	84	138	202	279	372	606	912
	2	35	67	111	168	233	311	505	760
	5	32	61	104	153	215	289	480	724
	10	27	54	94	143	200	274	455	700
	15	NA	46	84	130	186	258	432	666
15	0	49	91	151	223	312	420	684	1040
	2	39	72	122	186	260	350	570	865
	5	35	67	110	170	240	325	540	825
	10	30	58	103	158	223	308	514	795
	15	NA	50	93	144	207	291	488	760
	20	NA	NA	82	132	195	273	466	726
20	0	53	101	163	252	342	470	770	1190
	2	42	80	136	210	286	392	641	990
	5	38	74	123	192	264	364	610	945
	10	32	65	115	178	246	345	571	910
	15	NA	55	104	163	228	326	550	870
	20	NA	NA	91	149	214	306	525	832
30	0	56	108	183	276	384	529	878	1370
	2	44	84	148	230	320	441	730	1140
	5	NA	78	137	210	296	410	694	1080
	10	NA	68	125	196	274	388	656	1050
	15	NA	NA	113	177	258	366	625	1000
	20	NA	NA	99	163	240	344	596	960
	30	NA	NA	NA	NA	192	395	540	890
50	0	NA	120	210	310	443	590	980	1550
	2	NA	95	171	260	370	492	820	1290
	5	NA	NA	159	234	342	474	780	1230
	10	NA	NA	146	221	318	456	730	1190
	15	NA	NA	NA	200	392	407	705	1130
	20	NA	NA	NA	185	276	384	670	1080
	30	NA	NA	NA	NA	222	330	605	1010

For SI units, 1 in. = 25.4 mm, 1 ft. = 0.305 m, 1 in.^{2} = 645 mm^2, 1,000 Btu/hr = 0.293 kW

Table 8-10

Type B Double-Wall Vent [NFPA 54: Table 13.2(a)]

						Number of Appliances:				Two or More						
						Appliance Type:				Category I						
						Appliance Vent Connection:				Type B Double-Wall Connector						
Vent Connector Capacity																
		Type B Double-Wall Vent and Connector Diameter – *D* (in.)														
		3			4			5			6			7		
		Appliance Input Rating in Thousands of Btu per Hour														
Vent Height *H* (ft)	Connector Rise *R* (ft)	FAN		NAT	FAN		NAT	FAN		NAT	FAN		NAT	FAN		NAT
		Min	Max	Max	Min	Max	Max	Min	Max	Max	Min	Max	Max	Min	Max	Max
6	1	22	37	26	35	66	46	46	106	72	58	164	104	77	225	142
	2	23	41	31	37	75	55	48	121	86	60	183	124	79	253	168
	3	24	44	35	38	81	62	49	132	96	62	199	139	82	275	189
8	1	22	40	27	35	72	48	49	114	76	64	176	109	84	243	148
	2	23	44	32	36	80	57	51	128	90	66	195	129	86	269	175
	3	24	47	36	37	87	64	53	139	101	67	210	145	88	290	198
10	1	22	43	28	34	78	50	49	123	78	65	189	113	89	257	154
	2	23	47	33	36	86	59	51	136	93	67	206	134	91	282	182
	3	24	50	37	37	92	67	52	146	104	69	220	150	94	303	205
15	1	21	50	30	33	89	53	47	142	83	64	220	120	88	298	163
	2	22	53	35	35	96	63	49	153	99	66	235	142	91	320	193
	3	24	55	40	36	102	71	51	163	111	68	248	160	93	339	218
20	1	21	54	31	33	99	56	46	157	87	62	246	125	86	334	171
	2	22	57	37	34	105	66	48	167	104	64	259	149	89	354	202
	3	23	60	42	35	110	74	50	176	116	66	271	168	91	371	228
30	1	20	62	33	31	113	59	45	181	93	60	288	134	83	391	182
	2	21	64	39	33	118	70	47	190	110	62	299	158	85	408	215
	3	22	66	44	34	123	79	48	198	124	64	309	178	88	423	242
50	1	19	71	36	30	133	64	43	216	101	57	349	145	78	477	197
	2	21	73	43	32	137	76	45	223	119	59	358	172	81	490	234
	3	22	75	48	33	141	86	46	229	134	61	366	194	83	502	263
100	1	18	82	37	28	158	66	40	262	104	53	442	150	73	611	204
	2	19	83	44	30	161	79	42	267	123	55	447	178	75	619	242
	3	20	84	50	31	163	89	44	272	138	57	452	200	78	627	272

Common Vent Capacity															
Vent Height *H* (ft)	Type B Double-Wall Common Vent Diameter – *D* (in.)														
	4			5			6			7			8		
	Combined Appliance Input Rating in Thousands of Btu per Hour														
	FAN +FAN	FAN +NAT	NAT +NAT	FAN +FAN	FAN +NAT	NAT +NAT	FAN +FAN	FAN +NAT	NAT +NAT	FAN +FAN	FAN +NAT	NAT +NAT	FAN +FAN	FAN +NAT	NAT +NAT
6	92	81	65	140	116	103	204	161	147	309	248	200	404	314	260
8	101	90	73	155	129	114	224	178	163	339	275	223	444	348	290
10	110	97	79	169	141	124	243	194	178	367	299	242	477	377	315
15	125	112	91	195	164	144	283	228	206	427	352	280	556	444	365
20	136	123	102	215	183	160	314	255	229	475	394	310	621	499	405
30	152	138	118	244	210	185	361	297	266	547	459	360	720	585	470
50	167	153	134	279	244	214	421	353	310	641	547	423	854	706	550
100	175	163	NA	311	277	NA	489	421	NA	751	658	479	1025	873	625

Table 8-10

Type B Double-Wall Vent [NFPA 54: Table 13.2(a)] (continued)

	Number of Appliances:	Two or More
	Appliance Type:	Category I
	Appliance Vent Connection:	Type B Double-Wall Connector

Vent Connector Capacity

Vent Height *H* (ft)	Connector Rise *R* (ft)	Type B Double-Wall Vent and Connector Diameter – *D* (in.): 8			9			10			12			14		
		Appliance Input Rating in Thousands of Btu per Hour														
		FAN		NAT	FAN		NAT	FAN		NAT	FAN		NAT	FAN		NAT
		Min	Max	Max	Min	Max	Max	Min	Max	Max	Min	Max	Max	Min	Max	Max
6	1	92	296	185	109	376	237	128	466	289	174	764	496	223	1046	653
	2	95	333	220	112	424	282	131	526	345	180	897	616	230	1231	827
	3	97	363	248	114	463	317	134	575	386	NA	NA	NA	NA	NA	NA
8	1	100	320	194	118	408	248	138	507	303	186	822	516	238	1126	696
	2	103	356	230	121	454	294	141	564	358	192	952	644	244	1307	884
	3	105	384	258	123	492	330	143	612	402	198	1050	772	252	1445	1072
10	1	106	341	200	125	436	257	146	542	314	196	870	536	249	1195	730
	2	109	374	238	128	479	305	149	596	372	201	997	664	256	1371	924
	3	111	402	268	131	515	342	152	642	417	207	1095	792	263	1509	1118
15	1	110	389	214	134	493	273	162	609	333	214	967	568	272	1334	790
	2	112	419	253	137	532	323	165	658	394	221	1085	712	279	1499	1006
	3	115	445	286	140	565	365	167	700	444	228	1181	856	286	1632	1222
20	1	107	436	224	131	552	285	158	681	347	223	1051	596	291	1443	840
	2	110	463	265	134	587	339	161	725	414	230	1162	748	298	1597	1064
	3	113	486	300	137	618	383	164	764	466	237	1253	900	307	1726	1288
30	1	103	512	238	125	649	305	151	802	372	216	1217	632	286	1664	910
	2	105	534	282	129	679	360	155	840	439	223	1316	792	294	1802	1160
	3	108	555	317	132	706	405	158	874	494	231	1400	952	303	1920	1410
50	1	97	627	257	120	797	330	144	984	403	206	1479	689	273	2023	1007
	2	100	645	306	123	820	392	148	1014	478	213	1561	860	281	2139	1291
	3	103	661	343	126	842	441	151	1043	538	221	1631	1031	290	2242	1575
100	1	91	810	266	112	1038	341	135	1285	417	192	1923	712	254	2644	1050
	2	94	822	316	115	1054	405	139	1306	494	200	1984	888	263	2731	1346
	3	97	834	355	118	1069	455	142	1327	555	208	2035	1064	272	2811	1642

Common Vent Capacity

Vent Height *H* (ft)	Type B Double-Wall Common Vent Diameter – *D* (in.): 9			10			12			14			16		
	Combined Appliance Input Rating in Thousands of Btu per Hour														
	FAN +FAN	FAN +NAT	NAT +NAT	FAN +FAN	FAN +NAT	NAT +NAT	FAN +FAN	FAN +NAT	NAT +NAT	FAN +FAN	FAN +NAT	NAT +NAT	FAN +FAN	FAN +NAT	NAT +NAT
6	547	434	335	672	520	410	900	696	588	1284	990	815	1735	1336	1065
8	602	480	378	740	577	465	994	773	652	1423	1103	912	1927	1491	1190
10	649	522	405	800	627	495	1076	841	712	1542	1200	995	2093	1625	1300
15	753	612	465	924	733	565	1247	986	825	1794	1410	1158	2440	1910	1510
20	842	688	523	1035	826	640	1405	1116	916	2006	1588	1290	2722	2147	1690
30	979	808	605	1209	975	740	1658	1327	1025	2373	1892	1525	3220	2558	1990
50	1164	977	705	1451	1188	860	2024	1640	1280	2911	2347	1863	3964	3183	2430
100	1408	1215	800	1784	1502	975	2569	2131	1670	3732	3076	2450	5125	4202	3200

Table 8-10

Type B Double-Wall Vent [NFPA 54: Table 13.2(a)] (continued)

		Number of Appliances:	Two or More
		Appliance Type:	Category I
		Appliance Vent Connection:	Type B Double-Wall Connector

Vent Connector Capacity

Vent Height H (ft)	Connector Rise R (ft)	Type B Double-Wall Vent and Connector Diameter – D (in.)														
		16			18			20			22			24		
		Appliance Input Rating in Thousands of Btu per Hour														
		FAN		NAT	FAN		NAT	FAN		NAT	FAN		NAT	FAN		NAT
		Min	Max	Max	Min	Max	Max	Min	Max	Max	Min	Max	Max	Min	Max	Max
6	1	281	1371	853	346	1772	1080	NA	NA	NA	NA	NA	NA	NA	NA	NA
	2	287	1617	1081	352	2069	1370	NA	NA	NA	NA	NA	NA	NA	NA	NA
	3	NA	NA	NA	NA	NA	NA	NA	NA	NA	NA	NA	NA	NA	NA	NA
8	1	298	1478	910	365	1920	1150	NA	NA	NA	NA	NA	NA	NA	NA	NA
	2	305	1719	1150	372	2211	1460	471	2737	1800	560	3319	2180	662	3957	2590
	3	313	1902	1390	380	2434	1770	478	3018	2180	568	3665	2640	669	4373	3130
10	1	311	1570	955	379	2049	1205	NA	NA	NA	NA	NA	NA	NA	NA	NA
	2	318	1804	1205	387	2332	1535	486	2887	1890	581	3502	2280	686	4175	2710
	3	325	1989	1455	395	2556	1865	494	3169	2290	589	3849	2760	694	4593	3270
15	1	336	1760	1030	408	2317	1305	NA	NA	NA	NA	NA	NA	NA	NA	NA
	2	344	1978	1320	416	2579	1665	523	3197	2060	624	3881	2490	734	4631	2960
	3	351	2157	1610	424	2796	2025	533	3470	2510	634	4216	3030	743	5035	3600
20	1	357	1911	1095	430	2533	1385	NA	NA	NA	NA	NA	NA	NA	NA	NA
	2	365	2116	1395	438	2778	1765	554	3447	2180	661	4190	2630	772	5005	3130
	3	373	2287	1695	450	2984	2145	567	3708	2650	671	4511	3190	785	5392	3790
30	1	367	2183	1190	461	2891	1540	NA	NA	NA	NA	NA	NA	NA	NA	NA
	2	376	2366	1510	474	3110	1920	619	3840	2365	728	4861	2860	847	5606	3410
	3	384	2524	1830	485	3299	2340	632	4080	2875	741	4976	3480	860	5961	4150
50	1	350	2659	1315	435	3548	1665	NA	NA	NA	NA	NA	NA	NA	NA	NA
	2	359	2814	1685	447	3730	2135	580	4601	2633	709	5569	3185	851	6633	3790
	3	369	2951	2055	461	3893	2605	594	4808	3208	724	5826	3885	867	6943	4620
100	1	326	3490	1370	402	4707	1740	NA	NA	NA	NA	NA	NA	NA	NA	NA
	2	336	3606	1760	414	4842	2220	523	5982	2750	639	7254	3330	769	8650	3950
	3	346	3714	2150	426	4968	2700	539	6143	3350	654	7453	4070	786	8892	4810

Common Vent Capacity

Vent Height H (ft)	Type B Double-Wall Common Vent Diameter – D (in.)											
	18			20			22			24		
	Combined Appliance Input Rating in Thousands of Btu per Hour											
	FAN +FAN	FAN +NAT	NAT +NAT	FAN +FAN	FAN +NAT	NAT +NAT	FAN +FAN	FAN +NAT	NAT +NAT	FAN +FAN	FAN +NAT	NAT +NAT
6	2253	1732	1345	2838	2180	1660	3488	2677	1970	4206	3226	2390
8	2507	1936	1510	3162	2439	1860	3890	2998	2200	4695	3616	2680
10	2727	2113	1645	3444	2665	2030	4241	3278	2400	5123	3957	2920
15	3184	2484	1910	4026	3133	2360	4971	3862	2790	6016	4670	3400
20	3561	2798	2140	4548	3552	2640	5573	4352	3120	6749	5261	3800
30	4197	3326	2520	5303	4193	3110	6539	5157	3680	7940	6247	4480
50	5184	4149	3075	6567	5240	3800	8116	6458	4500	9837	7813	5475
100	6749	5509	4050	8597	6986	5000	10,681	8648	5920	13,004	10,499	7200

For SI units, 1 in. = 25.4 mm, 1 ft. = 0.305 m, 1 in.^{2} = 645 mm^2, 1,000 Btu/hr = 0.293 kW

Table 8-11

Type B Double-Wall Vent [NFPA 54: Table 13.2(b)]

	Number of Appliances:	Two or More
	Appliance Type:	Category I
	Appliance Vent Connection:	Single-Wall Metal Connector

Vent Connector Capacity

Vent Height *H* (ft)	Connector Rise *R* (ft)	Single-Wall Metal Vent Connector Diameter – *D* (in.)														
		3			4			5			6			7		
		Appliance Input Rating in Thousands of Btu per Hour														
		FAN		NAT	FAN		NAT	FAN		NAT	FAN		NAT	FAN		NAT
		Min	Max	Max	Min	Max	Max	Min	Max	Max	Min	Max	Max	Min	Max	Max
6	1	NA	NA	26	NA	NA	46	NA	NA	71	NA	NA	102	207	223	140
	2	NA	NA	31	NA	NA	55	NA	NA	85	168	182	123	215	251	167
	3	NA	NA	34	NA	NA	62	121	131	95	175	198	138	222	273	188
8	1	NA	NA	27	NA	NA	48	NA	NA	75	NA	NA	106	226	240	145
	2	NA	NA	32	NA	NA	57	125	126	89	184	193	127	234	266	173
	3	NA	NA	35	NA	NA	64	130	138	100	191	208	144	241	287	197
10	1	NA	NA	28	NA	NA	50	119	121	77	182	186	110	240	253	150
	2	NA	NA	33	84	85	59	124	134	91	189	203	132	248	278	183
	3	NA	NA	36	89	91	67	129	144	102	197	217	148	257	299	203
15	1	NA	NA	29	79	87	52	116	138	81	177	214	116	238	291	158
	2	NA	NA	34	83	94	62	121	150	97	185	230	138	246	314	189
	3	NA	NA	39	87	100	70	127	160	109	193	243	157	255	333	215
20	1	49	56	30	78	97	54	115	152	84	175	238	120	233	325	165
	2	52	59	36	82	103	64	120	163	101	182	252	144	243	346	197
	3	55	62	40	87	107	72	125	172	113	190	264	164	252	363	223
30	1	47	60	31	77	110	57	112	175	89	169	278	129	226	380	175
	2	51	62	37	81	115	67	117	185	106	177	290	152	236	397	208
	3	54	64	42	85	119	76	122	193	120	185	300	172	244	412	235
50	1	46	69	34	75	128	60	109	207	96	162	336	137	217	460	188
	2	49	71	40	79	132	72	114	215	113	170	345	164	226	473	223
	3	52	72	45	83	136	82	119	221	123	178	353	186	235	486	252
100	1	45	79	34	71	150	61	104	249	98	153	424	140	205	585	192
	2	48	80	41	75	153	73	110	255	115	160	428	167	212	593	228
	3	51	81	46	79	157	85	114	260	129	168	433	190	222	603	256

Common Vent Capacity

Vent Height *H* (ft)	Type B Double-Wall Common Vent Diameter – *D* (in.)											
	4			5			6			7		
	Combined Appliance Input Rating in Thousands of Btu per Hour											
	FAN +FAN	FAN +NAT	NAT +NAT	FAN +FAN	FAN +NAT	NAT +NAT	FAN +FAN	FAN +NAT	NAT +NAT	FAN +FAN	FAN +NAT	NAT +NAT
6	NA	78	64	NA	113	99	200	158	144	304	244	196
8	NA	87	71	NA	126	111	218	173	159	331	269	218
10	NA	94	76	163	137	120	237	189	174	357	292	236
15	121	108	88	189	159	140	275	221	200	416	343	274
20	131	118	98	208	177	156	305	247	223	463	383	302
30	145	132	113	236	202	180	350	286	257	533	446	349
50	159	145	128	268	233	208	406	337	296	622	529	410
100	166	153	NA	297	263	NA	469	398	NA	725	633	464

Table 8-11

Type B Double-Wall Vent [NFPA 54: Table 13.2(b)] (continued)

Number of Appliances:	Two or More
Appliance Type:	Category I
Appliance Vent Connection:	Single-Wall Metal Connector

Vent Connector Capacity

Vent Height *H* (ft)	Connector Rise *R* (ft)	Single-Wall Metal Vent Connector Diameter – *D* (in.)								
		8			9			10		
		Appliance Input Rating in Thousands of Btu per Hour								
		FAN		NAT	FAN		NAT	FAN		NAT
		Min	Max	Max	Min	Max	Max	Min	Max	Max
6	1	262	293	183	325	373	234	447	463	286
	2	271	331	219	334	422	281	458	524	344
	3	279	361	247	344	463	316	468	574	385
8	1	285	316	191	352	403	244	481	502	299
	2	293	253	228	360	450	292	492	560	355
	3	302	381	256	370	489	328	501	609	400
10	1	302	335	196	372	429	252	506	534	308
	2	311	369	235	381	473	302	517	589	368
	3	320	398	265	391	511	339	528	637	413
15	1	312	380	208	397	482	266	556	596	324
	2	321	411	248	407	522	317	568	646	387
	3	331	438	281	418	557	360	579	690	437
20	1	306	425	217	390	538	276	546	664	336
	2	317	453	259	400	574	331	558	709	403
	3	326	476	294	412	607	375	570	750	457
30	1	296	497	230	378	630	294	528	779	358
	2	307	521	274	389	662	349	541	819	425
	3	316	542	309	400	690	394	555	855	482
50	1	284	604	245	364	768	314	507	951	384
	2	294	623	293	376	793	375	520	983	458
	3	304	640	331	387	816	423	535	1013	518
100	1	269	774	249	345	993	321	476	1236	393
	2	279	788	299	358	1011	383	490	1259	469
	3	289	801	339	368	1027	431	506	1280	527

Common Vent Capacity

Vent Height *H* (ft)	Type B Double-Wall Common Vent Diameter – *D* (in.)								
	8			9			10		
	Combined Appliance Input Rating in Thousands of Btu per Hour								
	FAN +FAN	FAN +NAT	NAT +NAT	FAN +FAN	FAN +NAT	NAT +NAT	FAN +FAN	FAN +NAT	NAT +NAT
6	398	310	257	541	429	332	665	515	407
8	436	342	285	592	473	373	730	569	460
10	467	369	309	638	512	398	787	617	487
15	544	434	357	738	599	456	905	718	553
20	606	487	395	824	673	512	1013	808	626
30	703	570	459	958	790	593	1183	952	723
50	833	686	535	1139	954	689	1418	1157	838
100	999	846	606	1378	1185	780	1741	1459	948

For SI units, 1 in. = 25.4 mm, 1 ft. = 0.305 m, 1 in.^{2} = 645 mm^2, 1,000 Btu/hr = 0.293 kW

Table 8-12

Masonry Chimney [NFPA 54: Table 13.2(c)]

Number of Appliances:	Two or More
Appliance Type:	Category I
Appliance Vent Connection:	Type B Double-Wall Connector

Vent Connector Capacity

Vent Height *H* (ft)	Connector Rise *R* (ft)	Type B Double-Wall Vent Connector Diameter – *D* (in.)														
		3			4			5			6			7		
		Appliance Input Rating in Thousands of Btu per Hour														
		FAN		NAT	FAN		NAT	FAN		NAT	FAN		NAT	FAN		NAT
		Min	Max	Max	Min	Max	Max	Min	Max	Max	Min	Max	Max	Min	Max	Max
6	1	24	33	21	39	62	40	52	106	67	65	194	101	87	274	141
	2	26	43	28	41	79	52	53	133	85	67	230	124	89	324	173
	3	27	49	34	42	92	61	55	155	97	69	262	143	91	369	203
8	1	24	39	22	39	72	41	55	117	69	71	213	105	94	304	148
	2	26	47	29	40	87	53	57	140	86	73	246	127	97	350	179
	3	27	52	34	42	97	62	59	159	98	75	269	145	99	383	206
10	1	24	42	22	38	80	42	55	130	71	74	232	108	101	324	153
	2	26	50	29	40	93	54	57	153	87	76	261	129	103	366	184
	3	27	55	35	41	105	63	58	170	100	78	284	148	106	397	209
15	1	24	48	23	38	93	44	54	154	74	72	277	114	100	384	164
	2	25	55	31	39	105	55	56	174	89	74	299	134	103	419	192
	3	26	59	35	41	115	64	57	189	102	76	319	153	105	448	215
20	1	24	52	24	37	102	46	53	172	77	71	313	119	98	437	173
	2	25	58	31	39	114	56	55	190	91	73	335	138	101	467	199
	3	26	63	35	40	123	65	57	204	104	75	353	157	104	493	222
30	1	24	54	25	37	111	48	52	192	82	69	357	127	96	504	187
	2	25	60	32	38	122	58	54	208	95	72	376	145	99	531	209
	3	26	64	36	40	131	66	56	221	107	74	392	163	101	554	233
50	1	23	51	25	36	116	51	51	209	89	67	405	143	92	582	213
	2	24	59	32	37	127	61	53	225	102	70	421	161	95	604	235
	3	26	64	36	39	135	69	55	237	115	72	435	180	98	624	260
100	1	23	46	24	35	108	50	49	208	92	65	428	155	88	640	237
	2	24	53	31	37	120	60	51	224	105	67	444	174	92	660	260
	3	25	59	35	38	130	68	53	237	118	69	458	193	94	679	285

Common Vent Capacity

Vent Height *H* (ft)	Minimum Internal Area of Masonry Chimney Flue (in.^{2})														
	12			19			28			38			50		
	Combined Appliance Input Rating in Thousands of Btu per Hour														
	FAN +FAN	FAN +NAT	NAT +NAT	FAN +FAN	FAN +NAT	NAT +NAT	FAN +FAN	FAN +NAT	NAT +NAT	FAN +FAN	FAN +NAT	NAT +NAT	FAN +FAN	FAN +NAT	NAT +NAT
6	NA	74	25	NA	119	46	NA	178	71	NA	257	103	NA	351	143
8	NA	80	28	NA	130	53	NA	193	82	NA	279	119	NA	384	163
10	NA	84	31	NA	138	56	NA	207	90	NA	299	131	NA	409	177
15	NA	NA	36	NA	152	67	NA	233	106	NA	334	152	523	467	212
20	NA	NA	41	NA	NA	75	NA	250	122	NA	368	172	565	508	243
30	NA	NA	NA	NA	NA	NA	NA	270	137	NA	404	198	615	564	278
50	NA	NA	NA	NA	NA	NA	NA	NA	NA	NA	NA	NA	NA	620	328
100	NA	NA	NA	NA	NA	NA	NA	NA	NA	NA	NA	NA	NA	NA	348

Table 8-12

Masonry Chimney [NFPA 54: Table 13.2(c)] (continued)

		Number of Appliances:	Two or More							
		Appliance Type:	Category I							
		Appliance Vent Connection:	Single-Wall Metal Connector							
Vent Connector Capacity										
		Type B Double-Wall Vent Connector Diameter – *D* (in.)								
		8			9			10		
		Appliance Input Rating in Thousands of Btu per Hour								
Vent Height *H* (ft)	Connector Rise *R* (ft)	FAN Min	FAN Max	NAT Max	FAN Min	FAN Max	NAT Max	FAN Min	FAN Max	NAT Max
6	1	104	370	201	124	479	253	145	599	319
	2	107	436	232	127	562	300	148	694	378
	3	109	491	270	129	633	349	151	795	439
8	1	113	414	210	134	539	267	156	682	335
	2	116	473	240	137	615	311	160	776	394
	3	119	517	276	139	672	358	163	848	452
10	1	120	444	216	142	582	277	165	739	348
	2	123	498	247	145	652	321	168	825	407
	3	126	540	281	147	705	366	171	893	463
15	1	125	511	229	153	658	297	184	824	375
	2	128	558	260	156	718	339	187	900	432
	3	131	597	292	159	760	382	190	960	486
20	1	123	584	239	150	752	312	180	943	397
	2	126	625	270	153	805	354	184	1011	452
	3	129	661	301	156	851	396	187	1067	505
30	1	119	680	255	145	883	337	175	1115	432
	2	122	715	287	149	928	378	179	1171	484
	3	125	746	217	152	968	418	182	1220	535
50	1	115	798	294	140	1049	392	168	1334	506
	2	118	827	326	143	1085	433	172	1379	558
	3	121	854	357	147	1118	474	176	1421	611
100	1	109	907	334	134	1222	454	161	1589	596
	2	113	933	368	138	1253	497	165	1626	651
	3	116	956	399	141	1282	540	169	1661	705

Common Vent Capacity									
	Minimum Internal Area of Masonry Chimney Flue (in.^{2})								
	63			78			113		
	Combined Appliance Input Rating in Thousands of Btu per Hour								
Vent Height *H* (ft)	FAN +FAN	FAN +NAT	NAT +NAT	FAN +FAN	FAN +NAT	NAT +NAT	FAN +FAN	FAN +NAT	NAT +NAT
6	NA	458	188	NA	582	246	1041	853	NA
8	NA	501	218	724	636	278	1144	937	408
10	606	538	236	776	686	302	1226	1010	454
15	682	611	283	874	781	365	1374	1156	546
20	742	668	325	955	858	419	1513	1286	648
30	816	747	381	1062	969	496	1702	1473	749
50	879	831	461	1165	1089	606	1905	1692	922
100	NA	NA	499	NA	NA	669	2053	1921	1058

For SI units, 1 in. = 25.4 mm, 1 ft. = 0.305 m, 1 in.^{2} = 645 mm^2, 1,000 Btu/hr = 0.293 kW

Table 8-13

Masonry Chimney [NFPA 54: Table 13.2(d)]

Number of Appliances:	Two or More
Appliance Type:	Category I
Appliance Vent Connection:	Single-Wall Metal Connector

Vent Connector Capacity

Vent Height *H* (ft)	Connector Rise *R* (ft)	Single-Wall Metal Vent Connector Diameter – *D* (in.) / Appliance Input Rating Limits in Thousands of Btu per Hour: 3 FAN Min	3 FAN Max	3 NAT Max	4 FAN Min	4 FAN Max	4 NAT Max	5 FAN Min	5 FAN Max	5 NAT Max	6 FAN Min	6 FAN Max	6 NAT Max	7 FAN Min	7 FAN Max	7 NAT Max
6	1	NA	NA	21	NA	NA	39	NA	NA	66	179	191	100	231	271	140
	2	NA	NA	28	NA	NA	52	NA	NA	84	186	227	123	239	321	172
	3	NA	NA	34	NA	NA	61	134	153	97	193	258	142	247	365	202
8	1	NA	NA	21	NA	NA	40	NA	NA	68	195	208	103	250	298	146
	2	NA	NA	28	NA	NA	52	137	139	85	202	240	125	258	343	177
	3	NA	NA	34	NA	NA	62	143	156	98	210	264	145	266	376	205
10	1	NA	NA	22	NA	NA	41	130	151	70	202	225	106	267	316	151
	2	NA	NA	29	NA	NA	53	136	150	86	210	255	128	276	358	181
	3	NA	NA	34	97	102	62	143	166	99	217	277	147	284	389	207
15	1	NA	NA	23	NA	NA	43	129	151	73	199	271	112	268	376	161
	2	NA	NA	30	92	103	54	135	170	88	207	295	132	277	411	189
	3	NA	NA	34	96	112	63	141	185	101	215	315	151	286	439	213
20	1	NA	NA	23	87	99	45	128	167	76	197	303	117	265	425	169
	2	NA	NA	30	91	111	55	134	185	90	205	325	136	274	455	195
	3	NA	NA	35	96	119	64	140	199	103	213	343	154	282	481	219
30	1	NA	NA	24	86	108	47	126	187	80	193	347	124	259	492	183
	2	NA	NA	31	91	119	57	132	203	93	201	366	142	269	518	205
	3	NA	NA	35	95	127	65	138	216	105	209	381	160	277	540	229
50	1	NA	NA	24	85	113	50	124	204	87	188	392	139	252	567	208
	2	NA	NA	31	89	123	60	130	218	100	196	408	158	262	588	230
	3	NA	NA	35	94	131	68	136	231	112	205	422	176	271	607	255
100	1	NA	NA	23	84	104	49	122	200	89	182	410	151	243	617	232
	2	NA	NA	30	88	115	59	127	215	102	190	425	169	253	636	254
	3	NA	NA	34	93	124	67	133	228	115	199	438	188	262	654	279

Common Vent Capacity

Vent Height *H* (ft)	Combined Appliance Input Rating in Thousands of Btu per Hour: 12 FAN +FAN	12 FAN +NAT	12 NAT +NAT	19 FAN +FAN	19 FAN +NAT	19 NAT +NAT	28 FAN +FAN	28 FAN +NAT	28 NAT +NAT	38 FAN +FAN	38 FAN +NAT	38 NAT +NAT	50 FAN +FAN	50 FAN +NAT	50 NAT +NAT
6	NA	NA	25	NA	118	45	NA	176	71	NA	255	102	NA	348	142
8	NA	NA	28	NA	128	52	NA	190	81	NA	276	118	NA	380	162
10	NA	NA	31	NA	136	56	NA	205	89	NA	295	129	NA	405	175
15	NA	NA	36	NA	NA	66	NA	230	105	NA	335	150	NA	400	210
20	NA	NA	NA	NA	NA	74	NA	247	120	NA	362	170	NA	503	240
30	NA	NA	NA	NA	NA	NA	NA	NA	135	NA	398	195	NA	558	275
50	NA	NA	NA	NA	NA	NA	NA	NA	NA	NA	NA	NA	NA	612	325
100	NA	NA	NA	NA	NA	NA	NA	NA	NA	NA	NA	NA	NA	NA	NA

Table 8-13

Masonry Chimney [NFPA 54: Table 13.2(d)] (continued)

Number of Appliances: Two or More
Appliance Type: Category I
Appliance Vent Connection: Single-Wall Metal Connector

Vent Connector Capacity

Vent Height *H* (ft)	Connector Rise *R* (ft)	Single-Wall Metal Vent Connector Diameter – *D* (in.)								
		8			9			10		
		Appliance Input Rating Limits in Thousands of Btu per Hour								
		FAN		NAT	FAN		NAT	FAN		NAT
		Min	Max	Max	Min	Max	Max	Min	Max	Max
6	1	292	366	200	362	474	252	499	594	316
	2	301	432	231	373	557	299	509	696	376
	3	309	491	269	381	634	348	519	793	437
8	1	313	407	207	387	530	263	529	672	331
	2	323	465	238	397	607	309	540	766	391
	3	332	509	274	407	663	356	551	838	450
10	1	333	434	213	410	571	273	558	727	343
	2	343	489	244	420	640	317	569	813	403
	3	352	530	279	430	694	363	580	880	459
15	1	349	502	225	445	646	291	623	808	366
	2	359	548	256	456	706	334	634	884	424
	3	368	586	289	466	755	378	646	945	479
20	1	345	569	235	439	734	306	614	921	387
	2	355	610	266	450	787	348	627	986	443
	3	365	644	298	461	831	391	639	1042	496
30	1	338	665	250	430	864	330	600	1089	421
	2	348	699	282	442	908	372	613	1145	473
	3	358	729	312	452	946	412	626	1193	524
50	1	328	778	287	417	1022	383	582	1302	492
	2	339	806	320	429	1058	425	596	1346	545
	3	349	831	351	440	1090	466	610	1386	597
100	1	315	875	328	402	1181	444	560	1537	580
	2	326	899	361	415	1210	488	575	1570	634
	3	337	921	392	427	1238	529	589	1604	687

Common Vent Capacity

Vent Height *H* (ft)	Minimum Internal Area of Masonry Chimney Flue (in.^{2})								
	63			78			113		
	Combined Appliance Input Rating in Thousands of Btu per Hour								
	FAN +FAN	FAN +NAT	NAT +NAT	FAN +FAN	FAN +NAT	NAT +NAT	FAN +FAN	FAN +NAT	NAT +NAT
6	NA	455	187	NA	579	245	NA	846	NA
8	NA	497	217	NA	633	277	1136	928	405
10	NA	532	234	771	680	300	1216	1000	450
15	677	602	280	866	772	360	1359	1139	540
20	765	661	321	947	849	415	1495	1264	640
30	808	739	377	1052	957	490	1682	1447	740
50	NA	821	456	1152	1076	600	1879	1672	910
100	NA	NA	494	NA	NA	663	2006	1885	1046

For SI units, 1 in. = 25.4 mm, 1 ft. = 0.305 m, 1 in.^{2} = 645 mm^2, 1,000 Btu/hr = 0.293 kW

Table 8-14 continued

Table 8-14

Single-Wall Metal Pipe or Type B Asbestos Cement Vent [NFPA 54: Table 13.2(e)]

Number of Appliances:	**Two or More**
Appliance Type:	**Draft Hood-Equipment**
Appliance Vent Connection:	**Direct to Pipe or Vent**

Vent Connector Capacity

Total Vent Height *H* (ft)	Connector Rise *R* (ft)	Vent Connector Diameter – *D* (in.)					
		3	4	5	6	7	8
		Maximum Appliance Input Rating in Thousands of Btu per Hour					
6-8	1	21	40	68	102	146	205
	2	28	53	86	124	178	235
	3	34	61	98	147	204	275
15	1	23	44	77	117	179	240
	2	30	56	92	134	194	265
	3	35	64	102	155	216	298
30 and up	1	25	49	84	129	190	270
	2	31	58	97	145	211	295
	3	36	68	107	164	232	321

Common Vent Capacity

Total Vent Height *H* (ft)	Common Vent Diameter – *D* (in.)						
	4	5	6	7	8	10	12
	Combined Appliance Input Rating in Thousands of Btu per Hour						
6	48	78	111	155	205	320	NA
8	55	89	128	175	234	365	505
10	59	95	136	190	250	395	560
15	71	115	168	228	305	480	690
20	80	129	186	260	340	550	790
30	NA	147	215	300	400	650	940
50	NA	NA	NA	360	490	810	1190

For SI units, 1 in. = 25.4 mm, 1 ft. = 0.305 m, 1 in.^{2} = 645 mm^2, 1,000 Btu/hr = 0.293 kW

Table 8-15
Exterior Masonry Chimney [NFPA 54: Table 13.1(f)]

Number of Appliances:	Single
Appliance Type:	NAT
Appliance Vent Connection:	Type B Double-Wall Connector

INTENDED USE: Minimum allowable Input Rating of Space-Heating Appliance in Thousands of Btu per Hour

Vent Height *H* (ft)	Internal Area of chimney (in.^{2})							
	12	19	28	38	50	63	78	113
	Local 99% winter design temperature: 37°F or lower							
6	0	0	0	0	0	0	0	0
8	0	0	0	0	0	0	0	0
10	0	0	0	0	0	0	0	0
15	NA	0	0	0	0	0	0	0
20	NA	NA	123	190	249	184	0	0
30	NA	NA	NA	NA	NA	393	334	0
50	NA	NA	NA	NA	NA	NA	NA	579
	Local 99% winter design temperature: 27°F to 36°F							
6	0	0	68	116	156	180	212	266
8	0	0	82	127	167	187	214	263
10	0	51	97	141	183	201	225	265
15	NA	NA	NA	NA	233	253	274	305
20	NA	NA	NA	NA	NA	307	330	362
30	NA	NA	NA	NA	NA	419	445	485
50	NA	NA	NA	NA	NA	NA	NA	763
	Local 99% winter design temperature: 17°F to 26°F							
6	NA	NA	NA	NA	NA	215	259	349
8	NA	NA	NA	NA	197	226	264	352
10	NA	NA	NA	NA	214	245	278	358
15	NA	NA	NA	NA	NA	296	331	398
20	NA	NA	NA	NA	NA	352	387	457
30	NA	NA	NA	NA	NA	NA	507	581
50	NA	NA	NA	NA	NA	NA	NA	NA
	Local 99% winter design temperature: 5°F to 16°F							
6	NA	NA	NA	NA	NA	NA	NA	416
8	NA	NA	NA	NA	NA	NA	312	423
10	NA	NA	NA	NA	NA	289	331	430
15	NA	NA	NA	NA	NA	NA	393	485
20	NA	NA	NA	NA	NA	NA	450	547
30	NA	NA	NA	NA	NA	NA	NA	682
50	NA	NA	NA	NA	NA	NA	NA	972
	Local 99% winter design temperature: -10°F to 4°F							
6	NA	NA	NA	NA	NA	NA	NA	484
8	NA	NA	NA	NA	NA	NA	NA	494
10	NA	NA	NA	NA	NA	NA	NA	513
15	NA	NA	NA	NA	NA	NA	NA	586
20	NA	NA	NA	NA	NA	NA	NA	650
30	NA	NA	NA	NA	NA	NA	NA	805
50	NA	NA	NA	NA	NA	NA	NA	1003
	Local 99% winter design temperature -11°F or lower Not recommended for any vent configurations							

For SI units, 1 in. = 25.4 mm, 1 in.^{2} = 645 mm^2, 1 ft. = 0.305 m, 1,000 Btu/hr = 0.293 kW, °C = (°F - 32)/1.8

Note: See Figure G.2.4 for a map showing local 99 percent winter design temperatures in the United States.

Table 8-16

Exterior Masonry Chimney [NFPA 54: Table 13.2(f)]

				Number of Appliances:	Two or More			
				Appliance Type:	NAT + NAT			
				Appliance Vent Connection:	Type B Double-Wall Connector			
INTENDED USE: Combined Appliance Maximum Input Rating in Thousands of Btu per Hour								
Vent Height *H* (ft)	**Internal Area of Chimney (in.²)**							
	12	**19**	**28**	**38**	**50**	**63**	**78**	**113**
6	25	46	71	103	143	188	246	NA
8	28	53	82	119	163	218	278	408
10	31	56	90	131	177	236	302	454
15	NA	67	106	152	212	283	365	546
20	NA	NA	NA	NA	NA	325	419	648
30	NA	NA	NA	NA	NA	NA	496	749
50	NA	NA	NA	NA	NA	NA	NA	922
100	NA	NA	NA	NA	NA	NA	NA	NA

For SI units, 1 in. = 25.4 mm, 1 ft. = 0.305 m, 1 in.^{2} = 645 mm^2, 1,000 Btu/hr = 0.293 kW

Table 8-17

Exterior Masonry Chimney [NFPA 54: Table 13.2(g)]

	Number of Appliances:	Two or More
	Appliance Type:	NAT + NAT
	Appliance Vent Connection:	Type B Double-Wall Connector

INTENDED USE: Minimum Allowable Input Rating of Space-Heating Appliance in Thousands of Btu per Hour

Vent Height *H* (ft)	Internal Area of Chimney (in.^{2})							
	12	19	28	38	50	63	78	113
	Local 99% winter design temperature: 37°F or greater							
6	0	0	0	0	0	0	0	NA
8	0	0	0	0	0	0	0	0
10	0	0	0	0	0	0	0	0
15	NA	0	0	0	0	0	0	0
20	NA	NA	NA	NA	NA	184	0	0
30	NA	NA	NA	NA	NA	393	334	0
50	NA	NA	NA	NA	NA	NA	NA	579
100	NA	NA	NA	NA	NA	NA	NA	NA
	Local 99% winter design temperature: 27°F to 36°F							
6	0	0	68	NA	NA	180	212	NA
8	0	0	82	NA	NA	187	214	263
10	0	51	NA	NA	NA	201	225	265
15	NA	NA	NA	NA	NA	253	274	305
20	NA	NA	NA	NA	NA	307	330	362
30	NA	NA	NA	NA	NA	NA	445	485
50	NA	NA	NA	NA	NA	NA	NA	763
100	NA	NA	NA	NA	NA	NA	NA	NA
	Local 99% winter design temperature: 17°F to 26°F							
6	NA	NA	NA	NA	NA	NA	NA	NA
8	NA	NA	NA	NA	NA	NA	264	352
10	NA	NA	NA	NA	NA	NA	278	358
15	NA	NA	NA	NA	NA	NA	331	398
20	NA	NA	NA	NA	NA	NA	387	457
30	NA	NA	NA	NA	NA	NA	NA	581
50	NA	NA	NA	NA	NA	NA	NA	862
100	NA	NA	NA	NA	NA	NA	NA	NA
	Local 99% winter design temperature: 5°F to 16°F							
6	NA	NA	NA	NA	NA	NA	NA	NA
8	NA	NA	NA	NA	NA	NA	NA	NA
10	NA	NA	NA	NA	NA	NA	NA	430
15	NA	NA	NA	NA	NA	NA	NA	485
20	NA	NA	NA	NA	NA	NA	NA	547
30	NA	NA	NA	NA	NA	NA	NA	682
50	NA	NA	NA	NA	NA	NA	NA	NA
100	NA	NA	NA	NA	NA	NA	NA	NA
	Local 99% winter design temperature: 4°F or lower Not recommended for any vent configurations							

For SI units, 1 in. = 25.4 mm, 1 in.^{2} = 645 mm^2, 1 ft. = 0.305 m, 1,000 Btu/hr = 0.293 kW, °C = (°F - 32)/1.8

Note: See Figure G.2.4 for a map showing local 99 percent winter design temperatures in the United States.

Table 8-18
Exterior Masonry Chimney [NFPA 54: Table 13.2(h)]

					Number of Appliances:	Two or More		
					Appliance Type:	FAN + NAT		
					Appliance Vent Connection:	Type B Double-Wall Connector		
INTENDED USE: Combined Appliance Maximum Input Rating in Thousands of Btu per Hour								
Vent Height *H* (ft)	Internal Area of Chimney (in.^{2})							
	12	19	28	38	50	63	78	113
6	74	119	178	257	351	458	582	853
8	80	130	193	279	384	501	636	937
10	84	138	207	299	409	538	686	1010
15	NA	152	233	334	467	611	781	1156
20	NA	NA	250	368	508	668	858	1286
30	NA	NA	NA	404	564	747	969	1473
50	NA	NA	NA	NA	NA	831	1089	1692
100	NA	NA	NA	NA	NA	NA	NA	1921

For SI units, 1 in. = 25.4 mm, 1 ft. = 0.305 m, 1 in.^{2} = 645 mm^2, 1,000 Btu/hr = 0.293 kW

Table 8-19

Exterior Masonry Chimney [NFPA 54: Table 13.2(i)]

				Number of Appliances:	Two or More			
				Appliance Type:	FAN + NAT			
				Appliance Vent Connection:	Type B Double-Wall Connector			
INTENDED USE: Minimum Allowable Input Rating of Space-Heating Appliance in Thousands of Btu per Hour								
Vent Height *H* (ft)	Internal Area of Chimney (in.^{2})							
	12	19	28	38	50	63	78	113
	Local 99% winter design temperature: 37°F or greater							
6	0	0	0	0	0	0	0	0
8	0	0	0	0	0	0	0	0
10	0	0	0	0	0	0	0	0
15	NA	0	0	0	0	0	0	0
20	NA	NA	123	190	249	184	0	0
30	NA	NA	NA	334	398	393	334	0
50	NA	NA	NA	NA	NA	714	707	579
100	NA	NA	NA	NA	NA	NA	NA	1600
	Local 99% winter design temperature: 27°F to 36°F							
6	0	0	68	116	156	180	212	266
8	0	0	82	127	167	187	214	263
10	0	51	97	141	183	210	225	265
15	NA	111	142	183	233	253	274	305
20	NA	NA	187	230	284	307	330	362
30	NA	NA	NA	330	319	419	445	485
50	NA	NA	NA	NA	NA	672	705	763
100	NA	NA	NA	NA	NA	NA	NA	1554
	Local 99% winter design temperature: 17°F to 26°F							
6	0	55	99	141	182	215	259	349
8	52	74	111	154	197	226	264	352
10	NA	90	125	169	214	245	278	358
15	NA	NA	167	212	263	296	331	398
20	NA	NA	212	258	316	352	387	457
30	NA	NA	NA	362	429	470	507	581
50	NA	NA	NA	NA	NA	723	766	862
100	NA	NA	NA	NA	NA	NA	NA	1669
	Local 99% winter design temperature: 5°F to 16°F							
6	NA	78	121	166	214	252	301	416
8	NA	94	135	182	230	269	312	423
10	NA	111	149	198	250	289	331	430
15	NA	NA	193	247	305	346	393	485
20	NA	NA	NA	293	360	408	450	547
30	NA	NA	NA	377	450	531	580	682
50	NA	NA	NA	NA	NA	797	853	972
100	NA	NA	NA	NA	NA	NA	NA	1833
	Local 99% winter design temperature: -10°F to 4°F							
6	NA	NA	145	196	249	296	349	484
8	NA	NA	159	213	269	320	371	494
10	NA	NA	175	231	292	339	397	513
15	NA	NA	NA	283	351	404	457	586
20	NA	NA	NA	333	408	468	528	650
30	NA	NA	NA	NA	NA	603	667	805
50	NA	NA	NA	NA	NA	NA	955	1003
100	NA	NA	NA	NA	NA	NA	NA	NA
	Local 99% winter design temperature: -11°F or lower Not recommended for any vent configurations							

For SI units, 1 in. = 25.4 mm, 1 in.^{2} = 645 mm^2, 1 ft. = 0.305 m, 1,000 Btu/hr = 0.293 kW .

Note: See Figure G.2.4 for a map showing local 99 percent winter design temperatures in the United States.

CHAPTER 9

INSTALLATION OF SPECIFIC APPLIANCES

901.0 Scope.

Warm air heating systems, vented decorative appliances, floor furnaces, vented wall furnaces, unit heaters and room heaters shall comply with the general requirements of Chapter 3 of this code, in addition to the requirements of this chapter.

902.0 General.

(A) This chapter is applicable primarily to nonindustrial-type gas utilization appliances and installations and, unless specifically indicated, does not apply to industrial-type appliances and installations. Listed gas utilization appliances shall be installed in accordance with their listing and the manufacturer's instructions, or as elsewhere specified in this chapter. Unlisted appliances shall be installed as specified in this part as applicable to the equipment. For additional information concerning particular appliances and accessories, including industrial types, reference can be made to the standards listed in Chapter 17. [NFPA 54:10.1.1]

(B) Gas utilization appliances shall not be installed so its combustion, ventilation, and dilution air are obtained only from a bedroom or bathroom unless the bedroom or bathroom has the required volume in accordance with Section 701.2. [NFPA 54:10.1.2]

(C) Where the room size in comparison with the size of the appliance is to be calculated, the total volume of the appliance is determined from exterior dimensions and is to include fan compartments and burner vestibules, where used. Where the actual ceiling height of a room is greater than eight (8) feet (2,438 mm), the volume of the room is figured on the basis of a ceiling height of eight (8) ft. (2,438 mm). [NFPA 54:10.1.3]

902.1 Added or Converted Appliances. When additional or replacement appliances or equipment are installed or an appliance is converted to gas from another fuel, the location in which the appliances or equipment are to be operated shall be checked to verify the following:

(A) Air for combustion and ventilation is provided where required, in accordance with the provisions of Chapter 7. Where existing facilities are not adequate, they shall be upgraded to meet Chapter 7 specifications.

(B) The installation components and appliances meet the clearances to combustible material provisions of this code. It shall be determined that the installation and operation of the additional or replacement appliances do not render the remaining appliances unsafe for continued operation.

(C) The venting system is constructed and sized in accordance with the provisions of Chapter 8. Where the existing venting system is not adequate, it shall be upgraded to comply with Chapter 8. [NFPA 54:9.1.2]

902.2 Type of Gas(es). It shall be determined whether the appliance has been designed for use with the gas to which it will be connected. No attempt shall be made to convert the appliance from the gas specified on the rating plate for use with a different gas without consulting the installation instruction, the serving gas supplier, or the appliance manufacturer for complete instructions. [NFPA 54:9.1.3]

903.0 Air-Conditioning Appliances (Gas-Fired Air Conditioners and Heat Pumps).

903.1 Independent Gas Piping. Gas piping serving heating gas utilization appliances shall be permitted to also serve cooling appliances where heating and cooling appliances cannot be operated simultaneously. (See Section 1309.4.) [NFPA 54:10.2.1]

903.2 Connection of Gas Engine-Powered Air Conditioners. To protect against the effects of normal vibration in service, gas engines shall not be rigidly connected to the gas supply piping. [NFPA 54:10.2.2]

903.3 Clearances for Indoor Installation. The installation of air-conditioning appliances shall comply with the following requirements:

(A) Listed air-conditioning appliances installed in rooms that are large in comparison with the size of the appliance shall be installed with clearances per the terms of their listing and the manufacturer's instructions.

(B) Air-conditioning appliances installed in rooms that are NOT large (such as alcoves and closets) in comparison with the size of the equipment shall be listed for such installations and installed in accordance with the manufacturer's instructions. Listed clearances shall not be reduced by the protection methods described in Table 5-3, regardless of whether the enclosure is of combustible or noncombustible material.

(C) Unlisted air-conditioning appliances shall be installed with clearances from combustible material of not less than eighteen (18) inches (457 mm) above the appliance and at the sides, front, rear, and nine (9) inches (229 mm) from the draft hood.

(D) Air-conditioning appliances (listed and unlisted) installed in rooms that are large in comparison with the size of the appliances shall be permitted to be installed with reduced clearances to combustible material provided the combustible material or appliance is protected as described in Table 5-3 [see 5-3(2)].

(E) Where the furnace plenum is adjacent to plaster on metal lath or noncombustible material attached to combustible material, the clearance shall be measured to the surface of the plaster or other noncombustible finish where the clearance specified is two (2) inches (50 mm) or less.

(F) Listed air-conditioning appliances shall have the clearance from supply ducts within three (3) feet (914 mm) of the furnace plenum be not less than that specified from the furnace plenum. No clearance is necessary beyond this distance. [NFPA 54:10.2.3]

903.4 Assembly and Installation. Air-conditioning appliances shall be installed in accordance with the manufacturer's instructions. Unless the appliance is listed for installation on a combustible surface, such as a floor or roof, or unless the surface is protected in an approved manner, it shall be installed on a surface of noncombustible construction with noncombustible material and surface finish and with no combustible material against the underside thereof. [NFPA 54:10.2.4]

903.5 Furnace Plenums and Air Ducts. A furnace plenum supplied as a part of the air-conditioning appliance shall be installed in accordance with the manufacturer's instructions. Where a furnace plenum is not supplied with the appliance, any fabrication and installation instructions provided by the manufacturer shall be followed. The method of connecting supply and return ducts shall facilitate proper circulation of air. [NFPA 54:10.2.5]

903.6 Refrigeration Coils. (See Sections 904.7 and 904.8) [NFPA 54:10.2.6]

903.7 Switches in Electrical Supply Line. Means for interrupting the electrical supply to the air-conditioning appliance and to its associated cooling tower (if supplied and installed in a location remote from the air conditioner) shall be provided within sight of and not over fifty (50) feet (15 m) from the air-conditioner and cooling tower. [NFPA 54:10.2.7]

904.0 Central Heating Boilers and Furnaces.

904.1 Location. Central heating furnace and low-pressure boiler installations in bedrooms or bathrooms shall comply with one of the following:

(1) Central heating furnaces and low-pressure boilers may be installed in a closet located in the bedroom or bathroom, provided the closet is equipped with a listed, gasketed door assembly and a listed, self-closing device. The self-closing door assembly shall meet the requirements of section 904.1.1. The door assembly shall be installed with a threshold and bottom door seal and shall meet the requirements of section 904.1.2. All combustion air for such installations shall be obtained from the outdoors. The closet shall be for the exclusive use of the central heating furnace or low-pressure boiler.

(2) Central heating furnaces and low-pressure boilers shall be of the direct-vent type. [NFPA 54:10.3.1]

904.1.1 Self-Closing Doors. Self-closing doors shall swing easily and freely and shall be equipped with a self-closing device to cause the door to close and latch each time it is opened. The closing mechanism shall not have a hold-open feature. [NFPA 80:6.1.4.2.1, 6.1.4.2.2]

904.1.2 Gasketing. Gasketing on gasket doors or frames shall be furnished only in accordance with the published listings of the door, frame, or gasketing material manufacturer. [NFPA 80:6.4.8]

Exception: Where acceptable to the Authority Having Jurisdiction, gasketing of noncombustible or limited-combustible material (see NFPA 220, *Standard on Types of Building Construction*) shall be permitted to be applied to the frame, provided closing and latching of the door are not inhibited.

904.2 Clearance.

(A) Listed central heating furnaces and low-pressure boilers installed in rooms that are large in comparison with the size of the appliance shall be installed with clearances per the terms of their listings and the manufacturer's instructions. (See Sections 208.0 and 1004.0 for definition.) [NFPA 54:10.3.2.1]

(B) Central-heating furnaces and low-pressure boilers installed in rooms that are NOT large (such as alcoves and closets) in comparison with the size of the appliance shall be listed for such installations. Listed clearances shall not be reduced by the protection methods described in Table 5-3 and illustrated in Figures 9-1(a) through 9-1(c), regardless of whether the enclosure is of combustible or noncombustible material. [NFPA 54:10.3.2.2]

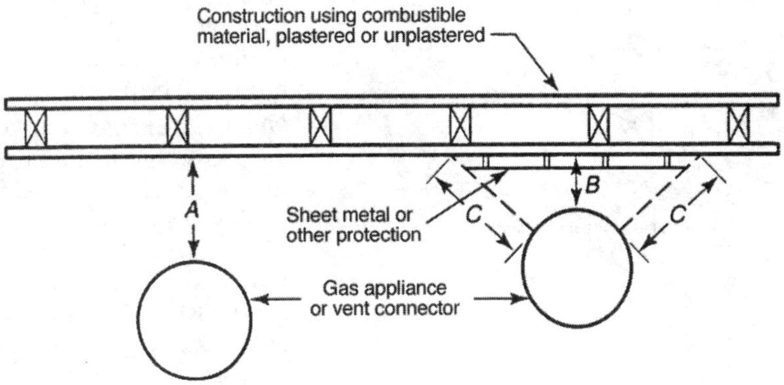

Notes:

A equals the clearance with no protection.

B equals the reduced clearance permitted. The protection applied to the construction using combustible material shall extend far enough in each direction to make C equal to A.

FIGURE 9-1(a) Extent of Protection Necessary to Reduce Clearance from Gas Equipment or Vent Connectors.

(C) Unlisted central-heating furnaces and low pressure boilers installed in rooms that are large in comparison with the size of the appliance shall be installed with clearances not less than those specified in Table 9-1. [NFPA 54:10.3.2.3]

(D) Central-heating furnaces and low-pressure boilers (listed and unlisted) installed in rooms that are large in comparison with the size of the appliance shall be permitted to be installed with reduced clearances to combustible material provided the combustible material or appliance is protected as described in Table 5-3. [NFPA 54:10.3.2.4]

(E) Front clearance shall be sufficient for servicing the burner and the furnace or boiler. [NFPA 54:10.3.2.5]

(F) Where the furnace plenum is adjacent to plaster on metal lath or noncombustible material attached to combustible material, the clearance shall be measured to the surface of the plaster or other noncombustible finish where the clearance specified is two (2) inches (50 mm) or less. [NFPA 54:10.3.2.6]

(G) The clearance to these appliances shall not interfere with combustion air, draft hood clearance and relief, and accessibility for servicing. (See Sections 304, 701.0, and 802.12.7.) [NFPA 54:10.3.2.7]

(H) Listed central heating furnaces shall have the clearance from supply ducts within three (3) feet (914 mm) of the furnace plenum be not less than that specified from the furnace plenum. No clearance is necessary beyond this distance. [NFPA 54:10.3.2.8]

(I) Unlisted central heating furnaces with temperature limit controls that cannot be set higher than 250°F (121°C) shall have the clearance from supply duct within six (6) feet (1,829 mm) of the furnace plenum be not less than six (6) inches (150 mm). No clearance is necessary beyond this distance. [NFPA 54:10.3.2.9]

(J) Central-heating furnaces other than those listed in 904.2 (h) or (i) shall have clearances from the supply ducts of not less than eighteen (18) inches (457 mm) from the furnace plenum for the first three (3) feet (914 mm), then six (6) inches (150 mm) for the next three (3) feet (914 mm) and one (1) inch (25 mm) beyond six (6) feet (1,829 mm). [NFPA 54:10.3.2.10]

904.3 Assembly and Installation. A central-heating boiler or furnace shall be installed in accordance with the manufacturer's instructions and shall be installed on a floor of noncombustible construction with noncombustible flooring and surface finish and with no combustible material against the underside thereof, or on fire-resistive slabs or arches having no combustible material against the underside thereof.

Exception:

(1) Appliances listed for installation on a combustible floor.

(2) Installation on a floor protected in an approved manner. [NFPA 54:10.3.3]

904.3.1 Under-floor Installation. Furnaces installed in an under-floor area of the building shall comply with the Section 904.3.1.1 through 904.3.1.3.

904.3.1.1 Supported by Ground. Where a furnace is supported by the ground, it shall be installed on a concrete slab not less than three (3) inches (76 mm) above the adjoining ground level.

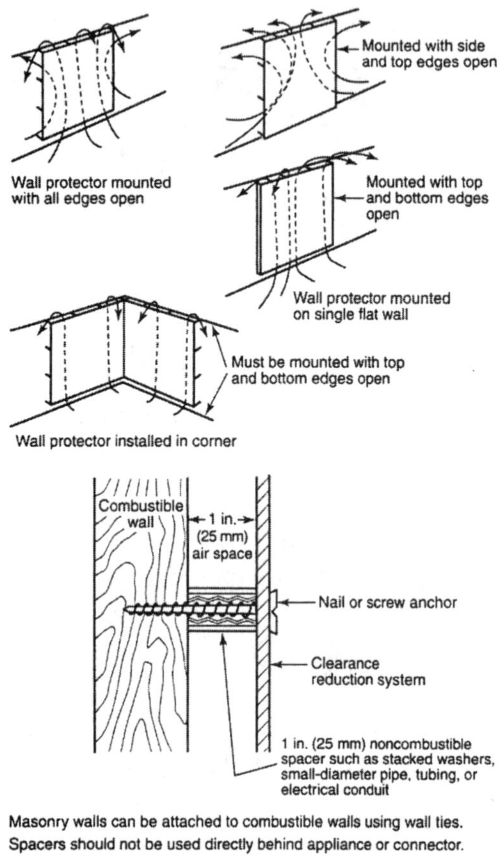

FIGURE 9-1(b) Wall Protector Clearance Reduction System.

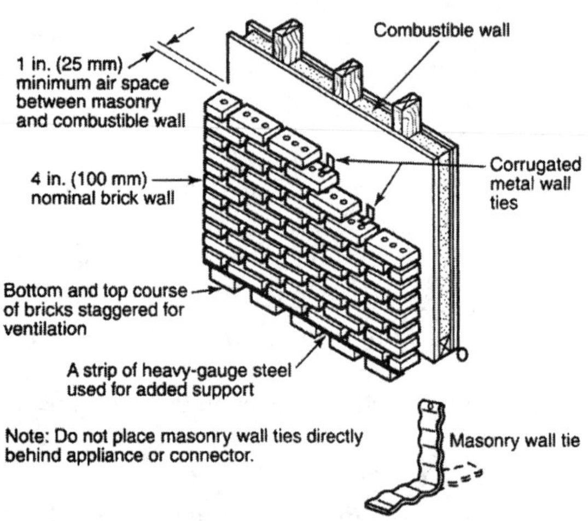

FIGURE 9-1(c) Masonry Clearance Reduction System.

904.3.1.2 Supported from Above. Where a furnace is supported from above, a minimum clearance of six (6) inches (152 mm) shall be provided from finished grade.

904.3.1.3 Excavation. Where excavation is necessary to install a furnace, it shall extend to a depth of six (6) inches (152 mm) below and twelve (12) inches (300 mm) on all sides of the furnace, except on the service side, which shall have thirty (30) inches (762 mm). If the depth of the excavation for either the furnace or passageway exceeds twelve (12) inches (300 mm), walls shall be lined with concrete or masonry four (4) inches (102 mm) above the adjoining ground level.

904.4 Temperature- or Pressure-Limiting Devices. Steam and hot water boilers, respectively, shall be provided with approved automatic limiting devices for shutting down the burner(s) to prevent boiler steam pressure or boiler water temperature from exceeding the maximum allowable working pressure or temperature. Safety limit controls shall not be used as operating controls. [NFPA 54:10.3.4]

904.5 Low-Water Cutoff. Hot water boilers installed above the radiation level and all steam boilers shall be provided with an automatic means to shut off the fuel supply to the burner(s) if the boiler water level drops to the lowest safe water line. [NFPA 54:10.3.5]

904.6 Steam Safety and Pressure-Relief Valves. Steam and hot water boilers shall be equipped, respectively, with listed or approved steam safety or pressure-relief valves of appropriate discharge capacity and conforming with ASME requirements. A shutoff valve shall not be placed between the relief valve and the boiler or on discharge pipes between such valves and the atmosphere.

(1) Relief valves shall be piped to discharge near the floor.

(2) The entire discharge piping shall be at least the same size as the relief valve discharge piping.

(3) Discharge piping shall not contain a threaded end connection at its termination point. [NFPA 54:10.3.6]

904.7 Furnace Plenums and Air Ducts.

(A) Furnace plenums and air ducts shall be installed in accordance with NFPA 90A, *Standard for the Installation of Air-Conditioning and Ventilating Systems*, or NFPA 90B, *Standard for the Installation of Warm-Air Heating and Air-Conditioning Systems*. [NFPA 54:10.3.7.1]

(B) A furnace plenum supplied as a part of a furnace shall be installed in accordance with the manufacturer's instructions. [NFPA 54:10.3.7.2]

(C) Where a furnace plenum is not supplied with the furnace, any fabrication and installation instructions provided by the manufacturer shall be followed. The method of connecting supply and return ducts shall facilitate proper circulation of air. [NFPA 54:10.3.7.3]

(D) Where a furnace is installed so supply ducts carry air circulated by the furnace to areas outside the space containing the furnace, the return air shall also be handled by a duct(s) sealed to the furnace casing and terminating outside the space containing the furnace. [NFPA 54:10.3.7.4]

904.8 Refrigeration Coils. The installation of refrigeration coils shall comply with the following requirements: [NFPA 54:10.3.8]

(A) A refrigeration coil shall not be installed in conjunction with a forced-air furnace where circulation of cooled air is provided by the furnace blower, unless the blower has sufficient capacity to overcome the external static resistance imposed by the duct system and cooling coil and the air throughput necessary for heating or cooling, whichever is greater. [NFPA 54:10.3.8(1)]

(B) Furnaces shall not be located upstream from cooling units, unless the cooling unit is designed or equipped so as not to develop excessive temperature or pressure. [NFPA 54:10.3.8(2)]

(C) Refrigeration coils shall be installed in parallel with or on the downstream side of central furnaces to avoid condensation in the heating element, unless the furnace has been specifically listed for downstream installation. With a parallel flow arrangement, the dampers or other means used to control flow of air shall be sufficiently tight to prevent any circulation of cooled air through the furnace. [NFPA 54:10.3.8(3)]

(D) Means shall be provided for disposal of condensate and to prevent dripping of condensate on the heating element. [NFPA 54:10.3.8(4)]

904.9 Cooling Units Used with Heating Boilers.

(A) Boilers, where used in conjunction with refrigeration systems, shall be installed so that the chilled medium is piped in parallel with the heating boiler with appropriate valves to prevent the chilled medium from entering the heating boiler. [NFPA 54:10.3.9.1]

(B) Where hot-water-heating boilers are connected to heating coils located in air-handling units where they can be exposed to refrigerated air circulation, such boiler piping systems shall be equipped with flow control valves or other automatic means to prevent gravity circulation of the boiler water during the cooling cycle. [NFPA 54:10.3.9.2]

904.10 Access to Appliances on Roofs.

904.10.1 Appliances located on roofs or other elevated locations shall be accessible. [NFPA 54-09:9.4.3.1]

904.10.2 Buildings of more than fifteen (15) feet (4,572 mm) in height shall have an inside means of access to the roof, unless other means acceptable to the Authority Having Jurisdiction are used. [NFPA 54-09:9.4.3.2]

904.10.3 The inside means of access shall be a permanent or foldaway inside stairway or ladder, terminating in an enclosure, scuttle, or trapdoor. Such scuttles or trapdoors shall be at least twenty-two (22) inches x twenty-four (24) inches (560 mm x 610 mm) in size, shall open easily and safely under all conditions, especially snow, and shall be constructed so as to permit access from the roof side unless deliberately locked on the inside. At least six (6) feet (1,829 mm) of clearance shall be available between the access opening and the edge of the roof or similar hazard, or rigidly fixed rails or guards a minimum of forty-two (42) inches (1,067 mm) in height shall be provided on the exposed side. Where parapets or other building structures are utilized in lieu of guards or rails, they shall be a minimum of forty-two (42) inches (1,067 mm) in height. [NFPA 54-09:9.4.3.3]

904.11 Appliances in Attics and Under-Floor Spaces.

904.11.1 Access. An attic or under-floor space in which an appliance is installed shall be accessible through an opening and passageway at least as large as the largest component of the appliance, and not less than twenty-two (22) inches x thirty (30) inches (560 mm x 760 mm). [NFPA 54-09:9.5.1]

904.11.2 Passageway Height. Where the height of the passageway is less than six (6) feet (1,829 mm), the distance from the passageway access to the appliance shall not exceed twenty (20) feet (6,096 mm) measured along the center line of the passageway. [NFPA 54-09:9.5.1.1]

904.11.3 Passageway. The passageway shall be unobstructed and shall have solid flooring not less than twenty-four (24) inches (610 mm) wide from the entrance opening to the appliance. [NFPA 54:9.5.1.2]

904.11.4 Work Surface. A level working platform or grade surface not less than thirty (30) inches (760 mm) by thirty (30) inches (760 mm) shall be provided in front of the service side of the appliance. [NFPA 54:9.5.2]

904.11.5 Lighting and Convenience Outlet. A permanent 120-volt receptacle outlet and a lighting fixture shall be installed near the appliance. The switch controlling the lighting fixture shall be located at the entrance to the passageway. [NFPA 54:9.5.3]

904.11.6 Furnace (Upright and Horizontal). Upright furnaces may be installed in an attic, furred, or under-floor space more than five (5) feet (1,524 mm) in height, provided the required listings and furnace and duct clearances are observed. Horizontal furnaces may be installed in an attic, furred, or under-floor space, provided the required listings and furnace and duct clearances are observed.

905.0 Clothes Dryers.

905.1 Clearance. The installation of clothes dryers shall comply with the following requirements: [NFPA 54:10.4.1]

(A) Listed Type 1 clothes dryers shall be installed with a minimum clearance of six (6) inches (150 mm) from adjacent combustible material. Clothes dryers listed for installation at reduced clearances shall be installed in accordance with their listing. Type 1 clothes dryers installed in closets shall be specifically listed for such installation. [NFPA 54:10.4.1(1)]

(B) Listed Type 2 clothes dryers shall be installed with clearances of not less than that shown on the marking plate and in the manufacturer's instructions. Type 2 clothes dryers designed and marked, "For use only in noncombustible locations," shall not be installed elsewhere. [NFPA 54:10.4.19(2)]

(C) Unlisted clothes dryers shall be installed with clearances to combustible material of not less than eighteen (18) inches (460 mm). Combustible floors under unlisted clothes dryers shall be protected in an approved manner. [NFPA 54:10.4.1(3)]

905.2 Exhausting to the Outdoors. Type 1 and Type 2 clothes dryers shall be exhausted to the outside air. [NFPA 54:10.4.2]

905.3 Provisions for Makeup Air.

(A) Makeup air shall be provided for Type 1 clothes dryers in accordance with the manufacturer's installation instructions. [NFPA 54:10.4.3.1]

(B) Provision for makeup air shall be provided for Type 2 clothes dryers, with a minimum free area (see Section 701.6) of one (1) in.^{2} (6.5 m^2) for each one thousand (1,000) Btu/h (2,200 mm^2/kW) total input rating of the dryer(s) installed. [NFPA 54:10.4.3.2]

905.4 Exhaust Ducts for Type 1 Clothes Dryers.

(A) A clothes dryer exhaust duct shall not be connected into any vent connector, gas vent, chimney, crawl space, attic, or other similar concealed space. [NFPA 54:10.4.4.1]

(B) Ducts for exhausting clothes dryers shall not be assembled with screws or other fastening means that extend into the duct and that would catch lint and reduce the efficiency of the exhaust system. [NFPA 54:10.4.4.2]

(C) Exhaust ducts shall be constructed of rigid metallic material. Transition ducts used to connect the dryer to the exhaust duct shall be listed for that application or installed in accordance with the clothes dryer manufacturer's installation instructions. [NFPA 54:10.4.4.3]

905.5 Exhaust Ducts for Type 2 Clothes Dryers.

(A) Exhaust ducts for Type 2 clothes dryers shall comply with Section 905.4. [NFPA 54:10.4.5.1]

(B) Exhaust ducts for Type 2 clothes dryers shall be constructed of sheet metal or other noncombustible material. Such ducts shall be equivalent in strength and corrosion resistance to ducts made of galvanized sheet steel not less than 0.0195 inch (0.5 mm) thick. [NFPA 54:10.4.5.2]

(C) Type 2 clothes dryers shall be equipped or installed with lint-controlling means. [NFPA 54:10.4.5.3]

(D) Exhaust ducts for Type 2 clothes dryers shall be installed with a minimum clearance of six (6) inches (150 mm) from adjacent combustible material. Where exhaust ducts for Type 2 clothes dryers are installed with reduced clearances, the adjacent combustible material shall be protected in accordance with Table 5-3. [NFPA 54:10.4.5.4]

(E) Where ducts pass through walls, floors, or partitions, the space around the duct shall be sealed with noncombustible material. [NFPA 54:10.4.5.5]

(F) Multiple installations of Type 2 clothes dryers shall be made in a manner to prevent adverse operation due to back pressures that might be created in the exhaust systems. [NFPA 54:10.4.5.6]

905.6 Multiple Family or Public Use. All clothes dryers installed for multiple-family or public use shall be equipped with approved safety shutoff devices and shall be installed as specified for a Type 2 clothes dryer under Section 905.5. [NFPA 54:10.4.6]

906.0 Conversion Burners.

Installation of conversion burners shall conform to ANSI Z21.8, *Standard for Installation of Domestic Gas Conversion Burners*. [NFPA 54:10.5]

907.0 Decorative Appliances for Installation in Vented Fireplaces.

907.1 Prohibited Installations. Decorative appliances for installation in vented fireplaces shall not be installed in bathrooms or bedrooms unless the appliance is listed and the bedroom or bathroom has the required volume in accordance with Section 701.2. [NFPA 54:10.6.1]

907.2 Installation. A decorative appliance for installation in a vented fireplace shall be installed only in a vented fireplace having a working chimney flue and constructed of noncombustible materials. These appliances shall not be thermostatically controlled.

(1) A listed decorative appliance for installation in a vented fireplace shall be installed in accordance with its listing and the manufacturer's instructions. [NFPA 54:10.6.2.1]

(2) A decorative appliance for installation in a vented fireplace where installed in a manufactured home shall be listed for installation in manufactured homes. [NFPA 54:10.6.2.2]

(3) An unlisted decorative appliance for installation in a vented fireplace shall be installed in a fireplace having a permanent free opening, based on appliance input rating and chimney height, equal to or greater than that specified in Table 9-2. [NFPA 54:10.6.2.3]

907.3 Fireplace Screens. A fireplace screen shall be installed with a decorative appliance for installation in a vented fireplace. [NFPA 54:10.6.3]

908.0 Gas Fireplaces, Vented.

908.1 Prohibited Installations. Vented gas fireplaces shall not be installed in bathrooms or bedrooms unless the appliance is listed and the bedroom or bathroom has the required volume in accordance with Section 701.2. [NFPA 54:10.7.1]

Exception: Direct-vent gas fireplaces.

908.2 Installation. The installation of vented gas fireplaces shall comply with the following requirements:

(A) Listed vented gas fireplaces shall be installed in accordance with their listing and the manufacturer's instructions and where installed in or attached to combustible material shall be specifically listed for such installation. [NFPA 54:10.7.2(1)]

(B) Unlisted vented gas fireplaces shall not be installed in or attached to combustible material. They shall have a clearance at the sides and rear of not less than eighteen (18) inches (460 mm). Combustible floors under unlisted vented gas fireplaces shall be protected in an approved manner. Unlisted appliances of other than the direct-vent type shall be equipped with a draft hood and shall be properly vented in accordance with Chapter 8. Appliances that use metal, asbestos, or ceramic material to direct radiation to the front of the appliance shall have a clearance of thirty-six (36) inches (910 mm) in front and, if constructed with a double back of metal or ceramic, shall be installed with a minimum clearance of eighteen (18) inches (460 mm) at the sides and twelve (12) inches (300 mm) at the rear. [NFPA 54:10.7.2(2)]

(C) Panels, grilles, and access doors that are required to be removed for normal servicing operations shall not be attached to the building. [NFPA 54:10.7.2(3)]

(D) Direct-vent gas fireplaces shall be installed with the vent-air intake terminal in the outdoors and in accordance with the manufacturer's instructions. [NFPA 54:10.7.2(4)]

908.3 Combustion and Circulating Air. Combustion and circulating air shall be provided in accordance with Section 701.0. [NFPA 54:10.7.3]

909.0 Non-Recirculating Direct Gas-Fired Industrial Air Heaters.

909.1 Application. Direct gas-fired industrial air heaters of the non-recirculating type shall be listed in accordance with ANSI Z83.4/CSA 3.7 *Non-Recirculating Direct Gas-Fired Industrial Air Heaters*. [NFPA 54:10.8.1]

909.2 Prohibited Installations.

(A) Non-recirculating direct gas-fired industrial air heaters shall not serve any area containing sleeping quarters. [NFPA 54:10.8.2.1]

(B) Non-recirculating direct gas-fired industrial air heaters shall not recirculate room air. [NFPA 54:10.8.2.2]

909.3 Installation. Installation of direct gas-fired industrial air heaters shall comply with the following requirements:

(A) Non-recirculating direct gas-fired industrial air heaters shall be installed in accordance with the manufacturer's instructions. [NFPA 54:10.8.3.1]

(B) Non-recirculating direct gas-fired industrial air heaters shall be installed only in industrial or commercial occupancies. [NFPA 54:10.8.3.2]

(C) Non-recirculating direct gas-fired industrial air heaters shall be permitted to provide fresh air ventilation. [NFPA 54:10.8.3.3]

(D) Non-recirculating direct gas-fired industrial air heaters shall be provided with access for removal of burners; replacement of motors, controls, filters, and other working parts; and for adjustment and lubrication of parts requiring maintenance. [NFPA 54:10.8.3.4]

909.4 Clearance from Combustible Materials. Non-recirculating direct gas-fired industrial air heaters shall be installed with a clearance from combustible materials of not less than that shown on the rating plate and the manufacturer's instructions. [NFPA 54:10.8.4]

909.5 Air Supply. All air to the non-recirculating direct gas-fired industrial air heater shall be ducted directly from outdoors. Where outside-air dampers or closing louvers are used, they shall be verified to be in the open position prior to main burner operation. [NFPA 54:10.8.5]

909.6 Atmospheric Vents, Gas Reliefs or Bleeds. Non-recirculating direct gas-fired industrial air heaters with valve train components equipped with atmospheric vents, gas reliefs, or bleeds shall have their vent lines, gas reliefs, or bleeds lead to a safe point outdoors. Means shall be employed on these lines to prevent water from entering and to prevent blockage from insects and foreign matter. An atmospheric vent line shall not be required to be provided on a valve train component equipped with a listed vent limiter. [NFPA 54:10.8.6]

909.7 Relief Openings. The design of the installation shall include adequate provisions to permit the non-recirculating direct gas-fired industrial air heater to operate at its rated airflow without overpressurizing the space served by the heater by taking into account the structure's designed infiltration rate, properly designed relief openings, or an interlocked powered exhaust system, or a combination of these methods. [NFPA 54:10.8.7]

(A) The structure's designed infiltration rate and the size of relief opening(s) shall be determined by approved engineering methods. [NFPA 54:10.8.7.1]

(B) Louver or counterbalanced gravity damper relief openings shall be permitted. Where motorized dampers or closeable louvers are used, they shall be proved to be in their open position prior to main burner operation. [NFPA 54:10.8.7.2]

909.8 Purging. Inlet ducting, when used, shall be purged with at least four air changes prior to an ignition attempt. [NFPA 54:10.8.8]

910.0 Recirculating Direct Gas-Fired Industrial Air Heaters.

910.1 Application. Direct gas-fired industrial air heaters of the recirculating type shall be designed certified to be in compliance with the Standard for *Recirculating Direct Gas-Fired Industrial Air Heaters,* ANSI Z83.18. Unlisted direct gas-fired industrial air heaters of the recirculating type shall not be installed. [NFPA 54:10.9.1]

910.2 Prohibited Installations.

(A) Recirculating direct gas-fired industrial air heaters shall not serve any area containing sleeping quarters. [NFPA 54:10.9.2.1]

(B) Recirculating direct gas-fired industrial air heaters shall not recirculate room air in buildings that contain flammable solids, liquids, or gases, explosive materials, or substances that can become toxic when exposed to flame or heat. [NFPA 54:10.9.2.2]

910.3 Installation. Installation of direct gas-fired industrial air heaters shall comply with the following requirements: [NFPA 54:10.9.3]

(A) Recirculating direct gas-fired industrial air heaters shall be installed in accordance with the manufacturer's instructions. [NFPA 54:10.9.3(1)]

(B) Recirculating direct gas-fired industrial air heaters shall be installed only in industrial or commercial occupancies. [NFPA 54:10.9.3(2)]

(C) Recirculating direct gas-fired industrial air heaters shall be permitted to provide fresh air ventilation only for the amount that exceeds the minimum ventilation air specified on the heater's rating plate to maintain the combustion level created by the heater in the space being served by the heater below 25 ppm for carbon monoxide, 3 ppm for nitrogen dioxide, and 5000 ppm for carbon dioxide. Where gas-powered fork trucks or other fossil-fueled appliance are utilized in the conditioned space, additional ventilation requirements for the facility must be addressed separately. [NFPA 54:10.9.3(3)]

(D) Recirculating direct gas-fired industrial air heaters shall be provided with access for removal of burners; replacement of motors, controls, filters, and other working parts; and for adjustment and lubrication of parts requiring maintenance. [NFPA 54:10.9.3(4)]

910.4 Clearance from Combustible Materials. Recirculating direct gas-fired industrial air heaters shall be installed with a clearance from combustible materials of not less than that shown on the rating plate and the manufacturer's instructions. [NFPA 54:10.9.4]

910.5 Air Supply. Ventilation air to the recirculating direct gas-fired industrial air heater shall be ducted directly from outdoors. Air to the recirculating direct gas-fired industrial air heater in excess of the

TABLE 9-1
Clearances to Combustible Material for Unlisted Furnaces, Boilers, and Air Conditioners Installed in Rooms That Are Large in Comparison with the Size of Appliances. [NFPA 54: Table 10.2.3(a)]

		Minimum Clearance (in.)					
	Appliance	Above and Sides of Furnace Plenum	Top of Boiler	Jacket Sides and Rear	Front	Draft Hood and Barometric Draft Regulator	Single-Wall Vent Connector
I	Automatically fired, forced air or gravity system, equipped with temperature limit control that cannot be set higher than 250°F (121°C)	6		6	18	6	18
II	Automatically fired heating boilers – steam boilers operating at not over fifteen (15) psi (103 kPa) and hot water boilers operating at 250°F (121°C) or less	6	6	6	18	18	18
III	Central heating boilers and furnaces, other than in I or II	18	18	18	18	18	18
IV	Air-conditioning appliance	18	18	18	18	18	18

Note:
See Section 903.3 for additional requirements for air-conditioning appliances and Section 904.1 for additional requirements for central heating boilers and furnaces.

TABLE 9-2
Free Opening Area of Chimney Damper for Venting Flue Gases from Unlisted Decorative Appliances for Installation in Vented Fireplaces. [NFPA 54: Table 10.6.2.3]

Chimney Height (ft.)	Minimum Permanent Free Opening (in.^{2})*						
	8	13	20	29	39	51	64
	Appliance Input Rating (Btu/h)						
6	7,800	14,000	23,200	34,000	46,400	62,400	80,000
8	8,400	15,200	25,200	37,000	50,400	68,000	86,000
10	9,000	16,800	27,600	40,400	55,800	74,400	96,400
15	9,800	18,200	30,200	44,600	62,400	84,000	108,800
20	10,600	20,200	32,600	50,400	68,400	94,000	122,200
30	11,200	21,600	36,600	55,200	76,800	105,800	138,600

Notes: 1 ft = 0.305 m, 1 in.^{2} = 645 mm^2, 1,000 Btu/h = 0.293 kW
*The first six (6) minimum permanent free openings (eight (8) in.^{2} to fifty-one (51) in.^{2}) correspond approximately to the cross-sectional areas of chimneys having diameters of three (3) in. through eight (8) in., respectively. The sixty four (64) in.^{2} opening corresponds to the cross-sectional area of standard eight (8) in. x eight (8) in. chimney tile.

minimum ventilation air specified on the heater's rating plate shall be taken from the building, ducted directly from outdoors, or a combination of both. Where outside-air dampers or closing louvers are used, they shall be verified to be in the open position prior to main burner operation. [NFPA 54:10.9.5]

910.6 Atmospheric Vents, Gas Reliefs or Bleeds. Recirculating direct gas-fired industrial air heaters with valve train components equipped with atmospheric vents, gas reliefs, or bleeds shall have their vent lines, gas reliefs, or bleeds lead to a safe point outdoors.

Means shall be employed on these lines to prevent water from entering and to prevent blockage from insects and foreign matter. An atmospheric vent line shall not be required to be provided on a valve train component equipped with a listed vent limiter. [NFPA 54:10.9.6]

910.7 Relief Openings. The design of the installation shall include adequate provisions to permit the recirculating direct gas-fired industrial air heater to operate at its rated airflow without overpressurizing the space served by the heater by taking into account the structure's designed infiltration rate, properly designed relief openings or an interlocked powered exhaust system, or a combination of these methods. [NFPA 54:10.9.7]

(A) The structure's designed infiltration rate and the size of relief opening(s) shall be determined by approved engineering methods. [NFPA 54:10.9.7.1]

(B) Louver or counterbalanced gravity relief openings shall be permitted. Where motorized dampers or closeable louvers are used, they shall be proved to be in their open position prior to main burner operation. [NFPA 54:10.9.7.2]

910.8 Purging. Inlet ducting, when used, shall be purged with at least four air changes prior to an ignition attempt. [NFPA 54:10.9.8]

911.0 Duct Furnaces.

911.1 Clearances. The installation of duct furnaces shall comply with the following clearance requirements:

(A) Listed duct furnaces shall be installed with clearances of at least six (6) inches (150 mm) between adjacent walls, ceilings, and floors of combustible material and the furnace draft hood. Furnaces listed for installation at lesser clearances shall be installed in accordance with their listings. In no case shall the clearance be such as to interfere with combustion air and accessibility. (See Sections 304 and 701.0.) [NFPA 54:10.10.1(1)]

(B) Unlisted duct furnaces shall be installed with clearances to combustible material in accordance with the clearances specified for unlisted furnaces and boilers in Table 9-1. Combustible floors under unlisted duct furnaces shall be protected in an approved manner. [NFPA 54:10.10.1(2)]

911.2 Installation of Duct Furnaces. Duct furnaces shall be installed in accordance with the manufacturer's instructions. [NFPA 54-09:10.10.2]

911.3 Access Panels. The ducts connected to duct furnaces shall have removable access panels on both the upstream and downstream sides of the furnace. [NFPA 54:10.10.3]

911.4 Location of Draft Hoods and Controls. The controls, combustion-air inlet, and draft hoods for duct furnaces shall be located outside the ducts. The draft hood shall be located in the same enclosure from which combustion air is taken. [NFPA 54:10.10.4]

911.5 Circulating Air. Where a duct furnace is installed so that supply ducts carry air circulated by the furnace to areas outside the space containing the furnace, the return air shall also be handled by a duct(s) sealed to the furnace casing and terminating outside the space containing the furnace. The duct furnace shall be installed on the positive-pressure side of the circulating air blower. [NFPA 54:10.10.5]

911.6 Duct Furnaces Used with Refrigeration Systems.

(A) A duct furnace shall not be installed in conjunction with a refrigeration coil where circulation of cooled air is provided by the blower.

> **Exception:** Where the blower has sufficient capacity to overcome the external static resistance imposed by the duct system, furnace, and the cooling coil and the air throughput necessary for heating or cooling, whichever is greater. [NFPA 54:10.10.6.1]

(B) Duct furnaces used in conjunction with cooling equipment shall be installed in parallel with or on the upstream side of cooling coils to avoid condensation within heating elements. With a parallel flow arrangement, the dampers or other means used to control the flow of air shall be sufficiently tight to prevent any circulation of cooled air through the unit.

> **Exception:** Where the duct furnace has been specifically listed for downstream installation. [NFPA 54:10.10.6.2]

(C) Where duct furnaces are to be located upstream from cooling units, the cooling unit shall be so designed or equipped as to not develop excessive temperatures or pressures. [NFPA 54:10.10.6.3]

(D) Where a duct furnace is installed downstream of an evaporative cooler or air washer, the heat exchanger shall be constructed of corrosion-resistant materials. Stainless steel, ceramic-coated steel, and an aluminum-coated steel in which the bond between the steel and the aluminum is an iron-aluminum alloy are considered to be corrosion resistant. Air washers operating with chilled water that deliver air below the dew point of the ambient air at the equipment are considered as refrigeration systems. [NFPA 54:9.10.6.4]

911.7 Installation in Commercial Garages and Aircraft Hangars. Duct furnaces installed in garages for more than three motor vehicles or in aircraft hangars shall be of a listed type and shall be installed in accordance with Sections 911.8 and 911.9. [NFPA 54:10.10.7]

911.8 Installation in Commercial Garages.

911.8.1 Parking Structures. Appliances installed in enclosed, basement, and underground parking structures shall be installed in accordance with NFPA 88A, *Standard for Parking Structures*. [NFPA 54:9.1.11.1]

911.8.2 Repair Garages. Appliances installed in repair garages shall be installed in a detached building or room, separated from repair areas by walls, partitions, floors, or floor-ceiling assemblies that are constructed so as to prohibit the transmission of vapors and having a fire resistance rating of not less than one (1) hour, and that have no openings in the wall separating the repair area within eight (8) feet (2,438 mm) of the floor. Wall penetrations shall be firestopped. Air for combustion purposes shall be obtained from outside the building. The heating room shall not be used for the storage of combustible materials.

Exceptions:

(1) Overhead heaters where installed not less than eight (8) feet (2,438 mm) above the floor shall be permitted.

(2) Heating appliances for vehicle repair areas where there is no dispensing or transferring of Class I or Class II flammable or combustible liquids or liquefied petroleum gas shall be installed in accordance with NFPA 30A, *Code for Motor Fuel Dispensing Facilities and Repair Garages*. [NFPA 54:9.1.11.2]

911.9 Installation in Aircraft Hangars. Heaters in aircraft hangars shall be installed in accordance with NFPA 409, *Standard on Aircraft Hangars*. [NFPA 54:9.1.12]

912.0 Floor Furnaces.

912.1 Installation. The installation of floor furnaces shall comply with the following requirements:

(A) Listed floor furnaces shall be installed in accordance with their listing and the manufacturer's instructions.

(B) Unlisted floor furnaces shall not be installed in combustible floors.

(C) Thermostats controlling floor furnaces shall not be located in a room or space that can be separated from the room or space in which the register of the floor furnace is located. [NFPA 54:10.11.1]

912.2 Temperature Limit Controls.

(A) Listed automatically operated floor furnaces shall be equipped with temperature limit controls.

(B) Unlisted automatically operated floor furnaces shall be equipped with a temperature limit control arranged to shut off the flow of gas to the burner in the event the temperature at the warm-air outlet register exceeds 350°F (177°C) above room temperature. [NFPA 54:10.11.2]

912.3 Combustion and Circulating Air. Combustion and circulating air shall be provided in accordance with Section 701.0. [NFPA 54:10.11.3]

912.4 Placement. The following provisions apply to furnaces that serve one story.

(A) Floors. Floor furnaces shall not be installed in the floor of any doorway, stairway landing, aisle, or passageway of any enclosure, public or private, or in an exitway from any such room or space. [NFPA 54:10.11.4(1)]

(B) Walls and Corners. The register of a floor furnace with a horizontal warm-air outlet shall not be placed closer than six (6) inches (150 mm) from the nearest wall. A distance of at least eighteen (18) in. (460 mm) from two adjoining sides of the floor furnace register to walls shall be provided to eliminate the necessity of occupants walking over the warm-air discharge. The remaining sides shall be a minimum of six (6) inches (150 mm) from a wall. Wall-register models shall not be placed closer than six (6) in. (150 mm) to a corner. [NFPA 54:10.11.4(2)]

(C) Draperies. The furnace shall be placed so that a door, drapery, or similar object cannot be nearer than twelve (12) inches (300 mm) to any portion of the register of the furnace. [NFPA 54:10.11.4(3)]

912.5 Bracing. The space provided for the furnace shall be framed with doubled joists and with headers not lighter than the joists. [NFPA 54:10.11.5]

912.6 Support. Means shall be provided to support the furnace when the floor register is removed. [NFPA 54:10.11.6]

912.7 Clearance. The lowest portion of the floor furnace shall have at least a six (6) inch (150 mm) clearance from the general ground level. A reduced clearance to a minimum of two (2) inches (50 mm) is permitted, provided the lower six (6) inches (150 mm) portion of the floor furnace is sealed by the manufacturer to prevent entrance of water. Where these clearances are not present, the ground below and to the sides shall be excavated to form a "basin-like" pit under the furnace so that the required clearance is provided beneath the lowest portion of the furnace. A twelve (12) inch (300 mm) clearance shall be provided on all sides except the control side, that shall have an eighteen (18) inch (460 mm) clearance. [NFPA 54:10.11.7]

912.8 Access. The space in which any floor furnace is installed shall be accessible by an opening in the foundation not less than twenty-four (24) inches x eighteen (18) inches (610 mm x 460 mm) or by a trap door not less than twenty-four (24) inches x twenty-four (24) inches (610 mm x 610 mm) in any cross-section thereof, and a passageway not less than twenty-four (24) inches x eighteen (18) inches (610 mm x 460 mm) in any cross-section thereof. [NFPA 54:10.11.8]

(B) Unlisted domestic hot plates and laundry stoves shall be installed with horizontal clearances to combustible material of not less than twelve (12) inches (300 mm). Combustible surfaces under unlisted domestic hot plates and laundry stoves shall be protected in an approved manner. [NFPA 54:10.14.2]

(C) The vertical distance between tops of all domestic hot plates and laundry stoves and combustible material shall be at least thirty (30) inches (760 mm). [NFPA 54:10.14.3]

916.0 Household Cooking Appliances.

916.1 Floor-Mounted Units.

(A) Clearance from Combustible Material. The clearances specified as follows shall not interfere with combustion air, accessibility for operation, and servicing.

(1) Listed floor-mounted household cooking appliances, where installed on combustible floors, shall be set on their own bases or legs and shall be installed in accordance with their listing and the manufacturer's instructions.

(2) Listed household cooking appliances with listed gas room heater sections shall be installed so that the warm-air discharge side shall have a minimum clearance of eighteen (18) inches (460 mm) from adjacent combustible material. A minimum clearance of thirty-six (36) inches (910 mm) shall be provided between the top of the heater section and the bottom of cabinets.

(3) Listed household cooking appliances that include a solid or liquid fuel-burning section shall be spaced from combustible material and otherwise installed in accordance with the standards applying to the supplementary fuel section of the appliance.

(4) Unlisted floor-mounted household cooking appliances shall be installed with at least six (6) inches (150 mm) clearance at the back and sides to combustible material. Combustible floors under unlisted appliances shall be protected in an approved manner. [NFPA 54:10.15.1.1]

(B) Vertical Clearance Above Cooking Top. Household cooking appliances shall have a vertical clearance above the cooking top of not less than thirty (30) inches (760 mm) to combustible material or metal cabinets. A minimum clearance of twenty-four (24) inches (610 mm) is permitted when one of the following is installed:

(1) The underside of the combustible material or metal cabinet above the cooking top is protected with not less than one-fourth (1/4) inch (6.4 mm) insulating millboard covered with sheet metal not less than 0.0122 inch (0.3 mm) thick.

(2) A metal ventilating hood of sheet metal not less than 0.0122 inch (0.3 mm) thick is installed above the cooking top with a clearance of not less than one-fourth (1/4) inch (6.4 mm) between the hood and the underside of the combustible material or metal cabinet, and the hood is at least as wide as the appliance and is centered over the appliance.

(3) A listed cooking appliance or microwave oven is installed over a listed cooking appliance and will conform to the terms of the upper appliance's listing and the manufacturer's instructions. [NFPA 54:10.15.1.2]

916.2 Built-In Units.

(A) Installation. Listed built-in household cooking appliances shall be installed in accordance with their listing and the manufacturer's instructions. The installation shall not interfere with combustion air, accessibility for operation, and servicing. Unlisted built-in household cooking appliances shall not be installed in or adjacent to combustible material. [NFPA 54-09:10.15.2.1]

(B) Vertical Clearance. Built-in top (or surface) cooking appliances shall have a vertical clearance above the cooking top of not less than thirty (30) inches (760 mm) to combustible material or metal cabinets. A minimum clearance of twenty-four (24) inches (610 mm) is permitted when one of the following is installed:

(1) The underside of the combustible material or metal cabinet above the cooking top is protected with not less than one-fourth (1/4) inch (6.4 mm) insulating millboard covered with sheet metal not less than 0.0122 inch (0.3 mm) thick.

(2) A metal ventilating hood of sheet metal not less than 0.0122 inch (0.3 mm) thick is installed above the cooking top with a clearance of not less than one-fourth (1/4) inch (6.4 mm) between the hood and the underside of the combustible material or metal cabinet, and the hood is at least as wide as the appliance and is centered over the appliance.

(3) A listed cooking appliance or microwave oven is installed over a listed cooking appliance and will conform to the terms of the upper appliance's listing and the manufacturer's instructions. [NFPA 54:10.15.2.2]

(C) Horizontal Clearance. The minimum horizontal distance from the center of the burner head(s) of a listed top (or surface) cooking appliance to vertical combustible walls extending above the top panel shall be not less than that distance specified by the permanent marking on the appliance. [NFPA 54:10.15.2.3]

(D) Level Installation. Built-in household cooking appliances shall be installed so that the cooking top, broiler pan, or oven racks are level. [NFPA 54:10.15.2.4]

917.0 Illuminating Appliances.

917.1 Clearances for Listed Appliances. Listed illuminating appliances shall be installed in accordance with their listing and the manufacturer's instructions. [NFPA 54:10.16.1]

917.2 Clearances for Unlisted Appliances.

(A) Enclosed Type.

(1) Unlisted enclosed illuminating appliances installed outdoors shall be installed with clearances in any direction from combustible material of not less than twelve (12) inches (300 mm).

(2) Unlisted enclosed illuminating appliances installed indoors shall be installed with clearances in any direction from combustible material of not less than eighteen (18) inches (460 mm).

(B) Open-Flame Type.

(1) Unlisted open-flame illuminating appliances installed outdoors shall have clearances from combustible material not less than that specified in Table 9-3. The distance from ground level to the base of the burner shall be at least seven (7) feet (2,133 mm) where installed within two (2) feet (609 mm) of walkways. Lesser clearances shall be permitted to be used where acceptable to the Authority Having Jurisdiction.

(2) Unlisted open-flame illuminating appliances installed outdoors shall be equipped with a limiting orifice or other limiting devices that will maintain a flame height consistent with the clearance from combustible material, as given in Table 9-3.

(3) Appliances designed for flame heights in excess of thirty (30) inch (760 mm) shall be permitted to be installed if acceptable to the Authority Having Jurisdiction. Such appliances shall be equipped with a safety shutoff device or automatic ignition.

(4) Unlisted open-flame illuminating appliances installed indoors shall have clearances from combustible material acceptable to the Authority Having Jurisdiction. [NFPA 54:10.16.2]

TABLE 9-3
Clearances for Unlisted Outdoor Open-Flame Illuminating Appliances
[NFPA 54: Table 10.16.2.2]

Flame Height Above Burner Head (in.)	Minimum Clearance from Combustible Material (ft.)[1]	
	Horizontal	Vertical
12	2	6
18	3	8
24	3	10
30	4	12

Notes:
[1] Measured from the nearest portion of the burner head.
1 in. =25.4 mm, 1 ft. = 0.305 m

917.3 Mounting on Buildings. Illuminating appliances designed for wall or ceiling mounting shall be securely attached to substantial structures in such a manner that they are not dependent on the gas piping for support. [NFPA 54:10.16.3]

917.4 Mounting on Posts. Illuminating appliances designed for post mounting shall be securely and rigidly attached to a post. Posts shall be rigidly mounted. The strength and rigidity of posts greater than three (3) feet (914 mm) in height shall be at least equivalent to that of a two and one-half (2-1/2) inch (64 mm) diameter post constructed of 0.064 in. (1.6 mm) thick steel or a one (1) in. Schedule 40 steel pipe. Posts three (3) feet (914 mm) or less in height shall not be smaller than a three-fourth (3/4) inch Schedule 40 steel pipe. Drain openings shall be provided near the base of posts where there is a possibility of water collecting inside them. [NFPA 54:10.16.4]

917.5 Gas Appliance Pressure Regulators. Where a gas appliance pressure regulator is not supplied with an illuminating appliance and the service line is not equipped with a service pressure regulator, an appliance pressure regulator shall be installed in the line serving one (1) or more illuminating appliances. [NFPA 54:10.16.5]

918.0 Incinerators, Commercial-Industrial.

Commercial-industrial-type incinerators shall be constructed and installed in accordance with NFPA 82, *Standard on Incinerators and Waste and Linen Handling Systems and Equipment.* [NFPA 54:10.17]

919.0 Infrared Heaters.

919.1 Support. Suspended-type infrared heaters shall be fixed in position independent of gas and electric supply lines. Hangers and brackets shall be of noncombustible material. Heaters subject to vibration shall be provided with vibration-isolating hangers. [NFPA 54:10.18.1]

919.2 Suspended Low-Intensity Infrared Tube Heaters. Suspended low-intensity infrared tube heaters shall be connected to the building piping system with a connector listed for the application in accordance with ANSI Z21.24/CGA 6.10, *Connectors for Gas Appliances.*

(A) The connector shall be installed in accordance with the tube heater installation instructions, and shall be in the same room as the appliance.

(B) Only one connector shall be used per appliance. [NFPA 54-09:9.6.1.3]

919.3 Clearance. The installation of infrared heaters shall comply with the following clearance requirements:

(A) Listed heaters shall be installed with clearances from combustible material in accordance with their listing and the manufacturer's instructions.

(B) Unlisted heaters shall be installed in accordance with clearances from combustible material acceptable to the Authority Having Jurisdiction.

(C) In locations used for the storage of combustible materials, signs shall be posted to specify the maximum permissible stacking height to maintain required clearances from the heater to the combustibles. [NFPA 54:10.18.2]

919.4 Combustion and Ventilation Air.

(A) Where unvented infrared heaters are used, natural or mechanical means shall be provided to supply and exhaust at least 4^3 ft./min/1,000 Btu/h (0.38 m^3/min/kW) input of installed heaters.

(B) Exhaust openings for removing flue products shall be above the level of the heaters. [NFPA 54:10.18.3]

919.5 Installation in Commercial Garages and Aircraft Hangars. Overhead heaters installed in garages for more than three motor vehicles or in aircraft hangars shall be of a listed type and shall be installed in accordance with Sections 911.8 and 911.9. [NFPA 54:10.18.4]

920.0 Open-Top Broiler Units.

920.1 Listed Units. Listed open-top broiler units shall be installed in accordance with the manufacturer's installation instructions. [NFPA 54:10.19.1]

920.2 Unlisted Units. Unlisted open-top broiler units shall be installed in accordance with the manufacturer's instructions but shall not be installed in combustible material. [NFPA 54:10.19.2]

920.3 Protection Above Domestic Units. Domestic open-top broiler units shall be provided with a metal ventilating hood not less than 0.0122 in. (0.3 mm) thick with a clearance of not less than 1/4 in. (6 mm) between the hood and the underside of combustible material or metal cabinets. A clearance of at least twenty-four (24) inches (610 mm) shall be maintained between the cooking top and the combustible material or metal cabinet, and the hood shall be at least as wide as the open-top broiler unit and centered over the unit. Listed domestic open-top broiler units incorporating an integral exhaust system and listed for use without a ventilating hood need not be provided with a ventilating hood if installed in accordance with 916.1(B)(1). [NFPA 54:10.19.3]

920.4 Commercial Units. Commercial open-top broiler units shall be provided with ventilation in accordance with Chapter 5. [NFPA 54:10.19.4]

921.0 Outdoor Cooking Appliances.

921.1 Listed Units. Listed outdoor cooking appliances shall be installed in accordance with their listing and the manufacturer's instructions. [NFPA 54:10.20.1]

921.2 Unlisted Units. Unlisted outdoor cooking appliances shall be installed outdoors with clearances to combustible material of not less than thirty-six (36) inches (910 mm) at the sides and back and not less than forty-eight (48) inches (1,220 mm) at the front. In no case shall the appliance be located under overhead combustible construction. [NFPA 54:10.20.2]

922.0 Pool Heaters.

922.1 Location. A pool heater shall be located or protected so as to minimize accidental contact of hot surfaces by persons. [NFPA 54:10.21.1]

922.2 Clearance. The installation of pool heaters shall comply with the following requirements:

(A) In no case shall the clearances be such as to interfere with combustion air, draft hood, or vent terminal clearance and relief, and accessibility for servicing.

(B) A listed pool heater shall be installed in accordance with its listing and the manufacturer's instructions.

(C) An unlisted pool heater shall be installed with a minimum clearance of twelve (12) inches (300 mm) on all sides and the rear. A combustible floor under an unlisted pool heater shall be protected in an approved manner. [NFPA 54:10.21.2]

922.3 Temperature or Pressure-Limiting Devices.

(A) An unlisted pool heater shall be provided with overtemperature protection or overtemperature and overpressure protection by means of an approved device(s).

(B) Where a pool heater is provided with over-temperature protection only and is installed with any device in the discharge line of the heater that can restrict the flow of water from the heater to the pool (such as a check valve, shutoff valve, therapeutic pool valving, or flow nozzles), a pressure-relief valve shall be installed either in the heater or between the heater and the restrictive device. [NFPA 54:10.21.3]

922.4 Bypass Valves. Where an integral bypass system is not provided as a part of the pool heater, a bypass line and valve shall be installed between the inlet and outlet piping for use in adjusting the flow of water through the heater. [NFPA 54:10.21.4]

922.5 Venting. A pool heater listed for outdoor installation shall be installed with the venting means supplied by the manufacturer and in accordance with the manufacturer's instructions. (See Sections 802.2.5, 802.2.6, 802.3.4, and 802.8.) [NFPA 54:10.21.5]

923.0 Refrigerators.

923.1 Clearance. Refrigerators shall be provided with clearances for ventilation at the top and back in accordance with the manufacturer's instructions. Where such instructions are not available, at least two (2) inches (50 mm) shall be provided between the back of the refrigerator and the wall and at least twelve (12) inches (300 mm) above the top. [NFPA 54:10.22.1]

923.2 Venting or Ventilating Kits Approved for Use With a Refrigerator. Where an accessory kit is used for conveying air for burner combustion or unit cooling to the refrigerator from areas outside the room in which it is located, or for conveying combustion products diluted with air containing waste heat from the refrigerator to areas outside the room in which it is located, the kit shall be installed in accordance with the refrigerator manufacturer's instructions. [NFPA 54:10.22.2]

924.0 Room Heaters.

924.1 Prohibited Installations. Unless specifically permitted by the Authority Having Jurisdiction, unvented room heaters shall not be installed as primary heat sources. Unvented room heaters shall not be permitted in spaces that do not have the required volume of indoor air as defined in section 701.2.

924.1.1 Unvented room heaters shall not be installed in bathrooms or bedrooms. [NFPA 54: 10.23.1] This subsection shall not apply to portable oil fired unvented heating appliances used as supplemental heating in Group S, Divisions 3, 4, and 5, and Group U Occupancies, and regulated by the Fire Code.

Exceptions:

(1) Where approved by the Authority Having Jurisdiction, one listed wall-mounted unvented room heater equipped with an oxygen depletion safety shutoff system shall be permitted to be installed in a bathroom provided that the input rating shall not exceed 6,000 Btu/hr. (1,760 W/hr) and combustion and ventilation air is provided as specified in Section 701.2. [NFPA 54:10.23.1]

(2) Where approved by the Authority Having Jurisdiction, one listed wall-mounted unvented room heater equipped with an oxygen depletion safety shutoff system shall be permitted to be installed in a bedroom provided that the input rating shall not exceed 10,000 Btu/hr. (2,930 W/hr.) and combustion and ventilation air is provided as specified in Section 701.2. [NFPA 54:10.23.1]

924.2 Installations in Institutions. Room heaters shall not be installed in the following occupancies: (1) residential board and care; and (2) health care. [NFPA 54:10.23.2]

924.3 Clearance. A room heater shall be placed so as not to cause a hazard to walls, floors, curtains, furniture, doors when open, and so on, and to the free movements of persons within the room. Heaters designed and marked, "For use in noncombustible fireplace only," shall not be installed elsewhere. Listed room heaters shall be installed in accordance with their listings and the manufacturer's instructions. In no case shall the clearances be such as to interfere with combustion air and accessibility. Unlisted room heaters shall be installed with clearances from combustible material not less than the following:

(A) Circulating Type. Room heaters having an outer jacket surrounding the combustion chamber, arranged with openings at top and bottom so that air circulates between the inner and outer jacket, and arranged without openings in the outer jacket to permit direct radiation, shall have clearance at sides and rear of not less than twelve (12) inches (300 mm).

(B) Radiating Type. Room heaters other than those of the circulating type described in Section 924.3(a) shall have clearance at sides and rear of not less than eighteen (18) inches (460 mm), except that heaters that make use of metal, asbestos, or ceramic material to direct radiation to the front of the heater shall have a clearance of thirty-six (36) inches (910 mm) in front and, if constructed with a double back of metal or ceramic, shall be permitted to be installed with a clearance of eighteen (18) inches (460 mm) at sides and twelve (12) inches (300 mm) at rear. Combustible floors under unlisted room heaters shall be protected in an approved manner. [NFPA 54:10.23.3]

924.4 Wall-Type Room Heaters. Wall-type room heaters shall not be installed in or attached to walls of combustible material unless listed for such installation. [NFPA 54:10.23.4]

925.0 Stationary Gas Engines.

The installation of gas engines shall conform with NFPA 37, *Standard for the Installation and Use of Stationary Combustion Engines and Gas Turbines.* [NFPA 54:10.24]

925.1 Stationary gas engines shall not be rigidly connected to the gas supply piping. [NFPA 54-09:10.24.1]

926.0 Gas-Fired Toilets.

926.1 Clearance. A listed gas-fired toilet shall be installed in accordance with its listing and the manufacturer's instructions, provided that the clearance shall in any case be sufficient to afford ready accessibility for use, cleanout, and necessary servicing. [NFPA 54:10.25.1]

926.2 Installation on Combustible Floors. Listed gas-fired toilets installed on combustible floors shall be listed for such installation. [NFPA 54-09:10.25.2]

926.3 Installation. Vents or vent connectors that are capable of being contacted during casual use of the room in which the toilet is installed shall be protected or shielded to prevent such contact. [NFPA 54:10.25.3]

927.0 Unit Heaters.

927.1 Support. Suspended-type unit heaters shall be safely and adequately supported with due consideration given to their weight and vibration characteristics. Hangers and brackets shall be of noncombustible material. [NFPA 54:10.26.1]

927.2 Clearance.

(A) Suspended-Type Unit Heaters. Suspended-type unit heaters shall comply with the following requirements:

(1) A listed unit heater shall be installed with clearances from combustible material of not less than eighteen (18) inches (460 mm) at the sides, twelve (12) inches (300 mm) at the bottom, and six (6) inches (150 mm) above the top where the unit heater has an internal draft hood, or one (1) inch (25 mm) above the top of the sloping side of a vertical draft hood. A unit heater listed for reduced clearances shall be installed in accordance with its listing and the manufacturer's instructions.

(2) Unlisted unit heaters shall be installed with clearances to combustible material of not less than eighteen (18) inches (460 mm).

(3) Clearances for servicing shall be in accordance with the manufacturer's recommendations contained in the installation instructions.

(B) Floor-Mounted-Type Unit Heaters. Floor-mounted-type unit heaters shall comply with the following requirements:

(1) A listed unit heater shall be installed with clearances from combustible material at the back and one side only of not less than six (6) inches (150 mm). Where the flue gases are vented horizontally, the six (6) inches (150 mm) clearance shall be measured from the draft hood or vent instead of the rear wall of the unit heater. A unit heater listed for reduced clearances shall be installed in accordance with its listing and the manufacturer's instructions.

(2) Floor-mounted-type unit heaters installed on combustible floors shall be listed for such installation.

(3) Combustible floors under unlisted floor-mounted unit heaters shall be protected in an approved manner.

(4) Clearances for servicing shall be in, accordance with the manufacturer's recommendations contained in the installation instructions. [NFPA 54:10.26.2]

927.3 Combustion and Circulating Air. Combustion and circulating air shall be provided in accordance with Section 701.0. [NFPA 54:10.26.3]

927.4 Ductwork. A unit heater shall not be attached to a warm-air duct system unless listed and marked for such installation. [NFPA 54:10.26.4]

927.5 Installation in Commercial Garages and Aircraft Hangars. Unit heaters installed in garages for more than three motor vehicles or in aircraft hangars shall be of a listed type and shall be installed in accordance with Sections 911.8 and 911.9. [NFPA 54:10.26.5]

928.0 Wall Furnaces.

928.1 Installation.

(A) Listed wall furnaces shall be installed in accordance with their listings and the manufacturer's instructions. Wall furnaces installed in or attached to combustible material shall be listed for such installation. [NFPA 54:10.27.1.1]

(B) Unlisted wall furnaces shall not be installed in or attached to combustible material. [NFPA 54:10.27.1.2]

(C) Vented wall furnaces connected to a Type B-W gas vent system listed only for a single story shall be installed only in single-story buildings or the top story of multistory buildings. Vented wall furnaces connected to a Type B-W gas vent system listed for installation in multistory buildings shall be permitted to be installed in single-story or multistory buildings. Type B-W gas vents shall be attached directly to a solid header plate that serves as a fire stop at that point and shall be permitted to be an integral part of the vented wall furnace. The stud space in which the vented wall furnace is installed shall be ventilated at the first ceiling level by installation of the ceiling plate spacers furnished with the gas vent. Fire stop spacers shall be installed at each subsequent ceiling or floor level penetrated by the vent. (See Figure 9-2 for Type B-W gas vent installation.) [NFPA 54:10.27.1.3]

(D) Direct-vent wall furnaces shall be installed with the vent-air intake terminal in the outside atmosphere. The thickness of the walls on which the furnace is mounted shall be within the range of wall thickness marked on the furnace and covered in the manufacturer's installation instructions. [NFPA 54:10.27.1.4]

(E) Panels, grilles, and access doors that are required to be removed for normal servicing operations shall not be attached to the building. For additional information on the venting of wall furnaces, consult Section 802.0 in this code. [NFPA 54:10.27.1.5]

928.2 Location. Wall furnaces shall be located so as not to cause a hazard to walls, floors, curtains, furniture, or doors. Wall furnaces installed between bathrooms and adjoining rooms shall not circulate air from bathrooms to other parts of the building. [NFPA 54:10.27.2]

928.3 Combustion and Circulating Air. Combustion and circulating air shall be provided in accordance with Section 701.0. [NFPA 54:10.27.3]

929.0 Appliances for Installation in Manufactured Housing.

Appliances installed in manufactured housing after the initial sale shall be listed for installation in manufactured housing, or approved, and shall be installed in accordance with the requirements of this code and the manufacturer's installation instructions. Appliances installed in the living space of manufactured housing shall be in accordance with the requirements of Section 701.0. [NFPA 54:10.30]

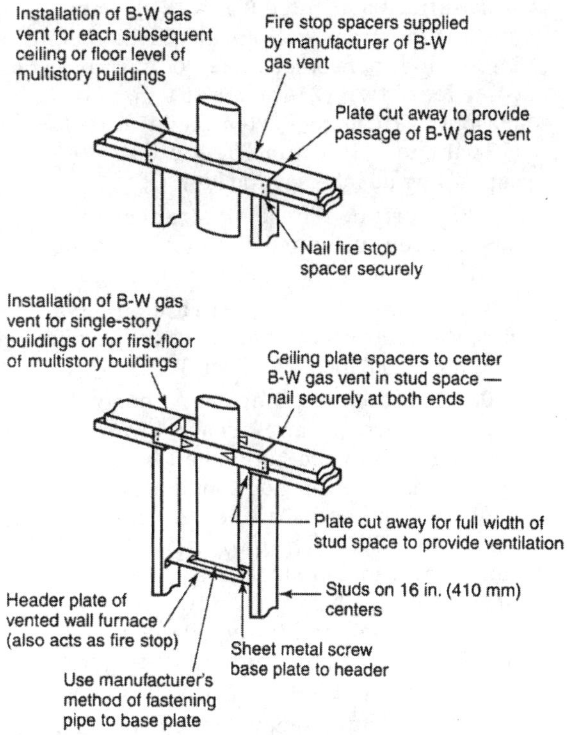

FIGURE 9-2 Installation of Type B-W Gas Vents for Vented Wall Furnaces.

930.0 Small Ceramic Kilns.

930.1 General. The provisions of this section apply to kilns used for ceramics that have a maximum interior volume of twenty (20) cubic feet (0.566 m^3) and are used for hobby or noncommercial purposes.

930.2 Installation. Kilns shall be installed in accordance with the manufacturer's instructions and the provisions of this code.

930.3 Fuel-Gas Controls. Fuel-gas controls shall comply with Sections 1312.1.4 and 306.0. Standing pilots shall not be used with gas-fired kilns.

930.4 Electrical Equipment. All electrical equipment used as part of, or in connection with, the installation of a kiln shall comply with the requirements set forth in the Electrical Code.

930.5 Installations Inside Buildings. In addition to other requirements specified in this section, interior installation shall meet the following requirements:

930.5.1 Kiln Clearances. The sides and tops of kilns shall be located a minimum of eighteen (18) inches (457 mm) from any noncombustible wall surface and three (3) feet (914 mm) from any combustible wall surface. Kilns shall be installed on noncombustible flooring consisting of at least two (2) inches (51 mm) of solid masonry or concrete extending at least twelve (12) inches (300 mm) beyond the base or supporting members of the kiln.

Exception: These clearances may be reduced, provided the kiln is installed in accordance with its listing.

In no case shall the clearance on the gas or electrical control side of a kiln be reduced to less than thirty (30) inches (762 mm).

930.5.2 Hoods. A canopy-type hood shall be installed directly above each kiln. The face opening area of the hood shall be equal to or greater than the top horizontal surface area of the kiln. The hood shall be constructed of not less than 0.024 inch (0.61 mm) (No. 24 U.S. gauge) galvanized steel or equivalent and be supported at a height of between twelve (12) inches (300 mm) and thirty (30) inches (76 mm) above the kiln by noncombustible supports.

Exception: Electric kilns installed with listed exhaust blowers may be used when marked as being suitable for the kiln and installed in accordance with manufacturer's instructions.

930.5.3 Gravity Ventilation Ducts. Each hood shall be connected to a gravity ventilation duct extending in a vertical direction to outside the building. This duct shall be of the same construction as the hood and shall have a minimum cross-sectional area of not less than one-fifteenth of the face opening area of the hood. The duct shall terminate a minimum of twelve (12) inches (300 mm) above any portion of a building within four (4) feet (1,219 mm) and terminate no less than four (4) feet (1,219 mm) from any openable windows or other openings into the building or adjacent property line. The duct opening to the outside shall be shielded, without reduction of duct area, to prevent entrance of rain into the duct. The duct shall be supported at each section by noncombustible supports.

930.5.4 Makeup Air. Provisions shall be made for air to enter the room in which a kiln is installed at a rate at least equal to the air being removed through the kiln hood.

930.5.5 Hood and Duct Clearances. Every hood and duct serving a fuel-burning kiln shall have a clearance from combustible construction of at least eighteen (18) inches (457 mm). This clearance may be reduced in accordance with Table 3-2.

930.6 Exterior Installations. Kilns shall be installed with minimum clearances as specified in Section 930.5.1. Wherever a kiln is located under a roofed area and is partially enclosed by more than two vertical wall surfaces, a hood and gravity ventilation duct shall be installed to comply with Sections 930.5.2, 930.5.3, and 930.5.5.

931.0 Compressed Natural Gas (CNG) Vehicular Fuel Systems.

The installation of compressed natural gas (CNG) fueling (dispensing) systems shall conform to NFPA 52, *Compressed Natural Gas (CNG) Vehicular Fuel Systems Code*. [NFPA 54-09:10.29]

932.0 Outdoor Open Flame Decorative Appliances.

932.1 Permanently fixed in place outdoor open flame decorative appliances shall be installed in accordance with the following. [NFPA 54-09:10.32]

932.1.1 Listed Units. Listed outdoor open flame decorative appliances shall be installed in accordance with the manufacturer's installation instructions. [NFPA 54-09:10.32.1]

932.1.2 Unlisted Units. Unlisted outdoor open flame decorative appliances shall be installed outdoors in accordance with the manufacturer's installation instructions and with clearances to combustible material of not less than thirty-six (36) inches (910 mm) from the sides. In no case shall the appliance be located under overhead combustible construction. [NFPA 54-09:10.32.2]

932.1.3 Connection to the Piping System. The connection to the gas piping system shall be in accordance with 1313.1 (1), (2), or (3) [NFPA 54-09:10.32.3]

933.0 Appliances for Installation in Manufactured Housing.

Appliances installed in manufactured housing after the initial sale shall be listed for installation in manufactured housing, or approved, and shall be installed in accordance with the requirements of this code and the manufacturer's installation instructions. Appliances installed in the living space of manufactured housing shall be in accordance with the requirements of Section 303.0 [NFPA 54-09:10.30]

934.0 Fuel Cell Power Plants.

Fuel cell power plants with a power output of less than 50 kW shall be listed and installed in accordance with the manufacturer's instructions. Fuel cell power plants with a power output of greater than 50 kW shall be installed in accordance with NFPA 853, *Standard for the Installation of Stationary Fuel Cell Power Systems*. [NFPA 54-09:10.31]

CHAPTER 10

STEAM AND HOT WATER BOILERS

1001.0 Scope.

The requirements of this chapter shall apply to the construction, installation, operation, repair, and alteration of all boilers and pressure vessels.

Exceptions:

(1) Listed and approved potable water heaters with a nominal capacity of one hundred and twenty (120) gallons (454 L) or less having a heat input of 200,000 Btu/h (58.62 kW) or less used for hot water supply at a pressure of one-hundred and sixty (160) psi (1,103 kPa) or less and at temperatures not exceeding 210°F (99°C), as regulated by the Plumbing Code.

(2) Pressure vessels used for unheated water supply, including those containing air that serves only as a cushion and is compressed by the introduction of water and tanks connected to sprinkler systems.

(3) Portable unfired pressure vessels and Interstate Commerce Commission (I.C.C.) containers.

(4) Containers for liquefied petroleum gases, bulk oxygen, and medical gas that are regulated by the Fire Code.

(5) Unfired pressure vessels in Groups B, F, H, M, R, S, and U Occupancies having a volume of five (5) cubic feet (0.14 m^3) or less and operated at pressures not exceeding two hundred and fifty (250) psi (1,724 kPa).

(6) Pressure vessels used in refrigeration systems that are regulated by Chapter 11 of this code.

(7) Pressure tanks used in conjunction with coaxial cables, telephone cables, power cables, and other similar humidity control systems.

(8) Any boiler or pressure vessel subject to regular inspection by federal inspectors or licensed by federal authorities.

(9) Boilers within the scope of NFPA 85, including associated fuel systems shall be designed and installed in accordance with NFPA 85, *Boiler and Combustion Systems Hazards Code.*

1002.0 Workmanship.

All equipment, appurtenances, devices, and piping shall be installed in a workmanlike manner in conformity with the provisions and intent of this chapter.

1003.0 Definitions.

Certain words and terms used in this chapter, unless clearly inconsistent with their context, shall mean as follows:

1003.1 **–A–**

AUTOMATIC BOILER – When applied to any class of boiler defined below, such boiler shall be equipped with certain controls and limit devices, as specified in Section 1009.0 and Table 10-3.

1003.2 **–B–**

BOILER – A closed vessel used for heating water or liquid, or for generating steam or vapor by direct application of heat from combustible fuels or electricity.

BURNER, AUTOMATIC BOILER – A device to convey fuel into the combustion chamber in proximity to its combustion-air supply so as to permit a stable controlled heat release compatible with the burner design and that is equipped with an ignition system to reliably ignite the entire heat release surface of the burner assembly.

1003.3 **–C–**

CONTINUOUS PILOT – A pilot that burns without turndown throughout the entire period that the boiler is in service, whether or not the main burner is firing.

1003.4 **–D–**

No definitions.

1003.5 **–E–**

No definitions.

1003.6 **–F–**

No definitions.

1003.7 **–G–**

No definitions.

1003.8 **–H–**

HOT WATER-HEATING BOILER – A boiler having a volume exceeding 120 gallons (454 L), a heat input exceeding 200,000 Btu/h (58.62 kW), or an operating temperature exceeding 210°F (99°C) that provides hot water to be used externally to itself.

1003.9 –I–

INTERLOCK – A device that senses a limit or off-limit condition or improper sequence of events and shuts down the offending or related piece of equipment or prevents proceeding in an improper sequence in order to prevent a hazardous condition from developing.

INTERMITTENT PILOT – A pilot that burns during light-off and while the main burner is firing, and that is shut off with the main burner.

INTERRUPTED PILOT – A pilot that burns during light-off and that is shut off during normal operation of the main burner.

1003.10 –J–

No definitions.

1003.11 –K–

No definitions.

1003.12 –L–

LOW-PRESSURE HOT WATER-HEATING BOILER – A boiler in which water is heated for the purpose of supplying heat at pressures not exceeding one hundred and sixty (160) psi (1,102 kPa) and temperatures not exceeding 250°F (121°C).

1003.13 –M–

MINIATURE BOILER – A power boiler having an internal shell diameter of sixteen (16) inches (406 mm) or less, a gross volume of five (5) cubic feet (142 L) or less, a heating surface of twenty (20) square feet (1.86 m^2) or less (not applicable to electric boilers), and not exceeding 100 psi (689 kPa).

1003.14 –N–

No definitions.

1003.15 –P–

PACKAGE BOILER – Any class of boiler defined herein and shall be a boiler equipped and shipped complete with fuel-burning equipment, automatic controls and accessories, and mechanical draft equipment, if used.

PILOT – A burner smaller than the main burner that is ignited by a spark or other independent and stable ignition source, and that provides ignition energy required to immediately light off the main burner.

POWER BOILER – A boiler in which steam is generated at pressures exceeding fifteen (15) psi (103 kPa).

POWER HOT WATER BOILER (HIGH TEMPERATURE WATER BOILER) – A boiler used for heating water or liquid to a pressure exceeding 160 psi (1,103 kPa) or to a temperature exceeding 250°F (121°C).

PRESSURE VESSEL (unfired) – A closed container, having a nominal internal diameter exceeding six (6) inches (152.4 mm) and a volume exceeding one and one-half (1-1/2) cubic feet (42 L), for liquids, gases, vapors subjected to pressures exceeding fifteen (15) psi (103 kPa), or steam under any pressure.

PURGE – The acceptable method of scavenging the combustion chamber, boiler passes, and breeching to remove all combustible gases.

1003.16 –Q–

No definitions.

1003.17 –R–

No definitions.

1003.18 –S–

STEAM-HEATING BOILER – A boiler operated at pressures not exceeding fifteen (15) psi (103 kPa) for steam.

1003.19 –T–

No definitions.

1003.20 –U–

No definitions.

1003.21 –V–

No definitions.

1003.22 –W–

No definitions.

1003.23 –X–

No definitions.

1003.24 –Y–

No definitions.

1003.25 –Z–

No definitions.

1004.0 Detailed Requirements.

1004.1 Safety Requirements. The construction of boilers and pressure vessels and the installation thereof shall conform to minimum requirements for safety from structural and mechanical failure and excessive pressures, as established by the Authority Having Jurisdiction in accordance with nationally recognized standards.

1004.2 Controls. Required electrical, mechanical, safety, and operating controls shall carry approval of an approved testing agency or be accepted by the Authority Having Jurisdiction. Electrical controls shall be of such design and construction as to be suitable for installation in the environment in which they are located.

1004.3 Gauges. All steam boilers shall be provided with a pressure gauge and a water level glass. All water boilers shall be provided with a pressure gauge and a temperature indicator.

1004.4 Stack Dampers. Stack dampers on boilers fired with oil or solid fuel shall not close off more than 80 percent of the stack area when closed, except on automatic boilers with prepurge, automatic draft control, and interlock. Operative dampers shall not be placed within any stack, flue, or vent of a gas-fired boiler, except on an automatic boiler with prepurge, automatic draft control, and interlock.

1004.5 Welding. Welding on pressure vessels shall be done by certified welders in conformity with nationally recognized standards.

1005.0 Expansion Tanks.

1005.1 General. All hot-water-heating systems shall be provided with an air expansion tank securely fastened to the structure. Supports shall be adequate to carry twice the weight of the tank filled with water without placing any strain on connecting piping.

All hot-water-heating systems incorporating hot water tanks or fluid relief columns shall be so installed as to prevent freezing under normal operating conditions.

1005.2 Systems with Open Expansion Tanks. Systems equipped with an open expansion tank to satisfy thermal water expansion shall be provided with an indoor overflow from the upper portion of the expansion tank in addition to an open vent. The indoor overflow shall be carried within the building to a suitable plumbing fixture or to the basement.

1005.3 Closed-Type Systems. Systems of the closed type shall have an airtight tank or other suitable air cushion that will be consistent with the volume and capacity of the system, and shall be suitably designed for a hydrostatic test pressure of two and one-half (2-1/2) times the allowable working pressure of the system. Expansion tanks for systems designed to operate at or above thirty (30) psig shall be constructed in accordance with nationally recognized standards approved by the Authority Having Jurisdiction. Provisions shall be made for draining the tank without emptying the system, except for pressurized tanks.

1005.4 Minimum Capacity of Closed-Type Tank. The minimum capacity of the closed-type expansion tank may be determined from Tables 10-1 and 10-2 or from the following formula:

$$Vt = \frac{(0.00041t - 0.0466)\, Vs}{\frac{(Pa}{(Pf} - \frac{Pa)}{Po)}}$$

WHERE:

Vt = minimum volume of expansion tank, gallons.

Vs = volume of system, not including expansion tank, gallons.

t = average operating temperature, °F.

Pa = atmospheric pressure, feet H_2O absolute.

Pf = fill pressure, feet H_2O absolute.

Po = maximum operating pressure, feet H_2O absolute.

TABLE 10-1
Expansion Tank Capacities for Gravity Hot Water Systems

Based on a two-pipe system with an average operating water temperature of 170°F (77°C), using cast-iron column radiation with a heat emission rate of 150 Btu per hour per square foot (473 W/m^2) equivalent direct radiation.

Square Feet of Installed Expansion Direct Radiation[1] (x 0.0929 for m^2)	Tank Capacity, Gallons (x 3.785 for L)
Up to 350	18
Up to 450	21
Up to 650	24
Up to 900	30
Up to 1,100	35
Up to 1,400	40
Up to 1,600	2 to 30
Up to 1,800	2 to 30
Up to 2,000	2 to 35
Up to 2,400	2 to 40

Note:

[1] For systems with more than 2,400 square feet (223 m^2) of installed equivalent direct water radiation, the required capacity of the cushion tank shall be increased on the basis of one (1) gallon (3.785 L) tank capacity per thirty-three (33) square feet (3.07 m^2) of additional equivalent direct radiation.

TABLE 10-2
Expansion Tank Capacities for Forced Hot Water Systems

Based on an average operating water temperature of 195°F (90.6°C), a fill pressure of twelve (12) psig (82.7 kPa), and a maximum operating pressure of thirty (30) psig (206.8 kPa).

System Volume,[1] Gallons (x 3.785 for L)	Tank Capacity, Gallons (x 3.785 for L)
100	15
200	30
300	45
400	60
500	75
1,000	150
2,000	300

[1] Includes volume of water in boiler, radiation, and piping, not including expansion tank.

1006.0 Safety or Relief Valve Discharge.

The discharge from relief valves shall be piped to within eighteen (18) inches (457 mm) of the floor or to an open receptacle, and when the operating temperature is in excess of 212°F (100°C), shall be equipped with a splash shield or centrifugal separator. When the discharge from safety valves would result in a hazardous discharge of steam inside the boiler room, such discharge shall be extended outside the boiler room. No valve of any description shall be placed between the safety or relief valve and the boiler, nor on the discharge pipe between the safety valve and the atmosphere. Discharges from relief valves on industrial boilers shall be discharged to an approved location.

1007.0 Shutoff Valves.

An approved manual shutoff valve shall be installed upstream of all control devices on the main burner of a gas-fired boiler. The takeoff point for the gas supply to the pilot shall be upstream of the gas shutoff valve of the main burner and shall be valved separately. A union or other approved means of disconnect shall be provided immediately downstream of these shutoff valves.

1008.0 Gas-Pressure Regulators.

An approved gas-pressure regulator shall be installed on gas-fired boilers where the gas supply pressure is higher than that at which the main burner is designed to operate. A separate approved gas-pressure regulator shall be installed to regulate the gas pressure to the pilot or pilots.

1009.0 Low-Water Cutoff.

Hot water heating boilers, other than manually fired, shall be equipped with a low-water cutoff, except that a coil-type boiler or a water-tube boiler that requires forced circulation to prevent overheating of the coils or tubes shall have a flow-sensing device installed in the outlet piping in lieu of the low-water cutoff. The required low-water cutoff or flow switch, as applicable, shall be mounted so as to prevent damage to the boiler and to permit testing of the fuel-supply cutoff without draining the heating system, except that such boilers used in Group R Occupancies of less than six dwelling units and Group U Occupancies need not be equipped with the low-water cutoff or flow switch.

1010.0 Combustion Regulators – Safety Valves.

The following requirements shall be retroactive:

1010.1 Every hot water-heating boiler, other than manually fired, shall be equipped with two temperature combustion regulators in series. Every steam-heating boiler, other than manually fired, shall be equipped with a pressure combustion regulator and a low-water cutoff. (See Section 1011.0.)

1010.2 Boilers and pressure vessels shall be provided with the required number, size, and capacity of safety or relief valves to ensure positive relief of overpressure in accordance with nationally recognized standards, as applicable. Valves so employed shall be constructed, sealed, and installed in accordance with nationally recognized standards, as applicable.

1011.0 Automatic Boilers.

Automatic boilers shall be equipped with controls and limit devices as set forth in Table 10-3. Automatic boilers shall also be equipped with the following gauges, as applicable: oil temperature, oil suction pressure, high and low gas pressure, stack temperature, and windbox pressure.

Except as otherwise specified, gas-fired boilers exceeding 400,000 Btu/h (117 kW) input shall conform to nationally recognized standards approved by the Authority Having Jurisdiction.

The Authority Having Jurisdiction shall have the authority to approve solid-fuel-fired boilers that can meet the safety requirements for automatic gas- or oil-fired boilers.

1012.0 Clearance for Access.

When boilers are installed or replaced, clearance shall be provided to allow access for inspection, maintenance, and repair. Passageways around all sides of boilers shall have an unobstructed width of not less than eighteen (18) inches (457 mm). Clearance for repair and cleaning may be provided through a door or access panel into another area, provided the opening is of sufficient size.

Exception: Subject to the approval of the Authority Having Jurisdiction, boilers may be installed with a side clearance of less than eighteen (18) inches (457 mm), provided that the lesser clearance does not inhibit inspection, maintenance, or repair.

Power boilers having a steam-generating capacity in excess of 5,000 pounds per hour (2,268 kg/h) or having a heating surface in excess of 1,000 square feet (93 m^2) or input in excess of 5,000,000 Btu/h (1,465 kW) shall have a minimum clearance of seven (7) feet (2,134 mm) from the top of the boiler to the ceiling.

Steam-heating boilers and hot-water-heating boilers that exceed one of the following limits—5,000,000 Btu/h input (1,465 kW); 5,000-pound-steam-per-hour (2,268 kg/h) capacity; or 1,000-square-foot (93 m^2) heating surface—and power boilers that do not exceed one of the following limits—5,000,000 Btu/h input (1,465 kW); 5,000-pound steam per hour (2,268 kg/h) capacity; or 1,000- square-foot (1,465 kW) heating surface—and all boilers with manholes on top of the boiler, except those described in the second and fourth paragraphs, shall have a minimum clearance of three (3) feet (914 mm) from the top of the boiler to the ceiling.

Package boilers, steam-heating boilers, and hot-water-heating boilers with no manhole on top of the shell and not exceeding one of the above limits shall have a minimum clearance of two (2) feet (610 mm) from the ceiling.

1013.0 Boiler Rooms and Enclosures.

Boiler rooms and enclosures and access thereto shall comply with Section 904.0 of this code and the Building Code.

1014.0 Mounting.

Equipment shall be set or mounted on a level base capable of supporting and distributing the weight contained thereon.

Boilers, tanks, and equipment shall be securely anchored to the structure.

Equipment requiring vibration isolation shall be installed as designed by a registered engineer and approved by the Authority Having Jurisdiction.

1015.0 Floors.

Boilers shall be mounted on floors of noncombustible construction unless listed for mounting on combustible flooring.

1016.0 Chimneys or Vents.

Boilers shall be connected to a chimney or vent, as provided for other fuel-burning equipment in Chapter 8 of this code.

1017.0 Drainage.

For heating or hot-water-supply boiler applications, the boiler room shall be equipped with a floor drain or other means suitable for disposing of the accumulation of liquid wastes incident to cleaning, recharging, and routine maintenance. No steam pipe shall be directly connected to any part of a plumbing or drainage system, nor shall any water having a temperature above 140°F (60°C) be discharged under pressure directly into any part of a drainage system. Pipes from boilers shall discharge by means of indirect waste piping, as determined by the Authority Having Jurisdiction or the boiler manufacturer's recommendations.

1018.0 Fuel Piping, Tanks, and Valves.

Tanks, piping, and valves for oil-burning appliances shall be installed in accordance with NFPA 31, *Standard for the Installation of Oil-Burning Equipment*. That portion of the oil-burning system supplied on boilers and covered within the scope of NFPA 85 shall be installed in accordance with NFPA 85.

1019.0 Air for Combustion and Ventilation.

Air for combustion and ventilation shall be provided in accordance with Chapter 7 of this code.

1020.0 Operating Adjustments and Instructions.

Hot water boiler installations, upon completion, shall have controls set, adjusted, and tested by the installing contractor. A complete control diagram of a permanent legible type, together with complete boiler operating instructions, shall be furnished by the installer for each installation.

1021.0 Inspections and Tests.

An installation for which a permit is required shall not be put into service until it has been inspected and approved by the Authority Having Jurisdiction.

It shall be the duty of the owner or his authorized representative to notify the Authority Having

Jurisdiction that the installation is ready for inspection and test. It also shall be the duty of the owner or his authorized representative to post in a conspicuous position on the installation a notice in substantially the following form: "Warning! This installation has not been inspected and approved by the Authority Having Jurisdiction and shall not be covered or concealed until so inspected and approved," and it shall be unlawful for anyone other than the Authority Having Jurisdiction to remove such notice. The Authority Having Jurisdiction shall require such tests as it deems necessary to determine that the installation complies with the provision of this section. Such tests shall be made by the owner or his authorized representative in the presence of the Authority Having Jurisdiction.

> **Exception:** On installations designed and supervised by a registered professional engineer, the Authority Having Jurisdiction shall have the authority to permit inspection and testing by such engineer.
>
> When the owner or his authorized representative requests inspection of a boiler prior to its installation, the Authority Having Jurisdiction shall make such inspection.

1022.0 Operating Permit.

It shall be unlawful to operate a boiler or pressure vessel without first obtaining a valid operating permit to do so from the Authority Having Jurisdiction. Such permit shall be displayed in a conspicuous place adjacent to the boiler or vessel. The operating permit shall not be issued until the equipment has been inspected and approved by the Authority Having Jurisdiction.

> **Exception:** The operation only of steam-heating boilers, low-pressure hot-water-heating boilers, hot water supply boilers, and pressure vessels in Group R Occupancies of less than six dwelling units and in Group U Occupancies.

1023.0 Maintenance Inspection.

The Authority Having Jurisdiction shall inspect all boilers and pressure vessels operated under permit at such intervals as deemed necessary, but not less frequently than noted below:

1023.1 Power boilers and miniature boilers shall be inspected externally annually. Where construction and operating conditions permit, they shall, in addition, be subject to inspection internally annually.

1023.2 Steam-heating boilers and hot-water-heating boilers shall be inspected externally annually. Where construction and operating conditions permit, they shall, in addition, be subject to inspection internally annually.

1023.3 Automatic steam-heating boilers shall be inspected externally biennially. Where construction and operating conditions permit, they shall, in addition, be subject to inspection internally biennially.

1023.4 Unfired pressure vessels shall be inspected externally biennially. When subject to corrosion and construction permits, they shall, in addition, be subject to inspection internally biennially.

Inspection of boilers and pressure vessels covered by insurance may be made by employees of the insuring company holding commissions from the National Board of Boiler and Pressure Vessel Inspectors, subject to approval of the Authority Having Jurisdiction. Approved insuring company inspectors shall make reports on prescribed forms on inspections authorized by the Authority Having Jurisdiction. The reports shall be filed in the Authority Having Jurisdiction office. Company inspectors shall notify the Authority Having Jurisdiction of suspension of insurance because of dangerous conditions, new insurance in effect, and discontinuance of insurance coverage.

1024.0 Operation and Maintenance of Boilers and Pressure Vessels.

Boilers and pressure vessels shall be operated and maintained in conformity with requirements for adequate protection of the public established by the Authority Having Jurisdiction in accordance with nationally recognized standards.

The Authority Having Jurisdiction shall notify the owner or authorized representative of defects or deficiencies that shall be promptly and properly corrected. If such corrections are not made, or if the operation of the boiler or pressure vessel is deemed unsafe by the Authority Having Jurisdiction, they shall have the authority to revoke the permit to operate the boiler or pressure vessel. If the operation of a boiler or pressure vessel is deemed by the Authority Having Jurisdiction to constitute an immediate danger, the pressure on such boiler or pressure vessel may be relieved at the owner's cost and the boiler or pressure vessel shall not thereafter be operated without approval of the Authority Having Jurisdiction.

TABLE 10-3
Controls and Limit Devices for Automatic Boilers

Boiler Group	Fuel	Fuel Input Range[1] (Inclusive), Btu/h	Type of Pilot[2]	Safety Control Timing (Nominal Maximum Time in Seconds)				Assured Fuel Supply Control[4]	Assured Air Supply Control[5]	Low Fire Start Up Control[6]	Pre-purging Control[7]	Hot Water Temperature and Low Water Limit Controls[8]	Steam Pressure and Low Water Limit Controls[9]	Approved Fuel Shutoff[10]	Control and Limit Device System Design[11]
				Trial for Pilot	Trial for Main Burner Frame: Direct Electric Ignition	Trial for Main Burner Frame: Flame Pilot	Main Burner Flame Failure[3]								
A	Gas	0-400,000	Any type	90	Not Required	90	90	Not Required	Required	Not Required	Not Required	Required	Required	Not Required	Required
B	Gas	400,001 - 2,500,000	Interrupted or Intermittent	15	15	15	2-4	Not Required	Required	Not Required	Not Required	Required	Required	Not Required	Required
C	Gas	2,500,001 - 5,000,000	Interrupted or Intermittent	15	15	15	2-4	Required	Required	Required	Required	Required	Required	Required	Required
D	Gas	Over 5,000,000	Interrupted	15	15	15	2-4	Required	Required	Required	Required	Required	Required	Required	Required
E	Oil	0-400,000	Any type	Not Required	90	90	90	Not Required	Required	Not Required	Not Required	Required	Required	Not Required	Required
F	Oil	400,001 - 1,000,000	Interrupted	Not Required	30	30	2-4	Required	Required	Not Required	Not Required	Required	Required	Not Required	Required
G	Oil	1,000,001 - 3,000,000	Interrupted	Not Required	15	15	2-4	Required	Required	Not Required	Not Required	Required	Required	Not Required	Required
H	Oil	Over 3,000,000	Interrupted	15	15	60	2-4	Required	Required	Required	Required	Required	Required	Required	Required
K	Electric	All	Not required	Not Required	Not Required	Not Required	Not Required	Not Required	Not Required	Not Required	Not Required	Required	Required	Not Required	Required
L	Gas, Oil and/or Coal	12,500,00 or more	Any	10 sec Per NFPA 85	Per NFPA 85	Per NFPA 85	Per NFPA 85	Per NFPA 85	Per NFPA 85	Per NFPA 85	Per NFPA 85	Per ASME Power Boiler Code, Section 1 and NFPA 85	Per ASME Power Boiler Code, Section 1 and NFPA 85	Per NFPA 85	Per NFPA 85
M	Heat Recovery Steam Generator	Any	None	Per NFPA 85	Per NFPA 85	Per NFPA 85	Per NFPA 85	Per NFPA 85	Per NFPA 85	Per NFPA 85	Per NFPA 85	Per ASME Boiler & Pressure Code & NFPA 85	Per ASME Boiler & Pressure Code & NFPA 85	Per NFPA 85	Per NFPA 85

FOOTNOTES FOR TABLE 10-3 (continued)

1 Fuel input shall be determined by one of the following:

(a) The maximum burner input as shown on the burner nameplate or as otherwise identified by the manufacturer.

(b) The nominal boiler rating, as determined by the building official, plus twenty-five percent (25%).

2 Automatic boilers shall have one flame failure device on each burner, which shall prove the presence of a suitable ignition source at the point where it will reliably ignite the main burner, except that boiler groups A, B, E, F, and G, which are equipped with direct electric ignition, shall monitor the main burner, and all boiler groups using interrupted pilots shall monitor only the main burner after the prescribed limited trial and ignition periods. Boiler group A, equipped with continuous pilot, shall accomplish 100 percent shutoff within ninety (90) seconds upon pilot flame failure. The use of intermittent pilots in boiler group C is limited to approved burner units.

3 In boiler groups B, C, and D a 90-second main burner flame failure limit may apply if continuous pilots are provided on manufacturer assembled boiler-burner units that have been approved by an approved testing agency as complying with nationally recognized standards approved by the building official. Boiler groups F and G equipped to re-energize their ignition systems within eight-tenth (0.8) second after main burner flame failure will be permitted thirty (30) seconds for group F or fifteen (15) seconds for group G to re-establish their main burner flames.

4 Boiler groups C and D shall have controls interlocked to accomplish a nonrecycling fuel shutoff upon high or low gas pressure, and boiler groups F, G, and H using steam or air for fuel atomization shall have controls interlocked to accomplish a nonrecycling fuel shutoff upon low atomizing steam or air pressure. Boiler groups F, G, and H equipped with a preheated oil system shall have controls interlocked to provide fuel shutoff upon low oil temperature.

5 Automatic boilers shall have controls interlocked to shut off the fuel supply in the event of draft failure if forced or induced draft fans are used or, in the event of low combustion airflow, if a gas power burner is used. Where a single motor directly driving both the fan and the oil pump is used, a separate control is not required.

6 Boiler groups C, D, and H, when firing in excess of four-hundred thousand (400,000) Btu per combustion chamber, shall be provided with low fire start of its main burner system to permit smooth light-off. This will normally be a rate of approximately one-third (1/3) of its maximum firing rate.

7 Boiler groups C, D, and H shall not permit pilot or main burner trial for ignition operation before a purging operation of sufficient duration to permit a minimum of four complete air changes through the furnace, including combustion chamber and the boiler passes. Where this is not readily determinable, five complete air changes of the furnace, including combustion chamber up to the first pass, will be considered equivalent. An atmospheric gas burner with no mechanical means of creating air movement or an oil burner that obtains two-thirds or more of the air required for combustion without mechanical means of creating air movement shall not require purge by means of four air changes, so long as its secondary air openings are not provided with means of closing. If such burners have means of closing secondary air openings, a time delay must be provided that puts these closures in a normally open position for four minutes before an attempt for ignition. An installation with a trapped combustion chamber shall, in every case, be provided with a mechanical means of creating air movement for purging.

8 Every automatic hot-water-heating boiler, low-pressure hot-water-heating boiler, and power hot water boiler shall be equipped with two high-temperature limit controls with a manual reset on the control, with the higher setting interlocked to shut off the main fuel supply, except that manual reset on the high-temperature limit control shall not be required on any automatic package boiler not exceeding four-hundred thousand (400,000) Btu/h input and that has been approved by an approved testing agency. Every automatic hot-water heating, power boiler, and package hot-water supply boiler shall be equipped with one low-water level limit control with a manual reset interlocked to shut off the fuel supply, so installed as to prevent damage to the boiler and to permit testing of the control without draining the heating system, except on boilers used in Group R Occupancies of less than six units and in Group U Occupancies and further, except that the low-water level limit control is not required on package hot-water supply boilers approved by a nationally recognized testing agency. However, a low-water flow limit control installed in the circulating water line may be used instead of the low-water level limit control for the same purpose on coil-type boilers.

9 Every automatic low-pressure steam-heating boiler, small power boiler, and power steam boiler shall be equipped with two high-steam pressure limit controls interlocked to shut off the fuel supply to the main burner with manual reset on the control, with the higher setting and two low-water-level limit controls, one of which shall be provided with a manual reset device and independent of the feed water controller. Coil-type flash steam boilers may use two high-temperature limit controls, one of which shall be manually reset in the hot water coil section of the boiler instead of the low-water level limit control.

10 Boiler groups C, D, and H shall use an approved automatic reset safety shutoff valve for the main burner fuel shutoff, which shall be interlocked to the programming control devices required. On oil burners where the safety shutoff valve will be subjected to pressures in excess of ten (10) psi when the burner is not firing, a second safety shutoff valve shall be provided in series with the first. Boiler groups C and D using gas in excess of one (1) pound-per-square-inch pressure or having a trapped combustion chamber or employing horizontal fire tubes shall be equipped with two approved safety shutoff valves, one of which shall be an automatic reset type, one of which may be used as an operating control, and both of which shall be interlocked to the limit-control devices required. Boiler groups C and D using gas in excess of one (1) pound per square inch pressure shall be provided with a permanent and ready means for making periodic tightness checks of the main fuel safety shutoff valves.

11 Control and limit device systems shall be grounded with operating voltage not to exceed one-hundred and fifty (150) volts, except that, upon approval by the building official, existing control equipment to be reused in an altered boiler control system may use two-hundred and twenty (220) volt single phase with one side grounded, provided such voltage is used for all controls. Control and limit devices shall interrupt the ungrounded side of the circuit. A readily accessible means of manually disconnecting the control circuit shall be provided with controls so arranged that when they are de-energized, the burner shall be inoperative.

CHAPTER 11

REFRIGERATION

1101.0 Scope.

Part I of this chapter covers refrigeration systems. Refrigeration systems, equipment, and devices, including the replacement of parts, alterations, and substitution of a different refrigerant, shall conform to the requirements of this chapter and other applicable provisions of this code.

Occupied spaces within refrigerated areas shall comply with this chapter and the applicable portions of the Building Code.

Part II covers cooling towers.

Part I – Refrigeration Systems.

1102.0 General.

Except as modified by this code, refrigeration systems shall comply with ASHRAE 15. In addition, ammonia refrigeration systems shall comply with IIAR 2.

1102.1 Refrigerants. The refrigerant used shall be of a type listed in Table 11-1 or as classified under Section 1103.0 as approved.

Exception: Lithium bromide absorption systems using water as the refrigerant.

1103.0 Refrigerant Classification.

Refrigerants shall be classified in accordance with the referenced standard for refrigerant designation and safety classification in Chapter 17.

1104.0 Classification of Refrigeration Systems.

Refrigeration systems shall be classified according to the degree of probability that a leakage of refrigerant could enter a normally occupied area.

1104.1 High-Probability System. Systems in which the basic design, or the location of components, is such that a leakage of refrigerant from a failed connection, seal, or component could enter the occupied space described in Section 1105.2.

1104.2 Low-Probability System. Systems that cannot be considered as high-probability system are classified as low-probability systems. This class includes systems whose joints and connections in the refrigerant circuit are effectively isolated from the occupied space described in Section 1105.2.

1105.0 Requirements for Refrigerant and Refrigeration System Use.

1105.1 System Selection. Refrigeration systems shall be limited in application in accordance with Table 11-2 and the requirements of this section.

1105.2 Volume of Occupied Space. The quantity of refrigerant in a single, independent circuit of a high-probability system shall not exceed the amounts shown in Table 11-1 based on the volume of the occupied space. The volume of the smallest, enclosed, occupied space shall be used to determine the permissible quantity of refrigerant in a system that is located in serves, or passes through such space. In accordance to this Section, occupied space shall include those rooms that are occupied occasionally for short periods of time such as storage rooms, equipment rooms other than refrigeration machinery rooms, or any room which can be entered with a door that can be closed after entry.

Exceptions:

(1) If the airflow to any enclosed space served by a portion of an air-duct system cannot be shut off or reduced below one-quarter of its maximum, the cubical contents of the entire space served by that portion of the air-duct system shall be used to determine the permissible quantity of refrigerant in the system.

(2) Refrigerated process or storage areas meeting the requirements of Section 1105.3.

1105.3 Refrigerated Process and Storage Areas. Refrigerant quantities in evaporators and piping within rooms or spaces used exclusively for processing or storage of materials under refrigerated conditions shall not be limited, provided that exiting is provided per the Building Code and that:

1105.3.1 The refrigerated room or space is equipped with a refrigerant vapor detection and alarm system complying with Section 1121.0.

1105.3.2 The refrigerated room or space is sealed from all other portions of the building by vapor-tight construction and tight-fitting, gasketed doors.

Exception: Adjoining refrigerated rooms.

1105.3.3 Where the quantity of a Group A2, B2, A3 or B3 refrigerant, other than ammonia, in an independent circuit would exceed 25% of the lower flammability limit if released to the surrounding room, (1) electrical equipment shall comply with the requirements of the electrical code for Class I, Division 2, or (2) the refrigerant vapor detection system required by Section 1105.3.1 shall automatically de-energize all electrical power within the space at vapor concentrations at or above 25% of the lower flammability limit.

1109.3 Storage. Storage of materials in a refrigeration machinery room shall be as permitted in the Fire Code.

1109.4 Emergency Control. A clearly identified switch of the break-glass type providing "off"-only control of electrically energized equipment and devices within the refrigeration machinery room shall be provided immediately adjacent to and outside of each refrigeration machinery room exit. In addition, emergency shutoff shall also be automatically activated when the concentration of refrigerant vapor exceeds 25 percent of the LFL.

1110.0 Refrigeration Piping, Containers, and Valves.

1110.1 General. Materials used in the construction and installation of refrigeration systems shall be suitable for the refrigerant, refrigerant oil, or brine in the system. Material or equipment that will deteriorate due to the chemical action of the refrigerant, the oil, or combination of both, shall not be installed.

1110.2 Nonferrous Materials. Copper and brass refrigeration piping, valves, fittings, and related parts used in the construction and installation of refrigeration systems shall be approved for the intended use.

1110.3 Ferrous Materials. Iron and steel refrigeration piping, valves, fittings, and related parts shall be approved for the intended use. Pipe more than two (2) inches (50 mm) iron pipe size shall be electric-resistance welded or seamless pipe.

1111.0 Erection of Refrigerant Piping.

1111.1 General. Piping and tubing shall be installed so as to prevent vibration and strains at joints and connections.

1111.2 Support. In addition to the requirements of Section 1106.2, piping and tubing shall be securely fastened to a permanent support within six (6) feet (1,829 mm) following the first bend in such tubing from the compressor and within two (2) feet (610 mm) of each subsequent bend or angle. Piping and tubing shall be supported at points not more than fifteen (15) feet (4,572 mm) apart.

1111.3 Protection from Damage. Refrigerant piping and tubing shall be installed so that it is not subject to damage from an external source. Soft annealed copper tubing shall not be larger than 1-3/8 inch (35 mm) nominal size. Mechanical joints shall not be made on tubing larger than 3/4 inch (20 mm) nominal size. Soft annealed copper tubing conveying refrigerant shall be enclosed in iron or steel piping and fittings, or in conduit, molding, or raceway that will properly protect the tubing against mechanical injury from an exterior source.

Exceptions:

(1) Tubing entirely within or tubing within five (5) feet (1,524 mm) of a refrigerant compressor when so located that it is not subject to external injury.

(2) Copper tubing serving a dwelling unit, when such tubing contains Group A1 refrigerant and is placed in locations not subject to damage from an external source.

1111.4 Visual Inspection. Refrigerant piping and joints shall be exposed to view for visual inspection and acceptance by the Authority Having Jurisdiction prior to being covered or enclosed.

Exception: Copper tubing enclosed in iron or steel piping conduit, molding, or raceway, provided there are no fittings or joints concealed therein.

1111.5 Prohibited Locations. Refrigerant piping shall not be located within a required exit.

1111.6 Underground Piping. Refrigerant piping placed underground shall be protected against corrosion.

1111.7 Joints. Iron or steel pipe joints shall be of approved threaded, flanged, or welded types. Exposed threads shall be tinned or coated with an approved corrosion inhibitor. Copper or brass pipe joints of iron pipe size shall be of approved threaded, flanged, or brazed types. Copper tubing joints and connections shall be approved flared, lapped, swaged, or brazed joints.

1111.8 Identification. Piping shall meet the reference standard for identification. The type of refrigerant, function, and pressure shall be indicated.

1112.0 Refrigerant Control Valves.

1112.1 Location. Stop valves shall be installed in the refrigerant piping of a refrigeration system at the following locations:

1112.1.1 At the inlet and outlet of a positive-displacement-type compressor, compressor unit, or condensing unit.

1112.1.2 At the refrigerant outlet from a liquid receiver.

1112.1.3 At the refrigerant inlet of a pressure vessel containing liquid refrigerant and having an internal gross volume exceeding three (3) cubic feet (108 m^3).

Exceptions:

(1) Systems with nonpositive-displacement compressors.

(2) Systems having a pump-out receiver for storage of the charge.

(3) Systems containing less than 110 pounds (49.9 kg) of Group A1 refrigerant.

(4) Self-contained systems that do not require a stop valve at the inlet of the receiver.

1112.2 Support. Stop valves installed in copper refrigerant lines of 3/4 inch (20 mm) or less outside diameter shall be securely supported independently of the tubing or piping.

1112.3 Access. Stop valves required by this section shall be readily accessible from the refrigeration machinery room floor or a level platform.

1112.4 Identification. Stop valves shall be identified by tagging in accordance with the reference standard for identification. A valve chart shall be mounted under glass at an approved location near the principal entrance to a refrigeration machinery room.

1113.0 Pressure-Limiting Devices.

1113.1 When Required. Pressure-limiting devices shall be provided on all systems operating above atmospheric pressure.

Exception: Factory-sealed systems containing less than twenty-two (22) pounds (9.98 kg) of Group A1 refrigerant listed by an approved agency.

1113.2 Setting. The maximum setting to which a pressure-limiting device may be set by use of the adjusting means provided shall not exceed the design pressure of the high side of a system not protected by a pressure-relief device or 90 percent of the pressure setting of the pressure-relief device installed on the high side of a system.

Exception: In systems using other than positive-displacement compressors that are protected by a pressure-relief device, the setting may be the design pressure of the high side of the system, provided the pressure-relief device is located in the low side, subject only to low-side pressure, and there is a permanent, unvalved relief path between the high side and the low side of the system.

1113.3 Connection. Pressure-limiting devices shall be connected between the pressure-imposing element and stop valve on the discharge side without intervening stop valves in the line leading to the pressure-limiting device.

1113.4 Operation. When the system is protected by a pressure-relief device, the pressure-limiting device shall stop the action of the pressure-imposing element at a pressure not more than 90 percent of the setting of the pressure-relief device.

1114.0 Pressure-Relief Devices.

1114.1 General. Refrigeration systems shall be protected by a pressure-relief device or other means designed to safely relieve pressure due to fire or abnormal conditions.

1114.2 Positive-Displacement Compressor. A positive-displacement compressor shall be equipped by the manufacturer with a pressure-relief device of adequate size and pressure setting to prevent rupture of the compressor or other component located between the compressor and the stop valve on the discharge side.

1114.3 Liquid-Containing Portions of Systems. Liquid-containing portions of systems, including piping, that can be isolated from pressure-relief devices required elsewhere and that can develop pressures exceeding their working design pressures due to temperature rise, shall be protected by the installation of pressure-relief devices.

1114.4 Evaporators. Evaporators located within eighteen (18) inches (457 mm) of a heating element or coil shall be fitted with a pressure-relief device.

Exceptions:

(1) Self-contained equipment.

(2) Factory-built coil assemblies specifically designed for installation on the discharge of forced-air heating units.

(3) If the connected and unvalved volume of the low side of the system meets the following:

$$V_1 > [W_1 - (V_2 - V_1) / V_{gt}]\, V_{gc}$$

(Equation 11-5)

WHERE:

V_1 = low-side volume, cubic feet.

V_2 = total volume of system, cubic feet.

W_1 = total weight of refrigerant in system, pounds.

V_{gt} = specific volume of refrigerant vapor at 110°F (43°C), cubic feet per pound.

V_{gc} = specific volume at critical temperature and pressure, cubic feet per pound.

(4) Reheat coils using the refrigerant for heat recovery.

1114.5 Actuation. Pressure-relief devices shall be direct-pressure actuated. Each part of a refrigeration system that can be valved off and that contains one or more pressure vessels having internal diameters greater than six (6) inches (150 mm) and containing liquid refrigerant shall be protected by a pressure-relief device.

1114.6 Stop Valves Prohibited. Stop valves shall not be located between a pressure-relief device and the portion of the system protected thereby.

1114.7 Location. Pressure-relief devices shall be connected as nearly as practical to the portion of the system protected thereby, above the liquid refrigerant level, and accessible for inspection and repair.

Exception: Fusible plugs on the high side may be located above or below the liquid refrigerant level.

1114.8 Materials. Seats and discs of pressure-relief devices shall be constructed of suitable material to resist corrosion or other chemical action caused by the refrigerant. Seats and discs of cast iron are prohibited.

1115.0 Pressure-Relief Device Settings.

1115.1 Pressure-Relief Valve Setting. Pressure-relief valves shall actuate at a pressure not exceeding the design pressure of the parts of the system protected.

1115.2 Rupture Member Setting. Rupture members used in lieu of, or in series with, a relief valve shall have a nominal rated rupture pressure not exceeding the design pressure of the parts of the system protected. Rupture members installed ahead of relief valves need not be larger, but shall not be smaller, than the relief-valve inlet.

1116.0 Marking of Pressure-Relief Devices.

1116.1 Pressure-Relief Valves. Pressure-relief valves for refrigerant-containing components shall be set and sealed by the manufacturer or by an approved assembler. Pressure-relief valves shall be marked by the manufacturer with the data required to show compliance with this chapter.

> **Exception:** Relief valves for systems with design pressures of fifteen (15) psig (103.35 kPa) or less may be marked by the manufacturer with pressure-setting capacity.

1116.2 Rupture Members. Rupture members for refrigerant pressure vessels shall be marked with the data required to show compliance with this chapter.

1116.3 Fusible Plugs. Fusible plugs shall be marked with the melting temperatures in °F (°C) to show compliance with this chapter.

1117.0 Overpressure Protection.

1117.1 General. Pressure vessels shall be provided with over-pressure protection as required by this section.

1117.2 Type of Protection. Pressure vessels with three (3) cubic feet (0.085 m³) internal gross volume or less may use a single pressure-relief device or a fusible plug. Pressure vessels over three (3) cubic feet (0.085 m³) but less than ten (10) cubic feet (0.283 m³) internal gross volume may use a single pressure-relief device but not a fusible plug. Pressure vessels of ten (10) cubic feet (0.283 m³) or more internal gross volume shall use a single rupture member or dual pressure-relief valve.

1117.3 Three-Way Valve Required. When dual pressure-relief valves are used, they shall be installed with a three-way valve to allow testing or repair.

> **Exception:** A single relief valve of the required relieving capacity may be used on low-side pressure vessels of ten (10) cubic feet (0.283 m³) or more when meeting the requirements of Section 1117.8, Exception 2.

1117.4 Parallel Pressure-Relief Devices. Two or more pressure-relief devices in parallel to obtain the required capacity shall be considered as one pressure-relief device. The discharge capacity shall be the sum of the capacities required for each pressure vessel being protected.

1117.5 Discharge Capacity. The minimum required discharge capacity of pressure-relief devices for each pressure vessel shall be determined by the following:

$$C = fDL \qquad \text{(Equation 11-6)}$$

WHERE:

C = minimum required discharge capacity of the relief device in pounds of air per minute.

D = outside diameter of vessel, feet.

L = length of vessel, feet.

f = factor dependent upon type of refrigerant from Table 11-3.

1117.6 Rating of Pressure-Relief Valves. Pressure-relief valves shall be of approved types and capacities. The rated discharge capacity of a pressure-relief valve shall be expressed in pounds of air per minute. Pipe and fittings between the pressure-relief valve and the parts of the system it protects shall have at least the area of the pressure-relief valve inlet.

1117.7 Rating of Rupture Members and Fusible Plugs. The rated discharge capacity of a rupture member or fusible plug discharging to atmosphere under critical flow conditions in pounds of air per minute shall be determined by the following formulas:

$$C = 0.8P_1d^2 \qquad \text{(Equation 11-7)}$$

$$d = 1.12\sqrt{C / P_1} \qquad \text{(Equation 11-8)}$$

WHERE:

C = rated discharge capacity of air, pounds per minute.

d = smallest internal diameter of the inlet pipe, retaining flanges, fusible plug, rupture member, inches.

For rupture members:

P_1 = (rated pressure psig x 1.1) + 14.7 (Equation 11-9)

For fusible plugs:

P_1 = absolute saturation pressure, corresponding to the stamped temperature melting point of the fusible plug or the critical pressure of the refrigerant used, whichever is smaller, psia.

1117.8 Discharge Location. Pressure-relief devices shall discharge to the atmosphere unless otherwise prohibited by this chapter at a location at least fifteen (15) feet (4,572 mm) above the adjoining grade level and at least twenty (20) feet (6,096 mm) from an opening into a building. The discharge termination shall be fitted with an approved diffuser directed to prevent spray of discharged refrigerant on personnel or entry of foreign material or water into the discharge piping. Discharge piping connected to the discharge side of a fusible plug or rupture member shall have provisions to prevent internal plugging of the pipe caused by the fusible plug or rupture-member function.

Exceptions:

(1) Systems containing less than 110 pounds (49.9 kg) of a Group A1 refrigerant.

(2) A pressure-relief valve may discharge into the low side of the system if the pressure-relief valve is of a type not affected by back pressure, provided the low side is equipped with pressure-relief devices of equal relieving capacity. The low-side pressure-relief device shall be set and discharged as required by this section. Fusible plugs or rupture members shall not be used for pressure relief into the low side.

1118.0 Discharge Piping.

The area of the discharge pipe from the pressure-relief device or fusible plug shall be at least as large as the outlet area of the pressure-relief device or fusible plug. A discharge pipe accepting discharge from more than one relief device or fusible plug shall be sized and have a maximum length of the common discharge not less than that required by the sum of the rated capacities of all relief valves discharging into the header at the lowest pressure setting of any of the relief valves discharging into the header.

The maximum length of the discharge piping permitted to be installed on the outlet of a pressure-relief device shall be determined by:

$$L = 9\,P^2 d^5 / 16C^2 \qquad \text{(Equation 11-10)}$$

WHERE:

C = minimum required discharge capacity, pounds of air per minute.

d = internal diameter of pipe, inches.

L = length of discharge pipe, feet.

For relief valves and rupture disks:

P = (rated pressure psig x 1.1) + 14.7
(Equation 11-11)

For fusible plugs $P=P_1$ where:

P_1 = absolute saturation pressure corresponding to the stamped temperature melting point of the fusible plug or the critical pressure of the refrigerant used, whichever is smaller, psia.

1119.0 Special Discharge Requirements.

1119.1 General. Systems containing other than Group A1 or B1 refrigerants shall discharge to atmosphere only through an approved flaring device.

Exceptions:

(1) Ammonia systems complying with Section 1120.0.

(2) Ammonia absorption systems serving a single dwelling unit.

(3) When the Authority Having Jurisdiction determines upon review of a rational engineering analysis that significant fire, health, or environmental hazard would not result from the proposed atmospheric release.

(4) Lithium bromide absorption system using water as the refrigerant.

1119.2 Design Requirements. Flaring devices shall be designed to incinerate the entire discharge. The products of refrigerant incineration shall not pose health or environmental hazards. Incineration shall be automatic upon initiation of discharge, shall be designed to prevent blow-back, and shall not expose structures or materials to threat of fire. Standby fuel, such as LPG, and standby power shall have the capacity to operate for 1.5 times the required time for complete incineration of the charge.

1119.3 Testing. Flaring systems shall be tested to demonstrate their safety and effectiveness. A report from an approved agency shall be submitted detailing the emission products from the system as installed.

1120.0 Ammonia Discharge.

Ammonia shall discharge into a tank of water that shall be used for no purpose except ammonia absorption. At least one (1) gallon (3.785 L) of fresh water shall be provided for each pound (454 g) of ammonia that will be released in one (1) hour from the largest relief device connected to the discharge pipe. The water used shall be prevented from freezing without the use of salt or chemicals. The tank shall be substantially constructed of not less than 1/8 inch (3.2 mm) or No. 10 MSG steel. The horizontal dimensions of the tank shall be equal to or less than one-half the height. The tank shall have a hinged cover or, if of the enclosed type, shall have a vent hole at the top. Pipe connections shall be through the top of the tank. The discharge pipe from the pressure-relief valves shall discharge ammonia in the center of the tank near the bottom but not more than thirty (30) feet (9,144 mm) below the surface of the water.

TABLE 11-1
Refrigerant Groups[1], Properties[2], and Allowable Quantities[3]
[ASHRAE 34: Table 1]

Refrig-erant	Chemical Formula	Chemical Name[4] (Composition for Blends)	Safety Group[1]	OEL[5] (ppm)	IDLH[6] (ppm)	Pounds per 1000 cf of Space[7]
R-11	CCl_3F	Trichlorofluoromethane	A1	C1000[8]	2,000	0.39
R-12	CCl_2F_2	Dichlorodifluoromethane	A1	1,000	15,000	5.6
R-13	$CClF_3$	Chlorotrifluoromethane	A1	1,000[10]	67,000	
R-13B1	$CBrF_3$	Bromotrifluoromethane	A1	1,000	40,000	
R-14	CF_4	Tetrafluoromethane (carbon tetrafluoride)	A1	1,000[10]	67,000	25
R-21	$CHCl_2F$	Dichlorofluoromethane	B1	10[14]	5,000	
R-22	$CHClF_2$	Chlorodifluoromethane	A1	1,000[14]	42,000[11]	13
R-23	CHF_3	Trifluoromethane	A1	1,000[10]		7.3
R-30	CH_2Cl_2	Dichloromethane (methylene chloride)	B2	C1000[8]	2,300	
R-32	CH_2F_2	Difluoromethane (methylene fluoride)	A2	1,000[13]		4.8
R-40	CH_3Cl	Chloromethane (methyl chloride)	B2	100	2,000	
R-50	CH_4	Methane	A3	1,000[10]		
R-113	CCl_2FCClF_2	1,1,2-trichloro-1, 2, 2 -trifluoroethane	A1	1,000	2,000	1.2
R-114	$CClF_2CClF_2$	1,2-dichloro-1,1,2,2 -tetraflouroethane	A1	1,000	15,000	8.7
R-115	$CClF_2CF_3$	Chloropentafluoroethane	A1	1,000[14]		47
R-116	CF_3CF_3	Hexafluoroethane	A1	1,000[10]		34
R-123	$CHCl_2CF_3$	2,2-dichloro-1,1,1, -trifluoroethane	B1	50[13]	4,000[11]	3.5
R-124	$CHClFCF_3$	2-chloro-1,1,1,2 –tetrafluoroethane	A1	1,000[13]		3.5
R-125	CHF_2CF_3	Pentafluoroethane	A1	1,000[13]		23
R-134a	CH_2FCF_3	1,1,1,2-tetrafluoroethane	A1	1,000[13]	50,000[11]	13
R-141b	CH_3CCl_2F	1,1-dichloro-1-fluoroethane	A1	500[13]		0.78
R-142b	CH_3CClF_2	1-chloro-1,1-difluoroethane	A2	1,000[13]		5.1
R-143a	CH_3CF_3	1,1,1-trifluoroethane	A2	1000[13]		4.5
R-152a	CH_3CHF_2	1,1-difluoroethane	A2	1,000[13]		2.0
R-170	CH_3CH_3	Ethane	A3	1,000	6,400	0.54
R-E170	CH_3OCH_3	Dimethyl ether	A3	1,000[10]		1.0
R-218	$CF_3CF_2CF_3$	Octafluoropropane	A1	1,000		43
R-227ea	CF_3CHFCF_3	1,1,1,2,3,3,3-heptafluoropropane	A1	1,000		36
R-236fa	$CF_3CH_2CF_3$	1,1,1,3,3,3-hexafluoropropane	A1	1,000[13]		21
R-245fa	$CHF_2CH_2CF_3$	1,1,1,3,3-pentafluoropropane	B1	300[13]		12
R-290	$CH_3CH_2CH_3$	Propane	A3	1,000	2,100	0.56
R-C318	$-(CF_2)_4-$	Octafluorocyclobutane	A1	1,000[10]		41
R-400	zeotrope	R-12/114 (50/50)	A1	1,000[10]		10
R-400	zeotrope	R-12/114 (60/40)	A1	1,000		11
R-401A	zeotrope	R-22/152a/124 (53.0/13.0/34.0)	A1	1,000[10]		6.6

TABLE 11-1
Refrigerant Groups[1], Properties[2], and Allowable Quantities[3] (continued)
[ASHRAE 34: Table 1]

Refrig-erant	Chemical Formula	Chemical Name[4] (Composition for Blends)	Safety Group[1]	OEL[5] (ppm)	IDLH[6] (ppm)	Pounds per 1000 cf of Space[7]
R-401B	zeotrope	R-22/152a/124 (61.0/11.0/28.0)	A1	1,000[10]		7.2
R-401C	zeotrope	R-22/152a/124 (33.0/15.0/52.0)	A1	1,000[10]		5.2
R-402A	zeotrope	R-125/290/22 (60.0/2.0/38.0)	A1	1,000[10]		8.5
R-402B	zeotrope	R-125/290/22 (38.0/2.0/60.0)	A1	1,000[10]		15
R-403A	zeotrope	R-290/22/218 (5.0/75.0/20.0)	A1	1,000[10]		7.6
R-403B	zeotrope	R-290/22/218 (5.0/56.0/39.0)	A1	1,000[10]		18
R-404A	zeotrope	R-125/143a/134a (44.0/52.0/4.0)	A1	1,000[10]		31
R-405A	zeotrope	R-22/152a/142b/C318 (45.0/7.0/5.5/42.5)		1,000[10]		16
R-406A	zeotrope	R-22/600a/142b (55.0/4.0/41.0)	A2	1,000[10]		4.7
R-407A	zeotrope	R-32/125/134a (20.0/40.0/40.0)	A1	1,000[10]		18
R-407B	zeotrope	R-32/125/134a (10.0/70.0/20.0)	A1	1,000[10]		20
R-407C	zeotrope	R-32/125/134a (23.0/25.0/52.0)	A1	1,000[10]		17
R-407D	zeotrope	R-32/125/134a (15.0/15.0/70.0)	A1	1,000[10]		15
R-407E	zeotrope	R-32/125/134a (25.0/15.0/60.0)	A1	1,000[10]		16
R-408A	zeotrope	R-125/143a/22 (7.0/46.0/47.0)	A1	1,000[10]		21
R-409A	zeotrope	R-22/124/142b (60.0/25.0/15.0)	A1	1,000[10]		7.1
R-409B	zeotrope	R-22/124/142b (65.0/25.0/10.0)	A1	1,000[10]		7.3
R-410A	zeotrope	R-32/125 (50.0/50.0)	A1	1,000[10]		25
R-410B	zoetrope	R-32.125 (45.0/55.0)	A1	1,000[10]		24
R-411A	zeotrope	R-1270/22/152a (1.5/87.5/11.0)	A2	990[10]		2.9
R-411B	zeotrope	R-1270/22/152a (3.0/94.0/3.0)	A2	980[10]		2.8
R-412A	zeotrope	R-22/218/143b (70.0/5.0/25.0)	A2	1,000[10]		5.1
R-413A	zeotrope	R-218/134a/600a (9.0/88.0/3.0)	A2	1,000		5.8
R-414A	zeotrope	R-22/124/600a/142b (51.0/28.5/4.0/16.5)	A1	1,000[10]		6.4
R-414B	zeotrope	R-22/124/600a/142b (50.0/39.0/1.5/9.5)	A1	1,000		6.0
R-415A	zeotrope	R-22/152a (82.0/18.0)	A2	1,000		12
R-415B	zeotrope	R-22/152a (25.0/75.0)	A2	1,000		9.3
R-416A	zeotrope	R-134a/124/600 (59.0/39.5/1.5)	A1	1,000		3.9
R-417A	zeotrope	R-125/134a/600 (46.6/50.0/3.4)	A1	1,000		3.5
R-418A	zeotrope	R-290/22/152a (1.5/96.0/2.5)	A2	1,000		13
R-419A	zeotrope	R-125/134a/E170 (77.0/19.0/4.0)	A2	1,000		19
R-420A	zeotrope	R-134a/142b (88.0/12.0)	A1	1,000		12
R-421A	zeotrope	R-125/134a (58.0/42.0)	A1	1,000		17
R-421B	zeotrope	R-125/134a (85.0/15.0)	A1	1,000		21

TABLE 11-1
Refrigerant Groups[1], Properties[2], and Allowable Quantities[3] (continued)
[ASHRAE 34: Table 1]

Refrig-erant	Chemical Formula	Chemical Name[4] (Composition for Blends)	Safety Group[1]	OEL[5] (ppm)	IDLH[6] (ppm)	Pounds per 1000 cf of Space[7]
R-422A	zeotrope	R-125/134a/600a (85.1/11.5/3.4)	A1	1,000		18
R-422B	zeotrope	R-125/134a/600a (55.0/42.0/3.0)	A1	1,000		16
R-422C	zeotrope	R-125/134a/600a (82.0/15.0/3.0)	A1	1,000		18
R-422D	zoetrope	R-125/134a/600a (65.1/31.5/3.4)	A1	1,000		16
R-423A	zeotrope	R-134a/227ea (52.5/47.5)	A1	1,000		19
R425A	zeotrope	R-32/134a/227ea (18.5/69.5/12.0)	A1	1,000		16
R426A	zoetrope	R-125/134a/600/601a (5.1/93.0/1.3/0.6)	A1	990		5.2
R427A	zoetrope	R-32/125/143a/134a (15.0/25.0/10.0/50.0)	A1	1,000		18
R428A	zoetrope	R-125/143a/290/600a (77.5/20.0/0.6/1.9)	A1	1,000		23
R-429A	zeotrope	R-E170/152a/600a (60.0/10.0/30.0)	A3	1,000		0.81
R-430A	zeotrope	R-152a/600a (76.0/24.0)	A3	1,000		1.3
R-431A	zeotrope	R-290/152a (71.0/29.0)	A3	1,000		0.69
R-432A	zeotrope	R-1270/E170 (80.0/20.0)	A3	710		0.13
R-433A	zeotrope	R-1270/290 (30.0/70.0)	A3	880		0.34
R-434A	zeotrope	R-125/143a/134a/600a (63.2/18.0/16.0/2.8)	A1	1,000		20
R-435A	zeotrope	R-E170/152a (80.0/20.0)	A3	1,000		1.1
R-436A	zeotrope	R-290/600a (56.0/44.0)	A3	1,000		0.5
R-436B	zeotrope	R-290/600a (52.0/48.0)	A3	1,000		0.5
R-437A	zeotrope	R-125/134a/600/601 (19.5/78.5/1.4/0.6)	A1	990		5
R-500	azeotrope	R-12/152a (73.8/26.2)	A1	1,000[10]	47,000[11]	7.6
R-501	azeotrope	R-22/12 (75.0/25.0)	A1	1,000		13
R-502	azeotrope	R-22/115(48.8/51.2)	A1	1,000[10]	65,000[11]	21
R-503	azeotrope	R-23/13 (40.1/59.9)		1,000	67,000	
R-504	azeotrope	R-32/115 (48.2/51.8)		1,000		29
R-507A	azeotrope	R-125/143a (50.0/50.0)	A1	1,000[10]		32
R-508A	azeotrope	R-23/116 (39.0/61.0)	A1	1,000[10]		14
R-508B	azeotrope	R-23/116 (46.0/54.0)	A1	1,000[10]		13
R509A	azeotrope	R-22/218 (44.0/56.0)	A1	1,000[10]		24
R-510A	azeotrope	R-E170/600a (88.0/12.0)	A3	1,000		0.87
R-600	$CH_3CH_2CH_2CH_3$	butane	A3	1,000	3,400	0.1
R-600a	$CH(CH_3)_2CH_3$	isobutane (2-methyl propane)	A3	1,000	3,400	0.6
R-601	$CH_3CH_2CH_2CH_2CH_3$	pentane	A3	600		0.2
R-601a	$(CH_3)_2CHCH_2CH_3$	2-methylbutane (isopentane)	A3	600		0.2

TABLE 11-1
Refrigerant Groups[1], Properties[2], and Allowable Quantities[3] (continued)
[ASHRAE 34: Table 1]

Refrig-erant	Chemical Formula	Chemical Name[4] (Composition for Blends)	Safety Group[1]	OEL[5] (ppm)	IDLH[6] (ppm)	Pounds per 1000 cf of Space[7]
R-611	$HCOOCH_3$	methyl formate	B2	100		
R-702	H_2	hydrogen	A3			
R-704	He	helium	A1			
R-717	NH_3	ammonia	B2	50[12]	300	0.014
R-718	H_2O	water	A1			footnote[15]
R-720	Ne	Neon	A1			
R-728	N_2	nitrogen	A1			
R-740	Ar	argon	A1			
R-744	CO_2	carbon dioxide	A1	5,000	40,000	4.5
R-764	SO_2	sulfur dioxide	B1	5	100	
R-1150	$CH_2{=}CH_2$	ethene (ethylene)	A3	200	5,200	0.38
R-1270	$CH_3CH{=}CH_2$	propene (propylene)	A3	500	3,400	0.1

Notes:

1 Refrigerant safety group designation is in accordance with Section 1103.0.

2 Refrigerant properties are those needed for this chapter.

3 Allowable quantities are for high-probability systems under Section 1104.0 only.

4 Chemical name shown is the preferred name. The popular name is shown in parenthesis.

5 OEL is the Occupational Exposure Limit. The value shown is the PEL designated in 29 CFR 1910.1000 unless otherwise indicated.

6 IDLH is that designated by NIOSH unless otherwise designated. If no value is shown, use the value listed under pounds per 1000cf of Space. Use the following formula to convert from pounds per 1000cf of Space to ppm. ppm = pounds per 1000cf of space/(0.000002557 x M), where M equals the molar mass of the refrigerant in g/mole.

7 Pounds of refrigerant in a high-probability system per 1000 cubic feet (28.3 kg/m^3) of occupied space. See Section 1104.0. This column does not apply to refrigerant machinery rooms or areas covered by Section 1107.0. If no value is listed use zero (0) unless sufficient data can be provided to determine the value as described in Section 7 of ASHRAE Standard 34-2007.

8 The OEL value shown is the TLV-C recommended by ACGIH.

9 The IDLH value shown is reduced from that designated by NIOSH in light of cardiac sensitization potential.

10 A PEL has not yet been established; the value given was determined in a consistent manner.

11 An IDLH has not yet been established; the value given was determined in a consistent manner.

12 OSHA PEL is 50 ppm; ACGIH TLV-TWA is 25 ppm.

13 The OEL value shown is the WEEL recommended by AIHA

14 The OEL value show is the ACGIH TLV-TWA

15 Quantity is unlimited when R-718 (water) is used as the refrigerant.

1 lb. = 0.454 kg, 1 ft^3 = 0.0283 m^3

TABLE 11-2
Permissible Refrigeration Systems[1]

Occupancy Group and Division	High-Probability System	Low-Probability System	Machinery Room
A–1	Group A1 only	Any	Any
A–2.1	Group A1 only	Any	Any
A–3	Group A1 only	Any	Any
A–4	Group A1 only	Any	Any
B	Group A1 only[2]	Any	Any
E–1	Group A1 only	Any	Any
E–2	Group A1 only	Any	Any
E–3	Group A1 only	Any	Any
F-1	Group A1 only[2]	Any	Any
F-2	Any[2]	Any	Any
H–1	Any	Any	Any
H–2	Any	Any	Any
H–3	Any	Any	Any
H–4	Group A1 only	Any	Any
H–5	Group A1 only	Any	Any
H–6	Group A1 only	Any	Any
H–7	Any	Any	Any
I–1.1	None	Any	Any
I–1.2	Group A1 only	Any	Any
I-2	Group A1 only	Any	Any
I-3	None	Any	Any
M	Group A1 only[2]	Any	Any
R–1	Group A1 only	Any	Any
R–2	Group A1 only	Any	Any
R–3	Group A1 only	Any	Any
S-1	Group A1 only[2]	Any	Any
S-2	Any[2]	Any	Any
S-3	Group A1 only	Any	Any
S-4	Group A1 only	Any	Any
S-5	Group A1 only	Any	Any
U-1	Any	Any	Any

Notes:
[1] See Section 1105.0.
[2] Any refrigerant may be used within a high-probability system when the room or space complies with Section 1105.3.

TABLE 11-3
Value of Factor Dependent for Equation 11–6

Refrigerant Number	*f*
11	1.0
12	1.6
22	1.6
113	1.0
114	1.6
115	2.5
123	1.0
134a	1.6
152a	1.0
500	1.6
502	2.5
717	0.5
744	1.0

TABLE 11-4
Field Leak Test Pressures in psig (kPa)

Refrigerant Number	High Side Water Cooled	High Side Air Cooled	Low Side
11	15	35	15
12	140	220	140
22	230	360	230
113	15	15	15
114	40	80	40
115	275	340	275
123	15	30	15
134a	150	250	150
152a	130	220	130
500	165	265	165
502	250	385	250
717	235	390	235
744*	N/A	N/A	N/A

Note: psig x 6.89476 = kPa, N/A—Not applicable.
*Special design required; test pressures typically exceed 1,000 psig (6890 kPa).

and insulation (including the insulation, jacket, and lap-seal adhesives) shall have a maximum peak heat release rate of 300 kW, a maximum total heat release of 50 MJ, a maximum total smoke release of five-thousand, three-hundred and eighty-two (5,382) square feet (500 m^2) and shall not generate flames that extend one (1) foot (305 mm) or more above the top of the vertical portion of the apparatus at any time during the test when tested in accordance with NFPA 274, *Standard Test Method to Evaluate Fire Performance Characteristics of Pipe Insulation.* Insulation coverings and linings shall not flame, glow, smolder, or smoke when tested in accordance with ASTM C 411, *Hot-Surface Performance of High Temperature Thermal Insulation,* at the temperature to which they are exposed in service. In no case shall the test temperature be below 250°F (121°C).

1201.2.1.9 Flashing Material. Flashings shall be lead, copper, galvanized iron, or other approved materials.

1201.2.1.10 Gaskets. Flange gaskets shall be metal, asbestos, or other approved materials.

1201.2.1.11 Hangers and Anchors. Hangers and anchors shall be suitable for the use intended.

1201.2.1.12 Sleeves. Sleeves shall be of steel, cast-iron or wrought-iron pipe, or tile.

1201.2.1.13 Standards. All piping, tubing, valves, joints, fittings, devices, and materials shall be free of defects and comply with nationally recognized standards.

1201.2.1.14 Marking. Materials and devices shall be suitably identified. In addition to the incised marking required in the standards, all hard-drawn copper tubing shall be marked in accordance with the applicable standard found in Chapter 17. Color coding shall be as follows:

Type L – Blue

Type K – Green

Type M – Red

Type ACR – Blue

Type DWV – Yellow

1201.2.2 Fabrication of Joints. Joints shall be made by the use of fittings except as otherwise permitted in this chapter.

1201.2.2.1 Screwed Joints. Threads on iron pipe size (IPS pipe) shall be standard taper pipe threads. All burrs shall be removed. Pipe ends shall be reamed or filed out to the full size of bore, and all chips shall be removed.

1201.2.2.2 Solder Joints. Surfaces to be joined by soldering shall be cleaned bright by manual or mechanical means. The joints shall be properly fluxed using a listed soldering flux. (See Soldering Fluxes in Referenced Standards, Chapter 17, Part II of this code). Tubing shall be reamed out to the full size of bore.

1201.2.2.3 Welded Joints. Welding shall be performed in accordance with nationally recognized standards by certified welders.

1201.2.2.4 Flanged Joints. Flanged joints shall be tightened evenly and provided with suitable nuts, bolts, and gaskets.

1201.2.2.5 Mechanical Joints. Mechanical joints shall comply with nationally recognized standards.

1201.2.3 Connections.

1201.2.3.1 Brass and Copper Piping. Joints in brass and copper piping shall be threaded, brazed, welded, flanged, or mechanical type.

1201.2.3.2 Cast-Iron Piping. Joints in cast-iron pipe shall be threaded, flanged, or mechanical type.

1201.2.3.3 Galvanized Wrought-Iron and Galvanized Steel Piping. Joints in galvanized wrought-iron and galvanized steel piping shall be threaded, flanged, or mechanical type.

1201.2.3.4 Black Wrought-Iron Piping. Joints in black wrought-iron piping shall be threaded, brazed, welded, flanged, or mechanical type, except that joints built into or embedded in concrete or masonry shall be welded.

1201.2.3.5 Black Steel Piping. Joints in black steel piping shall be threaded, brazed, welded, flanged, or mechanical type.

1201.2.3.6 Asbestos-Cement Piping. Joints in asbestos-cement piping shall be mechanical type and suitable for the service temperature intended.

1201.2.3.7 Copper Water Tubing. Joints in copper tubing shall be soldered, brazed or flared, except that joints under a building and in or under any concrete slab resting on the ground shall be silver brazed, or equal, and fittings shall be of wrought copper.

Mechanically formed tee fittings are also acceptable when brazed and installed in accordance with Section 1201.2.1.4. All solder joints shall be made with solders meeting the standard for solder metal found in Chapter 17. However, if steam pressures exceed fifteen (15) psig (103 kPa) or water pressures exceed thirty (30) psig (206 kPa), then 50 percent tin - 50 percent lead solder shall not be used. Solders and fluxes with a lead content which exceeds two-tenths (0.20) of 1 percent shall be prohibited in piping systems conveying potable water.

1201.2.3.8 Piping to Tubing. Joints connecting piping to tubing shall be made with adapter fittings connected as required in Sections 1201.2.3.1 through 1201.2.3.7.

1201.2.4 Changes in Direction. Changes in direction shall be made by the appropriate use of fittings, except that changes in direction in copper tubing may be made with bends having a radius not less than six (6) diameters of the tubing, provided that such bends are made by the use of forming equipment that does not deform or reduce appreciably the cross-sectional area of the tubing.

1201.2.5 Changes in Pipe Sizes. Where different sizes of pipe or pipe and fittings are to be connected, the proper size increasers or reducer fittings shall be used between the two sizes. When the branch is at least two sizes smaller than the main, weldolets or threadolets may be used in lieu of welding tees. Bushings shall not be used. Eccentric reducing fittings shall be used wherever necessary to provide free drainage of lines.

1201.2.6 Hangers and Supports. All piping and equipment shall be adequately supported to the satisfaction of the Authority Having Jurisdiction. Hot-water and steam piping shall be supported, anchored, and provided with swing joints, expansion loops or joints, or other means to avoid excessive strain on piping, equipment, or the building structure to the satisfaction of the Authority Having Jurisdiction.

1201.2.6.1 Vertical Piping—Attachment. Vertical piping and tubing shall be secured at sufficiently close intervals to keep the pipe in alignment and carry the weight of the pipe and contents.

1201.2.6.2 Horizontal Piping.

1201.2.6.2.1 Supports. Horizontal piping and tubing shall be supported at sufficiently close intervals to keep it in alignment and prevent excessive sagging.

1201.2.6.2.2 In Ground. Piping and tubing in the ground shall be laid on a firm bed for its entire length except when otherwise approved by the Authority Having Jurisdiction. Asbestos-cement piping shall be provided with adequate thrust blocking.

1201.2.7 Installation.

1201.2.7.1 Same Materials Required. All piping materials used, except valves and similar devices, shall be of a like material, except as otherwise acceptable to the Authority Having Jurisdiction.

1201.2.7.2 Wall Thickness.

1201.2.7.2.1 Piping shall be at least standard-weight brass or copper, Class 150 cast-iron, standard-weight wrought iron, ASTM Schedule 40 steel, or suitable asbestos-cement of adequate pressure rating.

1201.2.7.2.2 Tubing shall be at least Type K—for condensate return lines; Type L—for steam condenser cooling water lines, underground water lines, and aboveground water lines; Type M—for aboveground water lines not embedded in concrete or masonry.

1201.2.7.3 Piping Embedded in Structure. Piping shall not be built into or embedded in concrete or masonry, except where used for radiant panel heating or cooling. Black steel pipe, wrought-iron piping, or Type L copper tubing may be so embedded.

1201.2.7.4 Cutting Structure. Structural members shall not be seriously weakened or impaired by cutting or notching.

1201.2.7.5 Providing for Expansion, Contraction, and Settling. Piping shall be installed so that piping, connections, and equipment shall not be subjected to excessive strains or stresses, and provisions shall be made for expansion, contraction, shrinkage, and structural settlement.

1201.2.7.6 Circulation. Piping shall provide adequate circulation. Piping shall be graded so that all gases can move in the direction of the water flow to a vented section of the system. When sections of a piping system cannot be installed with the required grade, such sections shall be provided with automatic or manual air vents whose discharge is piped to an approved location. Steam traps shall be provided where required.

1201.2.7.7 Underground Piping.

1201.2.7.7.1 Cinders and Other Corrosive Material Fills. All piping passing through or under cinders or other corrosive fill materials shall be suitably protected from corrosion.

Exception: Where a soil analysis by an acceptable testing laboratory shows the soil to be free of materials that may corrode the pipe to be installed, the requirements for protective coatings may be waived.

1201.2.7.7.2 Beneath Buildings. Piping installed within a building and in, or under, a concrete floor slab resting on the ground shall be installed as follows:

Ferrous Piping. Ferrous piping shall be galvanized and covered with an approved protective coating.

Copper Tubing. Copper tubing shall be installed without joints, where possible.

Asbestos-Cement. Asbestos-cement pipe shall not be installed beneath any building.

1201.2.7.7.3 Outside of Buildings.

Black Wrought Iron and Black Steel. Black wrought-iron and black steel piping shall be protected against corrosion by an approved pipe wrapping.

Asbestos-Cement. Asbestos-cement piping shall be installed in accordance with the manufacturer's recommendations, but shall not be installed within two (2) feet (610 mm) of any building.

1201.2.7.7.4 Under Walls or Foundations. Piping passing under walls or foundations shall be protected from breakage.

1201.2.7.7.5 Openings into Buildings. Voids around piping passing through concrete or masonry floors or walls shall be appropriately sealed at the opening into the building. Sleeves shall be provided at such openings.

1201.2.7.8 Above ground Piping.

1201.2.7.8.1 Sleeves. Sleeves shall be provided to protect all piping through concrete and masonry walls.

1201.2.7.8.2 Insulation. The temperature of surfaces within normal reach of building occupants shall not exceed 140°F (60°C) unless they are protected by suitable insulation. Where sleeves are installed, any insulation shall continue full-sized through them.

1201.2.7.8.3 Lining. Combustible portions of unventilated spaces that contain piping or devices whose outside temperature, including insulation, exceeds 140°F (60°C), shall be lined with No. 24 gauge (0.021 inch) (0.53 mm) steel, or one-fourth (1/4) inch thick (6.4 mm) insulating millboard.

1201.2.7.8.4 Clearance. There shall be at least one (1) inch (25 mm) clearance from the structure around steam pipes.

1201.2.7.8.5 Exposed Piping. Exposed piping subject to excessive corrosion, erosion, or mechanical damage shall be suitably protected.

1201.2.7.8.6 Asbestos-Cement Piping. Asbestos-cement piping shall not be installed within any building.

1201.2.7.8.7 Roof and Wall Openings. Joints at the roof around pipes or appurtenances shall be made water-tight by the use of approved flashings or flashing material. Exterior wall openings shall be made water-tight.

1201.2.7.8.8 Drainage. Means shall be provided to drain all piping.

1201.2.7.8.9 Freezing. Where required, piping outside of a building or in an exterior wall shall be protected from freezing.

1201.2.7.9 Trenches and Tunnels.

1201.2.7.9.1 Protection of Structure. Trenches deeper than the footings of a building or structure and paralleling the same shall be at least 45 degrees therefrom, or approved per Section 105.0.

1201.2.7.9.2 Mechanical Equipment. Use of mechanical excavating equipment is prohibited within two (2) feet (610 mm) of existing piping or appurtenances.

1201.2.7.9.3 Tunneling and Driving. Tunnels shall, before backfilling, have a clear height of two (2) feet (610 mm) above the pipe and shall be limited in length to one-half the depth of the trench, with a maximum length of eight (8) feet (2,438 mm). When pipes are driven, the drive pipe shall be at least one size larger than the pipe to be laid.

1201.2.7.9.4 Backfilling. Excavations shall be completely backfilled as soon after inspection as practicable. Adequate precaution shall be taken to ensure proper compaction of backfill around piping without damage to such piping. Trenches shall be backfilled in thin layers to twelve (12) inches (304 mm) above the top of the piping with clean earth that shall not contain stones, boulders, cinderfill, or other materials that would damage, break the piping, or cause corrosive action. Mechanical devices, such as bulldozers, graders, etc., may then be used to complete backfill to grade. Fill shall be properly compacted. Suitable precautions shall be taken to ensure permanent stability for pipe laid in filled or made ground.

1201.2.8 Pressure Testing.

1201.2.8.1 Responsibility. The equipment, material, and labor necessary for inspection or test shall be furnished by the person to whom the permit is issued or by whom inspection is requested.

1201.2.8.2 Media. The piping shall be tested with water.

1201.2.8.3 Pressure Test. Piping shall be tested with a hydrostatic pressure of not less than 100 psig (689 kPa), but at least fifty (50) psig (345 kPa) greater than operating pressure. This pressure shall be maintained for at least thirty (30) minutes. Required tests shall be conducted by the owner or contractor in the presence of an authorized inspector. The piping being tested shall remain exposed to the inspector and shall not leak during the test.

1201.2.8.4 Moved Structures. Piping systems of a building and parts thereof that are moved from one foundation to another shall be completely tested as prescribed elsewhere in this section for new work, except that walls or floors need not be removed during such test when equivalent means of inspection are provided.

1201.2.8.5 Test Waived. No test or inspection shall be required where a system, or part thereof, is set up for exhibition purposes and has no connection with a water system.

1201.2.8.6 Exceptions. In cases where it would be impractical to provide the aforementioned tests, or for minor installations and repairs, the Authority Having Jurisdiction shall have the authority to make such inspection as it deems necessary.

1201.3 Those portions of the hot water piping systems in which the continuous pressure-temperature relationship does not exceed the following may be constructed of cross-linked polyethylene (PEX) tubing of SDR-9 conforming to specifications ASTM F876 and F877.

Temperature		Pressure	
°F	**(°C)**	**psi**	**(kPa)**
73	(23)	160	(1,103)
180	(82)	100	(689)
200	(93)	80	(550)

1201.3.1 Materials and Construction.

1201.3.1.1 PEX Tubing. Tubing shall be copper tube size, SDR-9 cross-linked polyethylene conforming to ASTM F876. PEX tubing and fittings shall be installed in accordance with the manufacturers' recommended installation instructions.

1201.3.1.2 Fittings. Fittings shall be manufactured and tested in accordance with the nationally recognized standards.

1201.3.1.3 Insulation. Coverings and insulation used for hot water pipes shall be of material suitable for the operating temperature of the system. The insulation, jackets, and lap-seal adhesives, including pipe coverings and linings, shall have a flame-spread index not greater than twenty-five (25) and a smoke-developed index not greater than fifty (50) when tested in accordance with NFPA 255, *Method of Test of Burning Characteristics of Building Materials*; or in accordance with ASTM E 84, *Surface Burning Characteristics of Building Materials*; or in accordance with the provisions of UL 723, *Test for Surface Burning Characteristics of Building Materials.* The specimen preparation and mounting procedures of ASTM E 2231, *Specimen Preparation and Mounting of Pipe and Duct Insulation Materials to Assess Surface Burning Characteristics* shall be used. Alternately, materials used for pipe coverings and insulation (including the insulation, jacket, and lap-seal adhesives) shall have a maximum peak heat release rate of 300 kW, a maximum total heat release of 50 MJ, a maximum total smoke release of 500 m^2 and shall not generate flames that extend one (1) foot (305 mm) or more above the top of the vertical portion of the apparatus at any time during the test when tested in accordance with NFPA 274, *Standard Test Method to Evaluate Fire Performance Characteristics of Pipe Insulation*. Insulation coverings and linings shall not flame, glow, smolder, or smoke when tested in accordance with ASTM C 411, *Hot-Surface Performance of High Temperature Thermal Insulation*, at the temperature to which they are exposed in service. In no case shall the test temperature be below 250°F (121°C).

1201.3.1.4 Hangers, Sleeves, and Anchors. Hangers, sleeves, and anchors shall be suitable for the use intended as recommended by the manufacturers' installation instructions.

1201.3.1.5 Standards. All piping, tubing, and fitting materials shall be free of defects and comply with nationally recognized standards approved by the Authority Having Jurisdiction.

1201.3.1.6 Marking. Materials and devices shall be suitably identified.

1201.3.2 Fabrication of Joints. All joining methods shall meet the performance requirements set forth in ASTM F877. Joints shall be made by one or more of the following methods:

1201.3.2.1 Crimp/Insert Fittings. Insert fittings of metal with crimp rings of copper may be used.

1201.3.2.2 Compression Fittings. Metallic fittings utilizing compression seals are acceptable.

1201.3.2.3 Cold Expansion Fittings. Cold expansion fittings utilizing a PEX reinforcing ring or metal compression sleeve may be used.

1201.3.2.4 Transition Fittings. Connections to other piping materials shall be made of approved types of special transition fittings.

1201.3.3 Changes in Direction. Changes in direction shall be made by the appropriate use of fittings or with pipe bends having a radius of not less than six (6) times the outside diameter of the tubing. No forming equipment or heating is required.

1201.4 Those portions of the hot water piping systems in which the continuous pressure temperature relationship does not exceed the following shall be permitted to be constructed of cross-linked polyethylene/aluminum/cross-linked polyethylene (PEX-AL-PEX) piping conforming to specification ASTM F1281.

Temperature		Pressure	
°F	(°C)	psi	(kPa)
73	(23)	200	(1,380)
180	(82)	125	(862)
200	(93)	100	(690)

1201.4.1 Materials and Construction.

1201.4.1.1. PEX-AL-PEX. Piping shall be crosslinked polyethylene/aluminum/cross-linked polyethylene conforming to ASTM F 1281. PEX-AL-PEX piping and fittings shall be installed in accordance with the manufacturers' recommended installation instructions.

1201.4.1.2 Fittings. Fittings shall be manufactured and tested in accordance with the nationally recognized standards.

1201.4.1.3 Insulation. Coverings and insulation used for hot water pipes shall be of materials suitable for the operating temperature of the system. The insulation, jackets, and lap-seal adhesives shall be tested as a composite developed rating of not more than fifty (50) when tested in accordance with Building Code standards.

1201.4.1.4 Hangers, Sleeves, and Anchors. Hangers, sleeves, and anchors shall be suitable for the use intended as recommended by the manufacturers' installation instructions.

1201.4.1.5 Standards. All piping and fitting materials shall be free of defects and comply with nationally recognized standards in Chapter 17.

1201.4.1.6 Markings. Materials and devices shall be suitably identified.

1201.4.2 Fabrication of Joints. All joining methods shall meet the performance requirements set forth in ASTM F1281. Joints shall be made by one or more of the following methods:

1201.4.2.1 Crimp/Insert Fittings. Insert fittings of metal with crimp rings of copper may be used.

1201.4.2.2 Mechanical Compression Fittings. Metallic fittings utilizing a split ring and compression nut are acceptable.

1201.4.2.3 Transition Fittings. Connections to other piping materials shall be made of approved types of special transition fittings.

1201.4.3 Changes in Direction. Changes in direction shall be made by fittings or with pipe bends having a radius of not less than five (5) times the outside diameter of the piping. No forming equipment or heating is required.

1201.5 Those portions of the hot water piping systems in which the continuous pressure/temperature relationship does not exceed the following shall be permitted to be constructed of polypropylene (PP) piping conforming to specification ASTM F2389, which is rated at or above these pressures and temperatures.

Temperature		Pressure	
°F	(°C)	psi	(kPa)
73	(23)	200	(1,380)
180	(82)	100	(690)
200	(93)	40	(345)

1201.5.1 Materials and Construction.

1201.5.1.1 Polypropylene Pipe. Pipe shall conform to ASTM F2389. (PP) pipe and fittings shall be installed in accordance with the manufacturers' recommended installation instructions.

1201.5.1.2 Fittings. Fittings shall be manufactured and tested in accordance with the ASTM F2389.

1201.5.1.3 Hangers, Sleeves and Anchors. Hangers, sleeves, and anchors shall be suitable for the use intended as recommended by the manufacturers' installation instructions.

1201.5.1.4 Standards. All pipe and fittings materials shall be free of defects and comply with nationally recognized standards approved by the Authority Having Jurisdiction.

1201.5.1.5 Marking. Materials and devices shall be suitably identified. (PP) pipe and fittings shall be marked in accordance with ASTM F2389.

1201.5.1.6 Heat-fusion Joints. Heat-Fusion for polypropylene (PP) pipe and fitting joints shall be installed with socket-type heat-fused polypropylene fittings, butt-fusion polypropylene fittings or pipe, or electrofusion polypropylene fittings. Joint surfaces shall be clean and free from moisture. The joint shall be undisturbed until cool.

1201.5.1.7 Mechanical and Compression Sleeve Joints. Mechanical and compression sleeve joints shall be installed in accordance with the manufacturer's instructions.

1201.5.1.8 Transition Fittings. Connections to other piping materials shall be made with approved types of transition fittings.

1201.6 Those portions of the hot water piping systems in which the continuous pressure/ temperature relationship does not exceed the following shall be permitted to be constructed of Raised Temperature Polyethylene (PE-RT) tubing conforming to specification ASTM F2623.

Temperature		Pressure	
°F	(°C)	psi	(kPa)
73	(23)	160	(1,103)
140	(60)	100	(690)
180	(82)	80	(552)

1201.6.1 Materials and Construction.

1201.6.1.1 PE-RT Tubing. Tubing shall be copper tube size, SDR 9 Raised Temperature Polyethylene conforming to ASTM F2623. PE-RT tubing and fittings shall be installed in accordance with the manufacturer's recommended installation instructions.

1201.6.1.2 Fittings. Fittings shall be manufactured and tested in accordance with the section 1201.6.1.6.

1201.6.1.3 Hangers, Sleeves and Anchors. Hangers, sleeves, and anchors shall be suitable for the use intended as recommended by the manufacturer's instructions.

1201.6.1.4 Standards. All Piping, tubing and fitting materials shall be free of defects and comply with recognized standards approved by the Authority Having Jurisdiction.

1201.6.1.5 Marking. Materials and devices shall be suitably identified.

1201.6.1.6 Fabrication of Joints. Fittings shall be manufactured and tested in accordance with the application for which they are intended. Joints shall be made by one or more of the following methods:

1201.6.1.7 Crimp/Insert Fittings. Insert fittings of metal or plastic with crimp rings of copper may be used.

1201.6.1.8 Compression Fittings. Metallic fittings utilizing compression seals are acceptable.

1201.6.1.9 Transition Fittings. Connections to other piping materials shall be made of approved types of special transition fittings.

1201.6.2 Changes in Direction. Changes in direction shall be made by the appropriate use of fittings or with pipe bends having a radius of not less than six (6) times the outside diameter of tubing. No forming equipment or heating is required.

1201.7 Those portions of the hot water piping systems in which the continuous pressure temperature relationship does not exceed the following shall be permitted to be constructed of Polyethylene/Aluminum/Polyethylene (PE-AL-PE) piping conforming to specification ASTM F1282.

Temperature		Pressure	
°F	(°C)	psi	(kPa)
73	(23)	200	(1,380)
140	(60)	150	(1,035)
180	(82)	100	(690)

1201.7.1 Materials and Construction.

1201.7.1.1 PE-AL-PE. Piping shall be PE-Al-PE conforming to ASTM F1282. PE-AL-PE piping and fittings shall be installed in accordance with the manufacturer's recommended installation instructions.

1201.7.1.2 Fittings. Fittings shall be manufactured and tested in accordance with the nationally recognized standards.

1201.7.1.3 Hangers, Sleeves, and Anchors. Hangers, sleeves, and anchors shall be suitable for the use intended as recommended by the manufacturers' instructions.

1201.7.1.4 Standards. All piping, tubing and fitting materials shall be free of defects and comply with nationally recognized standards in Chapter 17.

1201.7.1.5 Marking. Materials and devices shall be suitably identified.

1201.7.1.6 Fabrication of Joints. All joining methods shall meet the performance requirements set forth in ASTM F1282 or ASTM F1974. Joints shall be made by one or more of the following methods:

1201.7.1.7 Crimp/Insert Fittings. Insert fittings of metal or plastic with crimp rings of copper may be used.

1201.7.1.8 Mechanical Compression Fittings. Metallic fittings utilizing compression seals are acceptable.

1201.7.1.9 Transition Fittings. Connections to other piping materials shall be made of approved types of special transition fittings.

1201.7.2 Changes in Direction. Changes in direction shall be made by the appropriate use of fittings or with pipe bends having a radius of not less than six (6) times the outside diameter of the piping. No forming equipment or heating is required.

Part II – Hydronic Panel Heating Systems.

1202.0 Scope.

The purpose of this part is to establish and provide minimum standards for the protection of public health, welfare, and property by regulating and controlling the design and installation of panel heating systems.

1203.0 Installation.

1203.1 Panel systems shall be designed and installed in accordance with installation standards incorporated in Chapter 17 Panel Heating, and the requirements of this code.

1203.2 Piping to be embedded in concrete shall be pressure-tested prior to pouring concrete. During pouring, the pipe shall be maintained at the proposed operating pressure.

1204.0 Piping Materials.

1204.1 Panel(s). Piping for heating panels shall be standard-weight steel pipe, Type L copper tubing, or approved plastic pipe or tubing rated at 100 psi (689 kPa) at 180°F (82°C).

1204.2 Hot-Water Supply Lines. Piping for hot-water supply lines shall be installed according to the requirements in Chapter 10.

1205.0 Piping Joints.

Joints of pipe or tubing forming the panel that are embedded in a portion of the building, for example, concrete or plaster, shall be in accordance with the following:

1205.1 Steel pipe welded with electrical arc or oxygen/acetylene method.

1205.2 Copper tubing joined with brazing alloys having a melting point above 1,000°F (538°C).

1206.0 Heat Sources.

Heat sources for generating hot water for use in hydronic panel radiant heating systems shall include conventional fossil fuel, hot water boilers, electrical-resistance heated boilers, air/water or water/water heat pumps, or solar heat collector systems. The latter system may include booster or backup heating units.

Systems shall be protected by pressure-temperature relief valves as outlined in this code.

1207.0 Testing.

Approved piping or tubing installed as a portion of a radiant piping system that will be embedded in the walls, floors, or ceilings of the building it is designed to heat shall be tested for leaks by the hydrostatic test method by applying at least 100 psi (689 kPa) water pressure or 1.5 times the operating pressure, whichever is greater.

For metal piping, a pressure gauge shall be connected to the piping, and after the pressure has been raised, the hydrostatic pressure connection shall be discontinued and the systems under pressure shall remain at the test pressure for a sufficient period of time to determine whether any leaks exist in the system. Leaks shall be indicated by the pressure drop on the gauge. The minimum test period shall be thirty (30) minutes.

For flexible plastic piping, the test pressure shall be applied for a period of thirty (30) minutes. During this time, the system shall be maintained at the test pressure by the periodic addition of makeup water to compensate for the initial stretching of the pipe. The system shall then be visually inspected for tightness.

Tests for tightness of radiant piping systems shall be witnessed by the Authority Having Jurisdiction.

CHAPTER 13

FUEL GAS PIPING

Part I – Fuel Piping.

1301.0 Scope.

(A) Coverage of piping systems shall extend from the point of delivery to the connections with each gas utilization device. For other than undiluted liquefied petroleum gas systems, the point of delivery shall be considered the outlet of the service meter assembly or the outlet of the service regulator or service shutoff valve where no meter is provided. For undiluted liquefied petroleum gas systems, the point of delivery shall be considered the outlet of the final pressure regulator, exclusive of the line gas regulators, in the system. [NFPA 54:1.1.1.1(A)]

(B) Piping systems requirements shall include design, materials, components, fabrications, assembly, installation, testing, inspection, operation, and maintenance. [NFPA 54:1.1.1.1(C)]

(C) This code shall not apply to the following (reference standards for some of which appear in Chapter 17): [NFPA 54:1.1.1.2]

(1) Portable LP-Gas equipment of all types that is not connected to a fixed fuel piping system.

(2) Installation of farm equipment, such as brooders, dehydrators, dryers, and irrigation equipment.

(3) Raw material (feedstock) applications except for piping to special atmosphere generators.

(4) Oxygen-fuel gas cutting and welding systems.

(5) Industrial gas applications using such gases as acetylene and acetylenic compounds, hydrogen, ammonia, carbon monoxide, oxygen, and nitrogen.

(6) Petroleum refineries, pipeline compressor or pumping stations, loading terminals, compounding plants, refinery tank farms, and natural gas processing plants.

(7) Large integrated chemical plants or portions of such plants where flammable or combustible liquids or gases are produced by chemical reactions or used in chemical reactions.

(8) LP-Gas installations at utility gas plants.

(9) Liquefied natural gas (LNG) installations.

(10) Fuel-gas piping in electric utility power plants. [NFPA 54-09:1.1.1.2(10)]

(11) Proprietary items of equipment, apparatus, or instruments, such as gas-generating sets, compressors, and calorimeters.

(12) LP-Gas equipment for vaporization, gas mixing, and gas manufacturing.

(13) LP-Gas piping for buildings under construction or renovations that are not to become part of the permanent building piping system—that is, temporary fixed piping for building heat.

(14) Installation of LP-Gas systems for railroad switch heating.

(15) Installation of LP-Gas and compressed natural gas systems on vehicles.

(16) Gas piping, meters, gas-pressure regulators, and other appurtenances used by the serving gas supplier in the distribution of gas, other than undiluted LP-Gas.

1302.0 General.

The regulations of this chapter shall govern the installation of all fuel gas piping in or in connection with any building, structure or within the property lines of any premises up to five (5) psi (34.5 kPa), other than service pipe. Fuel oil piping systems shall be installed in accordance with NFPA 31.

1303.0 Definitions.

For the purposes of this code, these definitions shall apply to this chapter. Certain terms, phrases, words, and their derivatives shall be interpreted as set forth in this section, provided, however, that whenever the words "gas meters" appear, they shall be construed to also mean valves and those devices required for the regulation of pressure and the measurement of natural gas being dispensed for any building, structure, or premises.

1303.1 Appliance Fuel Connector – An assembly of listed semirigid or flexible tubing and fittings to carry fuel between a fuel-piping outlet and a fuel-burning appliance.

1303.2 Fuel Gas – Natural, manufactured, liquefied petroleum, or a mixture of these.

1303.3 Gas Piping – Any installation of pipe, valves, or fittings that is used to convey fuel gas, installed on any premises or in any building, but shall not include:

(1) Any portion of the service piping.

(2) Any approved piping connection six (6) feet (1,829 mm) or less in length between an existing gas outlet and a gas appliance in the same room with the outlet.

1303.4 Gas Piping System – Any arrangement of gas piping supplied by one (1) meter and each arrangement of gas piping serving a building, structure, or premises, whether individually metered or not.

1303.5 Liquefied Petroleum Gas (LP-Gas) Facilities – Liquefied petroleum gas facilities are tanks, containers, container valves, regulating equipment, meters, and/or appurtenances for the storage and supply of liquefied petroleum gas for any building, structure, or premises.

1303.6 Provision for Location of Point of Delivery – The location of the point of delivery shall be acceptable to the serving gas supplier. [NFPA 54:5.2]

1303.7 Quick-Disconnect Device – A hand-operated device that provides a means for connecting and disconnecting an appliance or an appliance connector to a gas supply and that is equipped with an automatic means to shut off the gas supply when the device is disconnected.

1303.8 Service Piping – The piping and equipment between the street gas main and the gas piping system inlet that is installed by, and is under the control and maintenance of, the serving gas supplier.

1303.9 Transition Gas Riser – Any listed or approved section or sections of pipe and fittings used to convey fuel gas and installed in a gas piping system for the purpose of providing a transition from below ground to aboveground.

1304.0 Inspection.

1304.1 Upon completion of the installation, alteration, or repair of any gas piping, and prior to the use thereof, the Authority Having Jurisdiction shall be notified that such gas piping is ready for inspection.

1304.2 All excavations required for the installation of underground piping shall be kept open until such time as the piping has been inspected and approved. If any such piping is covered or concealed before such approval, it shall be exposed upon the direction of the Authority Having Jurisdiction.

1304.3 The Authority Having Jurisdiction shall make the following inspections and shall either approve that portion of the work as completed, or shall notify the permit holder wherein the same fails to comply with this code.

1304.3.1 Rough Piping Inspection. This inspection shall be made after all gas piping authorized by the permit has been installed, before any such piping has been covered or concealed, or before any fixture or appliance has been attached thereto. This inspection shall include a determination that the gas piping size, material, and installation meet the requirements of this code.

1304.3.2 Final Piping Inspection. This inspection shall be made after all piping authorized by the permit has been installed and after all portions thereof that are to be covered or concealed are so concealed and before any fixtures, appliance, or shutoff valve has been attached thereto. This inspection shall be in accordance with Section 1316.1. Test gauges used in conducting tests shall comply with the following:

(1) ***Test Gauges.*** *Tests required by this code, that are performed utilizing dial gauges, shall be limited to gauges having the following pressure graduations or incrementations.*

(2) *Required pressure tests of ten (10) psi (69 kPa) or less shall be performed with gauges of 1/10 pound (0.7 kPa) incrementation or less.*

(3) *Required pressure tests exceeding ten (10) pounds (69 kPa) but less than one hundred (100) psi (689 kPa) shall be performed with gauges of one (1) psi (6.9 kPa) incrementation or less.*

(4) *Required pressure tests exceeding one hundred (100) psi (689 kPa) shall be performed with gauges incremented for two (2) percent or less of the required test pressure.*

(5) *Test gauges shall have a pressure range not greater than twice the test pressure applied.*

1304.4 In cases where the work authorized by the permit consists of a minor installation of additional piping to piping already connected to a gas meter, the foregoing inspections may be waived at the discretion of the Authority Having Jurisdiction. In this event, the Authority Having Jurisdiction shall make such inspection as deemed advisable in order to be assured that the work has been performed in accordance with the intent of this code.

1305.0 Certificate of Inspection.

1305.1 If, upon final piping inspection, the installation is found to comply with the provisions of this code, a certificate of inspection may be issued by the Authority Having Jurisdiction.

1305.2 A copy of the certificate of such final piping inspection shall be issued to the serving gas supplier supplying gas to the premises.

1305.3 It shall be unlawful for any serving gas supplier or person furnishing gas to turn on, or cause to be turned on, any fuel gas or any gas meter or meters until such certificate of final inspection, as herein provided, has been issued.

1306.0 Authority to Render Gas Service.

1306.1 It shall be unlawful for any person, firm, or corporation, excepting an authorized agent or employee of a person, firm, or corporation engaged in the business of furnishing or supplying gas and whose service pipes supply or connect with the particular premises, to turn on or reconnect gas service in or on any premises where and when gas service is, at the time, not being rendered.

1306.2 It shall be unlawful to turn on or connect gas in or on any premises unless all outlets are properly and securely connected to gas appliances or capped or plugged with screw joint fittings.

1307.0 Authority to Disconnect.

1307.1 The Authority Having Jurisdiction or the serving gas supplier is hereby authorized to disconnect any gas piping or appliance, or both, that shall be found not to conform to the requirements of this code or that may be found defective and in such condition as to endanger life or property.

1307.2 Where such disconnection has been made, a notice shall be attached to such gas piping or appliance, or both, that shall state the same has been disconnected, together with the reasons therefore.

1307.3 It shall be unlawful to remove or disconnect any gas piping or gas appliance without capping or plugging with a screw joint fitting the outlet from which said pipe or appliance was removed. All outlets to which gas appliances are not connected shall be left capped gastight on any piping system that has been installed, altered, or repaired.

Exception: When an approved listed quick-disconnect device is used.

1308.0 Temporary Use of Gas.

Where temporary use of gas is desired and the Authority Having Jurisdiction deems the use necessary, a permit may be issued for such use for a period of time not to exceed that designated by the Authority Having Jurisdiction, provided that such gas piping system otherwise conforms to the requirements of this code regarding material, sizing, and safety.

1309.0 Gas Piping System Design, Materials, and Components.

1309.1 Piping Plan.

1309.1.1 Installation of Piping System. Where required by the Authority Having Jurisdiction, a piping sketch or plan shall be prepared before proceeding with the installation. This plan shall show the proposed location of piping, the size of different branches, the various load demands, and the location of the point of delivery.

1309.1.2 Addition to Existing System. When additional appliances are being connected to a gas piping system, the existing piping shall be checked to determine whether it has adequate capacity (see Section 1309.4.3). If inadequate, the existing system shall be enlarged as required, or separate gas piping of adequate capacity shall be provided. [NFPA 54:5.1.2]

1309.2 Provision for Location of Point of Delivery. The location of the point of delivery shall be acceptable to the serving gas supplier. [NFPA 54:5.2]

1309.3 Interconnections Between Gas Piping Systems. [NFPA 54:5.3]

1309.3.1 Interconnections Supplying Separate Users. Where two (2) or more meters, or two (2) or more service regulators where meters are not provided, are located on the same premises and supply separate users, the gas piping systems shall not be interconnected on the outlet side of the meters or service regulators. [NFPA 54:5.3.1]

1309.3.2 Interconnections for Standby Fuels. Where a supplementary gas for standby use is connected downstream from a meter or a service regulator where a meter is not provided, a device to prevent backflow shall be installed. A three-way valve installed to admit the standby supply and at the same time shut off the regular supply shall be permitted to be used for this purpose. [NFPA 54:5.3.2]

1309.4 Sizing of Gas Piping Systems.

1309.4.1 General Considerations. Gas piping systems shall be of such size and so installed as to provide a supply of gas sufficient to meet the maximum demand and supply gas to each appliance inlet at not less than the minimum supply pressure required by the appliance. [NFPA 54-09:5.4.1]

1309.4.2 Maximum Gas Demand. The volume of gas to be provided (in cubic feet per hour) shall be determined directly from the manufacturer's input ratings of the appliance served. Where the input rating is not indicated, the gas supplier, the appliance manufacturer, or a qualified agency shall be contacted, or the rating from Table 13-1 shall be used for estimating the volume of gas to be supplied. The total connected hourly load shall be used as the basis for pipe sizing, assuming all appliances are operating at full capacity simultaneously. [NFPA 54:5.4.2]

Exception: Sizing shall be permitted to be based upon established load diversity factors.

Table 13-1
Approximate Gas Input for Typical Appliances
[NFPA 54: Table 5.4.2.1]

Appliance	Input Btu/h (approx.)
Space Heating Units	
Warm air furnace	
Single family	100,000
Multifamily, per unit	60,000
Hydronic boiler	
Single family	100,000
Multifamily, per unit	60,000
Space and Water Heating Units	
Hydronic boiler	
Single family	120,000
Multifamily, per unit	75,000
Water Heating Appliances	
Water heater, automatic storage	
30 to 40 gal tank	35,000
Water heater, automatic storage	
50 gal tank	50,000
Water heater, automatic instantaneous	
Capacity at 2 gal/minute	142,800
Capacity at 4 gal/minute	285,000
Capacity at 6 gal/minute	428,400
Water heater, domestic,	
Circulating or side-arm	35,000
Cooking Appliances	
Range, free standing, domestic	65,000
Built-in oven or broiler unit, domestic	25,000
Built-in top unit, domestic	40,000
Other Appliances	
Refrigerator	3,000
Clothes dryer, Type 1 (domestic)	35,000
Gas fireplace, direct vent	40,000
Gas log	80,000
Barbecue	40,000
Gas light	2,500

Note: 1 Btu/hr = 0.293 W

1309.4.3 Sizing Methods. Gas piping shall be sized in accordance with one of the following:

(1) Pipe sizing tables or sizing equations in this chapter.

(2) Other approved engineering methods acceptable to the Authority Having Jurisdiction.

(3) Sizing tables included in a listed piping system manufacturer's installation instructions. [NFPA 54:5.4.3]

1309.4.4 Allowable Pressure Drop. The design pressure loss in any piping system under maximum probable flow conditions, from the point of delivery to the inlet connection of the appliance, shall be such that the supply pressure at the appliance is greater than or equal to the minimum pressure required by the appliance. [NFPA 54-09:5.4.4]

1309.5 Acceptable Piping Materials and Joining Methods.

1309.5.1 General.

1309.5.1.1 Materials. Materials used for piping systems shall comply with the requirements of this chapter or shall be acceptable to the Authority Having Jurisdiction. [NFPA 54:5.6.1.1]

1309.5.1.2 Used Materials. Pipe, fittings, valves, or other materials shall not be used again unless they are free of foreign materials and have been ascertained to be adequate for the service intended. [NFPA 54:5.6.1.2]

1309.5.1.3 Other Materials. Material not covered by the standard specifications listed herein shall be investigated and tested to determine that it is safe and suitable for the proposed service and, in addition, shall be recommended for that service by the manufacturer and shall be acceptable to the Authority Having Jurisdiction. [NFPA 54:5.6.1.3]

1309.5.2 Metallic Pipe.

1309.5.2.1 Cast-iron pipe shall not be used. [NFPA 54:5.6.2.1]

1309.5.2.2 Steel and wrought-iron pipe shall be at least of standard weight (Schedule 40) and shall comply with one of the following standards:

(1) ASME B36.10, *Welded and Seamless Wrought-Steel Pipe*

(2) ASTM A 53, *Standard Specification for Pipe, Steel, Black and Hot-Dipped, Zinc-Coated Welded and Seamless*

(3) ASTM A 106, *Standard Specification for Seamless Carbon Steel Pipe for High-Temperature Service* [NFPA 54:5.6.2.2]

1309.5.2.3 Copper and brass pipe shall not be used if the gas contains more than an average of 0.3 grains of hydrogen sulfide per one-hundred (100) scf of gas (0.7 mg/100 L). [NFPA 54:5.6.2.3]

Threaded copper, brass, or aluminum alloy pipe shall not be used with gases corrosive to such material. [NFPA 54:5.6.2.4]

1309.5.2.4 Aluminum alloy pipe shall comply with ASTM B 241, *Standard Specification for Aluminum and Aluminum-Alloy Seamless Pipe and Seamless Extruded Tube* (except that the use of alloy 5456 is prohibited) and shall be marked at each end of each length indicating compliance. Aluminum alloy pipe shall be coated to protect against external corrosion where it is in contact with masonry, plaster, insulation, or is subject to repeated wettings by such liquids as water, detergents, or sewage. [NFPA 54:5.6.2.5]

Aluminum alloy pipe shall not be used in exterior locations or underground. [NFPA 54:5.6.2.6]

1309.5.3 Metallic Tubing. Seamless copper, aluminum alloy, or steel tubing shall not be used with gases corrosive to such material. [NFPA 54:5.6.3]

1309.5.3.1 Steel tubing shall comply with ASTM A 254, *Standard Specification for Copper-Brazed Steel Tubing.* [NFPA 54-09:5.6.3.1]

1309.5.3.2 Copper and brass tubing shall not be used if the gas contains more than an average of 0.3 grains of hydrogen sulfide per one-hundred (100) scf of gas (0.7 mg/100 L). Copper tubing shall comply with standard Type K or L of ASTM B 88, *Standard Specification for Seamless Copper Water Tube*, or ASTM B 280, *Standard Specification for Seamless Copper Tube for Air Conditioning and Refrigeration Field Service.* [NFPA 54:5.6.3.2]

1309.5.3.3 Aluminum alloy tubing shall comply with ASTM B 210, *Standard Specification for Aluminum and Aluminum-Alloy Drawn Seamless Tubes*, or ASTM B 241, *Standard Specification for Aluminum and Aluminum-Alloy Seamless Pipe and Seamless Extruded Tube.* Aluminum alloy tubing shall be coated to protect against external corrosion where it is in contact with masonry, plaster, insulation, or is subject to repeated wettings by such liquids as water, detergent, or sewage. Aluminum alloy tubing shall not be used in exterior locations or underground. [NFPA 54:5.6.3.3]

1309.5.3.4 Corrugated stainless steel tubing shall be tested and listed in compliance with the construction, installation, and performance requirements of IAS LC-1, *Standard for Fuel Gas Piping Systems Using Corrugated Stainless Steel Tubing.* [NFPA 54:5.6.3.4]

1309.5.4 Plastic Pipe, Tubing, and Fittings. Plastic pipe, tubing, and fittings used to supply fuel gas shall conform with ASTM D 2513, *Standard Specification for Thermoplastic Gas Pressure Pipe, Tubing, and Fittings.* Pipe to be used shall be marked "gas" and "ASTM D 2513." [NFPA 54:5.6.4.1]

Anodeless risers shall comply with the following: [NFPA 54:5.6.4.3]

1309.5.4.1 Factory-assembled anodeless risers shall be recommended by the manufacturer for the gas used and shall be leak-tested by the manufacturer in accordance with written procedures. [NFPA 54:5.6.4.3(1)]

1309.5.4.2 Service head adapters and field-assembled anodeless risers incorporating service head adapters shall be recommended by the manufacturer for the gas used by the manufacturer and shall be design-certified to meet the requirements of Category I of ASTM F1973, *Factory Assembled Anodeless Riser and Transition Fitting in Polyethylene (PE) Fuel Gas Distribution Systems*, and the Code of Federal Regulations, Title 49, Part 192.281(e). The manufacturer shall provide the user-qualified installation instructions as pre-scribed by the Code of Federal Regulations, Title 49, Part 192.283(b). [NFPA 54:5.6.4.3(2)]

1309.5.4.3 The use of plastic pipe, tubing, and fittings in undiluted liquefied petroleum gas piping systems shall be in accordance with NFPA 58, *Liquefied Petroleum Gas Code.* [NFPA 54:5.6.4.3(3)]

1309.5.5 Workmanship and Defects. Gas pipe, tubing, and fittings shall be clear and free from cutting burrs and defects in structure or threading and shall be thoroughly brushed and chip and scale blown. Defects in pipe, tubing, and fittings shall not be repaired. Defective pipe, tubing, and fittings shall be replaced. [NFPA 54:5.6.5]

1309.5.6 Protective Coating. Where in contact with material or atmosphere exerting a corrosive action, metallic piping and fittings coated with a corrosion-resistant material shall be used. External or internal coatings or linings used on piping or components shall not be considered as adding strength. [NFPA 54:5.6.6]

1309.5.7 Metallic Pipe Threads.

(A) Specifications for Pipe Threads. Metallic pipe and fitting threads shall be taper pipe threads and shall comply with ASME B1.20.1, *Standard for Pipe Threads, General Purpose (Inch)*. [NFPA 54:5.6.7.1]

(B) Damaged Threads. Pipe with threads that are stripped, chipped, corroded, or otherwise damaged shall not be used. Where a weld opens during the operation of cutting or threading, that portion of the pipe shall not be used. [NFPA 54:5.6.7.2]

(C) Number of Threads. Field threading of metallic pipe shall be in accordance with Table 13-2. [NFPA 54:5.6.7.3]

(D) Thread Compounds. Thread (joint) compounds (pipe dope) shall be resistant to the action of liquefied petroleum gas or to any other chemical constituents of the gases to be conducted through the piping. [NFPA 54:5.6.7.4]

TABLE 13-2
Specifications for Threading Metallic Pipe
[NFPA 54:5.6.7.3]

Iron Pipe Size (in.)	Approximate Length of Threaded Portion (in.)	Approximate No. of Threads to Be Cut
1/2	3/4	10
3/4	3/4	10
1	7/8	10
1-1/4	1	11
1-1/2	1	11
2	1	11
2-1/2	1-1/2	12
3	1-1/2	12
4	1-5/8	13

Note: 1 in. = 25.4 mm

1309.5.8 Metallic Piping Joints and Fittings. The type of piping joint used shall be suitable for the pressure-temperature conditions and shall be selected giving consideration to joint tightness and mechanical strength under the service conditions. The joint shall be able to sustain the maximum end force due to the internal pressure and any additional forces due to temperature expansion or contraction, vibration, fatigue, or the weight of the pipe and its contents. [NFPA 54:5.6.8]

1309.5.8.1 Pipe Joints. Pipe joints shall be threaded, flanged, brazed, press connected, or welded. Where nonferrous pipe is brazed, the brazing materials shall have a melting point in excess of 1,000°F (538°C). Brazing alloys shall not contain more than 0.05 percent phosphorus. [NFPA 54:5.6.8.1]

1309.5.8.2 Tubing Joints. Tubing joints shall either be made with approved gas tubing fittings, be brazed with a material having a melting point in excess of 1,000°F (538°C), or made by press-connect fittings complying with CSA LC-4. Brazing alloys shall not contain more than 0.05 percent phosphorus. [NFPA 54:5.6.8.2]

1309.5.8.3 Flared Joints. Flared joints shall be used only in systems constructed from nonferrous pipe and tubing where experience or tests have demonstrated that the joint is suitable for the conditions and where provisions are made in the design to prevent separation of the joints. [NFPA 54:5.6.8.3]

1309.5.8.4 Metallic Fittings (Including Valves, Strainers, Filters).

(1) Threaded fittings in sizes larger than four (4) inches (100 mm) shall not be used unless acceptable to the Authority Having Jurisdiction.

(2) Fittings used with steel or wrought-iron pipe shall be steel, brass, bronze, malleable iron, or cast-iron.

(3) Fittings used with copper or brass pipe shall be copper, brass, or bronze.

(4) Fittings used with aluminum alloy pipe shall be of aluminum alloy.

(5) Cast-Iron Fittings.
   (a) Flanges shall be permitted.
   (b) Bushings shall not be used.
   (c) Fittings shall not be used in systems containing flammable gas-air mixtures.
   (d) Fittings in sizes four (4) inches (100 mm) and larger shall not be used indoors unless approved by the Authority Having Jurisdiction.
   (e) Fittings in sizes six (6) inches (150 mm) and larger shall not be used unless approved by the Authority Having Jurisdiction.

(6) Aluminum Alloy Fittings. Threads shall not form the joint seal.

(7) Zinc-Aluminum Alloy Fittings. Fittings shall not be used in systems containing flammable gas-air mixtures.

(8) Special Fittings. Fittings such as couplings, proprietary-type joints, saddle tees, gland-type compression fittings, and flared, flareless, or compression-type tubing fittings shall be (1) used within the fitting manufacturer's pressure-temperature recommendations; (2) used within the service conditions anticipated with respect to vibration, fatigue, thermal expansion, or contraction; (3) installed or braced to prevent separation of the joint by gas pressure or external physical damage; and (4) acceptable to the Authority Having Jurisdiction. [NFPA 54:5.6.8.4]

1309.5.9 Flanges. Flanges shall comply with ASME B16.1, *Standard for Cast-Iron Pipe Flanges and Flanged Fittings;* ASME B16.20, *Standard for Ring-Joint Gaskets and Grooves for Steel Pipe Flanges;* or MSS SP-6, *Standard Finishes for Contact Faces of Pipe Flanges and Connecting-End Flanges of Valves and Fittings.* The pressure-temperature ratings shall equal or exceed that required by the application. [NFPA 54:5.6.10]

(A) Flange Facings. Standard facings shall be permitted for use under this code. Where one-hundred and fifty (150) psi (1,090 kPa) steel flanges are bolted to Class 125 cast-iron flanges, the raised face on the steel flange shall be removed. [NFPA 54:5.6.10.1]

(B) Lapped Flanges. Lapped flanges shall be used only aboveground or in exposed locations accessible for inspection. [NFPA 54:5.6.10.2]

1309.5.10 Flange Gaskets. The material for gaskets shall be capable of withstanding the design temperature and pressure of the piping system and the chemical constituents of the gas being conducted without change to its chemical and physical properties. The effects of fire exposure to the joint shall be considered in choosing the material. [NFPA 54:5.6.11]

(1) Acceptable materials include the following:
  - (a) Metal or metal-jacketed asbestos (plain or corrugated)
  - (b) Asbestos
  - (c) Aluminum O-rings and spiral-wound metal gaskets [NFPA 54:5.6.11.1]

(2) When a flanged joint is opened, the gasket shall be replaced. [NFPA 54:5.6.11.2]

(3) Full-face gaskets shall be used with all bronze and cast-iron flanges. [NFPA 54:5.6.11.3]

1309.6 Gas Meters.

1309.6.1 Capacity. Gas meters shall be selected for the maximum expected pressure and permissible pressure drop. [NFPA 54:5.7.1]

1309.6.2 Location.

(A) Gas meters shall be located in ventilated spaces readily accessible for examination, reading, replacement, or necessary maintenance. [NFPA 54:5.7.2.1]

(B) Gas meters shall not be placed where they will be subjected to damage, such as adjacent to a driveway, under a fire escape, in public passages, halls, or coal bins, or where they will be subject to excessive corrosion or vibration. [NFPA 54:5.7.2.2]

(C) Gas meters shall not be located where they will be subjected to extreme temperatures or sudden extreme changes in temperature.

Meters shall not be located in areas where they are subjected to temperatures beyond those recommended by the manufacturer. [NFPA 54:5.7.2.4]

1309.6.3 Supports. Gas meters shall be supported or connected to rigid piping so as not to exert a strain on the meters. Where flexible connectors are used to connect a gas meter to downstream piping at mobile homes in mobile home parks, the meter shall be supported by a post or bracket placed in a firm footing or by other means providing equivalent support. [NFPA 54:5.7.3]

1309.6.4 Meter Protection. Meters shall be protected against overpressure, back pressure, and vacuum, where such conditions are anticipated. [NFPA 54:5.7.4]

1309.6.5 Identification. Gas piping at multiple meter installations shall be marked by a metal tag or other permanent means attached by the installing agency, designating the building or the part of the building being supplied. [NFPA 54:5.7.5]

1309.7 Gas Pressure Regulators.

1309.7.1 Where Required. A line gas pressure regulator or gas appliance pressure regulator, as applicable, shall be installed where the gas supply pressure is higher than that at which the branch supply line or appliance is designed to operate or varies beyond design pressure limits. [NFPA 54:5.8.1]

1309.7.2 Listing. Line gas pressure regulators shall be listed in accordance with CSA Z21.80. [NFPA 54:5.8.2]

1309.7.3 Location. The gas pressure regulator shall be accessible for servicing. [NFPA 54:5.8.3]

1309.7.4 Regulator Protection. Pressure regulators shall be protected against physical damage. [NFPA 54:5.8.4]

1309.7.5 Venting.

(A) Line Gas Pressure Regulators.

(1) An independent vent to the outside of the building, sized in accordance with the regulator manufacturer's instructions, shall be provided where the location of a regulator is such that a ruptured diaphragm will cause a hazard. Where there is more than one (1) regulator at a location, each regulator shall have a separate vent to the outside, or if approved by the Authority Having Jurisdiction, the vent lines shall be permitted to be manifolded in accordance with accepted engineering practices to minimize back pressure in the event of diaphragm failure. (See NFPA 54:5.9.7 for information on properly locating the vent.) Materials for vent piping shall be in accordance with Section 1309.5.

Exception: A regulator and vent-limiting means combination listed as complying with CSA Z21.80, *Standard for Line Pressure Regulators,* shall be permitted to be used without a vent to the outdoors.

The following references were extracted from NFPA 54:

5.9.7 Vents. The discharge stacks, vents or outlet parts of all pressure-relieving and pressure-limiting devices shall be located so that gas is safely discharged into the outside atmosphere. Discharge stacks or vents shall be designed to prevent the entry of water, insects, or any other foreign material that could cause blockage. The discharge stack or vent line shall be at least the same size as the outlet of the pressure-relieving device.

(2) The vent shall be designed to prevent the entry of water, insects, or other foreign materials that could cause blockage.

(3) At locations where regulators might be submerged during floods, a special anti-flood-type breather vent fitting shall be installed, or the vent line shall be extended above the height of the expected flood waters.

(4) A regulator shall not be vented to the appliance flue or exhaust system. [NFPA 54:5.8.5.1]

(B) Gas Appliance Pressure Regulators. For venting of gas appliance pressure regulators (see Section 1311.0).

1309.7.6 Bypass Piping. Valved and regulated bypasses shall be permitted to be placed around gas line pressure regulators where continuity of service is imperative. [NFPA 54:5.8.6]

1309.7.7 Identification. Line pressure regulators at multiple regulator installations shall be marked by a metal tag or other permanent means designating the building or the part of the building being supplied. [NFPA 54:5.8.7]

1309.8 Back-Pressure Protection.

1309.8.1 Where to Install. Protective devices shall be installed as close to the appliance as practical where the design of appliance connected is such that air, oxygen, or standby gases could be forced into the gas supply system. [NFPA 54:5.10.1.1] Gas and air combustion mixers incorporating double diaphragm "zero" or "atmosphere" governors or regulators shall require no further protection unless connected directly to compressed air or oxygen at pressures of 5 psi (34 kPa) or more. [NFPA 54:5.10.1.2]

1309.8.2 Protective Devices. Protective devices shall include, but not be limited to, the following:

(1) Check valves.

(2) Three-way valves (of the type that completely closes one side before starting to open the other side).

(3) Reverse flow indicators controlling positive shutoff valves.

(4) Normally closed air-actuated positive shutoff pressure regulators. [NFPA 54:5.10.2]

1309.9 Low-Pressure Protection. A protective device shall be installed between the meter and the appliance if the operation of the appliance is such (i.e., a gas compressor) that it could produce a vacuum or a dangerous reduction in gas pressure at the meter. Such devices include, but are not limited to, mechanical, diaphragm-operated, or electrically operated low-pressure shutoff valves. [NFPA 54:5.11]

1309.10 Shutoff Valves. Shutoff valves shall be approved and shall be selected giving consideration to pressure drop, service involved, emergency use, and reliability of operation. Shutoff valves of size 1 inch National Pipe Thread and smaller shall be listed. [NFPA 54:5.12]

1309.11 Expansion and Flexibility.

1309.11.1 Design. Piping systems shall be designed to have sufficient flexibility to prevent thermal expansion or contraction from causing excessive stresses in the piping material, excessive bending or loads at joints, or undesirable forces or moments at points of connections to appliances and at anchorage or guide points. Formal calculations or model tests shall be required only where reasonable doubt exists as to the adequate flexibility of the system. [NFPA 54:5.13.1]

Flexibility shall be provided by the use of bends, loops, offsets, or couplings of the slip type. Provision shall be made to absorb thermal changes by the use of expansion joints of the bellows type, or by the use of "ball" or "swivel" joints. Expansion joints of the slip type shall not be used inside buildings or for thermal expansion. Where expansion joints are used, anchors or ties of sufficient strength and rigidity shall be installed to provide for end forces due to fluid pressure and other causes. [NFPA 54:5.13.1.1]

Pipe alignment guides shall be used with expansion joints, according to the recommended practice of the joint manufacturer. [NFPA 54:5.13.1.2]

1309.11.2 Special Local Conditions. Where local conditions include earthquake, tornado, unstable ground, or flood hazards, special consideration shall be given to increased strength and flexibility of piping supports and connections. [NFPA 54:5.13.2]

1310.0 Excess Flow Valve. Where automatic excess flow gas shutoff devices (valves) are installed, they shall be listed for the application and shall be sized according to the manufacturer's instructions. [NFPA 54-09:5.13]

1311.0 Venting of Gas Appliance Pressure Regulators.

Venting of gas appliance pressure regulators shall comply with the following requirements :

(1) Gas appliance pressure regulators requiring access to the atmosphere for successful operation shall be equipped with vent piping leading outdoors or, if the regulator vent is an integral part of the appliance, into the combustion chamber adjacent to a continuous pilot, unless constructed or equipped with a vent-limiting means to limit the escape of gas from the vent opening in the event of diaphragm failure.

(2) Vent-limiting means shall be employed on listed gas appliance pressure regulators only.

(3) In the case of vents leading outdoors, means shall be employed to prevent water from entering this piping and also to prevent blockage of vents by insects and foreign matter.

(4) Under no circumstances shall a regulator be vented to the appliance flue or exhaust system.

(5) In the case of vents entering the combustion chamber, the vent shall be located so the escaping gas will be readily ignited by the pilot and the heat liberated thereby will not adversely affect the normal operation of the safety shutoff system. The terminus of the vent shall be securely held in a fixed position relative to the pilot. For manufactured gas, the need for a flame arrester in the vent piping shall be determined.

(6) Vent line(s) from a gas appliance pressure regulator and bleed line(s) from a diaphragm-type valve shall not be connected to a common manifold terminating in a combustion chamber. Vent lines shall not terminate in positive-pressure-type combustion chambers. [NFPA 54:9.1.19]

1312.0 Gas Piping Installation.

1312.1 Piping Underground.

1312.1.1 Clearances. Underground gas piping shall be installed with sufficient clearance from any other underground structure to avoid contact therewith, to allow maintenance, and to protect against damage from proximity to other structures. In addition, underground plastic piping shall be installed with sufficient clearance or shall be insulated from any source of heat so as to prevent the heat from impairing the serviceability of the pipe. [NFPA 54:7.1.1]

1312.1.2 Protection Against Damage.

(A) Cover Requirements. Underground piping systems shall be installed with a minimum of eighteen (18) inches (460 mm) of cover. Where external damage to the pipe is not likely to result, the minimum cover shall be twelve (12) inches (300 mm). Where a minimum of twelve (12) inches (300 mm) of cover cannot be provided, the pipe shall be installed in conduit or bridged (shielded). [NFPA 54:7.1.2.1]

(B) Trenches. The trench shall be graded so that the pipe has a firm, substantially continuous bearing on the bottom of the trench. [NFPA 54:7.1.2.2]

(C) Backfilling. Where flooding of the trench is done to consolidate the backfill, care shall be exercised to see that the pipe is not floated from its firm bearing on the trench bottom. [NFPA 54:7.1.2.3]

1312.1.3 Protection Against Corrosion. Gas piping in contact with earth or other material that could corrode the piping shall be protected against corrosion in an approved manner. When dissimilar metals are joined underground, an insulating coupling or fitting shall be used. Piping shall not be laid in contact with cinders. Uncoated threaded or socket-welded joints shall not be used in piping in contact with soil or where internal or external crevice corrosion is known to occur. [NFPA 54:7.1.3]

1312.1.4 Protection Against Freezing. Where the formation of hydrates or ice is known to occur, piping shall be protected against freezing. [NFPA 54:7.1.4]

1312.1.5 Piping Through Foundation Wall. Underground piping, where installed through the outer foundation or basement wall of a building, shall be encased in a protective pipe. The space between the gas piping and the building shall be sealed to prevent entry of gas or water. [NFPA 54:7.1.5]

1312.1.6 Piping Underground Beneath Buildings. Where the installation of gas piping underground beneath buildings is unavoidable, the piping shall be encased in an approved conduit designed to withstand the superimposed loads. [NFPA 54:7.1.6]

The conduit shall extend into a normally usable and accessible portion of the building and, at the point where the conduit terminates in the building, the space between the conduit and the gas piping shall be sealed to prevent the possible entrance of any gas leakage. Where the end sealing is of a type that will retain the full pressure of the pipe, the conduit shall be designed for the same pressure as the pipe. The conduit shall extend at least four (4) inches (100 mm) outside the building, be vented above finished grade to the outside, and be installed so as to prevent the entrance of water and insects. [NFPA 54:7.1.6.1]

1312.1.7 Plastic Pipe.

(A) Connection of Plastic Piping. Plastic pipe shall be installed outside, underground only.

Exceptions:

(1) Plastic pipe shall be permitted to terminate above ground where an anodeless riser is used.

(2) Plastic pipe shall be permitted to terminate with a wall head adapter above-ground in buildings, including basements, where the plastic pipe is inserted in a piping material permitted for use in buildings. [NFPA 54:7.1.7.1]

(B) Connections made outside and underground between metallic and plastic piping shall be between metallic and plastic piping shall be made with fittings conforming to one of the following:

(1) ASTM D 2513, *Standard Specification for Thermoplastic Gas Pressure Pipe, Tubing, Fittings, Category I transition fittings.*

(2) ASTM F 1973, *Standard Specification for Factory Assembled Anodeless Risers and Transition Fittings in Polyethylene (PE) and Polyamide 11 (PA11) Fuel Gas Distribution Systems.*

(3) ASTM F 2509, *Standard Specification for Field-assembled Anodeless Riser Kits for Use on Outside Diameter Controlled Polyethylene Gas Distribution Pipe and Tubing.* [NFPA 54-09:7.1.7.2]

(C) An electrically continuous corrosion-resistant tracer wire (minimum AWG 14) or tape shall be buried with the plastic pipe to facilitate locating. One (1) end shall be brought above ground at a building wall or riser. [NFPA 54:7.1.7.3]

1312.2 Installation of Piping.

1312.2.1 Piping installed above ground shall be securely supported and located where it will be protected from physical damage (also see 1311.1.4). Where passing through an outside wall, the piping shall also be protected against corrosion by coating or wrapping with an inert material approved for such applications. Where piping is encased in a protective pipe sleeve, the annular space between the gas piping and the sleeve shall be sealed at the wall to prevent the entry of water, insects, or rodents. [NFPA 54:7.2.1]

1312.2.2 Building Structure.

(1) The installation of gas piping shall not cause structural stresses within building components to exceed allowable design limits. [NFPA 54:7.2.2.1]

(2) Approval shall be obtained before any beams or joists are cut or notched. [NFPA 54:7.2.2.2]

1312.2.3 Other Than Dry Gas. Drips, sloping, protection from freezing, and branch pipe connections, as provided for in Sections 1311.1.4,

1311.6.1, and 1311.8, shall be provided when other than dry gas is distributed and climactic conditions make such provisions necessary. [NFPA 54:7.2.3]

1312.2.4 Gas Piping to be Sloped. Piping for other than dry gas conditions shall be sloped not less than one-fourth (1/4) inch in fifteen (15) feet (7 mm in 4.6 m) to prevent traps. [NFPA 54:7.2.4]

1312.2.4.1 Ceiling Locations. Gas piping shall be permitted to be installed in accessible spaces between a fixed ceiling and a dropped ceiling, whether or not such spaces are used as a plenum. Valves shall not be located in such spaces.

Exception: Equipment shutoff valves required by this code shall be permitted to be installed in accessible spaces containing vented appliances.

1312.2.5 Prohibited Locations. Gas piping inside any building shall not be installed in or through a circulating air duct, clothes chute, chimney or gas vent, ventilating duct, dumbwaiter, or elevator shaft. [NFPA 54:7.2.5] This provision shall not apply to ducts used to provide combustion and ventilation air in accordance with Section 701.0 or to above-ceiling spaces as covered in 1311.2.4.1.

1312.2.6 Hangers, Supports, and Anchors.

(A) Piping shall be supported with metal pipe hooks, metal pipe straps, metal bands, metal brackets, or metal hangers, or building structural components, suitable for the size of piping, of adequate strength and quality, and located at intervals so as to prevent or damp out excessive vibration. Piping shall be anchored to prevent undue strains on connected appliances and shall not be supported by other piping. Pipe hangers and supports shall conform to the requirements of MSS SP-58, *Pipe Hangers and Supports - Materials, Design and Manufacture.* [NFPA 54:7.2.6.1]

(B) Spacings of supports in gas piping installations shall not be greater than shown in Table 13-3. Spacing of supports for CSST shall be in accordance with the CSST manufacturer's instruction. [NFPA 54:7.2.6.2]

(C) Supports, hangers, and anchors shall be installed so as not to interfere with the free expansion and contraction of the piping between anchors. All parts of the supporting appliances shall be designed and installed so they will not be disengaged by movement of the supported piping. [NFPA 54:7.2.6.3]

TABLE 13-3
Support of Piping
[NFPA 54: Table 7.2.6.2]

Steel Pipe, Nominal Size of Pipe (in.)	Spacing of Supports (ft.)	Nominal Size of Tubing Smooth-wall (in. O.D.)	Spacing of Supports (ft.)
1/2	6	1/2	4
3/4 or 1	8	5/8 or 3/4	6
1-1/4 or larger (horizontal)	10	7/8 or 1 (horizontal)	8
1-1/4 or larger (vertical)	Every floor level	1 or larger (vertical)	Every floor level

Note: 1 ft. = 0.305 m

1312.2.7 Removal of Pipe. Where piping containing gas is to be removed, the line shall be first disconnected from all sources of gas and then thoroughly purged with air, water, or inert gas before any cutting or welding is done. (See Section 1316.6) [NFPA 54:7.2.7]

1312.3 Concealed Piping Building.

1312.3.1 General. Gas piping in concealed locations shall be installed in accordance with this section. [NFPA 54:7.3.1]

1312.3.2 Connections. Where gas piping is to be concealed, unions, tubing fittings, right and left couplings, bushings, swing joints, and compression couplings made by combinations of fittings shall not be used. Connections shall be of the following type:

(1) Pipe fittings, such as elbows, tees, and couplings.

(2) Joining tubing by brazing (See Section 1309.5.8.2.).

(3) Fittings listed for use in concealed spaces that have been demonstrated to sustain, without leakage, any forces due to temperature expansion or contraction, vibration, or fatigue based on their geographic location, application, or operation.

(4) Where necessary to insert fittings in gas pipe that has been installed in a concealed location, the pipe shall be reconnected by welding, flanges, or the use of a ground joint union with the nut center-punched to prevent loosening by vibration. [NFPA 54:7.3.2]

1312.3.3 Piping in Partitions. Concealed gas piping shall not be located in solid partitions. [NFPA 54:7.3.3]

1312.3.4 Tubing in Partitions. This provision shall not apply to tubing that pierces walls, floors, or partitions. Tubing installed vertically and horizontally inside hollow walls or partitions without protection along its entire concealed length shall meet the following requirements: [NFPA 54:7.3.4]

1312.3.4.1 Tubing shall only be installed horizontally or vertically.

1312.3.4.2 A steel striker barrier not less than 0.0508 in. (1.3 mm) thick, or equivalent, shall be installed between the tubing and the finished wall and shall extend at least four (4) inches (100 mm) beyond concealed penetrations of plates, fire stops, wall studs, and similar construction features. [NFPA 54:7.3.4(1)]

1312.3.4.3 The tubing shall be installed in single runs and shall not be rigidly secured. [NFPA 54:7.3.4(2)]

1312.3.4.4 Where the wall or partition contains batt-type insulation the insulation shall not restrict the movement of the tubing.

1312.3.5 Piping in Floors. In industrial occupancies, gas piping in solid floors such as concrete shall be laid in channels in the floor and covered to permit access to the piping with a minimum of damage to the building. Where piping in floor channels could be exposed to excessive moisture or corrosive substances, the piping shall be protected in an approved manner. [NFPA 54:7.3.5.1]

Exception: In other than industrial occupancies and where approved by the Authority Having Jurisdiction, gas piping embedded in concrete floor slabs constructed with portland cement shall be surrounded with a minimum of 1-1/2 inch (38 mm) of concrete and shall not be in physical contact with other metallic structures, such as reinforcing rods or electrically neutral conductors. All piping, fittings, and risers shall be protected against corrosion in accordance with Section 1309.5.6. Piping shall not be embedded in concrete slabs containing quickset additives or cinder aggregate. [NFPA 54:7.3.5.2]

1312.4 Piping in Vertical Chases. (See Section 1302.0) Where gas piping exceeding five (5) psi (34 kPa) is located within vertical chases in accordance with NFPA 54 Section 5.5.1(2), the requirements of Sections 1311.4.1 through 1311.4.3 shall apply. [NFPA 54:7.4]

The following references were extracted from NFPA 54:

5.5.1 Maximum Design Operating Pressure. The maximum design operating pressure for piping systems located inside buildings shall not exceed 5 psi (34 kPa) unless one or more of the following conditions are met:

(1) The piping system is welded.

(2) The piping is located in a ventilated chase or otherwise enclosed for protection against accidental gas accumulation.

(3) The piping is located inside buildings or separate areas of buildings used exclusively for one of the following:

(a) Industrial processing or heating

(b) Research

(c) Warehousing

(d) Boiler or mechanical equipment rooms

(4) The piping is a temporary installation for building under construction.

(5) The piping serves appliances or equipment used for agricultural purposes.

(6) The piping system is an LP-Gas piping system with a design operating pressure greater than 20 psi and complies with NFPA 58: 5.5.2 Liquefied Petroleum Gas Systems. LP-Gas systems designed to operate below -5°F (-21°C) or with butane or a propane-butane mix shall be designed to either accommodate liquid LP-Gas or to prevent LP-Gas vapor from condensing back into liquid.

5.5.2 Liquefied Petroleum Gas Systems. LP-Gas systems designed to operate below -5 F (-21C) or with butane or a propane-butane mix shall be designed to either accommodate liquid LP-Gas or to prevent LP-Gas vapor from condensing back into a liquid.

1312.4.1 Pressure Reduction. (See Section 1302) Where pressure reduction is required in branch connections for compliance with NFPA 54 Section 5.5.1, such reduction shall take place either inside the chase or immediately adjacent to the outside wall of the chase. Regulator venting and downstream overpressure protection shall comply with Section 1309.7.5. The regulator shall be accessible for service and repair and vented in accordance with one (1) of the following:

(1) Where the fuel gas is lighter than air, regulators equipped with a vent-limiting means shall be permitted to be vented into the chase. Regulators not equipped with a vent-limiting means shall be permitted to be vented either directly to the outdoors or to a point within the top one (1) foot (305 mm) of the chase.

(2) Where the fuel gas is heavier than air, the regulator vent shall be vented only directly to the outdoors. [NFPA 54:7.4.1]

1312.4.2 Construction. Chase construction shall comply with local building codes with respect to fire resistance and protection of horizontal and vertical openings. [NFPA 54:7.4.2]

1312.4.3 Ventilation. A chase shall be ventilated to the outdoors and only at the top. The opening(s) shall have a minimum free area (in square inches) equal to the product of one-half (1/2) of the maximum pressure in the piping (in psi) times the largest nominal diameter of that piping (in inches), or the cross-sectional area of the chase, whichever is smaller. Where more than one (1) fuel gas piping system is present, the free area for each system shall be calculated and the largest area used. [NFPA 54:7.4.3]

1312.5 Gas Pipe Turns. Changes in direction of gas pipe shall be made by the use of fittings or factory bends. [NFPA 54:7.5]

1312.5.1 Metallic Pipe. Metallic pipe bends shall comply with the following:

(1) Bends shall be made only with bending equipment and procedures intended for that purpose.

(2) All bends shall be smooth and free from buckling, cracks, or other evidence of mechanical damage.

(3) The longitudinal weld of the pipe shall be near the neutral axis of the bend.

(4) Pipe shall not be bent through an arc of more than 90 degrees.

(5) The inside radius of a bend shall be not less than six (6) times the outside diameter of the pipe. [NFPA 54:7.5.1]

1312.5.2 Plastic Pipe. Plastic pipe bends shall comply with the following:

(1) The pipe shall not be damaged, and the internal diameter of the pipe shall not be effectively reduced.

(2) Joints shall not be located in pipe bends.

(3) The radius of the inner curve of such bends shall not be less than 25 times the inside diameter of the pipe.

(4) Where the piping manufacturer specifies the use of special bending equipment or procedures, such equipment or procedures shall be used. [NFPA 54:7.5.2]

1312.5.3 Elbows. Factory-made welding elbows or transverse segments cut therefrom shall have an arc length measured along the crotch of at least one (1) inch (25 mm) for pipe sizes two (2) inches (50 mm) and larger. [NFPA 54:7.5.3]

1312.6 Drips and Sediment Traps.

1312.6.1 Provide Drips Where Necessary. For other than dry gas conditions, a drip shall be provided at any point in the line of pipe where condensate could collect. Where required by the Authority Having Jurisdiction or the serving gas supplier, a drip shall also be provided at the outlet of the meter. This drip shall be so installed as to constitute a trap wherein an accumulation of condensate will shut off the flow of gas before it will run back into the meter. [NFPA 54:7.6.1]

1312.6.2 Location of Drips. All drips shall be installed only in such locations that they will be readily accessible to permit cleaning or emptying. A drip shall not be located where the condensate is likely to freeze. [NFPA 54:7.6.2]

1312.6.3 Sediment Traps. (See Section 1312.7.)

1312.7 Outlets.

1312.7.1 Location and Installation.

(1) The outlet fittings or piping shall be securely fastened in place. [NFPA 54:7.7.1.1]

(2) Outlets shall not be located behind doors. [NFPA 54:7.7.1.2]

(3) Outlets shall be located far enough from floors, walls, patios, slabs, and ceilings to permit the use of wrenches without straining, bending, or damaging the piping. [NFPA 54:7.7.1.3]

(4) The unthreaded portion of gas piping outlets shall extend not less than (1) inch (25 mm) through finished ceilings or indoor or outdoor walls. [NFPA 54:7.7.1.4]

(5) The unthreaded portion of gas piping outlets shall extend not less than two (2) inches (50 mm) above the surface off floors or outdoor patios or slabs. [NFPA 54:7.7.1.5]

(6) The provisions of Section 1311.7.1(4) and (5) shall not apply to listed quick disconnect devices of the flush-mounted type or listed gas convenience outlets. Such devices shall be installed in accordance with the manufacturer's installation instructions. [NFPA 54:7.7.1.6]

1312.7.2 Cap All Outlets.

(A) Each outlet, including a valve, shall be closed gastight with a threaded plug or cap immediately after installation and shall be left closed until the appliance is connected thereto. When an appliance is disconnected from an outlet and the outlet is not to be used again immediately, it shall be closed gas-tight. [NFPA 54:7.7.2.1]

Exceptions:

(1) Laboratory equipment installed in accordance with 1313.2(A) shall be permitted.

(2) The use of a listed quick-disconnect device with integral shutoff or listed gas convenience outlet shall be permitted.

(B) Equipment shutoff valves installed in fireplaces shall be removed and the piping capped gastight where the fireplace is used for solid-fuel burning. [NFPA 54:7.7.2.2]

1312.8 Branch Pipe Connection. When a branch outlet is placed on a main supply line before it is known what size pipe will be connected to it, the outlet shall be of the same size as the line that supplies it. [NFPA 54:7.8]

1312.9 Manual Gas Shutoff Valves. See Section 1313.4.

1312.9.1 Valves at Regulators. An accessible gas shutoff valve shall be provided upstream of each gas pressure regulator. Where two (2) gas pressure regulators are installed in series in a single gas line, a manual valve shall not be required at the second regulator. [NFPA 54:7.9.1]

1312.9.2 Valves Controlling Multiple Systems.

(A) Accessibility of Gas Valves. Main gas shutoff valves controlling several gas piping systems shall be readily accessible for operation and installed so as to be protected from physical damage. They shall be marked with a metal tag or other permanent means attached by the installing agency so that the gas piping systems supplied through them can be readily identified. [NFPA 54:7.9.2.1]

(B) Shutoff Valves for Multiple House Lines. In multiple tenant buildings supplied through a master meter, or through one (1) service regulator where a meter is not provided, or where meters or service regulators are not readily accessible from the appliance location, an individual shutoff valve for each apartment or tenant line shall be provided at a convenient point of general accessibility.

In a common system serving a number of individual buildings, shutoff valves shall be installed at each building. [NFPA 54:7.9.2.2]

1312.9.3 Emergency Shutoff Valves. An exterior shutoff valve to permit turning off the gas supply to each building in an emergency shall be provided. The emergency shutoff valves shall be plainly marked as such and their locations posted as required by the Authority Having Jurisdiction. [NFPA 54:7.9.2.3]

1312.9.4 Shutoff Valve for Laboratories. Each laboratory space containing two (2) or more gas outlets installed on tables, benches, or in hoods in educational, research, commercial and industrial occupancies shall have a single shutoff valve through which all such gas outlets are supplied. The shutoff valve shall be accessible and shall be located within the laboratory or located adjacent to the laboratory's egress door and shall be identified. [NFPA 54-09:7.9.2.4]

1312.10 Prohibited Devices. No device shall be placed inside the gas piping or fittings that will reduce the cross-sectional area or otherwise obstruct the free flow of gas, except where proper allowance in the piping system design has been made for such a device and where approved by the Authority Having Jurisdiction. [NFPA 54:7.10]

1312.11 Systems Containing Gas-Air Mixtures Outside the Flammable Range. Where gas-air mixing machines are employed to produce mixtures above or below the flammable range, they shall be provided with stops to prevent adjustment of the mixture to within or approaching the flammable range. [NFPA 54:7.11]

1312.12 Systems Containing Flammable Gas-Air Mixtures.

1312.12.1 Required Components. A central premix system with a flammable mixture in the blower or compressor shall consist of the following components:

(1) Gas-mixing machine in the form of an automatic gas-air proportioning device combined with a downstream blower or compressor.

(2) Flammable mixture piping, minimum Schedule 40 NPS.

(3) Automatic firecheck(s).

(4) Safety blowout(s) or backfire preventers for systems utilizing flammable mixture lines above 2-1/2 inch nominal pipe size or the equivalent. [NFPA 54:7.12.1]

1312.12.2 Optional Components. The following components shall also be permitted to be utilized in any type of central premix system: [NFPA 54:7.12.2]

(1) Flowmeter(s)

(2) Flame arrester(s)

1312.12.3 Additional Requirements. Gas-mixing machines shall have nonsparking blowers and shall be so constructed that a flashback will not rupture machine casings. [NFPA 54:7.12.3]

1312.12.4 Special Requirements for Mixing Blowers. A mixing blower system shall be limited to applications with minimum practical lengths of mixture piping, limited to a maximum mixture pressure of ten (10) inches water column (25 Pa) and limited to gases containing no more than 10 percent hydrogen.

The blower shall be equipped with a gas-control valve at its air entrance so arranged that gas is admitted to the airstream, entering the blower in proper proportions for correct combustion by the type of burners employed, the said gas-control valve being of either the zero governor or mechanical ratio valve type that controls the gas and air adjustment simultaneously. No valves or other obstructions shall be installed between the blower discharge and the burner or burners. [NFPA 54:7.12.4]

1312.12.5 Installation of Gas-Mixing Machines.

(A) The machine shall be located in a large, well-ventilated area or in a small detached building or cutoff room provided with room construction and explosion vents in accordance with sound engineering principles.

Such rooms or below finished grade installations shall have adequate positive ventilation. [NFPA 54:7.12.5.1]

(B) Where gas-mixing machines are installed in well-ventilated areas, the type of electrical equipment shall be in accordance with NFPA 70, *National Electrical Code,* for general service conditions unless other hazards in the area prevail. Where gas-mixing machines are installed in small detached buildings or cutoff rooms, the electrical equipment and wiring shall be installed in accordance with NFPA 70, *National Electrical Code,* for hazardous locations (Articles 500 and 501, Class I, Division 2). [NFPA 54:7.12.5.2]

(C) Air intakes for gas-mixing machines using compressors or blowers shall be taken from outdoors whenever practical. [NFPA 54:7.12.5.3]

(D) Controls for gas-mixing machines shall include interlocks and a safety shutoff valve of the manual reset type in the gas supply connection to each machine arranged to automatically shut off the gas supply in the event of high or low gas pressure. Except for open-burner installations only, the controls shall be interlocked so that the blower or compressor will stop operating following a gas supply failure. Where a system employs pressurized air, means shall be provided to shut off the gas supply in the event of air failure. [NFPA 54:7.12.5.4]

(E) Centrifugal gas-mixing machines in parallel shall be reviewed by the user and equipment manufacturer before installation, and means or plans for minimizing these effects of downstream pulsation and equipment overload shall be prepared and utilized as needed. [NFPA 54:7.12.5.5]

1312.12.6 Use of Automatic Firechecks, Safety Blowouts, or Backfire Preventers. Automatic firechecks and safety blowouts or backfire preventers shall be provided in piping systems distributing flammable air-gas mixtures from gas-mixing machines to protect the piping and the machines in the event of flashback, in accordance with the following: [NFPA 54:7.12.6]

(A) Approved automatic firechecks shall be installed upstream as close as practicable to the burner inlets following the firecheck manufacturer's instructions. [NFPA 54:7.12.6(1)]

(B) A separate manually operated gas valve shall be provided at each automatic firecheck for shutting off the flow of gas-air mixture through the firecheck after a flashback has occurred. The valve shall be located up stream as close as practical to the inlet of the automatic firecheck.

Caution:

These valves shall not be reopened after a flashback has occurred until the firecheck has cooled sufficiently to prevent reignition of the flammable mixture and has been reset properly. [NFPA 54:7.12.6(2)]

(C) A safety blowout or backfiring preventer shall be provided in the mixture line near the outlet of each gas-mixing machine where the size of the piping is larger than two and one-half (2-1/2) inches NPS, or equivalent, to protect the mixing equipment in the event of an explosion passing through an automatic firecheck. The manufacturer's instructions shall be followed when installing these devices, particularly after a disc has burst. The discharge from the safety blowout or backfire preventer shall be located or shielded so that particles from the ruptured disc cannot be directed toward personnel. Wherever there are interconnected installations of gas-mixing machines with safety blowouts or backfire preventers, provision shall be made to keep the mixture from other machines from reaching any ruptured disc opening. Check valves shall not be used for this purpose. [NFPA 54:7.12.6(3)]

(D) Large-capacity premix systems provided with explosion heads (rupture discs) to relieve excessive pressure in pipelines shall be located at and vented to a safe outdoor location. Provisions shall be provided for automatically shutting off the supply of gas-air mixture in the event of rupture. [NFPA 54:7.12.6(4)]

1312.13 Electrical Bonding and Grounding.

(A) Pipe and Tubing other than CSST. Each above-ground portion of a gas piping system other than CSST that is likely to become energized shall be electrically continuous and bonded to an effective ground-fault current path. Gas piping, other than CSST shall be considered to be bonded when it is connected to appliances that are connected to the appliance grounding conductor of the circuit supplying that appliance. [NFPA 54-09:7.13.1]

(B) CSST gas piping systems shall be bonded to the electrical service grounding electrode system at the point where the gas service enters the building. The bonding jumper shall not be smaller than 6 AWG copper wire. [NFPA 54-09: 7.13.2]

(C) Gas piping shall not be used as a grounding conductor or electrode. This does not preclude the bonding of metallic piping to a grounding system. [NFPA 54-09:7.13.3]

1312.14 Lightning Protection System. Where a lighting system is installed, the bonding of the gas piping shall be in accordance with NFPA 780, *Standard for Installation of Lightning Protection Systems.* [NFPA 54-09:7.13.4]

1312.15 Electrical Circuits. Electrical circuits shall not utilize gas piping or components as conductors.

Exception: Low-voltage (50V or less) control circuits, ignition circuits, and electronic flame detection device circuits shall be permitted to make use of piping or components as a part of an electric circuit. [NFPA 54:7.14]

1312.16 Electrical Connections.

(A) All electrical connections between wiring and electrically operated control devices in a piping system shall conform to the requirements of NFPA 70, *National Electrical Code.* (See Section 1311.13) [NFPA 54:7.15.1]

(B) Any essential safety control depending on electric current as the operating medium shall be of a type that will shut off (fail safe) the flow of gas in the event of current failure. [NFPA 54:7.15.2]

1313.0 Appliance Connections to Building Piping.

1313.1 Connecting Gas Appliances. Gas utilization appliances shall be connected to the building piping in compliance with Sections 1313.4 through 1313.6 by one of the following: [NFPA 54:9.6.1]

(1) Rigid metallic pipe and fittings. [NFPA 54:9.6.1(1)]

(2) Semirigid metallic tubing and metallic fittings. Aluminum alloy tubing shall not be used in exterior locations. [NFPA 54:9.6.1(2)]

(3) Listed flexible gas connectors in compliance with CSA Z21.24, *Standard for Connectors for Gas Appliances*. The connectors shall be used in accordance with the terms of their listing that are completely in the same room as the appliance. [NFPA 54:9.6.1(3)]

(4) CSST, where installed in accordance with the manufacturer's instructions. [NFPA 54:9.6.1(5)]

(5) Listed nonmetallic gas hose connectors in accordance with Section 1313.2. [NFPA 54:9.6.1(6)]

(6) Gas-fired food service (commercial cooking) appliances listed for use with casters or otherwise subject to movement for cleaning, and other large and heavy appliances that can be moved, shall be connected in accordance with the connector manufacturer's installation instructions using a listed appliance connector complying with CSA Z21.69, *Standard for Connectors for Movable Gas Appliances*. The commercial cooking appliance connector installation shall be configured in accordance with the manufacturer's installation instructions. [NFPA 54-09:9.6.1.1]

(7) Movement of appliances with casters shall be limited by a restraining device installed in accordance with the connector and appliance manufacturer's installation instructions. [NFPA 54-09:9.6.1.2]

The building gas supply manifold branch connection for commercial cooking appliances shall be oriented as follows:

(1) The building gas supply manifold branch connection shall be oriented vertically down (toward the floor).

(2) The bottom of the building gas supply manifold branch connection shall be located not less than 36 in. (914 mm) and not more than 42 in. (1,067 mm) above the floor.

(3) The building gas supply manifold branch connection shall be directly behind the appliance it serves and not obstructed by any appliance, equipment or anything else.

(8) In 1313.1(2), (3), and (5), the connector or tubing shall be installed so as to be protected against physical and thermal damage. Aluminum alloy

tubing and connectors shall be coated to protect against external corrosion where they are in contact with masonry, plaster, insulation, or are subject to repeated wettings by such liquids as water (except rainwater), detergents, or sewage. [NFPA 54:9.6.1(7)]

1313.2 Use of Nonmetallic Gas Hose Connectors. Listed nonmetallic gas hose connectors shall be used in accordance with the terms of their listing and as follows: [NFPA 54:9.6.2]

(A) Indoor. Indoor gas hose connectors shall be used only to connect laboratory, shop, and ironing equipment requiring mobility during operation. An equipment shutoff valve shall be installed where the connector is attached to the building piping. The connector shall be of minimum length and shall not exceed six (6) feet (1,829 mm). The connector shall not be concealed and shall not extend from one (1) room to another or pass through wall partitions, ceilings, or floors. [NFPA 54:9.6.2(1)]

(B) Outdoor. Outdoor gas hose connectors are permitted to connect portable outdoor gas-fired appliances. An equipment shutoff valve, a listed quick-disconnect device, or a listed gas convenience outlet shall be installed where the connector is attached to the supply piping and in such a manner to prevent the accumulation of water or foreign matter. This connection shall only be made in the outdoor area where the appliance is to be used. [NFPA 54:9.6.2(2)]. The connector length shall not exceed fifteen (15) feet (4,572 mm).

1313.3 Connection of Portable and Mobile Industrial Appliances.

(A) Where portable industrial gas utilization appliances or appliances requiring mobility or subject to vibration is connected to the building gas piping system by the use of a flexible hose, the hose shall be suitable and safe for the conditions under which it can be used. [NFPA 54:9.6.3.1]

(B) Where industrial gas utiliazation appliances requiring mobility is connected to the rigid piping by the use of swivel joints or couplings, the swivel joints or couplings shall be suitable for the service required, and only the minimum number required shall be installed. [NFPA 54:9.6.3.2]

(C) Where industrial gas utilization appliances subject to vibration is connected to the building piping system by the use of all metal flexible connectors, the connectors shall be suitable for the service required. [NFPA 54:9.6.3.3]

(D) Where flexible connections are used, they shall be of the minimum practical length and shall not extend from one (1) room to another or pass through any walls, partitions, ceilings, or floors. Flexible connections shall not be used in any concealed location. They shall be protected against physical or thermal damage and shall be provided with gas shutoff valves in readily accessible locations in rigid piping upstream from the flexible connections. [NFPA 54:9.6.3.4]

1313.4 Appliance Shutoff Valves and Connections. Appliances connected to a piping system shall have an accessible, approved manual shutoff valve with a nondisplaceable valve member or a listed gas convenience outlet installed within six (6) feet (1,829 mm) of the appliance it serves. Where a connector is used, the valve shall be installed up-stream of the connector. A union or flanged connection shall be provided downstream from this valve to permit removal of controls. Shutoff valves serving decorative gas appliances shall be permitted to be installed in fireplaces if listed for such use. [NFPA 54:9.6.4]

Exceptions:

(1) Shutoff valves shall be permitted to be accessibly located inside or under an appliance when such appliance can be removed without removal of the shutoff valve.

(2) Shutoff valves shall be permitted to be accessibly located inside wall heaters and wall furnaces listed for recessed installation where necessary maintenance can be performed without removal of the shutoff valve.

1313.5 Quick-Disconnect Devices. Quick-disconnect devices used to connect appliances to the building piping shall be listed. Where installed indoors, an approved manual shutoff valve with a nondisplaceable valve member shall be installed upstream of the quick-disconnect device. [NFPA 54:9.6.5]

1313.6 Gas Convenience Outlets. Appliances shall be permitted to be connected to the building piping by means of a listed gas convenience outlet, in conjunction with a listed appliance connector, used in accordance with the manufacturer's installation instructions. [NFPA 54:9.6.6] Gas convenience outlets shall be listed in accordance with CSA Z21.90, *Gas Convenience Outlets and Optional Enclosures* and installed in accordance with the manufacturer's installation instructions. [NFPA 54-09:9.6.6]

1313.7 Sediment Trap. Where a sediment trap is not incorporated as a part of the appliance, a sediment trap shall be installed as close to the inlet of the appliance as practical at the time of appliance installation. The sediment trap shall be either a tee fitting with a capped nipple in the bottom outlet as illustrated in Figure 13-1 or other device recognized as an effective sediment trap. Illuminating appliances, ranges, clothes dryers, decorative vented appliances for installation in vented fireplaces, gas fireplaces, and outdoor grills shall not be required to be so equipped. [NFPA 54:9.6.7]

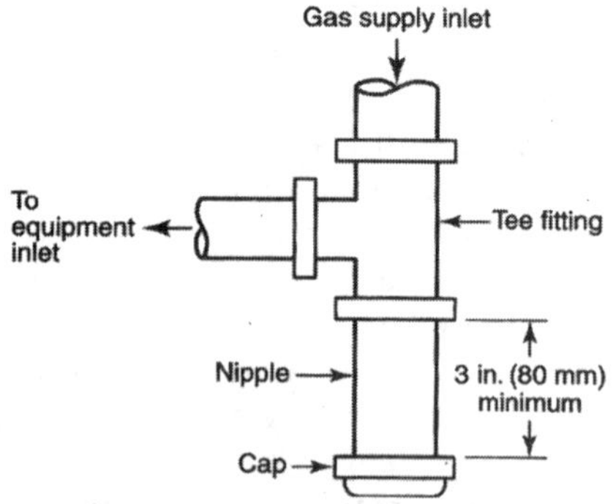

FIGURE 13-1 Method of Installing a Tee Fitting Sediment Trap.

1313.8 Installation of Piping. Piping shall be installed in a manner not to interfere with inspection, maintenance, or servicing of the appliance. [NFPA 54:9.6.8]

1314.0 Appliance Over Pressure Protection.

The maximum operating pressure for piping systems serving appliances designed to operate at 14 inches w.c. inlet pressure or less shall be 2 psig unless an over pressure protection device designed to limit pressure at the appliance to 2 psig upon failure of the line gas pressure regulator is installed.

1315.0 Liquefied Petroleum Gas Facilities and Piping.

Liquefied petroleum gas facilities shall comply with NFPA 58, *Liquefied Petroleum Gas Code.*

1316.0 Pressure Testing and Inspection.

1316.1 General.

1316.1.1 Prior to acceptance and initial operation, all piping installations shall be inspected and pressure-tested to determine that the materials, design, fabrication, and installation practices comply with the requirements of this code. [NFPA 54:8.1.1.1]

1316.1.2 Inspection shall consist of visual examination during or after manufacture, fabrication, assembly, or pressure tests as appropriate. Supplementary types of non-destructive inspection techniques, such as magnetic-particle, radiographic, and ultrasonic, shall not be required unless specifically listed herein or in the engineering design. [NFPA 54:8.1.1.2]

1316.1.3 Where repairs or additions are made following the pressure test, the affected piping shall be tested. Minor repairs and additions are not required to be pressure-tested provided that the work is inspected and connections are tested with a noncorrosive leak-detecting fluid or other leak-detecting methods approved by the Authority Having Jurisdiction. [NFPA 54:8.1.1.3]

1316.1.4 Where new branches are installed from the point of delivery to new appliance(s), only the newly installed branch(es) shall be required to be pressure-tested. Connections between the new piping and the existing piping shall be tested with a noncorrosive leak-detecting fluid or approved leak-detecting methods. [NFPA 54:8.1.1.4]

1316.1.5 A piping system shall be tested as a complete unit or in sections. Under no circumstances shall a valve in a line be used as a bulkhead between gas in one (1) section of the piping system and test medium in an adjacent section unless two (2) valves are installed in series with a valved "tell tale" located between these valves. A valve shall not be subjected to the test pressure unless it can be determined that the valve, including the valve-closing mechanism, is designed to safely withstand the pressure. [NFPA 54:8.1.1.5]

1316.1.6 Regulator and valve assemblies fabricated independently of the piping system in which they are to be installed shall be permitted to be tested with inert gas or air at the time of fabrication. [NFPA 54:8.1.1.6]

1316.1.7 Test Medium. The test medium shall be air, nitrogen, carbon dioxide, or an inert gas. OXYGEN SHALL NEVER BE USED. [NFPA 54:8.1.2]

1316.2 Test Preparation.

1316.2.1 Pipe joints, including welds, shall be left exposed for examination during the test. [NFPA 54:8.1.3.1]

Exception: Covered or concealed pipe end joints that have been previously tested in accordance with this code.

1316.2.2 Expansion joints shall be provided with temporary restraints, if required, for the additional thrust load under test. [NFPA 54:8.1.3.2]

1316.2.3 Appliances and equipment that are not to be included in the test shall be either disconnected from the piping or isolated by blanks, blind flanges, or caps. Flanged joints at which blinds are inserted to blank off other equipment during the test shall not be required to be tested. [NFPA 54:8.1.3.3]

1316.2.4 Where the piping system is connected to appliances, equipment, or equipment components designed for operating pressures of less than the test pressure, such appliances, equipment, or equipment components shall be isolated from the piping system by disconnecting them and capping the outlet(s). [NFPA 54:8.1.3.4]

1316.2.5 Where the piping system is connected to appliances, equipment, or equipment components designed for operating pressures equal to or greater than the test pressure, such appliances and equipment shall be isolated from the piping system by closing the individual appliance shutoff valve(s). [NFPA 54:8.1.3.5]

1316.2.6 Testing of piping systems shall be done with due regard for the safety of employees and the public during the test. Bulkheads, anchorage, and bracing suitably designed to resist test pressures shall be installed if necessary. Prior to testing, the interior of the pipe shall be cleared of all foreign material. [NFPA 54:8.1.3.6]

1316.3 Test Pressure.

1316.3.1 Test pressure shall be measured with a manometer or with a pressure-measuring device designed and calibrated to read, record, or indicate a pressure loss due to leakage during the pressure test period. The source of pressure shall be isolated before the pressure tests are made. Mechanical gauges used to measure test pressures shall have a range such that the highest end of the scale is not greater than five (5) times the test pressure. [NFPA 54:8.1.4.1]

1316.3.2 The test pressure to be used shall be no less than one and one-half (1-1/2) times the proposed maximum working pressure, but not less than three (3) psi (20 kPa), irrespective of design pressure. [NFPA 54:8.1.4.2]

1316.3.3 Test duration shall be not less than one-half (1/2) hour for each five-hundred (500) $ft.^3$ (14 m^3) of pipe volume or fraction thereof. When testing a system having a volume less than ten (10) $ft.^3$ (0.28 m^3) or a system in a single-family dwelling, the test duration shall be a minimum of ten (10) minutes. The duration of the test shall not be required to exceed twenty-four (24) hours. [NFPA 54:8.1.4.3]

1316.4 Detection of Leaks and Defects.

1316.4.1 The piping system shall withstand the test pressure specified without showing any evidence of leakage or other defects. Any reduction of test pressures, as indicated by pressure gauges, shall be deemed to indicate the presence of a leak unless such reduction can be readily attributed to some other cause. [NFPA 54:8.1.5.1]

1316.4.2 The leakage shall be located by means of an approved gas detector, a noncorrosive leak detection fluid, or other approved leak detection methods. Matches, candles, open flames, or other methods that provide a source of ignition shall not be used. [NFPA 54:8.1.5.2]

1316.4.3 Where leakage or other defects are located, the affected portion of the piping system shall be repaired or replaced and retested. (See Section 1316.1.3.) [NFPA 54:8.1.5.3]

1316.5 System and Appliance Leakage Test.

1316.5.1 Test Gases. Leak checks using fuel gas shall be permitted in piping systems that have been pressure-tested in accordance with Section 1316.0. [NFPA 54:8.2.1]

1316.5.2 Turning Gas On. During the process of turning gas on into a system of new gas piping, the entire system shall be inspected to determine that there are no open fittings or ends and that all valves at unused outlets are closed and plugged or capped. [NFPA 54-09:8.2.2]

1316.5.3 Test for Leakage. Immediately after the gas is turned on into a new system or into a system that has been initially restored after an interruption of service, the piping system shall be checked for leakage. Where leakage is indicated, the gas supply shall be shut off until the necessary repairs have been made. [NFPA 54:8.2.3]

1316.5.4 Placing Appliances in Operation. Gas utilization appliances shall not be placed in operation until after the piping system has been tested in accordance with Section 1316.5.3 and purged in accordance with Section 1316.6.2. [NFPA 54:8.2.4]

1316.6 Purging.

1316.6.1 Removal from Service. When gas piping is to be opened for servicing, addition, or modification, the section to be worked on shall be turned off from the gas supply at the nearest convenient point, and the line pressure vented to the outdoors or to ventilated areas of sufficient size to prevent accumulation of flammable mixtures. The remaining gas in this section of pipe shall be displaced with an inert gas as required by Table 13-4. [NFPA 54:8.3.1]

TABLE 13-4
Length of Piping Requiring Purging with Inert Gas for Servicing or Modification
[NFPA 54: Table 8.3.1]

Nominal Pipe Size (in.)	Length of Piping Requiring Purging (ft.)
2-1/2	> 50
3	> 30
4	> 15
6	> 10
8 or larger	Any length

Note: 1 in. = 25.4 mm, 1 ft. = 0.305 m

1316.6.2 Placing in Operation. When piping full of air is placed in operation, the air in the piping shall be displaced with fuel gas, except where such piping is required by Table 13-5 to be purged with an inert gas prior to introduction of fuel gas. The air can be safely displaced with fuel gas provided that a moderately rapid and continuous flow of fuel gas is introduced at one end of the line and air is vented out at the other end. The fuel-gas flow shall be continued without interruption until the vented gas is free of air. The point of discharge shall not be left unattended during purging. After purging, the vent shall then be closed. Where required by Table 13-5, the air in the piping shall first be displaced with an inert gas, and the inert gas shall then be displaced with fuel gas. [NFPA 54:8.3.2]

TABLE 13-5
Length of Piping Requiring Purging with Inert Gas Before Placing in Operation [NFPA 54: Table 8.3.2]

Nominal Pipe Size (in.)	Length of Piping Requiring Purging (ft.)
3	>30
4	>15
6	>10
8 or larger	Any length

Note: 1 ft. = 0.305 m

1316.6.3 Discharge of Purged Gases. The open end of piping systems being purged shall not discharge into confined spaces or areas where there are sources of ignition unless precautions are taken to perform this operation in a safe manner by ventilation of the space, control of purging rate, and elimination of all hazardous conditions. [NFPA 54:8.3.3]

1316.6.4 Placing Appliances in Operation. After the piping has been placed in operation, all appliances shall be purged and then placed in operation, as necessary. [NFPA 54:8.3.4]

1317.0 Interconnections between Gas-Piping Systems. [NFPA 54:5.3]

1317.1 Interconnections Supplying Separate Users. Where two (2) or more meters, or two (2) or more service regulators where meters are not provided, are located on the same premises and supply separate users, the gas piping systems shall not be interconnected on the outlet side of the meters or service regulators. [NFPA 54:5.3.1]

1317.2 Interconnections for Standby Fuels. Where supplementary gas for standby use is connected downstream from a meter, or a service regulator where a meter is not provided, a device to prevent backflow shall be installed. [NFPA 54:5.3.2.1] A three-way valve installed to admit the standby supply and at the same time shut off the regular supply shall be permitted to be used for this purpose. [NFPA 54:5.3.2.2]

1318.0 Required Gas Supply.

1318.1 The following regulations, as set forth in this section and in Section 1318.0, Required Gas Piping Size, shall be the standard for the installation of gas piping. All natural gas regulations and tables are based on the use of gas having a specific gravity of sixty hundredths (0.60), supplied at six (6) to eight (8) inches (152–203 mm) water column pressure at the outlet of the meter or regulator. Where the natural gas supplier cannot maintain a minimum inlet pressure of six (6) inches (152 mm) water column pressure, Table 13-7 allowing a maximum pressure drop of 0.5 inch (12.7 mm) water column shall be used. For undiluted liquefied petroleum gas, gas piping may be sized at eleven (11) inches (279 mm) water column pressure at the outlet of the meter or regulator and specific gravity of one and fifty hundredths (1.50).

Note: Where gas of a different specific gravity is to be delivered, the serving gas supplier should be contacted for specific gravity conversion factors to use in sizing piping systems from the pipe sizing tables in this chapter.

1318.2 The hourly volume of gas required at each piping outlet shall be taken as not less than the maximum hourly rating as specified by the manufacturer of the appliance or appliances to be connected to each such outlet.

1318.3 Where the gas appliances to be installed have not been definitely specified, Table 13-1 may be used as a reference to estimate requirements of typical appliances.

1318.3.1 To obtain the cubic feet per hour (L/sec.) of gas required, divide the input of the appliances by the average Btu (W) heating value per cubic foot (L) of the gas. The average Btu (W) per cubic foot (L) of the gas in the area of the installation may be obtained from the serving gas supplier.

1318.4 The size of the supply piping outlet for any gas appliance shall not be less than one-half (1/2) inch (15 mm).

1318.5 The minimum size of any piping outlet for a mobile home shall be three-quarter (3/4) inch (20 mm).

1319.0 Required Gas Piping Size.

1319.1 Pipe Sizing Methods. Where the pipe size is to be determined using any of the methods in Sections

1319.1.1 through 1319.1.3, the diameter of each pipe segment shall be obtained from the pipe sizing tables in Section 1319.2 or from the sizing equations in Section 1319.3. [NFPA 54:6.1]

1319.1.1 Longest Length Method. The pipe size of each section of gas piping shall be determined using the longest length of piping from the point of delivery to the most remote outlet and the load of the section. [NFPA 54:6.1.1]

1319.1.2 Branch Length Method. Pipe shall be sized as follows:

(A) The pipe size of each section of the longest pipe run from the point of delivery to the most remote outlet shall be determined using the longest run of piping and the load of the section.

(B) The pipe size of each section of branch piping not previously sized shall be determined using the length of piping from the point of delivery to the most remote outlet in each branch and the load of the section. [NFPA 54:6.1.2]

1319.1.3 Hybrid Pressure. The pipe size for each section of higher pressure gas piping shall be determined using the longest length of piping from the point of delivery to the most remote line pressure regulator. The pipe size from the line pressure regulator to each outlet shall be determined using the length of piping from the regulator to the most remote outlet served by the regulator. [NFPA 54:6.1.3]

1319.2 Tables for Sizing Gas Piping Systems. Tables 13-7 through 13-23 shall be used to size gas piping in conjunction with one of the methods described in Sections 1319.1.1 through 1319.1.3. [NFPA 54:6.2]

1319.3 Sizing Equations. The inside diameter of smooth wall pipe or tubing shall be determined by the sizing equations 13-1 and 13-2, using the equivalent pipe length determined by methods in Sections 1319.1.1 through 1319.1.3. [NFPA 54:6.4]

Equation 13-1 Low-Pressure Gas Formula [Less than 1.5 psi (10.3 kPa)] [NFPA 54:6.4.1]

$$D = \frac{Q^{0.381}}{19.17\left(\frac{\Delta H}{Cr \times L}\right)^{0.206}}$$

WHERE:

D = inside diameter of pipe, in.

Q = input rate appliance(s), cubic feet per hour at 60°F (16°C) and 30 in. (759 mm) mercury column

L = equivalent length of pipe, ft.

ΔH = pressure drop, in. water column (27.7 in. H_2O = 1 psi)

C_r and Y in accordance with Table 13-6

Equation 13-2 High-Pressure Gas Formula [1.5 psi (10.3 kPa) and above] [NFPA 54:6.4.2]

$$D = \frac{Q^{0.381}}{18.93\left[\frac{(P_1^2 - P_2^2) \cdot Y}{Cr \times L}\right]^{0.206}}$$

WHERE:

D = inside diameter of pipe, in.

Q = input rate of appliance(s)

P_1 = upstream pressure, psia (P_1 + 14.7)

P_2 = downstream pressure, psia (P_2 + 14.7)

L = equivalent length of pipe, ft.

C_r and Y in accordance with Table 13-6

TABLE 13-6
Cr and Y for Natural Gas and Undiluted Propane at Standard Conditions
[NFPA 54: Table 6.4.2]

	Formula Factors	
Gas	**Cr**	**Y**
Natural Gas	0.6094	0.9992
Undiluted Propane	1.2462	0.9910

Note: 1 ft.^{3} = 0.028 m^3, 1 ft. = 0.305 m, 1 in. w.c. = 0.249 kPa, 1 psi = 6.894 kPa, 1,000 Btu/h = 0.293 kW

1319.4 To determine the size of each section of pipe in any system within the range of the Table 13-7, proceed as follows:

(A) Measure the length of the pipe from the gas meter location to the most remote outlet on the system.

(B) Select the length in feet column and row showing that distance, or the next longer distance if the table does not give the exact length.

(C) Starting at the most remote outlet, find in the just selected the gas demand for that outlet. If the exact figure of demand is not shown, choose the next larger figure in the row.

(D) At the top of column in Table 13-7 will be found the correct size of pipe.

(E) Using this same row, proceed in a similar manner for each section of pipe serving this outlet. For each section of pipe, determine the total gas demand supplied by that section. Where gas piping sections serve both heating and cooling equipment and the installation prevents both units from operating simultaneously, only the larger of the two demand loads need to be used in sizing these sections.

(F) Size each section of branch piping not previously sized by measuring the distance from the gas meter location to the most remote outlet in that branch and follow the procedures of steps B, C, D, and E above.

Note: Size branch piping in the order of their distance from the meter location, beginning with the most distant outlet not previously sized.

1319.5 For conditions other than those covered by Section 1319.1, such as longer runs or greater gas demands, the size of each gas piping system shall be determined by standard engineering methods acceptable to the Authority Having Jurisdiction, and each such system shall be so designed that the total pressure drop between the meter or other point of supply and any outlet when full demand is being supplied to all outlets, will at no time exceed five-tenths (0.5) inches (12.7 mm) water column pressure.

1319.6 Where the gas pressure may be higher than fourteen (14) inches (356 mm) or lower than six (6) inches (152 mm) of water column, or when diversity demand factors are used, the design, pipe, sizing, materials, location, and use of such systems first shall be approved by the Authority Having Jurisdiction. Piping systems designed for pressures higher than the serving gas supplier's standard delivery pressure shall have prior verification from the gas supplier of the availability of the design pressure.

FIGURE 13-2 Example Illustrating Use of Tables 13-1 and 13-7

Problem: Determine the required pipe size of each section and outlet of the piping system shown in Figure 13-2. Gas to be used has a specific gravity of 0.60 and 1,100 Btu per cubic foot (11.4 Watt-Hour/L), delivered at eight (8) inch (203 mm) water column pressure.

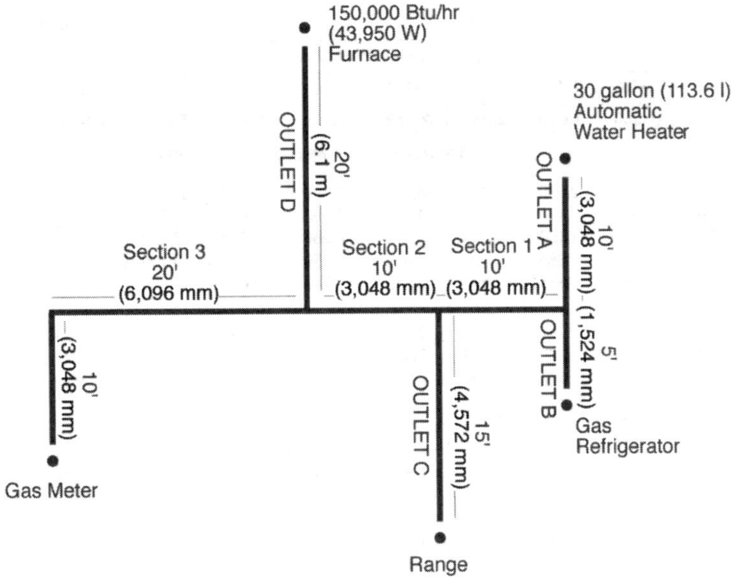

Solution:

(1) Maximum gas demand of outlet A —
32 cubic feet per hour (0.21 L/sec.) (from Table 13-1).
Maximum gas demand of outlet B —
3 cubic feet per hour (0.02 L/sec.) (from Table 13-1).
Maximum gas demand of outlet C —
59 cubic feet per hour (0.46 L/sec.) (from Table 13-1).
Maximum gas demand of outlet D —
136 cubic feet per hour (1.1 L/sec.) (150,000 Btu/hour [43,950 W])
divided by 1,100 Btu per cubic foot (11.4 Watt-hour/L)

(2) The length of pipe from the gas meter to the most remote outlet (outlet A) is 60 feet (18,288 mm).

(3) Using the length in feet column row marked 60 feet (18,288 mm) in Table 13-7:

Outlet A, supplying 32 cubic feet per hour (0.21 L/sec.), requires one-half (1/2) inch (15 mm) pipe.

Section 1, supplying outlets A and B, or 35 cubic feet per hour (0.24 L/sec.) requires one-half (1/2) inch (15 mm) pipe.

Section 2, supplying outlets A, B, and C, or 94 cubic feet per hour (0.7 L/sec.) requires three-quarter (3/4) inch (20 mm) pipe.

Section 3, supplying outlets A, B, C, and D, or 230 cubic feet per hour (1.8 L/sec.), requires one inch pipe.

(4) Using the column marked 60 feet (18,288 mm) in Table 13-7 (no column for actual length of 55 feet [16,764 mm]:

Outlet B supplying 3 cubic feet per hour (0.02 L/sec.), requires one-half (1/2) inch (15 mm) pipe.

Outlet C, supplying 59 cubic feet per hour (0.46 L/sec.), requires one-half (1/2) inch (15 mm) pipe.

(5) Using the column marked 60 feet (18,288 mm) in Table 13-7:

Outlet D, supplying 136 cubic feet per hour (1.1 L/sec.), requires (3/4) inch (20 mm) pipe.

Table 13-9
Schedule 40 Metallic Pipe [NFPA 54: Table 6.2(d)]

							Gas:	Natural	
							Inlet Pressure:	3.0 psi	
							Pressure Drop:	2.0 psi	
							Specific Gravity:	0.60	
	Pipe Size (in.)								
Nominal:	**1/2**	**3/4**	**1**	**1-1/4**	**1-1/2**	**2**	**2-1/2**	**3**	**4**
Actual ID:	**0.622**	**0.824**	**1.049**	**1.380**	**1.610**	**2.067**	**2.469**	**3.068**	**4.026**
Length (ft.)	**Capacity in Cubic Feet of Gas per Hour**								
10	2,350	4,920	9,270	19,000	28,500	54,900	87,500	155,000	316,000
20	1,620	3,380	6,370	13,100	19,600	37,700	60,100	106,000	217,000
30	1,300	2,720	5,110	10,500	15,700	30,300	48,300	85,400	174,000
40	1,110	2,320	4,380	8,990	13,500	25,900	41,300	73,100	149,000
50	985	2,060	3,880	7,970	11,900	23,000	36,600	64,800	132,000
60	892	1,870	3,520	7,220	10,800	20,800	33,200	58,700	120,000
70	821	1,720	3,230	6,640	9,950	19,200	30,500	54,000	110,000
80	764	1,600	3,010	6,180	9,260	17,800	28,400	50,200	102,000
90	717	1,500	2,820	5,800	8,680	16,700	26,700	47,100	96,100
100	677	1,420	2,670	5,470	8,200	15,800	25,200	44,500	90,800
125	600	1,250	2,360	4,850	7,270	14,000	22,300	39,500	80,500
150	544	1,140	2,140	4,400	6,590	12,700	20,200	35,700	72,900
175	500	1,050	1,970	4,040	6,060	11,700	18,600	32,900	67,100
200	465	973	1,830	3,760	5,640	10,900	17,300	30,600	62,400
250	412	862	1,620	3,330	5,000	9,620	15,300	27,100	55,300
300	374	781	1,470	3,020	4,530	8,720	13,900	24,600	50,100
350	344	719	1,350	2,780	4,170	8,020	12,800	22,600	46,100
400	320	669	1,260	2,590	3,870	7,460	11,900	21,000	42,900
450	300	627	1,180	2,430	3,640	7,000	11,200	19,700	40,200
500	283	593	1,120	2,290	3,430	6,610	10,500	18,600	38,000
550	269	563	1,060	2,180	3,260	6,280	10,000	17,700	36,100
600	257	537	1,010	2,080	3,110	5,990	9,550	16,900	34,400
650	246	514	969	1,990	2,980	5,740	9,150	16,200	33,000
700	236	494	931	1,910	2,860	5,510	8,790	15,500	31,700
750	228	476	897	1,840	2,760	5,310	8,470	15,000	30,500
800	220	460	866	1,780	2,660	5,130	8,180	14,500	29,500
850	213	445	838	1,720	2,580	4,960	7,910	14,000	28,500
900	206	431	812	1,670	2,500	4,810	7,670	13,600	27,700
950	200	419	789	1,620	2,430	4,670	7,450	13,200	26,900
1,000	195	407	767	1,580	2,360	4,550	7,240	12,800	26,100
1,100	185	387	729	1,500	2,240	4,320	6,890	12,200	24,800
1,200	177	369	695	1,430	2,140	4,120	6,570	11,600	23,700
1,300	169	353	666	1,370	2,050	3,940	6,290	11,100	22,700
1,400	162	340	640	1,310	1,970	3,790	6,040	10,700	21,800
1,500	156	327	616	1,270	1,900	3,650	5,820	10,300	21,000
1,600	151	316	595	1,220	1,830	3,530	5,620	10,000	20,300
1,700	146	306	576	1,180	1,770	3,410	5,440	9,610	19,600
1,800	142	296	558	1,150	1,720	3,310	5,270	9,320	19,000
1,900	138	288	542	1,110	1,670	3,210	5,120	9,050	18,400
2,000	134	280	527	1,080	1,620	3,120	4,980	8,800	18,000

Note: All table entries are rounded to 3 significant digits

Table 13-10
Schedule 40 Metallic Pipe [NFPA 54: Table 6.2(e)]

							Gas:	Natural	
							Inlet Pressure:	5.0 psi	
							Pressure Drop:	3.5 psi	
							Specific Gravity:	0.60	
	Pipe Size (in.)								
Nominal:	1/2	3/4	1	1-1/4	1-1/2	2	2-1/2	3	4
Actual ID:	0.622	0.824	1.049	1.380	1.610	2.067	2.469	3.068	4.026
Length (ft.)	Capacity in Cubic Feet of Gas per Hour								
10	3,190	6,430	11,800	24,200	36,200	69,700	111,000	196,000	401,000
20	2,250	4,550	8,320	17,100	25,600	49,300	78,600	139,000	283,000
30	1,840	3,720	6,790	14,000	20,900	40,300	64,200	113,000	231,000
40	1,590	3,220	5,880	12,100	18,100	34,900	55,600	98,200	200,000
50	1,430	2,880	5,260	10,800	16,200	31,200	49,700	87,900	179,000
60	1,300	2,630	4,800	9,860	14,800	28,500	45,400	80,200	164,000
70	1,200	2,430	4,450	9,130	13,700	26,400	42,000	74,300	151,000
80	1,150	2,330	4,260	8,540	12,800	24,700	39,300	69,500	142,000
90	1,060	2,150	3,920	8,050	12,100	23,200	37,000	65,500	134,000
100	979	1,980	3,620	7,430	11,100	21,400	34,200	60,400	123,000
125	876	1,770	3,240	6,640	9,950	19,200	30,600	54,000	110,000
150	786	1,590	2,910	5,960	8,940	17,200	27,400	48,500	98,900
175	728	1,470	2,690	5,520	8,270	15,900	25,400	44,900	91,600
200	673	1,360	2,490	5,100	7,650	14,700	23,500	41,500	84,700
250	558	1,170	2,200	4,510	6,760	13,000	20,800	36,700	74,900
300	506	1,060	1,990	4,090	6,130	11,800	18,800	33,300	67,800
350	465	973	1,830	3,760	5,640	10,900	17,300	30,600	62,400
400	433	905	1,710	3,500	5,250	10,100	16,100	28,500	58,100
450	406	849	1,600	3,290	4,920	9,480	15,100	26,700	54,500
500	384	802	1,510	3,100	4,650	8,950	14,300	25,200	51,500
550	364	762	1,440	2,950	4,420	8,500	13,600	24,000	48,900
600	348	727	1,370	2,810	4,210	8,110	12,900	22,900	46,600
650	333	696	1,310	2,690	4,030	7,770	12,400	21,900	44,600
700	320	669	1,260	2,590	3,880	7,460	11,900	21,000	42,900
750	308	644	1,210	2,490	3,730	7,190	11,500	20,300	41,300
800	298	622	1,170	2,410	3,610	6,940	11,100	19,600	39,900
850	288	602	1,130	2,330	3,490	6,720	10,700	18,900	38,600
900	279	584	1,100	2,260	3,380	6,520	10,400	18,400	37,400
950	271	567	1,070	2,190	3,290	6,330	10,100	17,800	36,400
1,000	264	551	1,040	2,130	3,200	6,150	9,810	17,300	35,400
1,100	250	524	987	2,030	3,030	5,840	9,320	16,500	33,600
1,200	239	500	941	1,930	2,900	5,580	8,890	15,700	32,000
1,300	229	478	901	1,850	2,770	5,340	8,510	15,000	30,700
1,400	220	460	866	1,780	2,660	5,130	8,180	14,500	29,500
1,500	212	443	834	1,710	2,570	4,940	7,880	13,900	28,400
1,600	205	428	806	1,650	2,480	4,770	7,610	13,400	27,400
1,700	198	414	780	1,600	2,400	4,620	7,360	13,000	26,500
1,800	192	401	756	1,550	2,330	4,480	7,140	12,600	25,700
1,900	186	390	734	1,510	2,260	4,350	6,930	12,300	25,000
2,000	181	379	714	1,470	2,200	4,230	6,740	11,900	24,300

Note: All table entries are rounded to 3 significant digits.

Table 13-11

Semi-Rigid Copper Tubing [NFPA 54: Table 6.2(f)]

								Gas:	**Natural**
								Inlet Pressure:	**Less than 2 psi**
								Pressure Drop:	**0.3 in. w.c.**
								Specific Gravity:	**0.60**
	Tube Size (in.)								
Nominal: K & L:	**1/4**	**3/8**	**1/2**	**5/8**	**3/4**	**1**	**1-1/4**	**1-1/2**	**2**
Nominal: ACR:	**3/8**	**1/2**	**5/8**	**3/4**	**7/8**	**1-1/8**	**1-3/8**	**–**	**–**
Outside:	**0.375**	**0.500**	**0.625**	**0.750**	**0.875**	**1.125**	**1.375**	**1.625**	**2.125**
Inside:*	**0.305**	**0.402**	**0.527**	**0.652**	**0.745**	**0.995**	**1.245**	**1.481**	**1.959**
Length (ft.)	**Capacity in Cubic Feet of Gas per Hour**								
10	20	42	85	148	210	448	806	1,270	2,650
20	14	29	58	102	144	308	554	873	1,820
30	11	23	47	82	116	247	445	701	1,460
40	10	20	40	70	99	211	381	600	1,250
50	NA	17	35	62	88	187	337	532	1,110
60	NA	16	32	56	79	170	306	482	1,000
70	NA	14	29	52	73	156	281	443	924
80	NA	13	27	48	68	145	262	413	859
90	NA	13	26	45	64	136	245	387	806
100	NA	12	24	43	60	129	232	366	761
125	NA	11	22	38	53	114	206	324	675
150	NA	10	20	34	48	103	186	294	612
175	NA	NA	18	31	45	95	171	270	563
200	NA	NA	17	29	41	89	159	251	523
250	NA	NA	15	26	37	78	141	223	464
300	NA	NA	13	23	33	71	128	202	420
350	NA	NA	12	22	31	65	118	186	387
400	NA	NA	11	20	28	61	110	173	360
450	NA	NA	11	19	27	57	103	162	338
500	NA	NA	10	18	25	54	97	153	319
550	NA	NA	NA	17	24	51	92	145	303
600	NA	NA	NA	16	23	49	88	139	289
650	NA	NA	NA	15	22	47	84	133	277
700	NA	NA	NA	15	21	45	81	128	266
750	NA	NA	NA	14	20	43	78	123	256
800	NA	NA	NA	14	20	42	75	119	247
850	NA	NA	NA	13	19	40	73	115	239
900	NA	NA	NA	13	18	39	71	111	232
950	NA	NA	NA	13	18	38	69	108	225
1,000	NA	NA	NA	12	17	37	67	105	219
1,100	NA	NA	NA	12	16	35	63	100	208
1,200	NA	NA	NA	11	16	34	60	95	199
1,300	NA	NA	NA	11	15	32	58	91	190
1,400	NA	NA	NA	10	14	31	56	88	183
1,500	NA	NA	NA	NA	14	30	54	84	176
1,600	NA	NA	NA	NA	13	29	52	82	170
1,700	NA	NA	NA	NA	13	28	50	79	164
1,800	NA	NA	NA	NA	13	27	49	77	159
1,900	NA	NA	NA	NA	12	26	47	74	155
2,000	NA	NA	NA	NA	12	25	46	72	151

NA means a flow of less than 10 cfh.
Note: All table entries are rounded to 3 significant digits.
*Table capacities are based on Type K copper tubing inside diameter (shown), which has the smallest inside diameter of the copper tubing products.

Table 13-12

Semi-Rigid Copper Tubing [NFPA 54: Table 6.2(g)]

							Gas:	Natural
							Inlet Pressure:	Less than 2 psi
							Pressure Drop:	0.5 in. w.c.
							Specific Gravity:	0.60

	Tube Size (in.)								
Nominal: K & L:	1/4	3/8	1/2	5/8	3/4	1	1-1/4	1-1/2	2
Nominal: ACR:	3/8	1/2	5/8	3/4	7/8	1-1/8	1-3/8	–	–
Outside:	0.375	0.500	0.625	0.750	0.875	1.125	1.375	1.625	2.125
Inside:*	0.305	0.402	0.527	0.652	0.745	0.995	1.245	1.481	1.959
Length (ft.)	Capacity in Cubic Feet of Gas per Hour								
10	27	55	111	195	276	590	1,060	1,680	3,490
20	18	38	77	134	190	406	730	1,150	2,400
30	15	30	61	107	152	326	586	925	1,930
40	13	26	53	92	131	279	502	791	1,650
50	11	23	47	82	116	247	445	701	1,460
60	10	21	42	74	105	224	403	635	1,320
70	NA	19	39	68	96	206	371	585	1,220
80	NA	18	36	63	90	192	345	544	1,130
90	NA	17	34	59	84	180	324	510	1,060
100	NA	16	32	56	79	170	306	482	1,000
125	NA	14	28	50	70	151	271	427	890
150	NA	13	26	45	64	136	245	387	806
175	NA	12	24	41	59	125	226	356	742
200	NA	11	22	39	55	117	210	331	690
250	NA	NA	20	34	48	103	186	294	612
300	NA	NA	18	31	44	94	169	266	554
350	NA	NA	16	28	40	86	155	245	510
400	NA	NA	15	26	38	80	144	228	474
450	NA	NA	14	25	35	75	135	214	445
500	NA	NA	13	23	33	71	128	202	420
550	NA	NA	13	22	32	68	122	192	399
600	NA	NA	12	21	30	64	116	183	381
650	NA	NA	12	20	29	62	111	175	365
700	NA	NA	11	20	28	59	107	168	350
750	NA	NA	11	19	27	57	103	162	338
800	NA	NA	10	18	26	55	99	156	326
850	NA	NA	10	18	25	53	96	151	315
900	NA	NA	NA	17	24	52	93	147	306
950	NA	NA	NA	17	24	50	90	143	297
1,000	NA	NA	NA	16	23	49	88	139	289
1,100	NA	NA	NA	15	22	46	84	132	274
1,200	NA	NA	NA	15	21	44	80	126	262
1,300	NA	NA	NA	14	20	42	76	120	251
1,400	NA	NA	NA	13	19	41	73	116	241
1,500	NA	NA	NA	13	18	39	71	111	232
1,600	NA	NA	NA	13	18	38	68	108	224
1,700	NA	NA	NA	12	17	37	66	104	217
1,800	NA	NA	NA	12	17	36	64	101	210
1,900	NA	NA	NA	11	16	35	62	98	204
2,000	NA	NA	NA	11	16	34	60	95	199

NA means a flow of less than 10 cfh.

Note: All table entries are rounded to 3 significant digits.

*Table capacities are based on Type K copper tubing inside diameter (shown), which has the smallest inside diameter of the copper tubing products.

Table 13-17
Semi-Rigid Copper Tubing [NFPA 54: Table 6.2(I)]

								Gas:	**Natural**
								Inlet Pressure:	**5.0 psi**
								Pressure Drop:	**3.5 psi**
								Specific Gravity:	**0.60**
	Tube Size (in.)								
Nominal: K & L:	**1/4**	**3/8**	**1/2**	**5/8**	**3/4**	**1**	**1-1/4**	**1-1/2**	**2**
Nominal: ACR:	**3/8**	**1/2**	**5/8**	**3/4**	**7/8**	**1-1/8**	**1-3/8**	**–**	**–**
Outside:	**0.375**	**0.500**	**0.625**	**0.750**	**0.875**	**1.125**	**1.375**	**1.625**	**2.125**
Inside:*	**0.305**	**0.402**	**0.527**	**0.652**	**0.745**	**0.995**	**1.245**	**1.481**	**1.959**
Length (ft.)	**Capacity in Cubic Feet of Gas per Hour**								
10	511	1,050	2,140	3,750	5,320	11,400	20,400	32,200	67,100
20	351	724	1,470	2,580	3,650	7,800	14,000	22,200	46,100
30	282	582	1,180	2,070	2,930	6,270	11,300	17,800	37,000
40	241	498	1,010	1,770	2,510	5,360	9,660	15,200	31,700
50	214	441	898	1,570	2,230	4,750	8,560	13,500	28,100
60	194	400	813	1,420	2,020	4,310	7,750	12,200	25,500
70	178	368	748	1,310	1,860	3,960	7,130	11,200	23,400
80	166	342	696	1,220	1,730	3,690	6,640	10,500	21,800
90	156	321	653	1,140	1,620	3,460	6,230	9,820	20,400
100	147	303	617	1,080	1,530	3,270	5,880	9,270	19,300
125	130	269	547	955	1,360	2,900	5,210	8,220	17,100
150	118	243	495	866	1,230	2,620	4,720	7,450	15,500
175	109	224	456	796	1,130	2,410	4,350	6,850	14,300
200	101	208	424	741	1,050	2,250	4,040	6,370	13,300
250	90	185	376	657	932	1,990	3,580	5,650	11,800
300	81	167	340	595	844	1,800	3,250	5,120	10,700
350	75	154	313	547	777	1,660	2,990	4,710	9,810
400	69	143	291	509	722	1,540	2,780	4,380	9,120
450	65	134	273	478	678	1,450	2,610	4,110	8,560
500	62	127	258	451	640	1,370	2,460	3,880	8,090
550	58	121	245	429	608	1,300	2,340	3,690	7,680
600	56	115	234	409	580	1,240	2,230	3,520	7,330
650	53	110	224	392	556	1,190	2,140	3,370	7,020
700	51	106	215	376	534	1,140	2,050	3,240	6,740
750	49	102	207	362	514	1,100	1,980	3,120	6,490
800	48	98	200	350	497	1,060	1,910	3,010	6,270
850	46	95	194	339	481	1,030	1,850	2,910	6,070
900	45	92	188	328	466	1,000	1,790	2,820	5,880
950	43	90	182	319	452	967	1,740	2,740	5,710
1,000	42	87	177	310	440	940	1,690	2,670	5,560
1,100	40	83	169	295	418	893	1,610	2,530	5,280
1,200	38	79	161	281	399	852	1,530	2,420	5,040
1,300	37	76	154	269	382	816	1,470	2,320	4,820
1,400	35	73	148	259	367	784	1,410	2,220	4,630
1,500	34	70	143	249	353	755	1,360	2,140	4,460
1,600	33	68	138	241	341	729	1,310	2,070	4,310
1,700	32	65	133	233	330	705	1,270	2,000	4,170
1,800	31	63	129	226	320	684	1,230	1,940	4,040
1,900	30	62	125	219	311	664	1,200	1,890	3,930
2,000	29	60	122	213	302	646	1,160	1,830	3,820

Note: All table entries are rounded to 3 significant digits.
*Table capacities are based on Type K copper tubing inside diameter (shown), which has the smallest inside diameter of the copper tubing products.

Table 13-18

Corrugated Stainless Steel Tubing (CSST) [NFPA 54-09: Table 6.2(m)]

											Gas:	**Natural**		
											Inlet Pressure:	**Less than 2 psi**		
											Pressure Drop:	**0.5 in. w.c.**		
											Specific Gravity:	**0.60**		
	Tube Size (EHD)*													
Flow Designation:	**13**	**15**	**18**	**19**	**23**	**25**	**30**	**31**	**37**	**39**	**46**	**48**	**60**	**62**
Length (ft)	**Capacity in Cubic Feet of Gas per Hour**													
5	46	63	115	134	225	270	471	546	895	1,037	1,790	2,070	3,660	4,140
10	32	44	82	95	161	192	330	383	639	746	1,260	1,470	2,600	2,930
15	25	35	66	77	132	157	267	310	524	615	1,030	1,200	2,140	2,400
20	22	31	58	67	116	137	231	269	456	536	888	1,050	1,850	2,080
25	19	27	52	60	104	122	206	240	409	482	793	936	1,660	1,860
30	18	25	47	55	96	112	188	218	374	442	723	856	1,520	1,700
40	15	21	41	47	83	97	162	188	325	386	625	742	1,320	1,470
50	13	19	37	42	75	87	144	168	292	347	559	665	1,180	1,320
60	12	17	34	38	68	80	131	153	267	318	509	608	1,080	1,200
70	11	16	31	36	63	74	121	141	248	295	471	563	1,000	1,110
80	10	15	29	33	60	69	113	132	232	277	440	527	940	1,040
90	10	14	28	32	57	65	107	125	219	262	415	498	887	983
100	9	13	26	30	54	62	101	118	208	249	393	472	843	933
150	7	10	20	23	42	48	78	91	171	205	320	387	691	762
200	6	9	18	21	38	44	71	82	148	179	277	336	600	661
250	5	8	16	19	34	39	63	74	133	161	247	301	538	591
300	5	7	15	17	32	36	57	67	95	148	226	275	492	540

Table 13-19

Corrugated Stainless Steel Tubing (CSST) [NFPA 54: Table 6.2(n)]

										Gas:	**Natural**		
										Inlet Pressure:	**Less than 2 psi**		
										Pressure Drop:	**3.0 in. w.c.**		
										Specific Gravity:	**0.60**		
	Tube Size (EHD)*												
Flow Designation:	**13**	**15**	**18**	**19**	**23**	**25**	**30**	**31**	**37**	**46**	**48**	**60**	**62**
Length (ft)	**Capacity in Cubic Feet of Gas per Hour**												
5	120	160	277	327	529	649	1,180	1,370	2,140	4,430	5,010	8,800	10,100
10	83	112	197	231	380	462	828	958	1,530	3,200	3,560	6,270	7,160
15	67	90	161	189	313	379	673	778	1,250	2,540	2,910	5,140	5,850
20	57	78	140	164	273	329	580	672	1,090	2,200	2,530	4,460	5,070
25	51	69	125	147	245	295	518	599	978	1,960	2,270	4,000	4,540
30	46	63	115	134	225	270	471	546	895	1,790	2,070	3,660	4,140
40	39	54	100	116	196	234	407	471	778	1,550	1,800	3,180	3,590
50	35	48	89	104	176	210	363	421	698	1,380	1,610	2,850	3,210
60	32	44	82	95	161	192	330	383	639	1,260	1,470	2,600	2,930
70	29	41	76	88	150	178	306	355	593	1,170	1,360	2,420	2,720
80	27	38	71	82	141	167	285	331	555	1,090	1,280	2,260	2,540
90	26	36	67	77	133	157	268	311	524	1,030	1,200	2,140	2,400
100	24	34	63	73	126	149	254	295	498	974	1,140	2,030	2,280
150	19	27	52	60	104	122	206	240	409	793	936	1,660	1,860
200	17	23	45	52	91	106	178	207	355	686	812	1,440	1,610
250	15	21	40	46	82	95	159	184	319	613	728	1,290	1,440
300	13	19	37	42	75	87	144	168	234	559	665	1,180	1,320

*EHD = Equivalent Hydraulic Diameter, which is a measure of the relative hydraulic efficiency between different tubing sizes. The greater the value of EHD, the greater the gas capacity of the tubing.

Notes:

(1) Table includes losses for four 90 degree bends and two end fittings. Tubing runs with larger numbers of bends and/or fittings shall be increased by an equivalent length of tubing to the following equation: $L = 1.3n$, where L is additional length (ft) of tubing and n is the number of additional fitting and/or bends.

(2) All table entries are rounded to 3 significant digits.

Table 13-20

Corrugated Stainless Steel Tubing (CSST) [NFPA 54: Table 6.2(o)]

										Gas:	**Natural**		
										Inlet Pressure:	**Less than 2 psi**		
										Pressure Drop:	**6.0 in. w.c.**		
										Specific Gravity:	**0.60**		
	Tube Size (EHD)*												
Flow Designation:	**13**	**15**	**18**	**19**	**23**	**25**	**30**	**31**	**37**	**46**	**48**	**60**	**62**
Length (ft)	**Capacity in Cubic Feet of Gas per Hour**												
5	173	229	389	461	737	911	1,690	1,950	3,000	6,280	7,050	12,400	14,260
10	120	160	277	327	529	649	1,180	1,370	2,140	4,430	5,010	8,800	10,100
15	96	130	227	267	436	532	960	1,110	1,760	3,610	4,100	7,210	8,260
20	83	112	197	231	380	462	828	958	1,530	3,120	3,560	6,270	7,160
25	74	99	176	207	342	414	739	855	1,370	2,790	3,190	5,620	6,400
30	67	90	161	189	313	379	673	778	1,250	2,540	2,910	5,140	5,850
40	57	78	140	164	273	329	580	672	1,090	2,200	2,530	4,460	5,070
50	51	69	125	147	245	295	518	599	978	1,960	2,270	4,000	4,540
60	46	63	115	134	225	270	471	546	895	1,790	2,070	3,660	4,140
70	42	58	106	124	209	250	435	505	830	1,660	1,920	3,390	3,840
80	39	54	100	116	196	234	407	471	778	1,550	1,800	3,180	3,590
90	37	51	94	109	185	221	383	444	735	1,460	1,700	3,000	3,390
100	35	48	89	104	176	210	363	421	698	1,380	1,610	2,850	3,210
150	28	39	73	85	145	172	294	342	573	1,130	1,320	2,340	2,630
200	24	34	63	73	126	149	254	295	498	974	1,140	2,030	2,280
250	21	30	57	66	114	134	226	263	447	870	1,020	1,820	2,040
300	19	27	52	60	104	122	206	240	409	793	936	1,660	1,860

*EHD = Equivalent Hydraulic Diameter, which is a measure of the relative hydraulic efficiency between different tubing sizes. The greater the value of EHD, the greater the gas capacity of the tubing.

Notes:

(1) Table includes losses for four 90 degree bends and two end fittings. Tubing runs with larger numbers of bends and/or fittings shall be increased by an equivalent length of tubing to the following equation: $L = 1.3n$, where L is additional length (ft) of tubing and n is the number of additional fittings and/or bends.

(2) All table entries are rounded to 3 significant digits.

Table 13-21

Corrugated Stainless Steel Tubing (CSST) [NFPA 54-09: Table 6.2(p)]

											Gas:	**Natural**		
											Inlet Pressure:	**2.0 psi**		
											Pressure Drop:	**1.0 psi**		
											Specific Gravity:	**0.60**		
	Tube Size (EHD)*													
Flow Designation:	**13**	**15**	**18**	**19**	**23**	**25**	**30**	**31**	**37**	**39**	**46**	**48**	**60**	**62**
Length (ft)	**Capacity in Cubic Feet of Gas per Hour**													
10	270	353	587	700	1100	1,370	2,590	2,990	4,510	5,037	9,600	10,700	18,600	21,600
25	166	220	374	444	709	876	1,620	1,870	2,890	3,258	6,040	6,780	11,900	13,700
30	151	200	342	405	650	801	1,480	1,700	2,640	2,987	5,510	6,200	10,900	12,500
40	129	172	297	351	567	696	1,270	1,470	2,300	2,605	4,760	5,380	9,440	10,900
50	115	154	266	314	510	624	1,140	1,310	2,060	2,343	4,260	4,820	8,470	9,720
75	93	124	218	257	420	512	922	1,070	1,690	1,932	3,470	3,950	6,940	7,940
80	89	120	211	249	407	496	892	1,030	1,640	1,874	3,360	3,820	6,730	7,690
100	79	107	189	222	366	445	795	920	1,470	1,685	3,000	3,420	6,030	6,880
150	64	87	155	182	302	364	646	748	1,210	1,389	2,440	2,800	4,940	5,620
200	55	75	135	157	263	317	557	645	1,050	1,212	2,110	2,430	4,290	4,870
250	49	67	121	141	236	284	497	576	941	1,090	1,890	2,180	3,850	4,360
300	44	61	110	129	217	260	453	525	862	999	1,720	1,990	3,520	3,980
400	38	52	96	111	189	225	390	453	749	871	1,490	1,730	3,060	3,450
500	34	46	86	100	170	202	348	404	552	783	1,330	1,550	2,740	3,090

*EHD = Equivalent Hydraulic Diameter, which is a measure of the relative hydraulic efficiency between different tubing sizes. The greater the value of EHD, the greater the gas capacity of the tubing.

Notes:

(1) Table does not include effect of pressure drop across the line regulator. Where regulator loss exceeds 3/4 psi, do not use this table. Consult with regulator manufacturer for pressure drops and capacity factors. Pressure drops across a regulator may vary with flow rate.

(2) CAUTION: Capacities shown in table may exceed maximum capacity for a selected regulator. Consult with regulator or tubing manufacturer for guidance.

(3) Table includes losses for four 90 degree bends and two end fittings. Tubing runs with larger numbers of bends and/or fittings shall be increased by an equivalent length of tubing according to the following equation: $L = 1.3n$, where L is additional length (ft) of tubing and n is the number of additional fittings and/or bends.

(4) All table entries are rounded to 3 significant digits.

Table 13-22

Corrugated Stainless Steel Tubing (CSST) [NFPA 54-09: Table 6.2(q)]

											Gas:	**Natural**		
											Inlet Pressure:	**5.0 psi**		
											Pressure Drop:	**3.5 psi**		
											Specific Gravity:	**0.60**		
	Tube Size (EHD)*													
Flow Designation:	**13**	**15**	**18**	**19**	**23**	**25**	**30**	**31**	**37**	**39**	**46**	**48**	**60**	**62**
Length (ft)	**Capacity in Cubic Feet of Gas per Hour**													
10	523	674	1,080	1,300	2,000	2,530	4,920	5,660	8,300	9,140	18,100	19,800	34,400	40,400
25	322	420	691	827	1,290	1,620	3,080	3,540	5,310	5,911	11,400	12,600	22,000	25,600
30	292	382	632	755	1,180	1,480	2,800	3,230	4,860	5,420	10,400	11,500	20,100	23,400
40	251	329	549	654	1,030	1,280	2,420	2,790	4,230	4,727	8,970	10,000	17,400	20,200
50	223	293	492	586	926	1,150	2,160	2,490	3,790	4,251	8,020	8,930	15,600	18,100
75	180	238	403	479	763	944	1,750	2,020	3,110	3,506	6,530	7,320	12,800	14,800
80	174	230	391	463	740	915	1,690	1,960	3,020	3,400	6,320	7,090	12,400	14,300
100	154	205	350	415	665	820	1,510	1,740	2,710	3,057	5,650	6,350	11,100	12,800
150	124	166	287	339	548	672	1,230	1,420	2,220	2,521	4,600	5,200	9,130	10,500
200	107	143	249	294	478	584	1,060	1,220	1,930	2,199	3,980	4,510	7,930	9,090
250	95	128	223	263	430	524	945	1,090	1,730	1,977	3,550	4,040	7,110	8,140
300	86	116	204	240	394	479	860	995	1,590	1,813	3,240	3,690	6,500	7,430
400	74	100	177	208	343	416	742	858	1,380	1,581	2,800	3,210	5,650	6,440
500	66	89	159	186	309	373	662	766	1,040	1,422	2,500	2,870	5,060	5,760

*EHD = Equivalent Hydraulic Diameter, which is a measure of the relative hydraulic efficiency between different tubing sizes. The greater the value of EHD, the greater the gas capacity of the tubing.

Notes:

(1) Table does not include effect of pressure drop across the line regulator. Where regulator loss exceeds 1 psi, do not use this table. Consult with regulator manufacturer for pressure drops and capacity factors. Pressure drops across regulator may vary with the flow rate.

(2) CAUTION: Capacities shown in table may exceed maximum capacity of selected regulator. Consult tubing manufacturer for guidance.

(3) Table includes losses for four 90 degree bends and two end fittings. Tubing runs with larger numbers of bends and/or fittings shall be increased by an equivalent length of tubing to the following equation: $L = 1.3n$, where L is additional length (ft) of tubing and n is the number of additional fittings and/or bends.

(4) All table entries are rounded to 3 significant digits.

Table 13-23

Polyethylene Plastic Pipe [NFPA 54: Table 6.2(r)]

					Gas:	**Natural**
					Inlet Pressure:	**Less than 2 psi**
					Pressure Drop:	**0.3 in. w.c.**
					Specific Gravity:	**0.60**
	Pipe Size (in.)					
Nominal OD:	**1/2**	**3/4**	**1**	**1-1/4**	**1-1/2**	**2**
Designation:	**SDR 9.33**	**SDR 11.00**	**SDR 11.00**	**SDR 10.00**	**SDR 11.00**	**SDR 11.00**
Actual ID:	**0.660**	**0.860**	**1.077**	**1.328**	**1.554**	**1.943**
Length (ft)	**Capacity in Cubic Feet of Gas per Hour**					
10	153	305	551	955	1,440	2,590
20	105	210	379	656	991	1,780
30	84	169	304	527	796	1,430
40	72	144	260	451	681	1,220
50	64	128	231	400	604	1,080
60	58	116	209	362	547	983
70	53	107	192	333	503	904
80	50	99	179	310	468	841
90	46	93	168	291	439	789
100	44	88	159	275	415	745
125	39	78	141	243	368	661
150	35	71	127	221	333	598
175	32	65	117	203	306	551
200	30	60	109	189	285	512
250	27	54	97	167	253	454
300	24	48	88	152	229	411
350	22	45	81	139	211	378
400	21	42	75	130	196	352
450	19	39	70	122	184	330
500	18	37	66	115	174	312

Note: All table entries are rounded to 3 significant digits.

Table 13-24
Polyethylene Plastic Pipe [NFPA 54: Table 6.2(s)]

					Gas:	Natural
					Inlet Pressure:	Less than 2 psi
					Pressure Drop:	0.5 in. w.c.
					Specific Gravity:	0.60
	Pipe Size (in.)					
Nominal OD:	1/2	3/4	1	1-1/4	1-1/2	2
Designation:	SDR 9.33	SDR 11.0	SDR 11.00	SDR 10.00	SDR 11.00	SDR 11.00
Actual ID:	0.660	0.860	1.077	1.328	1.554	1.943
Length (ft)	Capacity in Cubic Feet of Gas per Hour					
10	201	403	726	1,260	1,900	3,410
20	138	277	499	865	1,310	2,350
30	111	222	401	695	1,050	1,880
40	95	190	343	594	898	1,610
50	84	169	304	527	796	1,430
60	76	153	276	477	721	1,300
70	70	140	254	439	663	1,190
80	65	131	236	409	617	1,110
90	61	123	221	383	579	1,040
100	58	116	209	362	547	983
125	51	103	185	321	485	871
150	46	93	168	291	439	789
175	43	86	154	268	404	726
200	40	80	144	249	376	675
250	35	71	127	221	333	598
300	32	64	115	200	302	542
350	29	59	106	184	278	499
400	27	55	99	171	258	464
450	26	51	93	160	242	435
500	24	48	88	152	229	411

Note: All table entries are rounded to 3 significant digits.

Table 13-25
Polyethylene Plastic Pipe [NFPA 54: Table 6.2(t)]

					Gas:	Natural
					Inlet Pressure:	2.0 psi
					Pressure Drop:	1.0 psi
					Specific Gravity:	0.60
	Pipe Size (in.)					
Nominal OD:	**1/2**	**3/4**	**1**	**1-1/4**	**1-1/2**	**2**
Designation:	**SDR 9.33**	**SDR 11.0**	**SDR 11.00**	**SDR 10.00**	**SDR 11.00**	**SDR 11.00**
Actual ID:	**0.660**	**0.860**	**1.077**	**1.328**	**1.554**	**1.943**
Length (ft)	**Capacity in Cubic Feet of Gas per Hour**					
10	1,860	3,720	6,710	11,600	17,600	31,600
20	1,280	2,560	4,610	7,990	12,100	21,700
30	1,030	2,050	3,710	6,420	9,690	17,400
40	878	1,760	3,170	5,490	8,300	14,900
50	778	1,560	2,810	4,870	7,350	13,200
60	705	1,410	2,550	4,410	6,660	12,000
70	649	1,300	2,340	4,060	6,130	11,000
80	603	1,210	2,180	3,780	5,700	10,200
90	566	1,130	2,050	3,540	5,350	9,610
100	535	1,070	1,930	3,350	5,050	9,080
125	474	949	1,710	2,970	4,480	8,050
150	429	860	1,550	2,690	4,060	7,290
175	395	791	1,430	2,470	3,730	6,710
200	368	736	1,330	2,300	3,470	6,240
250	326	652	1,180	2,040	3,080	5,530
300	295	591	1,070	1,850	2,790	5,010
350	272	544	981	1,700	2,570	4,610
400	253	506	913	1,580	2,390	4,290
450	237	475	856	1,480	2,240	4,020
500	224	448	809	1,400	2,120	3,800
550	213	426	768	1,330	2,010	3,610
600	203	406	733	1,270	1,920	3,440
650	194	389	702	1,220	1,840	3,300
700	187	374	674	1,170	1,760	3,170
750	180	360	649	1,130	1,700	3,050
800	174	348	627	1,090	1,640	2,950
850	168	336	607	1,050	1,590	2,850
900	163	326	588	1,020	1,540	2,770
950	158	317	572	990	1,500	2,690
1,000	154	308	556	963	1,450	2,610
1,100	146	293	528	915	1,380	2,480
1,200	139	279	504	873	1,320	2,370
1,300	134	267	482	836	1,260	2,270
1,400	128	257	463	803	1,210	2,180
1,500	124	247	446	773	1,170	2,100
1,600	119	239	431	747	1,130	2,030
1,700	115	231	417	723	1,090	1,960
1,800	112	224	404	701	1,060	1,900
1,900	109	218	393	680	1,030	1,850
2,000	106	212	382	662	1,000	1,800

Note: All table entries are rounded to 3 significant digits.

Table 13-26
Polyethylene Plastic Tubing [NFPA 54: Table 6.2(u)]

	Gas:	Natural
	Inlet Pressure:	Less than 2.0 psi
	Pressure Drop:	0.3 in. w.c.
	Specific Gravity:	0.60
	Plastic Tubing Size (CTS)* (in.)	
Nominal OD:	1/2	3/4
Designation:	SDR 7.00	SDR 11.00
Actual ID:	0.445	0.927
Length (ft)	Capacity in Cubic Feet of Gas per Hour	
10	54	372
20	37	256
30	30	205
40	26	176
50	23	156
60	21	141
70	19	130
80	18	121
90	17	113
100	16	107
125	14	95
150	13	86
175	12	79
200	11	74
225	10	69
250	NA	65
275	NA	62
300	NA	59
350	NA	54
400	NA	51
450	NA	47
500	NA	45

* CTS = Copper tube size.
NA means a flow of less than 10 cfh.
Note: All table entries are rounded to 3 significant digits.

Table 13-27
Polyethylene Plastic Tubing [NFPA 54: Table 6.2(v)]

	Gas:	Natural
	Inlet Pressure:	Less than 2.0 psi
	Pressure Drop:	0.5 in. w.c.
	Specific Gravity:	0.60
	Plastic Tubing Size (CTS)* (in.)	
Nominal OD:	1/2	3/4
Designation:	SDR 7.00	SDR 11.00
Actual ID:	0.445	0.927
Length (ft)	Capacity in Cubic Feet of Gas per Hour	
10	72	490
20	49	337
30	39	271
40	34	232
50	30	205
60	27	186
70	25	171
80	23	159
90	22	149
100	21	141
125	18	125
150	17	113
175	15	104
200	14	97
225	13	91
250	12	86
275	11	82
300	11	78
350	10	72
400	NA	67
450	NA	63
500	NA	59

* CTS = Copper tube size.
NA means a flow of less than 10 cfh.
Note: All table entries are rounded to 3 significant digits.

Table 13-28

Schedule 40 Metallic Pipe [NFPA 54-09: Table 6.3(a)]

						Gas:	Undiluted Propane		
						Inlet Pressure:	10.0 psi		
						Pressure Drop:	1.0 psi		
						Specific Gravity:	1.50		
INTENDED USE: Pipe Sizing Between First Stage (High Pressure Regulator) and Second Stage (Low Pressure Regulator)									
	Pipe Size (in.)								
Nominal Inside:	1/2	3/4	1	1-1/4	1-1/2	2	2-1/2	3	4
Actual:	0.622	0.824	1.049	1.380	1.610	2.067	2.469	3.068	4.026
Length (ft.)	Capacity in Thousands of Btu per Hour								
10	3,320	6,950	13,100	26,900	40,300	77,600	124,000	219,000	446,000
20	2,280	4,780	9,000	18,500	27,700	53,300	85,000	150,000	306,000
30	1,830	3,840	7,220	14,800	22,200	42,800	68,200	121,000	246,000
40	1,570	3,280	6,180	12,700	19,000	36,600	58,400	103,000	211,000
50	1,390	2,910	5,480	11,300	16,900	32,500	51,700	91,500	187,000
60	1,260	2,640	4,970	10,200	15,300	29,400	46,900	82,900	169,000
70	1,160	2,430	4,570	9,380	14,100	27,100	43,100	76,300	156,000
80	1,080	2,260	4,250	8,730	13,100	25,200	40,100	70,900	145,000
90	1,010	2,120	3,990	8,190	12,300	23,600	37,700	66,600	136,000
100	956	2,000	3,770	7,730	11,600	22,300	35,600	62,900	128,000
125	848	1,770	3,340	6,850	10,300	19,800	31,500	55,700	114,000
150	768	1,610	3,020	6,210	9,300	17,900	28,600	50,500	103,000
175	706	1,480	2,780	5,710	8,560	16,500	26,300	46,500	94,700
200	657	1,370	2,590	5,320	7,960	15,300	24,400	43,200	88,100
250	582	1,220	2,290	4,710	7,060	13,600	21,700	38,300	78,100
300	528	1,100	2,080	4,270	6,400	12,300	19,600	34,700	70,800
350	486	1,020	1,910	3,930	5,880	11,300	18,100	31,900	65,100
400	452	945	1,780	3,650	5,470	10,500	16,800	29,700	60,600
450	424	886	1,670	3,430	5,140	9,890	15,800	27,900	56,800
500	400	837	1,580	3,240	4,850	9,340	14,900	26,300	53,700
550	380	795	1,500	3,070	4,610	8,870	14,100	25,000	51,000
600	363	759	1,430	2,930	4,400	8,460	13,500	23,900	48,600
650	347	726	1,370	2,810	4,210	8,110	12,900	22,800	46,600
700	334	698	1,310	2,700	4,040	7,790	12,400	21,900	44,800
750	321	672	1,270	2,600	3,900	7,500	12,000	21,100	43,100
800	310	649	1,220	2,510	3,760	7,240	11,500	20,400	41,600
850	300	628	1,180	2,430	3,640	7,010	11,200	19,800	40,300
900	291	609	1,150	2,360	3,530	6,800	10,800	19,200	39,100
950	283	592	1,110	2,290	3,430	6,600	10,500	18,600	37,900
1,000	275	575	1,080	2,230	3,330	6,420	10,200	18,100	36,900
1,100	261	546	1,030	2,110	3,170	6,100	9,720	17,200	35,000
1,200	249	521	982	2,020	3,020	5,820	9,270	16,400	33,400
1,300	239	499	940	1,930	2,890	5,570	8,880	15,700	32,000
1,400	229	480	903	1,850	2,780	5,350	8,530	15,100	30,800
1,500	221	462	870	1,790	2,680	5,160	8,220	14,500	29,600
1,600	213	446	840	1,730	2,590	4,980	7,940	14,000	28,600
1,700	206	432	813	1,670	2,500	4,820	7,680	13,600	27,700
1,800	200	419	789	1,620	2,430	4,670	7,450	13,200	26,900
1,900	194	407	766	1,570	2,360	4,540	7,230	12,800	26,100
2,000	189	395	745	1,530	2,290	4,410	7,030	12,400	25,400

Note: All table entries are rounded to 3 significant digits.

Table 13-29

Schedule 40 Metallic Pipe [NFPA 54-09: Table 6.3(b)]

						Gas:	Undiluted Propane	
						Inlet Pressure:	10.0 psi	
						Pressure Drop:	3.0 psi	
						Specific Gravity:	1.50	

INTENDED USE: Pipe Sizing Between First Stage (High Pressure Regulator) and Second Stage (Low Pressure Regulator)

	Pipe Size (in.)								
Nominal Inside:	**1/2**	**3/4**	**1**	**1-1/4**	**1-1/2**	**2**	**2-1/2**	**3**	**4**
Actual:	**0.622**	**0.824**	**1.049**	**1.380**	**1.610**	**2.067**	**2.469**	**3.068**	**4.026**
Length (ft.)	**Capacity in Thousands of Btu per Hour**								
10	5,890	12,300	23,200	47,600	71,300	137,000	219,000	387,000	789,000
20	4,050	8,460	15,900	32,700	49,000	94,400	150,000	266,000	543,000
30	3,250	6,790	12,800	26,300	39,400	75,800	121,000	214,000	436,000
40	2,780	5,810	11,000	22,500	33,700	64,900	103,000	183,000	373,000
50	2,460	5,150	9,710	19,900	29,900	57,500	91,600	162,000	330,000
60	2,230	4,670	8,790	18,100	27,100	52,100	83,000	147,000	299,000
70	2,050	4,300	8,090	16,600	24,900	47,900	76,400	135,000	275,000
80	1,910	4,000	7,530	15,500	23,200	44,600	71,100	126,000	256,000
90	1,790	3,750	7,060	14,500	21,700	41,800	66,700	118,000	240,000
100	1,690	3,540	6,670	13,700	20,500	39,500	63,000	111,000	227,000
125	1,500	3,140	5,910	12,100	18,200	35,000	55,800	98,700	201,000
150	1,360	2,840	5,360	11,000	16,500	31,700	50,600	89,400	182,000
175	1,250	2,620	4,930	10,100	15,200	29,200	46,500	82,300	167,800
200	1,160	2,430	4,580	9,410	14,100	27,200	43,300	76,500	156,100
250	1,030	2,160	4,060	8,340	12,500	24,100	38,400	67,800	138,400
300	935	1,950	3,680	7,560	11,300	21,800	34,800	61,500	125,400
350	860	1,800	3,390	6,950	10,400	20,100	32,000	56,500	115,300
400	800	1,670	3,150	6,470	9,690	18,700	29,800	52,600	107,300
450	751	1,570	2,960	6,070	9,090	17,500	27,900	49,400	100,700
500	709	1,480	2,790	5,730	8,590	16,500	26,400	46,600	95,100
550	673	1,410	2,650	5,450	8,160	15,700	25,000	44,300	90,300
600	642	1,340	2,530	5,200	7,780	15,000	23,900	42,200	86,200
650	615	1,290	2,420	4,980	7,450	14,400	22,900	40,500	82,500
700	591	1,240	2,330	4,780	7,160	13,800	22,000	38,900	79,300
750	569	1,190	2,240	4,600	6,900	13,300	21,200	37,400	76,400
800	550	1,150	2,170	4,450	6,660	12,800	20,500	36,200	73,700
850	532	1,110	2,100	4,300	6,450	12,400	19,800	35,000	71,400
900	516	1,080	2,030	4,170	6,250	12,000	19,200	33,900	69,200
950	501	1,050	1,970	4,050	6,070	11,700	18,600	32,900	67,200
1,000	487	1,020	1,920	3,940	5,900	11,400	18,100	32,000	65,400
1,100	463	968	1,820	3,740	5,610	10,800	17,200	30,400	62,100
1,200	442	923	1,740	3,570	5,350	10,300	16,400	29,000	59,200
1,300	423	884	1,670	3,420	5,120	9,870	15,700	27,800	56,700
1,400	406	849	1,600	3,280	4,920	9,480	15,100	26,700	54,500
1,500	391	818	1,540	3,160	4,740	9,130	14,600	25,700	52,500
1,600	378	790	1,490	3,060	4,580	8,820	14,100	24,800	50,700
1,700	366	765	1,440	2,960	4,430	8,530	13,600	24,000	49,000
1,800	355	741	1,400	2,870	4,300	8,270	13,200	23,300	47,600
1,900	344	720	1,360	2,780	4,170	8,040	12,800	22,600	46,200
2,000	335	700	1,320	2,710	4,060	7,820	12,500	22,000	44,900

Note: All table entries are rounded to 3 significant digits.

Table 13-30

Schedule 40 Metallic Pipe [NFPA 54-09: Table 6.3(c)]

Gas:	Undiluted Propane
Inlet Pressure:	2.0 psi
Pressure Drop:	1.0 psi
Specific Gravity:	1.50

INTENDED USE: Pipe Sizing Between 2 psi Service and Line Pressure Regulator

	Pipe Size (in.)								
Nominal	1/2	3/4	1	1-1/4	1-1/2	2	2-1/2	3	4
Actual ID:	0.622	0.824	1.049	1.380	1.610	2.067	2.469	3.068	4.026
Length (ft.)	Capacity in Thousands of Btu per Hour								
10	2,680	5,590	10,500	21,600	32,400	62,400	99,500	176,000	359,000
20	1,840	3,850	7,240	14,900	22,300	42,900	68,400	121,000	247,000
30	1,480	3,090	5,820	11,900	17,900	34,500	54,900	97,100	198,000
40	1,260	2,640	4,980	10,200	15,300	29,500	47,000	83,100	170,000
50	1,120	2,340	4,410	9,060	13,600	26,100	41,700	73,700	150,000
60	1,010	2,120	4,000	8,210	12,300	23,700	37,700	66,700	136,000
70	934	1,950	3,680	7,550	11,300	21,800	34,700	61,400	125,000
80	869	1,820	3,420	7,020	10,500	20,300	32,300	57,100	116,000
90	815	1,700	3,210	6,590	9,880	19,000	30,300	53,600	109,000
100	770	1,610	3,030	6,230	9,330	18,000	28,600	50,600	103,000
125	682	1,430	2,690	5,520	8,270	15,900	25,400	44,900	91,500
150	618	1,290	2,440	5,000	7,490	14,400	23,000	40,700	82,900
175	569	1,190	2,240	4,600	6,890	13,300	21,200	37,400	76,300
200	529	1,110	2,080	4,280	6,410	12,300	19,700	34,800	71,000
250	469	981	1,850	3,790	5,680	10,900	17,400	30,800	62,900
300	425	889	1,670	3,440	5,150	9,920	15,800	27,900	57,000
350	391	817	1,540	3,160	4,740	9,120	14,500	25,700	52,400
400	364	760	1,430	2,940	4,410	8,490	13,500	23,900	48,800
450	341	714	1,340	2,760	4,130	7,960	12,700	22,400	45,800
500	322	674	1,270	2,610	3,910	7,520	12,000	21,200	43,200
550	306	640	1,210	2,480	3,710	7,140	11,400	20,100	41,100
600	292	611	1,150	2,360	3,540	6,820	10,900	19,200	39,200
650	280	585	1,100	2,260	3,390	6,530	10,400	18,400	37,500
700	269	562	1,060	2,170	3,260	6,270	9,990	17,700	36,000
750	259	541	1,020	2,090	3,140	6,040	9,630	17,000	34,700
800	250	523	985	2,020	3,030	5,830	9,300	16,400	33,500
850	242	506	953	1,960	2,930	5,640	9,000	15,900	32,400
900	235	490	924	1,900	2,840	5,470	8,720	15,400	31,500
950	228	476	897	1,840	2,760	5,310	8,470	15,000	30,500
1,000	222	463	873	1,790	2,680	5,170	8,240	14,600	29,700
1,100	210	440	829	1,700	2,550	4,910	7,830	13,800	28,200
1,200	201	420	791	1,620	2,430	4,680	7,470	13,200	26,900
1,300	192	402	757	1,550	2,330	4,490	7,150	12,600	25,800
1,400	185	386	727	1,490	2,240	4,310	6,870	12,100	24,800
1,500	178	372	701	1,440	2,160	4,150	6,620	11,700	23,900
1,600	172	359	677	1,390	2,080	4,010	6,390	11,300	23,000
1,700	166	348	655	1,340	2,010	3,880	6,180	10,900	22,300
1,800	161	337	635	1,300	1,950	3,760	6,000	10,600	21,600
1,900	157	327	617	1,270	1,900	3,650	5,820	10,300	21,000
2,000	152	318	600	1,230	1,840	3,550	5,660	10,000	20,400

Note: All table entries are rounded to 3 significant digits.

Table 13-31

Schedule 40 Metallic Pipe [NFPA 54-09: Table 6.3(d)]

							Gas:	Undiluted Propane
							Inlet Pressure:	11.0 in. w.c.
							Pressure Drop:	0.5 in. w.c.
							Specific Gravity:	1.50

INTENDED USE: Pipe Sizing Between Single or Second Stage (Low Pressure Regulator) and Appliance									
	Pipe Size (in.)								
Nominal Inside:	1/2	3/4	1	1-1/4	1-1/2	2	2-1/2	3	4
Actual ID:	0.622	0.824	1.049	1.380	1.610	2.067	2.469	3.068	4.026
Length (ft.)	Capacity in Thousands of Btu per Hour								
10	291	608	1,150	2,350	3,520	6,790	10,800	19,100	39,000
20	200	418	787	1,620	2,420	4,660	7,430	13,100	26,800
30	160	336	632	1,300	1,940	3,750	5,970	10,600	21,500
40	137	287	541	1,110	1,660	3,210	5,110	9,030	18,400
50	122	255	480	985	1,480	2,840	4,530	8,000	16,300
60	110	231	434	892	1,340	2,570	4,100	7,250	14,800
80	101	212	400	821	1,230	2,370	3,770	6,670	13,600
100	94	197	372	763	1,140	2,200	3,510	6,210	12,700
125	89	185	349	716	1,070	2,070	3,290	5,820	11,900
150	84	175	330	677	1,010	1,950	3,110	5,500	11,200
175	74	155	292	600	899	1,730	2,760	4,880	9,950
200	67	140	265	543	814	1,570	2,500	4,420	9,010
250	62	129	243	500	749	1,440	2,300	4,060	8,290
300	58	120	227	465	697	1,340	2,140	3,780	7,710
350	51	107	201	412	618	1,190	1,900	3,350	6,840
400	46	97	182	373	560	1,080	1,720	3,040	6,190
450	42	89	167	344	515	991	1,580	2,790	5,700
500	40	83	156	320	479	922	1,470	2,600	5,300
550	37	78	146	300	449	865	1,380	2,440	4,970
600	35	73	138	283	424	817	1,300	2,300	4,700
650	33	70	131	269	403	776	1,240	2,190	4,460
700	32	66	125	257	385	741	1,180	2,090	4,260
750	30	64	120	246	368	709	1,130	2,000	4,080
800	29	61	115	236	354	681	1,090	1,920	3,920
850	28	59	111	227	341	656	1,050	1,850	3,770
900	27	57	107	220	329	634	1,010	1,790	3,640
950	26	55	104	213	319	613	978	1,730	3,530
1,000	25	53	100	206	309	595	948	1,680	3,420
1,100	25	52	97	200	300	578	921	1,630	3,320
1,200	24	50	95	195	292	562	895	1,580	3,230
1,300	23	48	90	185	277	534	850	1,500	3,070
1,400	22	46	86	176	264	509	811	1,430	2,930
1,500	21	44	82	169	253	487	777	1,370	2,800
1,600	20	42	79	162	243	468	746	1,320	2,690
1,700	19	40	76	156	234	451	719	1,270	2,590
1,800	19	39	74	151	226	436	694	1,230	2,500
1,900	18	38	71	146	219	422	672	1,190	2,420
2,000	18	37	69	142	212	409	652	1,150	2,350

Note: All table entries are rounded to 3 significant digits.

Table 13-32

Semi-Rigid Copper Tubing [NFPA 54-09: Table 6.3(e)]

Gas:	Undiluted Propane
Inlet Pressure:	10.0 psi
Pressure Drop:	1.0 psi
Specific Gravity:	1.50

INTENDED USE: Tube Sizing Between First Stage (High Pressure Regulator) and Second Stage (Low Pressure Regulator)

	Tube Size (in.)								
Nominal: K & L:	1/4	3/8	1/2	5/8	3/4	1	1-1/4	1-1/2	2
Nominal: ACR:	3/8	1/2	5/8	3/4	7/8	1-1/8	1-3/8	–	–
Outside:	0.375	0.500	0.625	0.750	0.875	1.125	1.375	1.625	2.125
Inside:*	0.305	0.402	0.527	0.652	0.745	0.995	1.245	1.481	1.959
Length (ft.)	Capacity in Thousands of Btu per Hour								
10	513	1,060	2,150	3,760	5,330	11,400	20,500	32,300	67,400
20	352	727	1,480	2,580	3,670	7,830	14,100	22,300	46,300
30	283	584	1,190	2,080	2,940	6,290	11,300	17,900	37,200
40	242	500	1,020	1,780	2,520	5,380	9,690	15,300	31,800
50	215	443	901	1,570	2,230	4,770	8,590	13,500	28,200
60	194	401	816	1,430	2,020	4,320	7,780	12,300	25,600
70	179	369	751	1,310	1,860	3,980	7,160	11,300	23,500
80	166	343	699	1,220	1,730	3,700	6,660	10,500	21,900
90	156	322	655	1,150	1,630	3,470	6,250	9,850	20,500
100	147	304	619	1,080	1,540	3,280	5,900	9,310	19,400
125	131	270	549	959	1,360	2,910	5,230	8,250	17,200
150	118	244	497	869	1,230	2,630	4,740	7,470	15,600
175	109	225	457	799	1,130	2,420	4,360	6,880	14,300
200	101	209	426	744	1,060	2,250	4,060	6,400	13,300
250	90	185	377	659	935	2,000	3,600	5,670	11,800
300	81	168	342	597	847	1,810	3,260	5,140	10,700
350	75	155	314	549	779	1,660	3,000	4,730	9,840
400	70	144	292	511	725	1,550	2,790	4,400	9,160
450	65	135	274	480	680	1,450	2,620	4,130	8,590
500	62	127	259	453	643	1,370	2,470	3,900	8,120
550	59	121	246	430	610	1,300	2,350	3,700	7,710
600	56	115	235	410	582	1,240	2,240	3,530	7,350
650	54	111	225	393	558	1,190	2,140	3,380	7,040
700	51	106	216	378	536	1,140	2,060	3,250	6,770
750	50	102	208	364	516	1,100	1,980	3,130	6,520
800	48	99	201	351	498	1,060	1,920	3,020	6,290
850	46	96	195	340	482	1,030	1,850	2,920	6,090
900	45	93	189	330	468	1,000	1,800	2,840	5,910
950	44	90	183	320	454	970	1,750	2,750	5,730
1,000	42	88	178	311	442	944	1,700	2,680	5,580
1,100	40	83	169	296	420	896	1,610	2,540	5,300
1,200	38	79	161	282	400	855	1,540	2,430	5,050
1,300	37	76	155	270	383	819	1,470	2,320	4,840
1,400	35	73	148	260	368	787	1,420	2,230	4,650
1,500	34	70	143	250	355	758	1,360	2,150	4,480
1,600	33	68	138	241	343	732	1,320	2,080	4,330
1,700	32	66	134	234	331	708	1,270	2,010	4,190
1,800	31	64	130	227	321	687	1,240	1,950	4,060
1,900	30	62	126	220	312	667	1,200	1,890	3,940
2,000	29	60	122	214	304	648	1,170	1,840	3,830

Note: All table entries are rounded to 3 significant digits.
*Table capacities are based on Type K copper tubing inside diameter (shown), which has the smallest inside diameter of the copper tubing products.

Table 13-33

Semi-Rigid Copper Tubing [NFPA 54-09: Table 6.3(f)]

							Gas:		**Undiluted Propane**	
							Inlet Pressure:		**11.0 in. w.c.**	
							Pressure Drop:		**0.5 in. w.c.**	
							Specific Gravity:		**1.50**	
INTENDED USE: Tube Sizing Between Single or Second Stage (Low Pressure Regulator) and Appliance										
		Tube Size (in.)								
Nominal:	**K & L:**	**1/4**	**3/8**	**1/2**	**5/8**	**3/4**	**1**	**1-1/4**	**1-1/2**	**2**
	ACR:	**3/8**	**1/2**	**5/8**	**3/4**	**7/8**	**1-1/8**	**1-3/8**	**–**	**–**
Outside:		**0.375**	**0.500**	**0.625**	**0.750**	**0.875**	**1.125**	**1.375**	**1.625**	**2.125**
Inside:*		**0.305**	**0.402**	**0.527**	**0.652**	**0.745**	**0.995**	**1.245**	**1.481**	**1.959**
Length (ft.)		**Capacity in Thousands of Btu per Hour**								
10		45	93	188	329	467	997	1,800	2,830	5,890
20		31	64	129	226	321	685	1,230	1,950	4,050
30		25	51	104	182	258	550	991	1,560	3,250
40		21	44	89	155	220	471	848	1,340	2,780
50		19	39	79	138	195	417	752	1,180	2,470
60		17	35	71	125	177	378	681	1,070	2,240
70		16	32	66	115	163	348	626	988	2,060
80		15	30	61	107	152	324	583	919	1,910
90		14	28	57	100	142	304	547	862	1,800
100		13	27	54	95	134	287	517	814	1,700
125		11	24	48	84	119	254	458	722	1,500
150		10	21	44	76	108	230	415	654	1,360
175		NA	20	40	70	99	212	382	602	1,250
200		NA	18	37	65	92	197	355	560	1,170
250		NA	16	33	58	82	175	315	496	1,030
300		NA	15	30	52	74	158	285	449	936
350		NA	14	28	48	68	146	262	414	861
400		NA	13	26	45	63	136	244	385	801
450		NA	12	24	42	60	127	229	361	752
500		NA	11	23	40	56	120	216	341	710
550		NA	11	22	38	53	114	205	324	674
600		NA	10	21	36	51	109	196	309	643
650		NA	NA	20	34	49	104	188	296	616
700		NA	NA	19	33	47	100	180	284	592
750		NA	NA	18	32	45	96	174	274	570
800		NA	NA	18	31	44	93	168	264	551
850		NA	NA	17	30	42	90	162	256	533
900		NA	NA	17	29	41	87	157	248	517
950		NA	NA	16	28	40	85	153	241	502
1,000		NA	NA	16	27	39	83	149	234	488
1,100		NA	NA	15	26	37	78	141	223	464
1,200		NA	NA	14	25	35	75	135	212	442
1,300		NA	NA	14	24	34	72	129	203	423
1,400		NA	NA	13	23	32	69	124	195	407
1,500		NA	NA	13	22	31	66	119	188	392
1,600		NA	NA	12	21	30	64	115	182	378
1,700		NA	NA	12	20	29	62	112	176	366
1,800		NA	NA	11	20	28	60	108	170	355
1,900		NA	NA	11	19	27	58	105	166	345
2,000		NA	NA	11	19	27	57	102	161	335

NA means a flow of less than 10,000 Btu/hr.

Note: All table entries are rounded to 3 significant digits.

*Table capacities are based on Type K copper tubing inside diameter (shown), which has the smallest inside diameter of the copper tubing products.

Table 13-34

Semi-Rigid Copper Tubing [NFPA 54-09: Table 6.3(g)]

Gas:	**Undiluted Propane**
Inlet Pressure:	**2.0 psi**
Pressure Drop:	**1.0 psi**
Specific Gravity:	**1.50**

INTENDED USE: Tube Sizing Between 2 psi Service and Line Pressure Regulator

Tube Size (in.)										
Nominal:	**K & L:**	**1/4**	**3/8**	**1/2**	**5/8**	**3/4**	**1**	**1-1/4**	**1-1/2**	**2**
	ACR:	**3/8**	**1/2**	**5/8**	**3/4**	**7/8**	**1-1/8**	**1-3/8**	**–**	**–**
Outside:		**0.375**	**0.500**	**0.625**	**0.750**	**0.875**	**1.125**	**1.375**	**1.625**	**2.125**
Inside:*		**0.305**	**0.402**	**0.527**	**0.652**	**0.745**	**0.995**	**1.245**	**1.481**	**1.959**
Length (ft.)		**Capacity in Thousands of Btu per Hour**								
10		413	852	1,730	3,030	4,300	9,170	16,500	26,000	54,200
20		284	585	1,190	2,080	2,950	6,310	11,400	17,900	37,300
30		228	470	956	1,670	2,370	5,060	9,120	14,400	29,900
40		195	402	818	1,430	2,030	4,330	7,800	12,300	25,600
50		173	356	725	1,270	1,800	3,840	6,920	10,900	22,700
60		157	323	657	1,150	1,630	3,480	6,270	9,880	20,600
70		144	297	605	1,060	1,500	3,200	5,760	9,090	18,900
80		134	276	562	983	1,390	2,980	5,360	8,450	17,600
90		126	259	528	922	1,310	2,790	5,030	7,930	16,500
100		119	245	498	871	1,240	2,640	4,750	7,490	15,600
125		105	217	442	772	1,100	2,340	4,210	6,640	13,800
150		95	197	400	700	992	2,120	3,820	6,020	12,500
175		88	181	368	644	913	1,950	3,510	5,540	11,500
200		82	168	343	599	849	1,810	3,270	5,150	10,700
250		72	149	304	531	753	1,610	2,900	4,560	9,510
300		66	135	275	481	682	1,460	2,620	4,140	8,610
350		60	124	253	442	628	1,340	2,410	3,800	7,920
400		56	116	235	411	584	1,250	2,250	3,540	7,370
450		53	109	221	386	548	1,170	2,110	3,320	6,920
500		50	103	209	365	517	1,110	1,990	3,140	6,530
550		47	97	198	346	491	1,050	1,890	2,980	6,210
600		45	93	189	330	469	1,000	1,800	2,840	5,920
650		43	89	181	316	449	959	1,730	2,720	5,670
700		41	86	174	304	431	921	1,660	2,620	5,450
750		40	82	168	293	415	888	1,600	2,520	5,250
800		39	80	162	283	401	857	1,540	2,430	5,070
850		37	77	157	274	388	829	1,490	2,350	4,900
900		36	75	152	265	376	804	1,450	2,280	4,750
950		35	72	147	258	366	781	1,410	2,220	4,620
1,000		34	71	143	251	356	760	1,370	2,160	4,490
1,100		32	67	136	238	338	721	1,300	2,050	4,270
1,200		31	64	130	227	322	688	1,240	1,950	4,070
1,300		30	61	124	217	309	659	1,190	1,870	3,900
1,400		28	59	120	209	296	633	1,140	1,800	3,740
1,500		27	57	115	201	286	610	1,100	1,730	3,610
1,600		26	55	111	194	276	589	1,060	1,670	3,480
1,700		26	53	108	188	267	570	1,030	1,620	3,370
1,800		25	51	104	182	259	553	1,000	1,570	3,270
1,900		24	50	101	177	251	537	966	1,520	3,170
2,000		23	48	99	172	244	522	940	1,480	3,090

Note: All table entries are rounded to 3 significant digits.
*Table capacities are based on Type K copper tubing inside diameter (shown), which has the smallest inside diameter of the copper tubing products.

Table 13-35

Corrugated Stainless Steel Tubing (CSST) [NFPA 54-09: Table 6.3(h)]

Gas:	Undiluted Propane
Inlet Pressure:	11.0 in. w.c.
Pressure Drop:	0.5 in w.c.
Specific Gravity:	1.50

INTENDED USE: CSST Sizing Between Single or Second Stage (Low Pressure) Regulator and Appliance Shutoff Valve

	Tube Size (EHD)*													
Flow Designation:	13	15	18	19	23	25	30	31	37	39	46	48	60	62
Length (ft)	Capacity in Thousands of Btu per Hour													
5	72	99	181	211	355	426	744	863	1,420	1,638	2,830	3,270	5,780	6,550
10	50	69	129	150	254	303	521	605	971	1,179	1,990	2,320	4,110	4,640
15	39	55	104	121	208	248	422	490	775	972	1,620	1,900	3,370	3,790
20	34	49	91	106	183	216	365	425	661	847	1,400	1,650	2,930	3,290
25	30	42	82	94	164	192	325	379	583	762	1,250	1,480	2,630	2,940
30	28	39	74	87	151	177	297	344	258	698	1,140	1,350	2,400	2,680
40	23	33	64	74	131	153	256	297	449	610	988	1,170	2,090	2,330
50	20	30	58	66	118	137	227	265	397	548	884	1,050	1,870	2,080
60	19	26	53	60	107	126	207	241	359	502	805	961	1,710	1,900
70	17	25	49	57	99	117	191	222	330	466	745	890	1,590	1,760
80	15	23	45	52	94	109	178	208	307	438	696	833	1,490	1,650
90	15	22	44	50	90	102	169	197	286	414	656	787	1,400	1,550
100	14	20	41	47	85	98	159	186	270	393	621	746	1,330	1,480
150	11	15	31	36	66	75	123	143	217	324	506	611	1,090	1,210
200	9	14	28	33	60	69	112	129	183	283	438	531	948	1,050
250	8	12	25	30	53	61	99	117	163	254	390	476	850	934
300	8	11	23	26	50	57	90	107	147	234	357	434	777	854

*EHD = Equivalent Hydraulic Diameter, which is a measure of the relative hydraulic efficiency between different tubing sizes. The greater the value of EHD, the greater the gas capacity of the tubing.

Notes:

(1) Table includes losses for four 90 degree bends and two end fittings. Tubing runs with larger numbers of bends and/or fittings shall be increased by an equivalent length of tubing to the following equation: $L = 1.3n$, where L is additional length (ft) of tubing and n is the number of additional fittings and/or bends.

(2) All table entries are rounded to 3 significant digits.

Table 13-36

Corrugated Stainless Steel Tubing (CSST) [NFPA 54-09: Table 6.3(i)]

Gas:	Undiluted Propane
Inlet Pressure:	2.0 psi
Pressure Drop:	1.0 psi
Specific Gravity:	1.50

INTENDED USE: CSST Sizing Between 2 psi Service and Line Pressure Regulator

	Tube Size (EHD)*													
Flow Designation:	13	15	18	19	23	25	30	31	37	39	46	48	60	62
Length (ft)	Capacity in Thousands of Btu per Hour													
10	426	558	927	1,110	1,740	2,170	4,100	4,720	7,130	7,958	15,200	16,800	29,400	34,200
25	262	347	591	701	1,120	1,380	2,560	2,950	4,560	5,147	9,550	10,700	18,800	21,700
30	238	316	540	640	1,030	1,270	2,330	2,690	4,180	4,719	8,710	9,790	17,200	19,800
40	203	271	469	554	896	1,100	2,010	2,320	3,630	4,116	7,530	8,500	14,900	17,200
50	181	243	420	496	806	986	1,790	2,070	3,260	3,702	6,730	7,610	13,400	15,400
75	147	196	344	406	663	809	1,460	1,690	2,680	3,053	5,480	6,230	11,000	12,600
80	140	189	333	393	643	768	1,410	1,630	2,590	2,961	5,300	6,040	10,600	12,200
100	124	169	298	350	578	703	1,260	1,450	2,330	2,662	4,740	5,410	9,530	10,900
150	101	137	245	287	477	575	1,020	1,180	1,910	2,195	3,860	4,430	7,810	8,890
200	86	118	213	248	415	501	880	1,020	1,660	1,915	3,340	3,840	6,780	7,710
250	77	105	191	222	373	448	785	910	1,490	1,722	2,980	3,440	6,080	6,900
300	69	96	173	203	343	411	716	829	1,360	1,578	2,720	3,150	5,560	6,300
400	60	82	151	175	298	355	616	716	1,160	1,376	2,350	2,730	4,830	5,460
500	53	72	135	158	268	319	550	638	1,030	1,237	2,100	2,450	4,330	4,880

*EHD = Equivalent Hydraulic Diameter, which is a measure of the relative hydraulic efficiency between different tubing sizes. The greater the value of EHD, the greater the gas capacity of the tubing.

Notes:

(1) Table does not include effect of pressure drop across the line regulator. Where regulator loss exceeds 1 psi, do not use this table. Consult with regulator manufacturer for pressure drops and capacity factors. Pressure drops across a regulator may vary with flow rate.

(2) CAUTION: Capacities shown in table may exceed maximum capacity for a selected regulator. Consult with regulator or tubing manufacturer for guidance.

(3) Table includes losses for four 90 degree bends and two end fittings. Tubing runs with larger numbers of bends and/or fittings shall be increased by an equivalent length of tubing to the following equation: $L = 1.3n$, where L is additional length (ft) of tubing and n is the number of additional fittings and/or bends.

(4) All table entries are rounded to 3 significant digits.

Table 13-37

Corrugated Stainless Steel Tubing (CSST) [NFPA 54-09: Table 6.3(j)]

Gas:	**Undiluted Propane**
Inlet Pressure:	**5.0 psi**
Pressure Drop:	**3.5 psi**
Specific Gravity:	**1.50**

	Tube Size (EHD)*													
Flow Designation:	**13**	**15**	**18**	**19**	**23**	**25**	**30**	**31**	**37**	**39**	**46**	**48**	**60**	**62**
Length (ft)	**Capacity in Thousands of Btu per Hour**													
10	826	1,070	1,710	2,060	3,150	4,000	7,830	8,950	13,100	14441	28,600	31,200	54,400	63,800
25	509	664	1,090	1,310	2,040	2,550	4,860	5,600	8,400	9339	18,000	19,900	34,700	40,400
30	461	603	999	1,190	1,870	2,340	4,430	5,100	7,680	8564	16,400	18,200	31,700	36,900
40	396	520	867	1,030	1,630	2,030	3,820	4,400	6,680	7469	14,200	15,800	27,600	32,000
50	352	463	777	926	1,460	1,820	3,410	3,930	5,990	6717	12,700	14,100	24,700	28,600
75	284	376	637	757	1,210	1,490	2,770	3,190	4,920	5539	10,300	11,600	20,300	23,400
80	275	363	618	731	1,170	1,450	2,680	3,090	4,770	5372	9,990	11,200	19,600	22,700
100	243	324	553	656	1,050	1,300	2,390	2,760	4,280	4830	8,930	10,000	17,600	20,300
150	196	262	453	535	866	1,060	1,940	2,240	3,510	3983	7,270	8,210	14,400	16,600
200	169	226	393	464	755	923	1,680	1,930	3,050	3474	6,290	7,130	12,500	14,400
250	150	202	352	415	679	828	1,490	1,730	2,740	3124	5,620	6,390	11,200	12,900
300	136	183	322	379	622	757	1,360	1,570	2,510	2865	5,120	5,840	10,300	11,700
400	117	158	279	328	542	657	1,170	1,360	2,180	2498	4,430	5,070	8,920	10,200
500	104	140	251	294	488	589	1,050	1,210	1,950	2247	3,960	4,540	8,000	9,110

*EHD = Equivalent Hydraulic Diameter, which is a measure of the relative hydraulic efficiency between different tubing sizes. The greater the value of EHD, the greater the gas capacity of the tubing.

Notes:

(1) Table does not include effect of pressure drop across the line regulator. Where regulator loss exceeds 1 psi, do not use this table. Consult with regulator manufacturer for pressure drops and capacity factors. Pressure drops across a regulator may vary with flow rate.

(2) CAUTION: Capacities shown in table may exceed maximum capacity for a selected regulator. Consult with regulator or tubing manufacturer for guidance.

(3) Table includes losses for four 90 degree bends and two end fittings. Tubing runs with larger numbers of bends and/or fittings shall be increased by an equivalent length of tubing to the following equation: $L = 1.3n$, where L is additional length (ft) of tubing and n is the number of additional fittings and/or bends.

(4) All table entries are rounded to 3 significant digits.

Table 13-38

Polyethylene Plastic Pipe [NFPA 54-09: Table 6.3(k)]

Gas:	**Undiluted Propane**
Inlet Pressure:	**11.0 in. w.c.**
Pressure Drop:	**0.5 in. w.c.**
Specific Gravity:	**1.50**

INTENDED USE: PE Sizing Between Integral 2-Stage Regulator at Tank or 2nd Stage (Low Pressure Regulator) and Building						
	Pipe Size (in.)					
Nominal OD:	**1/2**	**3/4**	**1**	**1-1/4**	**1-1/2**	**2**
Designation:	**SDR 9.33**	**SDR 11.0**	**SDR 11.00**	**SDR 10.00**	**SDR 11.00**	**SDR 11.00**
Actual ID:	**0.660**	**0.860**	**1.077**	**1.328**	**1.554**	**1.943**
Length (ft)	**Thousands of Btu per Hour**					
10	340	680	1,230	2,130	3,210	5,770
20	233	468	844	1,460	2,210	3,970
30	187	375	677	1,170	1,770	3,180
40	160	321	580	1,000	1,520	2,730
50	142	285	514	890	1,340	2,420
60	129	258	466	807	1,220	2,190
70	119	237	428	742	1,120	2,010
80	110	221	398	690	1,040	1,870
90	103	207	374	648	978	1,760
100	98	196	353	612	924	1,660
125	87	173	313	542	819	1,470
150	78	157	284	491	742	1,330
175	72	145	261	452	683	1,230
200	67	135	243	420	635	1,140
250	60	119	215	373	563	1,010
300	54	108	195	338	510	916
350	50	99	179	311	469	843
400	46	92	167	289	436	784
450	43	87	157	271	409	736
500	41	82	148	256	387	695

Note: All table entries are rounded to 3 significant digits.

Table 13-39

Polyethylene Plastic Pipe [NFPA 54-09: Table 6.3(I)]

			Gas:	Undiluted Propane		
			Inlet Pressure:	2.0 psi		
			Pressure Drop:	1.0 psi		
			Specific Gravity:	1.50		
INTENDED USE: PE Pipe Sizing Between 2 psi Service and Line Pressure Regulator						
	Pipe Size (in.)					
Nominal OD:	1/2	3/4	1	1-1/4	1-1/2	2
Designation:	SDR 9.33	SDR 11.0	SDR 11.00	SDR 10.00	SDR 11.00	SDR 11.00
Actual ID:	0.660	0.860	1.077	1.328	1.554	1.943
Length (ft)	Capacity in Cubic Feet of Gas per Hour					
10	3,130	6,260	11,300	19,600	29,500	53,100
20	2,150	4,300	7,760	13,400	20,300	36,500
30	1,730	3,450	6,230	10,800	16,300	29,300
40	1,480	2,960	5,330	9,240	14,000	25,100
50	1,310	2,620	4,730	8,190	12,400	22,200
60	1,190	2,370	4,280	7,420	11,200	20,100
70	1,090	2,180	3,940	6,830	10,300	18,500
80	1,010	2,030	3,670	6,350	9,590	17,200
90	952	1,910	3,440	5,960	9,000	16,200
100	899	1,800	3,250	5,630	8,500	15,300
125	797	1,600	2,880	4,990	7,530	13,500
150	722	1,450	2,610	4,520	6,830	12,300
175	664	1,330	2,400	4,160	6,280	11,300
200	618	1,240	2,230	3,870	5,840	10,500
250	548	1,100	1,980	3,430	5,180	9,300
300	496	994	1,790	3,110	4,690	8,430
350	457	914	1,650	2,860	4,320	7,760
400	425	851	1,530	2,660	4,020	7,220
450	399	798	1,440	2,500	3,770	6,770
500	377	754	1,360	2,360	3,560	6,390
550	358	716	1,290	2,240	3,380	6,070
600	341	683	1,230	2,140	3,220	5,790
650	327	654	1,180	2,040	3,090	5,550
700	314	628	1,130	1,960	2,970	5,330
750	302	605	1,090	1,890	2,860	5,140
800	292	585	1,050	1,830	2,760	4,960
850	283	566	1,020	1,770	2,670	4,800
900	274	549	990	1,710	2,590	4,650
950	266	533	961	1,670	2,520	4,520
1,000	259	518	935	1,620	2,450	4,400
1,100	246	492	888	1,540	2,320	4,170
1,200	234	470	847	1,470	2,220	3,980
1,300	225	450	811	1,410	2,120	3,810
1,400	216	432	779	1,350	2,040	3,660
1,500	208	416	751	1,300	1,960	3,530
1,600	201	402	725	1,260	1,900	3,410
1,700	194	389	702	1,220	1,840	3,300
1,800	188	377	680	1,180	1,780	3,200
1,900	183	366	661	1,140	1,730	3,110
2,000	178	356	643	1,110	1,680	3,020

Note: All table entries are rounded to 3 significant digits.

Table 13-40

Polyethylene Plastic Tubing [NFPA 54-09: Table 6.3(m)]

	Gas:	Undiluted Propane
	Inlet Pressure:	11.0 in. w.c.
	Pressure Drop:	0.5 in. w.c.
	Specific Gravity:	1.50
INTENDED USE: PE Tube Sizing Between Integral 2-Stage Regulator at Tank or Second Stage (Low Pressure Regulator) and Building		
	Plastic Tubing Size (CTS) (in.)	
Nominal OD:	1/2	3/4
Designation:	SDR 7.00	SDR 11.00
Actual ID:	0.445	0.927
Length (ft)	Thousands of Btu per Hour	
10	121	828
20	83	569
30	67	457
40	57	391
50	51	247
60	46	314
70	42	289
80	39	269
90	37	252
100	35	238
125	31	211
150	28	191
175	26	176
200	24	164
225	22	154
250	21	145
275	20	138
300	19	132
350	18	121
400	16	113
450	15	106
500	15	100

Note: All table entries are rounded to 3 significant digits.

Part II—Fuel Supply: Manufactured/Mobile Home Parks and Recreational Vehicle Parks.

1320.0 General.

All fuel gas piping systems serving manufactured homes, accessory buildings, or structures and communities shall be designed and constructed in accordance with any applicable provisions of NFPA 54, *National Fuel Gas Code,* and NFPA 58, *Liquefied Petroleum Gas Code.* NFPA 31, *Standard for the Installation of Oil-Burning Equipment,* shall apply to oil-fuel-burning systems and shall conform to the criteria of the Authority Having Jurisdiction. [NFPA 501A:4.1.1]

1321.0 Gas Supply Connections.

Gas supply connections at sites, where provided from an underground gas supply piping system, shall be located and arranged to permit attachment to a manufactured home occupying the site in a worklike manner. For the installation of liquefied petroleum gas (LP-Gas) storage systems, the applicable provisions of NFPA 58, *Liquefied Petroleum Gas Code,* shall be followed. [NFPA 501A:4.1.2]

1321.1 Location of Gas Supply Connection. The gas supply to the manufactured home shall be located within four (4) feet (1,219 mm) of the manufactured home stand.

> **Exception:** The above requirements shall not apply to gas supply connections for manufactured homes located on all-weather wood, concrete, or concrete block foundation systems or on foundations constructed in accordance with the local building code or, in the absence of a local code, with a recognized model building code. [NFPA 501A:4.1.3]

1322.0 Single and Multiple Manufactured Homesite Fuel Supply Systems.

1322.1 Gas Piping Installations.

1322.1.1 Gas Supply Connections. Underground Gas Piping. Gas supply connections at sites, where provided from an underground gas supply piping system, shall be located and arranged to permit attachment in a worklike manner to a manufactured home occupying the site. For the installation of LP-Gas storage systems, the provisions of NFPA 58, *Liquefied Petroleum Gas Code,* shall be followed.

1322.1.2 Underground gas piping system installations shall comply with any applicable building code and Sections 1322.1.2.1 and 1322.1.2.2. [NFPA 501A:4.2.1]

1322.1.2.1 Underground gas piping shall not be installed beneath that portion of a manufactured homesite reserved for the location of a manufactured home or manufactured home accessory building or structure unless installed in the open-ended gastight conduit of Section 1322.1.2.2. [NFPA 501A:4.2.1.1]

1322.1.2.2 The open-ended gastight conduit shall conform to the following: [NFPA 501A:4.2.1.2]

(1) The conduit shall be not less than Schedule 40 pipe that is approved for underground installation beneath buildings.

(2) The interior diameter of the conduit shall be not less than 0.5 inch (12.7 mm) larger than the outside diameter of the gas piping.

(3) The conduit shall extend to a point not less than four (4) inches (100 mm) beyond the outside wall of the manufactured home, accessory building, or structure, and the outer ends shall not be sealed.

(4) Where the conduit terminates within a manufactured home, accessory building, or structure, it shall be readily accessible, and the space between the conduit and the gas piping shall be sealed to prevent leakage of gas into the building.

1323.0 Manufactured Homesite Gas Shutoff Valve.

Each manufactured homesite shall have a listed gas shutoff valve installed upstream of the manufactured homesite gas outlet. The gas shutoff valve shall be located on the outlet riser at a height of not less than six (6) inches (150 mm) above finished grade. A gas shutoff valve shall not be located under any manufactured home. The outlet shall be equipped with a cap or plug to prevent discharge of gas whenever the manufactured homesite outlet is not connected to a manufactured home.

> **Exception:** All gas shutoff valves for manufactured homes located on foundations constructed in accordance with the local building code or, in the absence of a local code, with a recognized model building code. [NFPA 501A:4.2.2]

1324.0 Gas Meters.

1324.1 Support of Meters. Where installed, gas meters shall be adequately supported by a post or bracket placed on a firm footing or other means providing equivalent support and shall not depend on the gas outlet riser for support. [NFPA 501A:4.2.3.1]

1324.2 Location of Meters. Each gas meter shall be installed in an accessible location and shall be provided with unions or other fittings so that the meter is removed easily and replaced in an upright position. Meters shall not be installed in unventilated or inaccessible locations or closer than three (3) feet (0.91 m) to sources of ignition. [NFPA 501A:4.2.3.2]

1324.3 Meter Shutoff Valve or Cock. All gas meter installations shall be provided with shutoff valves or cocks located adjacent to and on the inlet side of the meters. In the case of a single meter installation utilizing an LP-Gas container, the container service valve shall be permitted to be used in lieu of the shutoff valve or cock. All gas meter installations shall be provided with test tees located adjacent to and on the outlet side of the meters. [NFPA 501A:4.2.4]

1325.0 Multiple Manufactured Homesite Fuel Distribution and Supply Systems. [NFPA 501A:4.3]

1326.0 Manufactured Home Community Natural Gas Distribution Systems.

All underground metallic fuel piping systems shall comply with the cathodic protection requirements of 49 CFR 191 and 192. [NFPA 501A:4.3.1]

1327.0 Manufactured Home Community LP-Gas Supply Systems.

Where ten (10) or more customers are served by one LP-Gas supply system, the installation of the gas supply system shall be in accordance with 49 CFR 192, *Transportation of Natural and Other Gas by Pipeline: Minimum Federal Safety Standards.* [NFPA 501A:4.3.2.1] Other liquefied petroleum gas supply systems and the storage and handling of LP-Gas shall be in accordance with NFPA 58, *Liquefied Petroleum Gas Code.* [NFPA 501A:4.3.2.2]

1328.0 Installation of Cathodic Protection Systems.

Where required by the federal standard cited in Section 1326.0, cathodic protection shall be installed for corrosion control of buried or submerged metallic gas piping. [NFPA 501A:4.3.3]

1329.0 Required Gas Supply.

The minimum hourly volume of gas required at each manufactured homesite outlet or any section of the manufactured home community gas piping system shall be calculated as shown in Table 13-41. [NFPA 501A:4.3.4.1]

1330.0 Gas Pipe Sizing and Pressure.

1330.1 The size of each section of a gas piping system shall be determined in accordance with NFPA 54, *National Fuel Gas Code,* or by other standard engineering methods acceptable to the Authority Having Jurisdiction. [NFPA 501A: 4.3.5.1]

1330.2 Where all connected appliances are operated at their rated capacity, the supply pressure shall be not less than 4 oz./in.^{2} (7 in. water column [1,743 Pa]). The gas supply pressure shall not exceed 8 oz./in.^{2} (14 in. water column [3,486 Pa]). [NFPA 501A:4.3.5.2]

1331.0 Gas Piping Materials.

1331.1 Metal. Metal gas pipe shall be standard weight wrought iron or steel (galvanized or black), yellow brass containing not more than 75 percent copper, or internally tinned or treated copper of iron pipe size. Galvanizing shall not be considered protection against corrosion.

Seamless copper or steel tubing shall be permitted to be used with gases not corrosive to such material. Steel tubing shall comply with ASTM A 539, *Standard Specification for Electric-Resistance-Welded Coiled Steel Tubing for Gas and Fuel Oil Lines,* or ASTM A 254, *Standard Specification for Copper Brazed Steel Tubing.* Copper tubing shall comply with ASTM B 88, *Specification for Seamless Copper Water Tubing (Type K or Type L),* or ASTM B 280, *Specification for Seamless Copper Tubing for Air Conditioning and Refrigeration Field Service.* Copper tubing (unless tin lined) shall not be used if the gas contains more than an average of 0.3 grains of hydrogen sulfide per 100 standard ft.^{3} of gas (0.7 mg/100 L) [NFPA 501A:4.3.6.1]

1331.2 Protection Coatings for Metal Gas Piping. All buried or submerged metallic gas piping shall be protected from corrosion by approved coatings or wrapping materials. All gas pipe protective coatings shall be approved types, shall be machine applied, and shall conform to recognized standards. Field wrapping shall provide equivalent protection and is restricted to those short sections and fittings that are necessarily stripped for threading or welding. Risers shall be coated or wrapped to a point at least six (6) inches (152 mm) aboveground. [NFPA 501A:4.3.6.2]

1331.3 Plastic. Plastic piping shall only be used underground and shall meet the requirements of ASTM D 2513, *Thermoplastic Gas Pressure Pipe, Tubing, and Fittings,* or ASTM D 2517, *Reinforced Epoxy Resin Gas Pressure Pipe and Fittings,* as well as the design pressure and design limitations of 49 CFR 192.123, and shall otherwise conform to the installation requirements thereof. [NFPA 501A:4.3.6.3]

TABLE 13-41
Demand Factors for Use in Calculating Gas Piping Systems in Manufactured Home Communities
[NFPA 501A: Table 4.3.4.1]

No. of Manufactured Homesites	Btu/h per Manufactured Homesite
1	125,000
2	117,000
3	104,000
4	96,000
5	92,000
6	87,000
7	83,000
8	81,000
9	79,000
10	77,000
11–20	66,000
21–30	62,000
31–40	58,000
41–60	55,000
Over 60	50,000

Note: In extreme climate areas, additional capacities shall be considered.
1 Btu/h = 0.29 Watt

1332.0 Gas Piping Installation.

1332.1 Minimum Burial Below Ground Level and Clearances. All gas piping installed below ground level shall have a minimum earth cover of eighteen (18) inches (457 mm) and shall be installed with at least twelve (12) inches (305 mm) of clearance in any direction from any other underground utility system. [NFPA 501A:4.3.7.1]

1332.2 Metallic Gas Piping.

1332.2.1 All metallic gas piping systems shall be installed in accordance with approved plans and specifications, including provisions for cathodic protection. Each cathodic protection system shall be designed and installed to conform to the provisions of 49 CFR 192. [NFPA 501A:4.3.7.2.1, 4.3.7.2.2]

1332.2.2 Where the cathodic protection system is designed to protect only the gas piping system, the gas piping system shall be electrically isolated from all other underground metallic systems or installations. Where only the gas piping system is cathodically protected against corrosion, a dielectric fitting shall be used in the manufactured home gas connection to insulate the manufactured home from the underground gas piping system. [NFPA 501A: 4.3.7.2.3, 4.3.7.2.4]

1332.2.3 Where a cathodic protection system is designed to provide all underground metallic systems and installations with protection against corrosion, all such systems and installations shall be electrically bonded together and protected as a whole. [NFPA 501A:4.3.7.2.5]

1332.3 Plastic Gas Piping. Plastic gas piping shall only be used underground and shall be installed with an electrically conductive wire for locating the pipe. The wire used to locate the plastic pipe shall be copper, not smaller in size than No. 18 AWG, with insulation approved for direct burial. Every portion of a plastic gas piping system consisting of metallic pipe shall be cathodically protected against corrosion. [NFPA 501A:4.3.7.3]

1332.4 Gas Piping System Shutoff Valve. A readily accessible and identifiable shutoff valve controlling the flow of gas to the entire manufactured home community gas piping system shall be installed near the point of connection to the service piping or to the supply connection of an LP-Gas container. [NFPA 501A:4.3.7.4]

1333.0 Liquefied Petroleum Gas Appliances.

LP-Gas appliances shall be installed in accordance with the applicable provisions of NFPA 58, *Liquefied Petroleum Gas Code.* [NFPA 501A:4.3.8]

1334.0 Oil Supply.

The following three methods of supplying oil to an individual manufactured homesite shall be permitted:

(1) Supply from an outside underground tank (see Section 1335.6).

(2) Supply from a centralized oil distribution system designed and installed in accordance with accepted engineering practices and in compliance with NFPA 31, *Standard for the Installation of Oil-Burning Equipment.*

(3) Supply from an outside aboveground tank (see Section 1335.6). [NFPA 501A:4.3.9]

1334.1 Minimum Oil Supply Tank Size. Oil supply tanks shall have a minimum capacity equal to 20 percent of the average annual oil consumption. [NFPA 501A:4.3.10]

1334.2 Oil Supply Connections — General. Oil supply connections at manufactured home stands, where provided from a centralized oil distribution system, shall be located and arranged to permit attachment in a worklike manner to a manufactured home utilizing the stand. The installation of such facilities shall meet the provisions of NFPA 31, *Standard for the Installation of Oil-Burning Equipment*, and particularly Section 5.8 thereof. [NFPA 501A:4.3.11]

1335.0 Fuel Supply Systems Installation.

1335.1 Flexible Gas Connector. Each gas supply connector shall be listed for outside manufactured home use, shall be not more than six (6) feet (1,829 mm) in length, and shall have a capacity rating adequate to supply the connected load.

Exception: All gas supply connections for manufactured homes located on an all-weather wood, concrete, or concrete block foundation system or on a foundation constructed in accordance with the local building code or, in the absence of a local code, with a recognized model building code.

1335.2 Use of Approved Pipe and Fittings of Extension. Where it is necessary to extend the manufactured home inlet to permit connection of the six (6) feet (1,829 mm) listed connector to the site gas outlet, the extension shall be of approved materials of the same size as the manufactured home inlet and shall be adequately supported at no more than four (4) feet (1,219 mm) intervals to the manufactured home. [NFPA 501A:4.4.2]

1335.3 Mechanical Protection. All gas outlet risers, regulators, meters, valves, or other exposed equipment shall be protected against accidental damage. [NFPA 501A:4.4.3]

1335.4 Special Rules on Atmospherically Controlled Regulators. Atmospherically controlled regulators shall be installed in such a manner that moisture cannot enter the regulator vent and accumulate above the diaphragm. Where the regulator vent is obstructed due to snow and icing conditions, shields, hoods, or other suitable devices shall be provided to guard against closing of the vent opening. [NFPA 501A:4.4.4]

1335.5 Fuel-Gas Piping Test. The manufactured home fuel-gas piping system shall be tested only with air before it is connected to the gas supply. The manufactured home gas piping system shall be subjected to a pressure test with all appliance shutoff valves in their closed positions. [NFPA 501A:4.4.5]

1335.5.1 The fuel-gas piping test shall consist of air pressure at not less than ten (10) inches water column or more than fourteen (14) inches water column (6 oz./in.^{2} to 8 oz./in.^{2} [2,490 Pa to 3,486 Pa]). The system shall be isolated from the air pressure source and shall maintain this pressure for not less than ten (10) minutes without perceptible leakage. Upon satisfactory completion of the test, the appliance valves shall be opened, and the gas appliance connectors shall be tested with soapy water or bubble solution while under the pressure remaining in the piping system. Solutions used for testing for leakage shall not contain corrosive chemicals. Pressure shall be measured with either a manometer, slope gauge, or gauge that is calibrated in either water inches or psi with increments of either 1/10 inch (2.5 mm) or 1/10 psi (0.6 kPa gauge), as applicable. Upon satisfactory completion of the test, the manufactured home gas supply connector shall be installed, and the connections shall be tested with soapy water or bubble solution. [NFPA 501A:4.4.5.1.1 - 4.4.5.1.6]

WARNING

Do not overpressurize the fuel-gas piping system. Damage to valves, regulators, and appliances can occur due to pressurization beyond the maximums specified. [NFPA 501A:4.4.5.2]

1335.5.2 Gas appliance vents shall be visually inspected to ensure that they have not been dislodged in transit and are connected securely to the appliance. [NFPA 501A:4.4.5.3]

1335.6 Oil Tanks. No more than one 660 gallon (2,500 L) tank or two tanks with aggregate capacity of 660 gallons (2,500 L) or less shall be connected to one oil-burning appliance. Two supply tanks, where used, shall be cross-connected and provided with a single fill and single vent, as described in NFPA 31, *Standard for the Installation of Oil-Burning Equipment*, and shall be on a common slab and rigidly secured, one to the other. Tanks having a capacity of 660 gallons (2,500 L) or less shall be securely supported by rigid, noncombustible supports to prevent settling, sliding, or lifting. [NFPA 501A:4.4.6]

1335.6.1 Oil supply tanks shall be installed in accordance with the applicable provisions of NFPA 31, *Standard for the Installation of Oil-Burning Equipment.* [NFPA 501A:4.4.6.1]

1335.6.2 A tank with a capacity no larger than 60 gallons (230 L) shall be permitted to be a DOT-5 shipping container (drum) and so marked, or a tank meeting the provisions of UL 80, Steel Inside Tank for Oil Burner Fuel. Tanks other than DOT-5 shipping containers having a capacity of not more than 660 gallons (2500 L) shall meet the provisions of UL 80, *Steel Inside Tank for Oil Burner Fuel.* Pressure tanks shall be built in accordance with Section VIII, Pressure Vessels, of the ASME *Boiler and Pressure Vessel Code.* [NFPA 501A:4.4.6.2]

1335.6.3 Tanks, as described in Sections 1335.6 and 1335.6.2, that are adjacent to buildings shall be located not less than (10) feet (3,048 mm) from a property line that is permitted to be built upon. [NFPA 501A:4.4.6.3]

1335.6.4 Tanks with a capacity no larger than 660 gallons (2,500 L) shall be equipped with an open vent no smaller than 1.5 inch (38 mm) iron pipe size; tanks with a 500 gallons (1,900 L) or less capacity shall have a vent of 1.25 inch (32 mm) iron pipe size. [NFPA 501A:4.4.6.4]

1335.6.5 Tanks shall be provided with a means of determining the liquid level. [NFPA 501A:4.4.6.5]

1335.6.6 The fill opening shall be a size and in a location that permits filling without spillage. [NFPA 501A:4.4.6.6]

1335.6.7 In flood hazard areas, tanks shall be elevated to or above the design flood elevation or they shall be designed, constructed, installed, and anchored to resist all flood-related and other loads during the design flood, or lesser floods, without release of contents into floodwaters or infiltration by floodwaters into the contents.

1336.0 Manufactured Home Accessory Building Fuel Supply Systems.

Fuel-gas supply systems installed in a manufactured home accessory building or structure shall comply with the applicable provisions of NFPA 54, *National Fuel Gas Code*, and NFPA 58, *Liquefied Petroleum Gas Code*. Fuel-oil supply systems shall comply with the applicable provisions of NFPA 31, *Standard for the Installation of Oil-Burning Equipment.* [NFPA 501A:4.5]

1337.0 Community Building Fuel Supply Systems in Manufactured Home Communities.

1337.1 Fuel-Gas Piping and Appliance Installations. Fuel-gas piping and appliances installed within a permanent building in a manufactured home community shall comply with nationally recognized appliance and fuel-gas piping codes and standards adopted by the Authority Having Jurisdiction. Where the state or other political subdivision does not assume jurisdiction, such fuel gas piping and appliance installations shall be designed and installed in accordance with the appropriate provisions of NFPA 54, *National Fuel Gas Code*, or NFPA 58, *Liquefied Petroleum Gas Code*. [NFPA 501A:4.6.1]

1337.2 Oil Supply Systems in Manufactured Home Communities. Oil-burning equipment and installations within a manufactured home community shall be designed and constructed in accordance with the applicable codes and standards adopted by the Authority Having Jurisdiction. Where the state or other political subdivision does not assume jurisdiction, such installation shall be designed and constructed in accordance with the applicable provisions of NFPA 31, *Standard for the Installation of Oil-Burning Equipment.* [NFPA 501A:4.6.2]

1337.3 Oil-Burning Equipment and Installation. Oil-burning equipment and installations within a building constructed in a manufactured home community in accordance with the local building code or a nationally recognized building code shall comply with nationally recognized codes and standards adopted by the Authority Having Jurisdiction. Where the state or other political subdivision does not assume jurisdiction, such oil-burning equipment and installation shall be designed and installed in accordance with the appropriate provisions of NFPA 31, *Standard for the Installation of Oil-Burning Equipment.* [NFPA 501A:4.6.3]

1338.0 Recreational Vehicle Park Fuel-Gas Equipment and Installations.

1338.1 All fuel-gas equipment and installations shall comply with Part II of this chapter, except as otherwise permitted or required by this code.

CHAPTER 14

PROCESS PIPING

1401.0 General.

The regulations of this chapter shall govern the installation of process piping in or in conjunction with a building or structure or located upon the premises.

1402.0 Definitions.

For the purposes of this chapter, certain terms, phrases, words, and their derivatives shall be interpreted as set forth in this section.

APPROVED. See Section 203.0.

EMERGENCY ALARM SYSTEM – A system intended to provide the indication and warning of abnormal conditions and summon appropriate aid.

EMERGENCY CONTROL STATION – An approved location on the premises where signals from emergency equipment are received.

FABRICATION AREA (FAB AREA) – An area within a Group H Occupancy semiconductor fabrication facility and related research and development areas in that there are processes involving hazardous production materials. Such areas are allowed to include ancillary rooms or areas such as dressing rooms and offices that are directly related to the fab area processes.

HAZARDOUS PROCESS PIPING (HPP) – A process material piping or tubing conveying a liquid or gas that has a degree-of-hazard rating in health, flammability, or reactivity of Class 3 or 4, as ranked by the Fire Code.

HIGH PURITY PIPING – A form of process piping but is usually specified for critical clean applications in the semiconductor, pharmaceutical, biotechnology, chemical, fiber optics, food, and dairy industries.

HPM STORAGE ROOM – A room used for the storage or dispensing of hazardous production material (HPM) and that is classified as a Group H, Division 1 or Division 2 Occupancy.

NONHAZARDOUS PROCESS PIPING (NPP) – Production material piping or tubing conveying a liquid or gas that is not classified as hazardous production material piping.

PROCESS PIPING – Piping or tubing that conveys liquid or gas, which is used directly in research, laboratory or production processes, and that is not regulated under the mechanical or plumbing code.

SERVICE CORRIDOR – A fully enclosed passage used for transporting hazardous production materials and purposes other than required exiting.

USE (MATERIAL) – The placing in action or making available for service by opening or connecting any container utilized for confinement of material, whether a solid, liquid, or gas.

1403.0 Permit.

It shall be unlawful to install, alter, or repair or cause to be installed, altered, or repaired any process material piping without first obtaining a permit.

Permits for process piping shall show the total number of outlets to be provided for on each system and such other information as may be required by the Authority Having Jurisdiction.

Fees for process piping permits are included in Table 1-1.

1404.0 Plans Required.

Plans, engineering calculations, diagrams, and other data shall be submitted in one or more sets with each application for a permit. The Authority Having Jurisdiction may require plans, computations, and specifications to be prepared and designed by an engineer licensed by the state to practice as such.

When plans or other data are submitted for review, a plan review fee shall be paid, as provided in Section 115.3.

1405.0 Workmanship.

Process piping shall not be strained or bent, nor shall tanks, vessels, vats, appliances, or cabinets be supported by or develop strain or stress on the piping.

1406.0 Inspections.

1406.1 General. Upon completion of the installation, alteration, or repair of process piping, and prior to the use thereof, the Authority Having Jurisdiction shall be notified that such piping is ready for inspection.

Excavations required for the installation of underground piping shall be kept open until such time as the piping has been inspected and approved. If any such piping is covered or concealed before such approval, it shall be exposed upon the direction of the Authority Having Jurisdiction.

1406.2 Required Inspections. The Authority Having Jurisdiction shall make the following inspections and shall either approve that portion of the work as completed or shall notify the permit holder wherein the same fails to comply with this code.

1406.2.1 Rough Piping Inspection. This inspection shall be made after all process piping authorized by the permit has been installed and before any such piping has been covered or concealed. This inspection shall include a determination that the piping size, material, and installation meet the requirements of this code.

1406.2.2 Final Piping Inspection. This inspection shall be made after all piping authorized by the permit has been installed and after all portions thereof that are to be covered or concealed are so concealed. This inspection shall include a pressure test, at which time the piping shall stand a pressure of not less than 1.5 times the maximum designed operating pressure when hydraulic testing is conducted or 110 percent when testing is conducted pneumatically. Test pressures shall be held for a length of time satisfactory to the Authority Having Jurisdiction, but in no case for less than thirty (30) minutes with no perceptible drop in pressure. HPM drain, waste, and vent piping shall be tested in accordance with the Plumbing Code. Tests shall be made in the presence of the Authority Having Jurisdiction. Necessary apparatus for conducting tests shall be furnished by the permit holder.

1406.3 Other Inspections. In addition to the inspections required by this section, the Authority Having Jurisdiction may require a special inspector, as specified in the Building Code, during installation of piping systems. In cases where the work authorized was installed in accordance with plans and specifications prepared by an engineer, the Authority Having Jurisdiction may require a final signed report stating that the work was installed in accordance with the approved plans and specifications and the applicable provisions of this chapter.

1407.0 Piping and Tubing.

1407.1 General. Process piping and tubing shall comply with this subsection and shall be installed in accordance with nationally recognized standards. Piping and tubing systems shall be metallic unless the material being transported is incompatible with such system.

1407.2 Hazardous Process Piping (HPP).

1407.2.1 General. HPP supply piping or tubing in service corridors shall be exposed to view. HPP piping shall be identified in accordance with nationally recognized standards to indicate the material being transported. All liquid HPP piping shall have an approved means for directing any spilled materials to an approved containment or drainage system.

All liquid HPP waste or drainage systems shall be installed in accordance with the Plumbing Code.

1407.2.2 Installation in Exit Corridors and Above Other Occupancies. Hazardous process supply pipe shall not be located within exit corridors, within any portion of a means of egress required to be enclosed in fire-resistive construction, or in concealed spaces in or above areas not classified as Group H Occupancies, except as permitted by this subsection.

Hazardous production material piping and tubing may be installed within the space defined by the walls of exit corridors and the floor or roof above, or in concealed spaces above other occupancies under the following conditions:

1407.2.2.1 Automatic sprinklers shall be installed within the space, unless the space is less than six (6) inches (152 mm) in least dimension.

1407.2.2.2 Ventilation at not less than six (6) air changes per hour shall be provided. The space shall not be used to convey air from any other area.

1407.2.2.3 When the piping or tubing is used to transport HPP liquids, a receptor shall be installed below such piping or tubing. The receptor shall be designed to collect any discharge or leakage and drain it to an approved location. The one (1) hour enclosure shall not be used as part of the receptor.

1407.2.2.4 All HPP supply piping and tubing and HPP nonmetallic waste lines shall be separated from the exit corridor and from any occupancy other than a semiconductor fabrication facility classified as a Group H Occupancy by construction, as required for walls or partitions that have a fire-protection rating of not less than one (1) hour. When gypsum wallboard is used, joints on the piping side of the enclosure need not be taped, provided the joints occur over framing members. Access openings into the enclosure shall be protected by approved fire assemblies.

1407.2.2.5 Readily accessible manual or automatic remotely activated fail-safe emergency shutoff valves shall be installed on piping and tubing other than waste lines at the following locations:

1407.2.2.5.1 At branch connections into the fabrication area.

1407.2.2.5.2 At entries into exit corridors. Excess flow valves shall be installed as required by the Fire Code.

1407.2.2.6 Electrical wiring and equipment located in the piping space shall be approved for Class I, Division 2, Hazardous Locations.

Exception: Occasional transverse crossing of the corridors by supply piping that is enclosed within a ferrous pipe or tube for the width of the corridor need not comply with Sections 1407.2.2.1 through 1407.2.2.6.

1407.3 Special Requirements for HPP Gases.

1407.3.1 General. In addition to other requirements of this section, HPP gases shall comply with this subsection and the Fire Code.

1407.3.2 Special Provisions.

1407.3.2.1 Excess Flow Control. Where HPP supply gas is carried in pressurized piping, a fail-safe system shall shut off flow due to a rupture in the piping. Where the piping originates from outside the building, the valve shall be located outside the building as close to the bulk source as practical.

1407.3.2.2 Piping and Tubing Installation. Piping and tubing shall be installed in accordance with approved standards. Supply piping for hazardous production materials having a health hazard ranking of 3 or 4 shall have welded connections throughout, unless an exhausted enclosure is provided.

Exception: Material that is incompatible with ferrous piping may be installed in nonmetallic piping with approved connections.

1407.3.2.3 Gas-Detection System. When hazardous production material gas is used or dispensed and the physiological warning properties for the gas are at a higher level than the accepted permissible exposure limit for the gas, a continuous gas-monitoring system shall be provided to detect the presence of a short-term hazard condition. When dispensing occurs and flammable gases or vapors may be present in quantities in excess of 20 percent of the lower explosive limit, a continuous gas-monitoring system shall be provided. The monitoring system shall be connected to the emergency control station.

CHAPTER 15

SOLAR SYSTEMS

See Section 1206.0, Heat Sources, and the *Uniform Solar Energy Code*, published by the International Association of Plumbing and Mechanical Officials. The *Uniform Solar Energy Code* provides requirements that may be adopted as part of the code by the Authority Having Jurisdiction.

CHAPTER 16

STATIONARY POWER PLANTS

1601.0 Scope Stationary Fuel Cell Power Plants.

Stationary fuel cell power plants shall be tested in accordance with CSA Z21.83, *American National Standard/Canadian Gas Association Standard for Fuel Cell Power Plants,* and shall be installed in accordance with NFPA 853, *Standard for the Installation of Stationary Fuel Cell Power Plants* and the manufacturer's installation instructions.

1602.0 Stationary Engine Generators.

Stationary engine generators shall be tested in accordance with UL 2200, *Standard for Stationary Engine Generator Assemblies,* and shall be installed in accordance with NFPA 37, *Standard for the Installation and Use of Stationary Combustion Engines and Gas Turbines* and the manufacturer's installation instructions.

CHAPTER 17
STANDARDS TABLE 17-1
Standards for Equipment and Materials

STANDARD NUMBER	STANDARD TITLE	APPLICATION	REFERENCED SECTION
ACCA Manual D-2002	Residential Duct Systems	Ducts	302.10, 304.1, 403.1, 503.1, 505.2, 601.1, 601.2, 602.4, 606.1, Table 6-2-P-1, 924.2
ACCA Manual J-2003	Residential Load Calculations	Ducts	113.2, 210.0, 403.1, 403.5.1, 601.2, Table 6-2-A, 1201.3.5.3,
ACCA Manual N Fifth Edition	Commercial Load Calculation	Ducts	113.2, 210.0, 403.1, 403.5.1, 601.2, Table 6-2-A, 1201.3.5.3
ACCA Manual Q	Low Pressure Low Velocity Duct Systems Design	Ducts	203.0, Table 4-1, 601.2, 605.0(E), 1002.0(7)
ARI 700-2006*	Specification for Fluorocarbon Refrigerant	Refrigerants	1102.0
ASCE 25-2006	Earthquake Actuated Automatic Gas Shutoff Devices	Earthquake Valves, Shutoff Valves	Ch. 13
ASHRAE 15-2007*	Safety Refrigeration Systems	Refrigeration Systems	Ch. 11, 1102.0
ASHRAE 34-2007*	Designation and Safety Classification of Refrigerants	Refrigeration Classifications	1102.0, 1103.0, 1133.0 Table 11-1
ASHRAE 62.1-2007*	Ventilation for Acceptable Indoor Air Quality	Indoor Air Quality Ventilation	Ch. 4
ASHRAE 129-1997* (RA 2002)	Measuring Air-Change Effectiveness	Used As An Alternate Sizing Value	Table 4-2
ASHRAE Climatic Conditions	ASHRAE Handbook – Fundamentals, Chapter 27 Table 1A & B, for the US 2001	Design	Ch. 4, Ch. 6
ASHRAE Handbook	Fundamentals, Climatic Conditions for the United States; Chapter 27, Table 1, and B, 2001	Climatic Conditions	Ch. 4
ASHRAE Volume 1980	Panel Heating Systems	Design	1202.0, 1203.0
ASME A13.1-2007*	Scheme for the Identification of Piping Systems	Piping	1111.8, 1112.4, 1309.6.5, 1201.2.1.14, 1309.7.7

STANDARD NUMBER	STANDARD TITLE	APPLICATION	REFERENCED SECTION
ASME A112.4.1-1993*	Water Heater Relief Valve Drain Tubes Valve	Drain tubes, Relief Valve Drains, Water Heater Drain	225.0, 802.2.2.5
ASME A112.18.6- 2003*	Flexible Water Connectors	Product; Connectors Flexible Connectors, Water Connectors	Tables 3-4, 3-5 and 3-6, 919.2, 1312.1.7
ASME A112.20.1-2004*	Qualification of Installers of High Purity Piping Systems	Certification: High Purity Piping, Installers	Ch. 14
ASME B1.20.1-1983 (R2006)*	Pipe Threads, General Purpose, Inch	Threads	1309.5.7(A), 1201.2.2.1
ASME B1.20.3-1976 (R2003)*	Dryseal Pipe Threads, Inch Joints	Threads	1309.5.8.4(6)
ASME B16.1-2005	Cast-Iron Pipe Flanges and Flanged Fittings, Classes 25, 125, 250	Cast Fittings, Cast Iron Fittings, Copper Alloy Fittings, Flanged Fittings, Pressure Fittings, Solder Fittings	1309.5.9
ASME B16.18-2001 (R2005)	Cast Copper Alloy Solder Joint Pressure Fittings	Cast Fittings, Copper Alloy Fittings, Pressure Fittings, Solder Fittings	1201.2.1.4
ASME B16.20-1998 (R2004)*	Metallic Gaskets for Pipe Flanges; Spiral-Wound, and Jacketed	Pipe Flanges	1309.5.9
ASME B16.22-2001 (R2005)*	Wrought Copper and Copper Alloy Solder Joint Pressure Fittings	Copper Alloy Fittings, Pressure Fittings, Solder Fittings, Wrought Fittings	1201.2.1.4.2
ASME B16.23-2002 (R2006)*	Cast Copper Alloy Solder Joint Drainage Fittings - DWV	Cast Fittings, Copper Alloy Fittings, Drainage Fittings, Solder Fittings	309.0
ASME B16.24-2006*	Cast Copper Alloy Pipe Flanged Fittings Classes 150, 300, 400, 600, 900, 1500, and 2500	Cast Fittings, Copper Alloy Fittings, Flanged Fittings	111.7, 1309.5.8.1 1309.5.10
ASME B16.26-2006*	Cast Copper Alloy Fittings for Flared Copper Tubes	Cast Fittings, Copper Alloy, Flared tube Fittings	1201.2.1.4
ASME B16.29-2007*	Wrought Copper and Wrought Copper Alloy Solder Joint Drainage Fittings	Copper Alloy Fittings, Drainage Fittings, Solder Fittings, Wrought Fittings	309.0

STANDARD NUMBER	STANDARD TITLE	APPLICATION	REFERENCED SECTION
ASME B16.33-(R2007)*	Manually Operated Metallic Gas Valves for Use in Gas Piping Systems up to 125 psi (Sizes ½ NPS - NPS 24)	Gas Valve, Manual Valve	1312.9
ASME B16.50-2001*	Wrought Copper and Copper Alloy Braze-Joint Pressure Fittings	Braze Joint Fittings, Copper Alloy Fittings, Copper Fittings, Pressure Fittings, Wrought Fittings	Ch. 12
ASME B31.5-2006*	Refrigeration Piping	Refrigeration Piping, Piping	1110.2, 1110.3
ASME B36.10M-2004*	Welded and Seamless Wrought Steel Pipe	Seamless Pipe, Steel Pipe, Welded Pipe, Wrought Steel Pipe	1309.5.2.2
ASME Section IV	Rules for Construction of Heating Boilers	Boiler Construction, Heating Boilers	1001.0
ASME Section VIII	Rules for Construction of Pressure Vessels	Pressure Vessel Construction, Pressure Vessels	1335.6.2, 1001.0
ASTM A 53-2007*	Pipe, Steel, Black and Hot-Dipped, Zinc-Coated, Welded, and Seamless	Black Steel Pipe Seamless Steel Pipe, Steel Pipe, Welded Steel Pipe, Zinc Coated Steel Pipe	1309.5.2.2(2)
ASTM A 254-97 (R2002)	Copper-Brazed Steel Tubing	Steel Pipe	1331.1
ASTM A 312-2007	Seamless, and Welded and Heavy Cold Worked of Austenitic Stainless Steel Pipes	Carbon Steel Fittings, Pipe Fittings, Pipe Nipples, Stainless Steel Fittings	Ch.12, Ch. 14
ASTM A 653-2007	Steel Sheet, Zinc-Coated (Galvanized) or Zinc-Iron Alloy-Coated (Galvannealed) by the Hot-Dip Process	Alloyed Coating, Bake Hardenable Steel, High Strength Low Alloy Minimized Spangle Coating, Sheet Steel, Spangle, Steel, Steel Sheet, Structural Steel	209.0, Appendix A
ASTM A 733-2003	Welded and Seamless Carbon Steel and Pipe Coatings, Pipe Nipples, Stainless Steel Fittings	Carbon Steel Fittings, Austenitic Stainless Steel Pipe Nipples	Ch. 12, Ch. 14

STANDARD NUMBER	STANDARD TITLE	APPLICATION	REFERENCED SECTION
ASTM A 861-2004	High-Silicon Iron Pipe and Fittings	Corrosion Resistant, Fittings, High-Silicon Iron, Hubless, Hub/Plain-End, No-hub	Ch. 12, Ch. 14
ASTM B 32-2004	Solder Metal	Bar, Flux, Flux Cored Solder, Ingot, Leadsilver Alloys, Lead-tin Alloys, Lead-tin-silver Alloys, Powder, Ribbon	1201.2.3.7
ASTM B 42-2002[e1]	Seamless Copper Pipe, Standard Sizes	Copper Pipe, Extra Strong, Regular, Standard Sizes	Ch. 12, Ch. 13
ASTM B 43-1998 (R2004)	Seamless Red Brass Pipe, Standard Sizes	Cooper Alloy UNS No. C23000, Red Brass Pipe	1111.7
ASTM B 75-2002	Seamless Copper Tube	Seamless Copper Tube, Seamless Tube, Tube	Ch.12, Ch. 13
ASTM B 88-2003	Seamless Copper Water Tube, (K, L, M)	Copper Tube, Seamless, UNS No. C10200, UNS No. C12000, UNS No. C12200, Water Tube	Ch. 12, 1309.5.3.2, 1331.1
ASTM B 135-2002, ASTM B 135M-2000 (2006)	Seamless Brass Tube	Seamless Brass Tube	Ch. 12, Ch. 13
ASTM B 210-2004/B210M-2005	Aluminum and Aluminum-Alloy Drawn Seamless Tubes/Aluminum and Aluminum-Alloy Drawn Seamless Tubes Metric	Product	1309.5.3.3
ASTM B 241/241M-2002	Aluminum and Aluminum-Alloy Seamless Pipe and Seamless Extruded Tube	Product	1309.5.2.4, 1309.5.3.3
ASTM B 280-2003	Seamless Copper Tube for Air Conditioning and Refrigeration Field Service	Air Conditioning Tube, Copper Tube, Field Service Tube, Refrigeration Tube, Seamless Copper Tube	Ch. 11, 1309.5.3.2, 1331.1
ASTM B 302-2007	Threadless Copper Pipe (TP)	Brazed-Joint Pipe Fittings, Copper Pipe, Threadless Copper Pipe, Threadless Pipe, Threadless Pipe (TP)	Ch. 12, Ch. 13

STANDARD NUMBER	STANDARD TITLE	APPLICATION	REFERENCED SECTION
ASTM B 306-2002	Copper Drainage Tube (DWV)	Copper Tube, Drainage Tube	309.0
ASTM B 370-2003	Copper Sheet and Strip for Building Construction	Building Construction, Coils For Building Construction, Copper Sheet, Copper Strip, Down Spouts, Flashing, Gutters, Roofing, Sheet/Strip In Flat Lengths	508.1.1
ASTM B 447-2007	Welded Copper Tube	Copper Tube, Copper Welded Tube, Welded Tube	1201.2.1.1, 1201.2.1.2, 1201.2.3.7, 1309.5.2.3, 1309.5.3.2
ASTM B 587-2006	Welded Brass Tube	Brass Tube, Welded Brass Tube	1201.2.1.1, 1201.2.1.3, 1201.2.1.4.1, 1201.2.3.1, 1309.5.2.3, 1309.5.3.2
ASTM B 687-1999 (R2005)[e1]	Brass, Copper, and Chromium-Plated Pipe Nipples	Chromium Plating, Copper, Copper Alloy Fittings, Em Tube, Pipe Nipples	1313.7
ASTM B 813-2000[e1]	Liquid and Paste Fluxes for Soldering of Copper Aggressiveness Test, and Copper Alloy Tube	Copper and Copper Alloys, Corrosiveness, Flushing, Liquid and Paste Fluxes, Resistivity, Sodering, Spreading Factor, Tarnish	1201.2.1.7
ASTM B 828-2002	Making Capillary Joints in Copper and Copper Alloy Tube and Fittings	Assembly, Capillary, Cleaning, Copper, Corrosivity, Fittings, Flux	1201.2.3.7
ASTM C 411-2005	Hot-Surface Performance of High Temperature Thermal Insulation	Block Board, Cracking, Delamination, Hot-Surface Performance, Pipe Thermal Insulation, Surface Analysis-Building, Temperature Tests-Insulation, Thermal Insulating Materials	1201.2.1.8, 1201.3.1.3, 1201.4.1.3

STANDARD NUMBER	STANDARD TITLE	APPLICATION	REFERENCED SECTION
ASTM C 518-2004	Steady-State Thermal Transmission Properties by Means of the Heat Flow Meter Apparatus	Certification: Calibration, Error Analysis, Heat Flow Meter Apparatus, Heat Flux, Instrument Verification, Thermal Conductivity, Thermal Resistance, Thermal Testing, Tunnel Test	Table 6-6A
ASTM D 93-2007	Standard Test Methods for Flash Point by Pensky-Martens Closed Cup Tester	Certification	503.3
ASTM D 2513-2007b*	Thermoplastic Gas Pressure Pipe, Tubing and Fittings	Product	1309.5.4, 1312.1.7(B)(1)
ASTM E 84-2007b	Surface Burning Characteristics of Building Materials	Certification: 25 Foot Tunnel, Flame Spread, Flame Spread Index, Smoke Developed, Smoke Developed Index, Steel Sheet, Surface Burning Test	602.2, 605.0(E), 1201.2.1.8, 1201.3.1.3, 1201.4.1.3, Table 3-2, footnote 6
ASTM E 814-2006	Fire Tests of Through-Penetration Fire Stops	Certification: Fire Testing-building Materials, Fire Testing-Hose Stream Test, Structural Analysis/Applications	507.2.5, 507.2.6, 928.1 (c),
ASTM E 2231-2007a	Specimen Preparation and Mounting of Pipe and Duct Insulation Materials to Assess Surface Burning Characteristics	Certification: IEEE, Insulation, Pipe, Preparation	605.0, 1201.2.1.8, 1201.4.1.3
ASTM E 2336-2004	Standard Test Method for Fire-Resisting Grease Duct Enclosure Systems	Field Fabricated Grease Duct	507.2.6
ASTM F 876-2006*	Crosslinked Polyethylene (PEX) Tubing	Piping	1201.3, 1201.3.1.1, 1201.3.2
ASTM F 877-2005^{e1}*	Crosslinked Polyethylene (PEX) Plastic Hot- and Cold-Water Distribution Systems	Piping	1201.3, 1201.3.1.1, 1201.3.2
ASTM F 1281-2007*	Crosslinked Polyethyene/Aluminum/Crosslinked Polyethylene (PEX-AL-PEX) Pressure Pipe	Composite, Crosslinked PE, PEX-AL-PEX, Pipe, Pressure	1201.4, 1201.4.1.1, 1201.4.2
ASTM F 1282-2006*	The Standard for Polyethylene/Aluminum/Polyethylene (PE-AL-PE) Composite Pressure Pipe	Composite Pressure Pipe	1201.7, 1201.7.1.1, 1201.7.1.6

STANDARD NUMBER	STANDARD TITLE	APPLICATION	REFERENCED SECTION
CSA 1.8-2006*	Gas Food Service Equipment	Food Service Equipment	913.1, 913.3(A), (B)
CSA 2.14a 2007*	Gas-Fired Construction Heaters	Product	307.1, 924.0, 927.0
CSA 2.22-2007*	Decorative Gas Appliances, Vented	Appliances, Decorative Appliances	203.0, 224.0, Table 8-1, 907.2, 907.3
CSA 2.3-2007*	Gas-Fired Central Furnaces	Central Furnaces	208.0, Table 3-2, 608.5, 902.0(A), 911.1(A), 911.6(B)
CSA 2.26b-2004*	Decorative Gas Appliances for Installation in Solid-Fuel Burning Fireplaces	Decorative Appliances, Decorative Fireplace Appliances	907.0, 1313.4
CSA 2.6-2006*	Gas Unit Heaters and Gas-Fired Duct Furnaces	Product	901.0, 911.0
CSA 2.7-1998 (R2004)*	Installation of Domestic Gas Conversion Burners	Conversion Burner Installation, Gas Burners	802.10.2.4, 802.12.1, 906.0, C.4.0,
CSA 2.7-M1998*	Domestic Gas Conversion Burners	Conversion Burners, Domestic, Gas Burners	802.10.2.4, 802.12.1, C.4.0
CSA 2.32-2007*	Gas-Fired Vented Heating Appliances	Gas Appliances, Vented Heaters	306.0, 307.1
CSA 2.34b-2004*	Gas-Fired Tube Type and Low Intensity Infrared Heaters	Infrared Heaters, Low Intensity Heaters	919.0
CSA 2.35b-2007*	Gas-Fired High Intensity Infrared Heaters	High Intensity Heaters, Infrared Heaters	919.0
CSA 3.7b-2006*	Non-Recirculating Direct Gas-Fired Industrial Air Heaters	Air heaters, Non-Recirculating, Non-Recirculating Heaters	909.0
CSA 4.1-2006*	Gas Water Heaters, Volume I, Water Heaters with Input Ratings of 75,000 Btu per Hour or Less	Gas Water Heaters	225.0, 517.1.2, 802.2.2.5, Table 13-1
CSA 4.3a-2007*	Gas Water Heaters, Volume III, Water Heaters with Input Ratings Above 75,000 Btu per Hour, Circulating and Instantaneous Water Heaters	Gas Water Heaters	225.0, 517.1.2, 802.2.2.5, Table 13-1
CSA 4.4b-2001 (R2003)*	Relief Valves for Hot Water Supply Systems	Hot Water, Pressure Relief, Relief Valves, Valves	904.6, 1006.0, 1007.0, 1010.2, 1206.0

STANDARD NUMBER	STANDARD TITLE	APPLICATION	REFERENCED SECTION
CSA 4.7-2006*	Gas-Fired Pool Heaters	Gas Fired Heaters, Pool Heaters, Spa Heaters	922.0
CSA 4.9b-2007*	Gas-Fired Low Pressure Steam and Hot Water Boilers	Gas Fired Boilers, Hot Water Boilers, Low Pressure Boilers, Steam Boilers	214.0, 904.0, 904.1.1, 904.2(E), 1003.1, Table 13-1
CSA 6.3b-2007*	Gas Appliance Pressure Regulators	Appliance Regulators, Gas Refrigerators, Pressure Regulators	917.5, 1309.7.5(B) 1311.0(1), (2)
CSA 6.5-2005*	Automatic Valves for Gas Appliances	Appliance Valves, Automatic Gas Valves, Valves	917.2 (B)(3), B.4.0
CSA 6.8-2005*	Pilot Gas Filters	Filters, Pilot Filters	1311.0
CSA 6.9A-2005*	Quick-Disconnect Devices for Use with Fuel Gas	Quick Disconnect	1303.7
CSA 6.10b-2006*	Connectors for Gas Appliances	Connectors, Gas Connectors	1312.1.7, Tables 3-4, 3-5 and 3-6
CSA 6.16b-2006*	Connectors for Moveable Gas Appliances	Appliance Connectors, Connectors	1312.1.7, Tables 3-4, 3-5 and 3-6
CSA 6.22a- 2005*	Line Pressure Regulators	Pressure Regulators, Regulators	1309.7, 1309.7.5(A),
CSA 6.26-2005*	Fuel Gas Piping Systems Using Corrugated Stainless Steel Tubing (CSST) Fuel Gas	CSST, Fuel Gas Tubing, Stainless Steel Tubing	1309.5.3.4
CSA A7.1b-2006*	Gas Clothes Dryers - Volume I, Type I	Clothes Dryers, Gas	504.3, 701.1.1, 701.1.4, 802.2.2.4, 905.0, 1312.7
CSA 7.2a-2006*	Gas Clothes Dryers - Volume II, Type 2	Clothes Dryers, Gas	504.3, 701.1.4, 905.0, 1312.7
CSA 8.4a-2006*	Gas Hose Connectors for Portable Outdoor Gas-Fired Appliances	Appliance Connectors, Gas Connectors, Hose Connectors, Indirect Water Heaters, Pool Heaters, Portable Appliance Connectors	1312.1.7

STANDARD NUMBER	STANDARD TITLE	APPLICATION	REFERENCED SECTION
CSA America FC 1-2004*	Fuel Cell Power Plants	Fuel Cell Power Plants	1601.0
CSA F379.1-88 (R2006)	Solar Domestic Hot Water Systems (Liquid to Liquid Heat Transfer)	Solar Heating Systems	1206.0, Ch. 15
CSA LC4-2007*	Press-Connect Copper and Copper Alloy Fittings For Use in Fuel Gas Distribution Systems	Copper Alloy Fittings	1309.5.8.2
CSA Z21.1-a-2007*	Household Gas Cooking Appliances	Household Cooking Appliances, Gas Fired	305.0, Table 3-2, 916.0, Table 13-1, 1313.1
CSA Z21.11.2b-2007*	Gas-Fired Unvented Room Heaters, Volume II	Room Heaters, Unvented Heaters	924.1.1
CSA Z21.12-1994 (R2000)*	Draft Hoods	Draft Hoods, Hoods	205.0, 206.0, 701.1.3, 802.2.2.5, 802.12, B.6.0
CSA Z21.20-2007*	Automatic Gas Ignition Systems and Components (same as CSA C22-2 No. 99	Automatic Ignition, Gas Ignition, Ignition	917.2(B)(3), B.4.0
CSA Z21.23b-2005	Gas Appliance Thermostats	Thermostats	912.1(C)
CSA Z21.24-2006*	Connectors for Gas Appliances (same as CSA 6.10)	Connectors For Gas	919.2
CSA Z21.42a-2004*	Gas-Fired Illuminating Appliances	Illuminating Appliances	203.0, Table 13-1
CSA Z21.61-1983 (R2004)*	Gas-Fired Toilets	Toilets	926.0
CSA Z21.71a-2005 (R2007)*	Automatic Intermittent Ignition Systems for Field Installation	Automatic Pilot Ignition, Pilot Ignition	1003.9, Table 10-3
CSA Z21.90-2001 (R2005)	Gas Convenience Outlets (same as CSA 6-24)		1312.7
CSA Z83.18a-2005*	Recirculating Direct Gas-Fired Industrial Air Heaters	Air Heaters, Industral Heaters, Recirculating	910.1
CSA 2.33-a-2007*	Vented Gas Fireplace Heaters	Fire Place Heaters	802.5.8
EPA 202-1996	Determination of Condensable Particulate Emissions for Stationary Sources	Miscellaneous	516.2.9
IAPMO IS 13-2006	Installation Protectively Coated Pipe	Pipe Coatings	1312.1.3

STANDARD NUMBER	STANDARD TITLE	APPLICATION	REFERENCED SECTION
IAPMO PS 33-2007a	Flexible PVC Hose for Pools, Hot Tubs, Spas and Jetted Bathtubs	Flexible Hose, Hose, PVC Hose	309.5
IAPMO PS 120-2004	Flashing and Stand Combination for Air Conditioning Units (Residential or Commercial Unit Curb)	Air Conditioning Flashing Stand	1106.2
IAS 1-1991	U.S. Requirements for Indirect Water Heaters for Use With External Heat Source	Indirect Water Heaters	225.0, 517.1.2, Table 13-1
I-B-R 1st Edition Aug, 2002	Rating Procedure for Heating Boilers	Certification	Ch. 10
IIAR2-1999	Mechanical Refrigeration, Equipment, Design, and Installation of Ammonia Systems	Ammonia Refrigeration Systems	1102.0
ISO Guide 65:1996	General Requirements for Bodies Operating Product Certification Systems	Certification	214.0, Ch. 3
MSS SP-6-2007	Standard Finishes for Contact Faces of Pipe Flanges and Connecting-End Flanges of Valves and Fittings	Butterfly Valves, Pipe Hangers, Pipe Supports, Valves	1309.10
MSS SP-58-2002*	Pipe Hangers and Supports - Materials, Design, and Manufacture	Pipe Hangers, Pipe Supports	1312.2.6(A)
MSS SP-67-2002a*	Butterfly Valves	Butterfly Valves,Valves	Ch. 12, Ch. 13, Ch. 14
MSS SP-69-2003*	Pipe Hangers and Supports - Selection and Application	Pipe Hangers, Pipe Supports	1312.2.6(A)
MSS SP-80-2003*	Bronze Gate, Globe, Angle, and Check Valves	Angle Valves, Bronze Valves, Check Valves, Gate Valves, Globe Valves, Valves	1309.8.2
MSS SP-89-2003*	Pipe Hangers and Supports – Fabrication and Installation Practices	Installing Pipe Hanger, Installing Piping Supports	1201.2.1.11, 1311.2.6(A)
MSS SP-104-2003*	Wrought Copper Solder Joint Pressure Fittings	Pressure Fittings, Solder Fittings, Wrought Copper Fittings	1201.2.2.2, Ch. 13, Ch. 14
MSS SP-106-2003*	Cast Copper Alloy Flanges and Flanged Fittings, Class 125, 150, and 300	Cast Copper Alloy Fittings, Copper Alloy Fittings, Copper Alloy Fittings, Flanged Fittings	1111.0
MSS SP-109-1997 (R2006)*	Welded Fabricated Copper Solder Joint Pressure Fittings	Copper Fittings, Fabricated Fittings, Pressure Fittings, Solder Fittings, Welded Pipe	1201.2.2.2, Ch. 13, Ch. 14

STANDARD NUMBER	STANDARD TITLE	APPLICATION	REFERENCED SECTION
NFPA 10-2007*	Portable Fire Extinguishers	Fire Extinguishing	513.11
NFPA 12-2008	Carbon Dioxide Extinguishing Systems	Fire Extinguishing	513.2.2.1
NFPA 13-2007	Installation of Sprinkler Systems	Fire Extinguishing	513.2.2.1
NFPA 17-2002*	Dry Chemical Extinguishing Systems	Fire Extinguishing	513.2.2.1, 513.3
NFPA 17A-2002*	Wet Chemical Extinguishing Systems	Fire Extinguishing	513.2.2.1, 513.3
NFPA 30-2008	Flammable and Combustible Liquids Code	Combustible Liquids, Flammable Liquids	204.0
NFPA 31-2006*	Installation of Oil Burning Equipment	Oil Burning Equipment	302.3, 1334.0, 1335.6, 1335.6.1
NFPA 37-2006	Standard for the Installation and Use of Stationary Combustion Engines and Gas Turbines	Generators	925.0, 1602.0
NFPA 52-2006	Compressed Natural Gas (CNG) Vehicular Fuel Systems	CNG Vehicle Fuel Systems	1301.0
NFPA 54/Z223.1-2006*	National Fuel Gas Code	Fuel Gas Code	516.2.1, 1309.7.5(A)(1)
NFPA 58-2004*	Liquefied Petroleum Gas Code	LPG Code	303.7, 516.2.1, 1309.5.4.3, 1313.0
NFPA 69-2008	Explosion Prevention Systems	Explosion Prevention	505.1.2.
NFPA 70-2008	National Electrical Code	Electrical Code	207.0, 511.1.6, 512.2.6, 516.2.7, 516.2.9(D)
NFPA 80-2007	Fire Doors and Fire Windows	Fire Doors	510.7.5
NFPA 82-2007	Standard on Incinerators and Waste and Linen Handling Systems and Equipment	Incinerator Shutes	918.0
NFPA 85-2004	Boiler and Combustion Systems Hazard Code	Boiler Hazard Code, Fuel Cell Power Plants Fuel Cells	1001.0(9), 1018.0, Table 10-3
NFPA 88A-2007	Parking Structures	Building	911.8.1
NFPA 90A-2002	The Installation of Air-Conditioning and Ventilating Systems	HVAC	Ch. 6, 904.7
NFPA 90B-2006	The Installation of Warm Air Heating and Air-Conditioning Systems	HVAC	Ch. 6, 904.7(A)

STANDARD NUMBER	STANDARD TITLE	APPLICATION	REFERENCED SECTION
NFPA 91-2004	Exhaust Systems for Air Conveying of Vapors, Gases, Mists, and Noncombustible Particulate Solids	Product Conveying Ducts	506.5.4
NFPA 96-2008	Ventilation Control and Fire Protection of Commercial Cooking Operations	Commercial Cooking	Ch. 5, Part II
NFPA 211-2006	Chimneys, Fireplaces, Vents, and Solid Fuel Burning Appliances	Chimneys, Fireplaces, Solid Fuel Appliances	517.7.1, 517.7.2, 517.7.3, 801.1, 801.2, 802.5.1.2, 802.5.1.3, 802.5.4.2
NFPA 255-2006	Method of Test of Burning Characteristics of Building Materials	Certification: Surface Burning Test (ASTM E84 and UL 723)	602.2, 605.0, 1201.2.1.8, 1201.3.1.3 1201.4.1.3, Table 3-1, footnote 6
NFPA 262-2007*	Flame Travel and Smoke of Wires and Cables for Use in Air-Handling Spaces	Certification: Surface Burning Test, Wire, Cable	602.2(5)
NFPA 274-2003*	Standard Test Method to Evaluate Fire Performance Characteristics of Pipe Insulation.	Pipe Insulation	1201.2.1.8, 1201.3.1.3, 1201.4.1.3
NFPA 409-2004	Standard on Aircraft Hangars	Building	911.9
NFPA 501A-2005	Fire Safety Criteria for Manufactured Home Installations Sites and Communities	Building	Ch. 13, Part II
NFPA 654-2006	Prevention of Fire and Dust Explosion from the Manufacturing, Processing, and Handling of Combustible Particulate Solids	Dust Explosion Prevention, Gutters	506.4
NFPA 780	Standard for the Installation of Lightning Protection Systems	Lightning Protection System	1312.14
NFPA 853-2007*	Installation of Stationary Fuel Cell Power Systems	Fuel Cell Power Plants, Fuel Cells	1601.0
PPI TR-4-2007b	Recommended Hydrostatic Strengths and Design Stresses for Thermoplastic Pipe and Fittings Compounds	Certification: Pipe and Fittings Compounds, Thermoplastic Compounds	Ch. 12, Ch. 14
SAE J512-97	Automotive Tube Fittings	Tube Fittings	Ch. 12
SMACNA-2005	HVAC Duct Construction Standards Metal and Flexible 3rd Edition	Ducts, Metal and Flexible	506.2, 602.1, 602.5, Tables 6-1 and 6-2
SMACNA-2003	Fibrous Glass Duct Construction Standard 7th Edition	Fiberglass Ducts	602.3
UL 17-1994	Vent or Chimney Connector Dampers for Oil-Fired Appliances	Vent Dampers	1005.4
UL 21-2007*	LP-Gas Hose	Hose, LPG Hose	1312.2
UL 33-2003*	Heat Responsive Links for Fire-Protection Service	Fusible Links	606.0
UL 51-2002	Power-Operated Pumps for Anhydrous Ammonia and LP-Gas	Ammonia Pumps, LPG Pumps, Pumps	516.2.1, 1309.5.4.3, 1313.0

STANDARD NUMBER	STANDARD TITLE	APPLICATION	REFERENCED SECTION
UL 80-2007*	Steel Tanks for Oil-Burner Fuels and Other Combustible Liquids	Oil Tanks, Tanks	302.3, 1018.0, 1335.6.2
UL 103-2001*	Factory-Built Chimneys for Residential Type and Building Heating Appliances	Chimneys, Heating Appliance Chimneys, Residential Chimneys	802.5.1
UL 125-2007*	Valves for Anhydrous Ammonia and LP-Gas (Other than Safety Relief)	Ammonia Valves, LPG Valves, Valves	516.2.1, 1309.5.4.3, 1313.0
UL 127-1996*	Factory-Built Fireplaces	Fireplaces	802.5.1.3, 802.5.4.1, 802.10.13
UL 132-2007*	Safety Relief Valves for Anhydrous Ammonia and LP-Gas	Ammonia Relief Valves, LPG Valves, Relief Valves	516.2.1, 1309.5.4.3, 1313.0
UL 144-1999*	LP-Gas Regulators	Low Pressure Boilers, Regulators	1301.0
UL 174-2004*	Household Electric Storage Tank Water Heaters	Water Heaters, Electric	225.0, 517.1.2, 802.2.2.5, 803.1.8.1, Table 13-1
UL 180-2003	Liquid-Level Indicating Gauges for Oil Burner Fuels	Gauges, Level Gauges	302.3, 1018.0, Ch. 13
UL 181-2005*	Factory-Made Air Ducts and Air Connectors	Air Connectors, Air Ducts	602.5, A.6.501
UL 181A-2005*	Closure Systems for Use with Rigid Air Ducts	Air Ducts	602.4
UL 181B-2005*	Closure Systems for Use with Flexible Air Ducts and Air Connectors	Air Connectors, Air Ducts	602.4, A.6.507(A)(4), A.6.508(2)
UL 197-2003*	Commercial Electric Cooking Appliances	Appliances, Commercial Cooking, Electric Appliances	507.1.1, 508.2.11
UL 207-2001	Refrigerant-Containing Components and Accessories, Non Electrical	Refrigerant Components	Ch. 11
UL 252-2003*	Compressed Gas Regulators	Regulators	Ch. 13
UL 268A-1998*	Smoke Detectors for Duct Application	Smoke Detectors	609.0
UL 296-2003*	Oil Burners	Burners	302.3, 1019.0, 1302.0, 1318.0, 1332.0(2), 1332.2, 1333.6, 1333.6.1, 1334.0, 1335.2, 1335.3
UL 300-2005*	Fire Testing of Fire Extinguishing Systems for Protection of Restaurant Cooking Equipment	Certification: Restaurant Fire Protection	513.2.2

STANDARD NUMBER	STANDARD TITLE	APPLICATION	REFERENCED SECTION
UL 343-1997	Pumps for Oil-Burning Appliances	Oil Pumps, Pumps	302.3, 1019.0, 1302.0, 1318.0, 1332.0(2), 1332.2, 1333.6, 1333.6.1, 1334.0, 1335.2, 1335.3
UL 352-2006	Constant-Level Oil Valves	Oil Valves,Valves	Appendix C
UL 353-1994	Limit Controls	Controls	912.2
UL 378-2006*	Draft Equipment	Draft Equipment	1003.15
UL 391-2006*	Solid-Fuel and Combination-Fuel Central and Supplementary Furnaces	Furnaces, Solid Fuel	Ch. 9
UL 399-1993*	Drinking-Water Coolers	Drinking Fountains, Water Coolers	Ch. 9
UL 404-2005*	Gauges for Compressed Gas Service Indicating Pressure	Gauges, Compressed Gas	Ch. 13
UL 412-2004*	Refrigeration Unit Coolers	Refrigeration	Ch. 11
UL 429-1999	Electrically Operated Valves	Valves	Ch. 12, 1309.9, Ch.14
UL 441-1996*	Gas Vents	Vents	802.6
UL 443-2006	Steel Auxiliary Tanks for Oil-Burner Fuel	Oil Tanks, Tanks	302.3, 1019.0, 1302.0, 1318.0, 1332.0(2), 1332.2, 1333.6, 1333.6.1, 1334.0, 1335.2, 1335.3
UL 471-2006	Commercial Refrigerators and Freezers	Freezers, Refrigerators	923.0, Ch. 11
UL 525-2004	Flame Arresters	Flame Arrestors	Ch. 9
UL 536-1997*	Flexible Metallic Hose	Hose, Metallic Hose	Ch. 13
UL 555-2006	Fire Dampers	Dampers	606.2
UL 555C-2006*	Ceiling Dampers	Dampers	606.0
UL 555S-1999	Smoke Dampers	Dampers	606.0
UL 565-1998	Liquid-Level Gauges and Indicators for Anhydrous Ammonia and LP-Gas	Gauges, LP Gas	516.2.1, 1309.5.4.3, 1309.5.9(D), 1313.0
UL 569-1995*	Pigtails and Flexible Hose Connectors for LP-Gas	Gas Connectors, Connectors	516.2.2, 1309.5.4.3, 1309.5.9(D), 1313.0
UL 641-1995	Type L Low-Temperature Venting Systems	Product	302.3

STANDARD NUMBER	STANDARD TITLE	APPLICATION	REFERENCED SECTION
UL 710-1995	Exhaust Hoods for Commercial Cooking Equipment	Exhaust Hoods, Hoods	508.6, 513.2.2
UL 723-2007*	Test for Surface Burning Characteristics of Building	Surface Burning Testing	508.2.8, 508.3, 1201.2.1.8
UL 726-1995	Oil-Fired Boiler Assemblies	Boiler, Oil Fired	302.3, 1320.0
UL 727-2006	Oil-Fired Central Furnaces	Furnaces, Central, Oil Fired	302.3, 1320.0
UL 729-2003	Oil-Fired Floor Furnaces	Furnaces, Floor, Oil Fired Furnaces	302.3, 912.0
UL 730-2003	Oil-Fired Wall Furnaces	Furnaces, Wall, Oil Fired	302.3, 928.0
UL 731-1995	Oil-Fired Heater Units	Heaters, Oil Fired	302.3, 927.0
UL 732-1995*	Oil-Fired Storage Tank Water Heaters	Water Heaters, Oil Fired	302.3, 1320.0
UL 733-1993	Oil-Fired Air Heaters and Direct-Fired Heaters	Water Heaters, Direct Fired, Oil Fired	302.3, 1320.0
UL 737-2007	Fireplace Stoves	Fireplace Stoves	208.0
UL 791-2006	Residential Incinerators	Incinerators	Table 8-2
UL 795-2006	Commercial-Industrial Gas Heating Equipment	Heating Equipment, Gas Fired	Ch. 10
UL 834-2004*	Heating Water Supply, and Power Boilers-Electric	Boilers, Electric Ranges Boilers	Ch. 10
UL 842-1997	Valves for Flammable Fluids	Valves	Ch. 13
UL 858-2005*	Household Electric Ranges	Electric Ranges, Ranges	Ch. 9
UL 867-2000	Electrostatic Air Cleaners	Filters	311.2
UL 896-1993	Oil-Burning Stoves	Stoves, Oil Fired	1320.0
UL 900-2004*	Air Filter Units	Air Filters, Filters	311.2

STANDARD NUMBER	STANDARD TITLE	APPLICATION	REFERENCED SECTION
UL 907-1994	Fireplace Accessories	Fireplace Accessories	Ch. 9
UL 921-2006*	Commercial Electric Dishwashers	Dishwashers	Ch. 5, Part II
UL 959-2001*	Medium Heat Appliance Factory-Built Chimneys	Chimneys	802.5
UL 984-1996	Hermetic Refrigerant Motor-Compressors	Compressors, Refrigeration	Ch. 11
UL 1026-2007*	Electric Household Cooking and Food Serving Appliances	Cooking Appliances, Electric Appliances	916.0
UL 1046-2000	Grease Filters for Exhaust Ducts	Filters-Grease	509.1
UL 1206-2003*	Electric Commercial Clothes-Washing Equipment	Clothes Washers, Commercial Clothes Washers	Ch. 5
UL 1240-2005*	Electric Commercial Clothes-Drying Equipment	Clothes Dryers, Commercial	905.0
UL 1453-2004*	Electric Booster and Commercial Storage	Electric Water Heaters, Water Heaters	302.3, 303.8.1.6, 308.0
UL 1479-2003	Fire Tests of Through-Penetration Firestops	Firestops	Ch. 12
UL 1482-1996	Solid-Fuel Type Room Heaters	Room Heaters, Solid Fuel Heaters	924.0
UL 1738-1993	Venting Systems for Gas-Burning Appliances, Categories II, III, and IV	Gas Vent Systems, Vent Systems	802.0
UL 1746-2007*	External Corrosion Protection Systems for Steel Underground Storage Tanks	Tanks	204.0, Ch. 10
UL 1777- 2007*	Chimney Liners	Chimneys, Liners	803.1.6, 803.2.19
UL 1812-1995	Ducted Heat Recovery Ventilators	Heat Recovery Ventilators	Ch. 6
UL 1815-2001	Nonducted Heat Recovery Ventilators	Heat Recovery Ventilators	Ch. 6
UL 1820-2004*	Fire Test of Pneumatic Tubing for Flame and Smoke Characteristics	Surface Burning Test, Pneumatic Tubing	602.2(7)
UL 1887-2004*	Fire Test of Plastic Sprinkler Pipe for Visible Flame and Smoke Characteristics	Surface Burning Test, Fire Sprinkler Pipe	602.2(6)
UL 1963-2005*	Refrigerant Recovery / Recycling Equipment	Refrigerant Equipment	Ch. 11
UL 1978-2005*	Grease Ducts	Ducts, Grease	508.2.11
UL 1995-2005*	Heating and Cooling Equipment	HVAC, Electric	307.0

STANDARD NUMBER	STANDARD TITLE	APPLICATION	REFERENCED SECTION
UL 2021-1997	Fixed and Location-Dedicated Electric Room Heaters	Product, Heating, Electric	Ch. 9
UL 2043-1996	Fire Test for Heat and Visible Smoke Release For Discrete Products and Their Accessories Installed in Air-Handling Spaces	Surface Burning Test, Discrete Products	602.2(8)
UL 2157-1997	Electric Clothes Washing Machines and Extractors	Clothes Washers, Extractors	Ch. 5
UL 2158-1997	Electric Clothes Dryers	Clothes Dryers, Electric	905.0
UL 2200,1998	Stationary Engine Generator Assemblies	Assemblies	1602.0
UL 2221-2001	Tests of Fire Resistive Grease Duct Enclosure Assemblies	Duct Wrap, Grease Duct Enclosure	507.2.5

Although these standards are referenced in Chapter 17, some of the materials and equipment in the standards may not be acceptable for use under the provisions of this code.

* *ANSI Approved Standards*

ABBREVIATIONS IN CHAPTER 17

ACCA	Air Conditioning Contractors of America Association, Inc., 2800 Shirlington Road, Suite 300, Arling ton, VA 22206; 703-575-4477
ANSI	American National Standards Institute, Inc., 25 W. 42nd Street, 4th Floor, New York, NY 10036; 202-293-8020
AHRI	Air-Conditioning, Heating, and Refrigeration Institute, 2111 Wilson Blvd., Suite 500, Arlington, VA 22201; 703-524-8800
ASCE	American Society of Civil Engineers, 1801 Alexander Bell Drive, Reston, VA 20191-4400; 800-548-2723
ASHRAE	American Society of Heating, Refrigerating, and Air-Conditioning Engineers, Inc., 1791 Tullie Circle, NE, Atlanta, GA 30329-2305; 404-636-8400
ASME	American Society of Mechanical Engineers, Three Park Avenue, New York, NY 10016; 800-843-2763
ASTM	American Society for Testing and Materials, 100 Barr Harbor Drive, West Conshohocken, PA 19428-2959; 610-832-9500
AWS	American Welding Society, 550 N.W. LeJeune Road, Miami, FL, 33126; 305-443-9353
AWWA	American Water Works Association, 6666 W. Quincy Ave., Denver, CO 80235; 303-794-7711
CGSB	Canadian General Standards Board, Place du Portage III, 6B1, 11 Laurier Street, Gatineau, Quebec, Canada K1A 1G6; 819-956-0425
CGA	Canadian Gas Association, 350 Sparks Street, Suite 809, Ottawa, Ontario, Canada K1R 7S8; 613-748-0057
CSA	Canadian Standards Association, 5060 Spectrum Way, Mississauga, Ontario, Canada L4W 5N6; 416-747-4000
EPA	Environmental Protection Agency, 1200 Pennsylvania Ave., N.W., Washington, DC 20460; 202-272-0167
IAPMO	International Association of Plumbing and Mechanical Officials, 5001 E. Philadelphia Street, Ontario, CA 91761-2816; 909-472-4100
IAS	International Approval Services, A Division of CSA, 8501 East Pleasant Valley Road Cleveland, OH 44131; 216-524-4990
I-B-R	Institute of Boiler and Radiator Manufacturers, 35 Russo Place, Berkeley Heights, NJ 07922; 908-464-8200
ISO	International Organization for Standardization, 1 ch. de la Voie-Creuse, CH-1211 Geneva 20, Switzerland; +41 22 749 01 011
MSS	Manufacturers Standardization Society, 127 Park St. NE, Vienna, VA 22180; 703-281-6613
NFPA	NFPA International, 1 Batterymarch Park, Quincy, MA 02169; 617-770-3000
PPI	Plastic Pipe Institute, 105 Recker Court, Suite 825, Irving, TX 75062; 469-499-1044
SAE	Society of Automotive Engineers, 400 Commonwealth Drive, Warrendale, PA 15096-0001; 724-776-4841
SMACNA	Sheet Metal and Air Conditioning Contractors National Association, 4201 Lafayette Center Drive, Chantilly, VA 20151-1209; 703-803-2980
UL	Underwriters Laboratories Inc., 333 Pfingsten Road, Northbrook, IL 60062-2096; 847-272-8800

APPENDICES

The appendices are intended to supplement the provisions of the installation requirements of this code. The definition in Chapter 2 are also applicable to the appendices.

TABLE OF CONTENTS

APPENDIX A

UNIFORM MECHANICAL CODE STANDARD NO. 2-2
STANDARD FOR GALVANIZED SHEET METALS

This standard has been deleted from UMC.

The Tentative Specification A525-64T of the American Society for Testing and Materials has been withdrawn and replaced by A653/A653M-03 Standard Specification for Steel Sheet, Zinc-Coat (Galvanized) or Zinc-Iron Alloy-Coat (Galvanized) by the Hot Dip Process

See Section 209.0 and 603.0 of the *Uniform Mechanical Code*.

APPENDIX A

UNIFORM MECHANICAL CODE STANDARD NO. 6-2
STANDARD FOR METAL DUCTS

Based on the first edition, 1995 HVAC Duct Construction Standards, Metal and Flexible, published by the Sheet Metal and Air Conditioning Contractors National Association.
See Sections 506.2, 602.1, and 602.5 of the *Uniform Mechanical Code.*

A.6.201 Scope.

The performance criteria and requirements herein contemplate a duct that is a structural assembly having the capacity to support occupant health and safety while minimizing its own contribution to property damage under emergency conditions. Ducts can supply fresh or treated air in support of life and health, can convey products of combustion away from a fire zone, can maintain a pressure differential that facilitates evacuation and reduces the spread of fire and smoke, and can facilitate firefighter access to a fire source.

A.6.202 Definitions.

Closure – The joining components on duct surfaces that are subjected to a pressure differential across the duct wall, including mechanical locks and metallic and nonmetallic materials used as fillers or sealers whether on straight sections of ducts or fittings or used in connection of ducts or fittings to other apparatus. Closure shall also pertain to penetrations of duct walls (such as by screws, rivets, pipes, electrical conduits, etc.).

Closure Failure – Any structural failure at a closure and also any loss of adhesive bond, creep, displacement, or other deterioration of closure materials and methods that results in an increase in air leakage above an allowable level and any behavior contributing to accelerated aging of the closure systems.

Duct Assembly – The completed composite connected assembly of straight sections, fittings, and accessories attached to or occurring in the perimeter of an air passageway that is used for supply air, exhaust air, or recirculated air mode. The passageway need not be continuous.

Note: Where apparatus, ceiling plenums, room boundaries, shafts, air terminals, etc., form a part of the passageways, such are not subject to the performance requirements in this standard, but are not prohibited from being made subject to similar requirements.

Metal Duct – A duct constructed of galvanized steel, uncoated steel, stainless steel, or aluminum material of commercial grade. Accessories and reinforcements are not required by this standard to be of the same material as the duct wall, but they must be electrolytically compatible and shall not reduce the performance levels for strength and tight closure of ducts.

Structural Failure – Any visible and measurable rupture, collapse, buckling, or other permanent deformation or separation of the elements in the duct assembly.

A.6.203 Performance Requirements for Ducts.

(A) Pressure. Structural failure or closure failure of a duct assembly shall not occur at 150 percent of the designed operating pressure classification when supported at the maximum recommended support spacing.

(B) Support Spacing. Structural failure or closure failure of duct assembly shall not occur when supported at twice the maximum recommended support spacing and atmospheric pressure is maintained inside the ducts.

(C) Deflection, Longitudinal. The maximum allowable deflection of rectangular duct walls at design pressure classification shall not exceed the following:

Duct Dimension		Deflection	
12" down	(305 mm)	1/2"	(12.7 mm)
13" to 18"	(330 to 457 mm)	5/8"	(15.9 mm)
19" to 24"	(483 to 610 mm)	3/4"	(19.1 mm)
over 24"	(over 610 mm)	1"	(25.4 mm)

The deflection reference plane is that passing through two adjacent corners of the duct. For cross-broken ducts, the measurement point shall be on the centerline of the duct at a point 1/8 of the duct width or reinforcement interval or joint spacing if unreinforced, whichever is greater, from the intersection of the cross-breaks. For cross-broken ducts, the allowable wall deflection shall be increased by 50 percent of the deflection "set" measured at atmospheric pressure level in the duct.

(D) Deflection, Transverse. The maximum allowable deflection for rectangular duct transverse joint systems and intermediate reinforcements shall be 1/4 inch (6.4 mm) for duct widths of 100 inches

(2,540 mm) or less and shall be 0.3 percent of the span for greater widths. In any case, the allowable deflection shall not be construed as permitting such members to exceed 60 percent of their yield stress.

(E) Allowable Stress. Loads imposed by fabricated duct weight designed pressure differential across the duct wall, bending moment created by supports, and 1/2 pound per square foot (2.4 kg/m^2) of insulation weight shall not exceed 60 percent of the yield stress of the metals in the duct and its connections. Other internal and external loads are not accounted for by this standard.

(F) Cyclical Loading Resistance. Duct assemblies shall be capable of withstanding a daily on-off cycle of operation without structural failure.

(G) Air-Tightness. Closures for ducts shall be substantially air-tight.

A.6.204 Support Requirements for Ducts.

(A) Method and Interval. The support spacing interval and method of contact with the duct assembly shall be selected to avoid structural failure or closure failure in the duct.

(B) Allowable Stress in Supports. Supports shall not be subjected to more than 60 percent of their yield stress.

(C) Safety Factor. The upper attachment of a duct suspension (hanging) system to a building structure shall have a safety factor of at least four (4) based on ultimate failure (pullout) load.

(D) Loading. Supports shall account for loads identified in Section A. 6.203(A).

(E) Fastening Loads. The location and arrangement of all fastenings in the support system shall be selected to avoid failure in tensile, compressive, shear, and flexure modes.

(F) Corrosion Control. The support system shall not cause corrosion by galvanic (electrolytic) action at points of contact with the duct.

(G) Character of Support System. The support system shall not penetrate enclosures containing fluids or electrical wiring in such a manner as to create a hazard, nor shall they penetrate a load-bearing member in such a manner as to cause excessive loading on the member.

A.6.205 Complying Systems.

Tables 6-2-A through 6-2-Q typify duct construction conforming with this standard. Throughout the tables in this standard, the minimum thickness in inches for galvanized sheet gauge numbers is used.

TABLE 6-2-A
Rectangular Duct Hangers, Minimum Size

Maximum Half of Duct Perimeter	Pair at 10 ft. Spacing		Pair at 8 ft. Spacing		Pair at 5 ft. Spacing		Pair at 4 ft. Spacing	
	Strap	Wire/Rod	Strap	Wire/Rod	Strap	Wire/Rod	Strap	Wire/Rod
P=30" 2	1" x 0.030 (1" x 22 ga.)	10 ga. (0.135")	1" x 0.030 (1" x 22 ga.)	10 ga. (0.135")	1" x 0.030 (1" x 22 ga.)	12 ga. (0.106")	1" x 0.030 (1" x 22 ga.)	12 ga. (0.106")
P=72" 2	1" x 0.047 (1" x 18 ga.)	3/8"	1" x 0.036 (1" x 20 ga.)	1/4"	1" x 0.030 (1" x 22 ga.)	1/4"	1" x 0.030 (1" x 22 ga.)	1/4"
P=96" 2	1" x 0.058 (1" x 16 ga.)	3/8"	1" x 0.047 (1" x 18 ga.)	3/8"	1" x 0.036 (1" x 20 ga.)	3/8"	1" x 0.030 (1" x 22 ga.)	1/4"
P=120" 2	1-1/2" x 0.058 (1-1/2" x 16 ga.)	1/2"	1" x 0.058 (1" x 16 ga.)	3/8"	1" x 0.047 (1" x 18 ga.)	3/8"	1" x 0.036 (1" x 20 ga.)	1/4"
P=168" 2	1-1/2" x 0.058 (1-1/2" x 16 ga.)	1/2"	1" x 0.058 (1-1/2" x 16 ga.)	1/2"	1" x 0.058 (1" x 16 ga.)	3/8"	1" x 0.047 (1" x 18 ga.)	3/8"
P=192" 2	—	1/2"	1-1/2" x 0.058 (1-1/2" x 16 ga.)	1/2"	1" x 0.058 (1" x 16 ga.)	3/8"	1" x 0.058 (1" x 16 ga.)	3/8"
P=193" up 2			Special analysis required					

When straps are lap-joined, use these minimum fasteners:

1 x 0.047", 0.036", 0.030 - two No. 10 or one 1/4 bolt
1" x 0.058" – two 1/4" dia.
1-1/2" x 0.058" - two 3/8" dia.
Place fasteners in series, not side by side.

Single Hanger Maximum Allowable Load

Strap	Wire or Rod (Dia.)
1" x 0.030"–260 lbs.	0.106"–80 lbs.
1" x 0.036"–320 lbs.	0.135"–120 lbs.
1" x 0.047"–420 lbs.	0.162"–160 lbs.
1" x 0.058"–700 lbs.	1/4"–270 lbs.
1-1/2" x 0.058"–1,100 lbs.	3/8"–680 lbs.
	1/2"–1,250 lbs.
	5/8"–2,000 lbs.
	3/4"–3,000 lbs.

Notes:

1. Dimensions other than gauge are in inches.
2. Tables allow for duct weight 1 lb./sf (4.8 kg/m^2), insulation weight normal reinforcement, and trapeze weight, but no external loads.
3. Straps are galvanized steel; other materials are uncoated steel.
4. Allowable loads for P/2 assume that ducts are 0.058 (16 ga.) (1.47 mm) maximum, except that when maximum duct dimension (w) is over 60" (1,524 mm), then P/2 maximum is 1.25 w.
5. For trapeze sizes see Table 6-2-B.
6. Twelve [0.081 in. (2.05 mm)], 10 [0.102 in. (2.59 mm)], or 8 [0.128 in. (3.25 mm)] ga. wire is steel of black-annealed, bright basic, or galvanized type.

TABLE 6-2-B
Allowable Loads for Trapeze Angles (for Rectangular Ducts)

inches x 25.4 = mm
pounds x 0.454 = kg
feet x 304.8 = mm

Length	1 x 1 x 0.058" (16 ga.)	1 x 1 x 1/8"	1-1/2 x 1-1/2 x 0.058" (16 ga.)	1-1/2 x 1-1/2 x 1/8"	1-1/2 x 1-1/2 x 3/16"	1-1/2 x 1-1/2 x 1/4" or 2 x 2 x 1/8"	2 x 2 x 3/16"	2 x 2 x 1/4"	2-1/2 x 2-1/2 x 3/16"	2-1/2 x 2-1/2 x 1/4"
18"	80	150	180	350	510	650	940	1,230	1,500	1,960
24"	75	150	180	350	510	650	940	1,230	1,500	1,960
30"	70	150	180	350	510	650	940	1,230	1,500	1,960
36"	60	130	160	340	500	620	920	1,200	1,480	1,940
42"	40	110	140	320	480	610	900	1,190	1,470	1,930
48"	—	80	110	290	450	580	870	1,160	1,440	1,900
54"	—	40	70	250	400	540	840	1,120	1,400	1,860
60"	—	—	—	190	350	490	780	1,060	1,340	1,800
66"	—	—	—	100	270	400	700	980	1,260	1,720
72"	—	—	—	—	190	320	620	900	1,180	1,640
78"	—	—	—	—	80	210	500	790	1,070	1,530
84"	—	—	—	—	—	80	380	660	940	1,400
96"	—	—	—	—	—	—	—	320	600	1,060
108"	—	—	—	—	—	—	—	—	150	610

Note: Loads above assume that a hanger rod is 6" max. distance from duct side.

TABLE 6-2-C
Round Duct Hangers

Duct Diameter	Maximum Spacing	Wire Diameter	Rod	Strap
10" down	12'	One 12 ga.	1/4"	1" x 0.030 (22 ga.)
11 to 18"	12'	Two 12 ga. or one 8 ga.	1/4"	1" x 0.030 (22 ga.)
19 to 24"	12'	Two 10 ga.	1/4"	1" x 0.030 (22 ga.)
25 to 36"	12'	Two 8 ga.	3/8"	1" x 0.036 (20 ga.)
37 to 50"	12'	fi	Two 3/8"	Two 1" x 0.036 (20 ga.)
51 to 80"	12'	fi	Two 3/8"	Two 1" x 0.047 (18 ga.)
81 to 94"	12'	fi	Two 3/8"	Two 1" x 0.058 (16 ga.)

Table 6-2-D
Flat Oval Duct Construction

Duct Width	Inches (Gauge)		
	Spiral Lock Seam Duct	Longitudinal Seam Duct	Fittings
To 24"	0.024 (24)	0.036 (20)	0.036 (20)
25" to 36"	0.030 (22)	0.036 (20)	0.036 (20)
37" to 48"	0.030 (22)	0.047 (18)	0.047 (18)
49" to 60"	0.036 (20)	0.047 (18)	0.047 (18)
61" to 70"	0.036 (20)	0.058 (16)	0.058 (16)
71" and up	0.047 (18)	0.058 (16)	0.058 (16)

TABLE 6-2-E
Round Duct Gauge Selection for Galvanized Steel

	Maximum 2" W.G. (498 Pa) Static Positive		Maximum 10" W.G. (2.5 kPa) Static Positive		Maximum 2" W.G. (498 Pa) Static Negative	
Duct Diameter, Inches	Spiral Seam, Inches (Gauge)	Longitudinal Seam, Inches (Gauge)	Spiral Seam, Inches (Gauge)	Longitudinal Seam, Inches (Gauge)	Spiral Seam, Inches (Gauge)	Longitudinal Seam, Inches (Gauge)
3 thru 8	0.016 (28)	0.016 (28)	0.019 (26)	0.024 (24)	0.016 (28)	0.024 (24)
9 thru 14	0.016 (28)	0.019 (26)	0.019 (26)	0.024 (24)	0.019 (26)	0.024 (24)
15 thru 26	0.019 (26)	0.024 (24)	0.024 (24)	0.030 (22)	0.024 (24)	0.030 (22)
27 thru 36	0.024 (24)	0.030 (22)	0.030 (22)	0.036 (20)	0.030 (22)	0.036 (20)
37 thru 50	0.030 (22)	0.036 (20)	0.036 (20)	0.036 (20)	0.036 (20)	0.047 (18)
51 thru 60	0.036 (20)	0.047 (18)	0.047 (18)	0.047 (18)	0.047 (18)	0.058 (16)
61 thru 84	0.047 (18)	0.058 (16)	0.047 (18)	0.058 (16)	0.058 (16)	0.070 (14)

SPIRAL AND LONGITUDINAL SEAMS

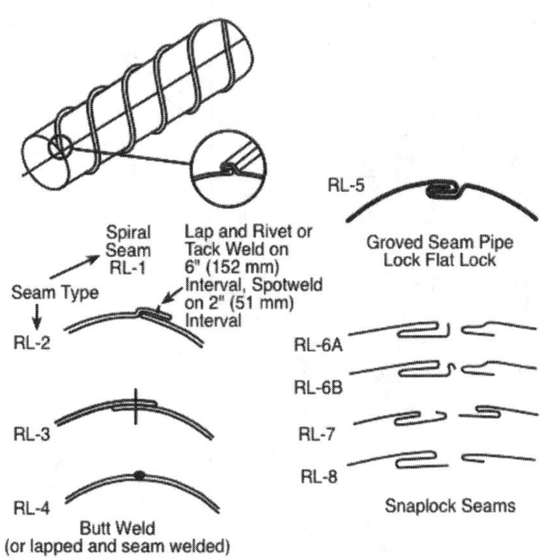

Pressure Class	Seam Type Permitted
Positive	
To +10" w.g. (2488 Pa)	RL-1, 4, 5 (2*, 3*) (51 mm, 76 mm)
To + 3" w.g. (746 Pa)	RL-1, 2, 3, 4, 5
To + 2" w.g. (498 Pa)	All
Negative	
To - 2" w.g. (498 Pa)	RL-1, 2, 3, 4, 5
To - 1" w.g. (249 Pa)	All

*Acceptable if spotwelded on 1" (25.4 mm) intervals.

TRANSVERSE JOINTS

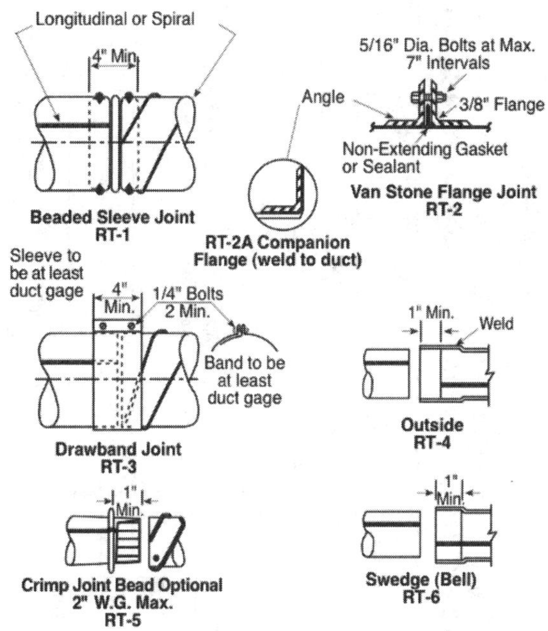

On joints RT-1, 2, 3, 4 and 6 screws must be used at uniform intervals 15" maximum along the circumference; three screws minimum on 14" or less diameter.

Minimum flange sizes for RT-2 and 2A:
1" x 1" x 10 GA on 14" maximum diameter
1-1/2" x 1-1/2" x 1/8" diameter over 14"

Longitudinal or spiral seam duct is acceptable for all joints except RT-4 and 6 (for longitudinal only)

(Inches x 25.4 = mm)

TABLE 6-2-G Rectangular Duct Transverse Reinforcement for Duct Material Thicknesses from 0.030" (22 ga.) (0.76 mm) to 0.019" (26 ga.) (0.48 mm)

water gauge x 248.8 = Pa
inches x 25.4 = mm

Pressure Class (Water Gauge) →	1/2"	1/2"	1"	1"	2"	2"	1/2"	1/2"	1"	1"	2"	2"	1/2"	1/2"	1"	1"	2"	2"
Duct Thickness Inches (Gauge) →	0.030" (22)	0.030" (22)	0.030" (22)	0.030" (22)	0.030" (22)	0.030" (22)	0.024" (24)	0.024" (24)	0.024" (24)	0.024" (24)	0.024" (24)	0.024" (24)	0.019" (26)	0.019" (26)	0.019" (26)	0.019" (26)	0.019" (26)	0.019" (26)
Duct Dimensions ↓																		
7" dn.																		
8-10"																		
11, 12"																	A/5D	A/8D
13, 14"											A/5D	A/8D			A/10D	A/8D	A/5D	A/4D
15, 16"					A/10D	A/8D			A/10D	A/8D	A/5D	A/8D			A/5D	A/8D	A/5D	A/4D
17, 18"					A/10D	A/8D			A/10D	A/8D	A/5D	A/8D			A/5D	A/8D	A/5D	A/4D
19, 20"			A/10D	A/8D	A/5D	B/8			A/10D	A/8D	A/5D	A/4D	A/10D	A/8D	A/5D	A/8D	A/5D	A/4D
21, 22"			A/10	A/8	A/5	B/8	A/10	A/8	A/10	A/8	A/5	A/4	A/10	A/8	A/5	A/4	A/5	A/4
23, 24"			B/10	B/8	B/5	C/8	A/10	A/8	B/10	B/8	B/5	B/4	A/10	A/8	A/5	A/4	B/5	B/4
25, 26"	A/10	A/8	B/10	B/8	B/5	C/8	A/10	A/8	A/5	B/8	B/5	B/4	A/10	A/8	A/5	A/4	B/5	B/4
27, 28"	B/10	B/8	C/10	C/8	C/5	B/4	B/10	B/8	B/5	C/8	C/5	B/4	B/5	B/8	B/5	B/4	B/2-1/2	B/4
29, 30"	B/10	B/8	C/10	C/8	C/5	C/4	B/10	B/8	B/5	C/8	C/5	C/4	B/5	B/8	B/5	B/4	C/2-1/2	C/4
31-36"	C/10	C/8	C/5	D/8	D/5	D/4	C/5	C/8	C/5	C/4	D/2-1/2	D/4	C/5	C/4	C/5	C/4		
37-42"	D/5	D/8	D/5	D/4	E/5	E/4	D/5	D/8	D/5	D/4	E/2-1/2	E/4	D/5	C/4	D/2-1/2	D/4		
43-48"	D/5	D/8	E/5	D/4	E/2-1/2	E/4	D/5	D/4	E/5	D/4	E/2-1/2	E/2	D/5	D/4	D/2-1/2	D/4		
49-54"	D/5	D/4	E/5	E/4	F/2-1/2	F/2	D/5	D/4	E/2-1/2	E/4	F/2-1/2	F/2	D/5	D/4				
55-60"	E/5	E/4	F/5	F/4	G/2-1/2	G/2	E/5	E/4	F/2-1/2	F/4			E/2-1/2	E/4				
61-72"	F/5	F/4	G/2-1/2	G/4	H/2-1/2	H/2	F/2-1/2	F/4										
73-84"	H/5	G/4					G/2-1/2	G/4										
85-96"	H/2-1/2	H/4																
97" up																		

Standing reinforcement is not required on duct sizes above this heavy line. Flat slips and drives or another choice may be used. Flat slips and drives must not be less than one gauge below duct gauge and not less than 24 gauge.

← Not Required →

Not Allowed

This number is the spacing interval between reinforcements, in feet.

This letter denotes the type of reinforcement to be selected from joint Tables 6-2-I, 6-2-J, or 6-2-K.

Note: Flat slips and drives may be backed with stiffeners from Table No. 6-2-J and thereby qualified for this table. Where the superscript D occurs, the flat drive is accepted as Class A. See Table No. 6-2-K.

TABLE 6-2-H Rectangular Duct Transverse Reinforcement for Duct Material Thicknesses from 0.058" (16 ga.) (1.47 mm) to 0.036" (20 ga.) (0.91 mm)

water gauge x 248.8 = Pa
inches x 25.4 = mm

Pressure Class (Water Gauge) →	1/2"	1/2"	1"	1"	2"	2"	1/2"	1/2"	1"	1"	2"	2"	1/2"	1/2"	1"	1"	2"	2"
Duct Thickness Inches (Gauge) →	0.058" (16)	0.058" (16)	0.058" (16)	0.058" (16)	0.058" (16)	0.058" (16)	0.047" (18)	0.047" (18)	0.047" (18)	0.047 (18)	0.047" (18)	0.047" (18)	0.036" (20)	0.036" (20)	0.036" (20)	0.036" (20)	0.036" (20)	0.036" (20)
Duct Dimensions																		
7" dn.																		
8-10"																		
11, 12"																		
13, 14"																		
15, 16"																		
17, 18"																		
19, 20"																	B/10	B/8
21, 22"											B/10	B/8			A/10	A/8	B/10	B/8
23, 24"											C/10	C/8			B/10	B/8	B/10	C/8
25, 26"					C/10	C/8					C/10	C/8			B/10	B/8	C/10	C/8
27, 28"					C/10	C/8			C/10	C/8	C/10	C/8	B/10	B/8	C/10	C/8	C/5	C/8
29, 30"					D/10	D/8			C/10	C/8	D/10	D/8	B/10	B/8	C/10	C/8	C/5	D/8
31-36"			D/10	D/8	E/10	E/8	C/10	C/8	D/10	D/8	D/5	E/8	C/10	C/8	D/10	D/8	D/5	D/4
37-42"	D/10	D/8	E/10	E/8	E/5	E/8	D/10	D/8	E/10	D/8	E/5	E/4	D/10	D/8	D/5	D/8	E/5	E/4
43-48"	E/10	D/8	F/10	F/8	F/5	G/8	E/10	D/8	E/5	E/8	F/5	E/4	E/10	D/8	E/5	D/4	F/5	E/4
49-54"	E/10	E/8	G/10	F/8	G/5	F/4	E/10	E/8	E/5	F/8	G/5	F/4	D/5	E/8	E/5	E/4	F/2-1/2	F/4
55-60"	F/10	F/8	F/5	G/8	H/5	G/4	F/10	F/8	F/5	G/8	H/5	G/4	E/5	F/8	F/5	F/4	G/2-1/2	G/4
61-72"	H/10	G/8	H/5	G/4	I/5	H/4	F/5	G/8	H/5	G/4	H/2-1/2	H/4	F/5	F/4	G/2-1/2	G/4	H/2-1/2	H/2
73-84"	H/5	H/8	I/5	H/4	I/2-1/2	J/4	H/5	G/4	I/5	H/4	I/2-1/2	J/4	H/5	G/4	H/2-1/2	H/4	I/2-1/2	I/2
85-96"	H/5	I/8	J/5	I/4	J/2-1/2	K/4	H/5	H/4	I/2-1/2	I/4	J/2-1/2	J/2	H/5	H/4	I/2-1/2	I/2	J/2-1/2	J/2
97" up	H/5	H/4	J/2-1/2	J/2	K/2-1/2	K/2	H/5	H/4	J/2-1/2	J/2	K/2-1/2	K/2						

Not Required (0.036" (20) columns, above the heavy line)

Standing reinforcement is not required on duct sizes above this heavy line. Flat slips and drives or another choice may be used. Flat slips and drives must not be less than one gauge below duct gauge and not less than 24 gauge.

Note: Flat slips and drives may be backed with stiffeners from Table No. 6-2-J and thereby qualified for this table.

This number is the spacing interval between reinforcements, in feet.

This letter denotes the type of reinforcement to be selected from joint Tables Nos. 6-2-I, 6-2-J or 6-2-K.

TABLE 6-2-1
Transverse Joint Reinforcement for Tables 6-2-G and 6-2-H

Minimum Rigidity Class		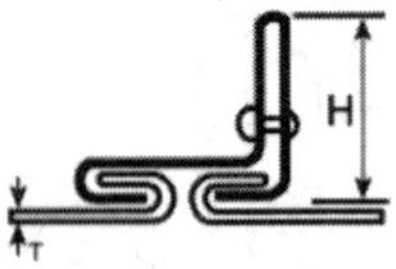 T-2 STANDING DRIVE SLIP		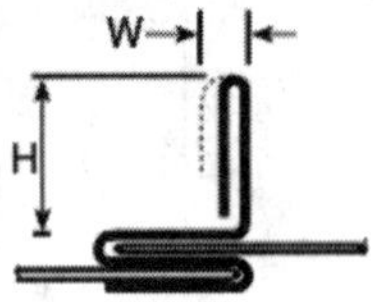 T-10 STANDING S		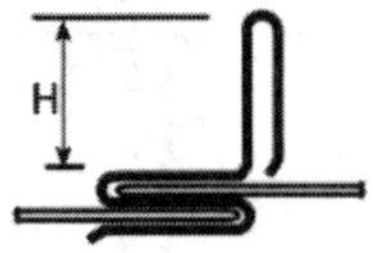 T-11 STANDING S		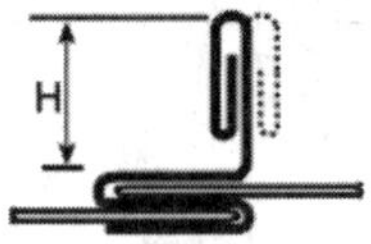 T-12 STANDING S		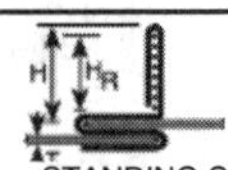 STANDING S (BAR REINFORCED) T-13 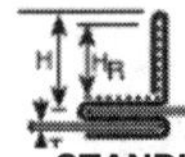 STANDING S (ANGLE REINFORCED) T-14	
	EI*	H x T (x 25.4 for mm)	WT/LF (x 1.488 for kg/m)	H x T (x 25.4 for mm)	WT/LF (x 1.488 for kg/m)	H x T (x 25.4 for mm)	WT/LF (x 1.488 for kg/m)	H x T (x 25.4 for mm)	WT/LF (x 1.488 for kg/m)	H x T + h (x 25.4 for mm)	WT/LF (x 1.488 for kg/m)
A	0.5	↑		↑		1/2 x 0.019" (1/2 x 26 ga.)	0.5	↑		↑	
B	1.0	1-1/8 x 0.019" (1-1/8 x 26 ga.)	0.5	Not given		1/2 x 0.030" (1/2 x 22 ga.)	0.7				
C	2.5	1-1/8 x 0.030" (1-1/8 x 22 ga.)	0.8	1 x 0.019" (1 x 26 ga.)	0.6 0.6	1 x 0.019" (1 x 26 ga.)	0.6 0.6	Not given			
D	0.5	Not given		1 x 0.024" (1 x 24 ga.)	0.7	1 x 0.024" (1 x 24 ga.)	0.7	1-1/8 x 0.019" (1-1/8 x 26 ga.)	0.7		
E	10			1-1/8 x 0.036" (1-1/8 x 20 ga.) w = 3/16"	0.9	Not given		1-1/8 x 0.047" (1-1/8 x 18 ga.)	1.4	Not given	
F	15			1-5/8 x 0.030" (1-5/8 x 22 ga.) w = 3/16"	1.0			1-1/2 x 0.024" (1-1/2 x 24 ga.) 1-1/2 x 1/8 bar	1.0	1-1/2 x 0.024" (1-1/2 x 24 ga.) 1-1/2 x 1/8 bar	1.5
G	25			1-5/8 x 0.047" (1-5/8 x 18 ga.) w = 3/16"	1.5			1-1/2 x 0.047" (1-1/2 x 18 ga.)	1.7	1-1/2 x 0.030" (1-1/2 x 22 ga.) 1-1/2 x 1/8 bar	1.6
H	50			Not given				Not given		1-1/2 x 0.036" (1-1/2 x 20 ga.) 1-1/2 x 1-1/2 x 3/16	2.9
I	75									2 x 0.036" (2 x 20 ga.) 2 x 2 x 1/8	2.9
J	100									2 x 0.036" (2 x 20 ga.) 2 x 2 x 3/16	3.7
K	150									Not given	
L	200	↓		↓		↓		↓		↓	

Note:

* Nominal EI is number listed times 10^5.

TABLE 6-2-I *continued*

Minimum Rigidity Class		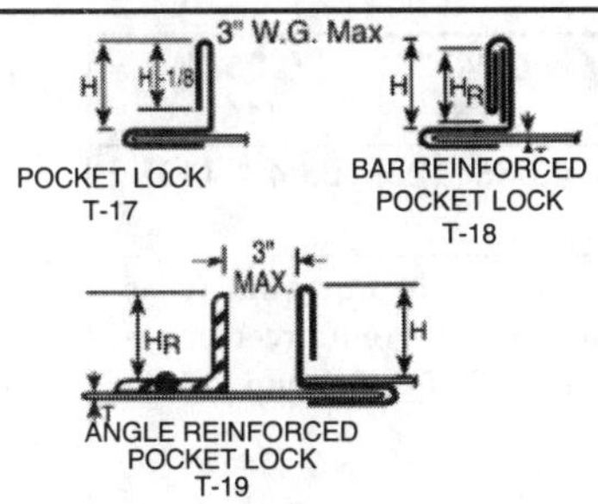			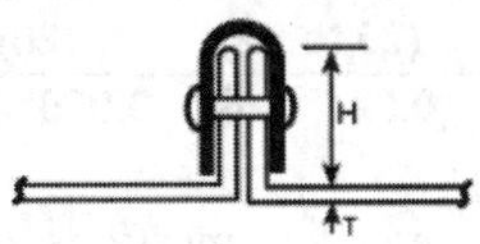 T-10 CAPPED FLANGE			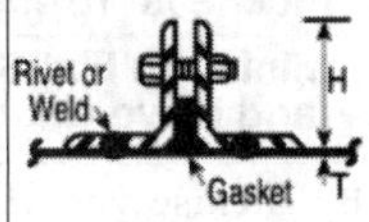 T-22 COMPANION ANGLES		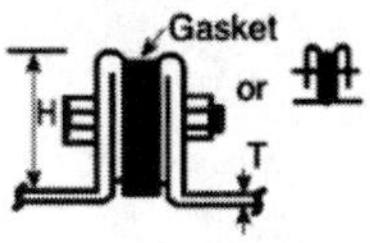 T-23 FLANGED	
	EI*	H	LOCK T, DUCT T, h (x 25.4 for mm)	WT LF (x 1.488 for kg/m)	H x T (x 25.4 for mm)	U	WT LF (x 1.488 for kg/m)	H x T (x 25.4 for mm)	WT LF (x 1.488 for kg/m)	H x T (x 25.4 for mm)	WT LF (x 1.488 for kg/m)
A	0.5		↑					↑		↑	
B	1.0				3/4 x 0.019" (26 ga.)	0.024" (24 ga.)	0.4				
C	2.5	T-17 1"	0.019" (26 ga.) lock on 0.019" (26 ga.)	0.6	1 x 0.024" (24 ga.)	0.024" (24 ga.)	0.5			1 x 0.024" (24 ga.)	1.0
D	5.0	T-17 1"	0.024" (24 ga.) lock on 0.024" (24 ga.)	0.7	1 x 0.030" (22 ga.)	0.030" (22 ga.)	0.6			1 x 0.030" (22 ga.)	1.0
E	10	T-18 1"	0.030" (22 ga.) lock 1 x 1/8 bar	1.4	1-1/2 x 0.024" (24 ga.)	0.030" (22 ga.)	0.8	two 1 x 1/8	1.7	1 x 0.058" (16 ga.) 1-1/2 x 0.024" (24 ga.)	1.0
F	15	T-17 1-1/2"	0.030" (22 ga.) lock	1.0	1-1/2 x 0.036" (20 ga.)	0.036" (20 ga.)	1.0	↑		1-1/2 x 0.030" (22 ga.) 1-1/2 x 0.036" (20 ga.)	1.0 1.0
G	25	T-18 1-1/2"	0.030" (22 ga.) lock 1-1/2 x 1/8 bar	1.6	↑	↑		two 1-1/4 x 1/8	2.1	1-1/2 x 0.047" (18 ga.)	1.5
H	50	T-19 1-1/2"	0.036" (20 ga.) lock 1-1/2 x 3/16 angle	2.9	2 x 0.058" (16 ga.)	0.036" (20 ga.)	1.5	two 1-1/2 x 1/8	2.6	2 x 0.047" (18 ga.)	1.5
I	75	T-10 1-1/2"	0.036" (20 ga.) lock 2 x 1/8 angle	2.8	2 x 0.058" (16 ga.) *			two* 1-1/2 x 3/16	3.7	2 x 0.058" (16 ga.)	2.0
J	100	T-19 1-1/2"	0.036" (20 ga.) lock 2 x 3/16 angle	3.5				two* 1-1/2 x 1/4	4.7	2 x 0.047" (18 ga.)	
K	150	T-19 1-1/2"	0.036" (20 ga.) lock 2-1/2 x 3/16 angle	4.2				two* 2 x 3/16	4.9	With tie rods	
L	200		Not given		↓			two* 2 x 1/4	6.5	↓	

Note:

* Nominal EI is number listed times 10^5.

TABLE 6-2-I *continued*

Duct Thickness	Inches (Gauge)	0.019" to 0.030" (26 to 22)	0.036" (20)	0.047" (18)	0.058" (16)
Minimum Flat Slip and Drive Gauge		0.024" (24)	0.030" (22)	0.036" (20)	0.047" (18)

By themselves, flat-type slip connections and lap joints do not qualify as reinforcement. When backed by a stiffener from Table 6-2-J, the reinforcement is graded by the stiffener. See exception for joint T-1 in Table 6-2-K. T-5, T-6, and T-7 are restricted to 2" W.G. maximum service pressure.

Drive Slip
T-1

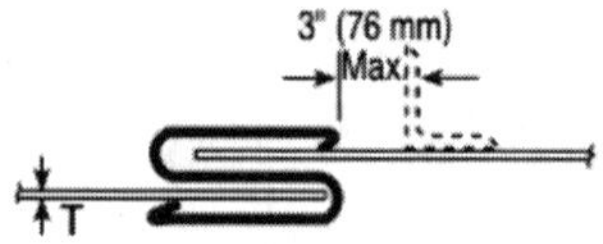

Hemmed "S" Slip
T-6

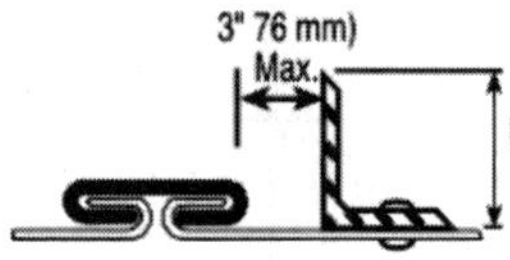

Drive Slip-Reinforced
T-3

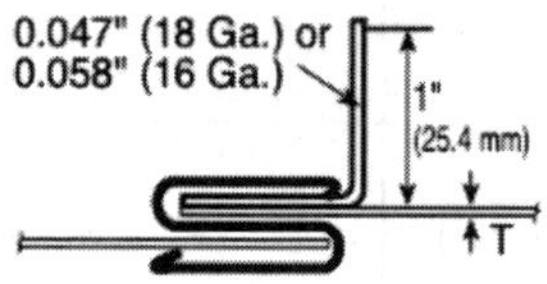

Reinforced "S" Slip
T-7

Lap
T-4

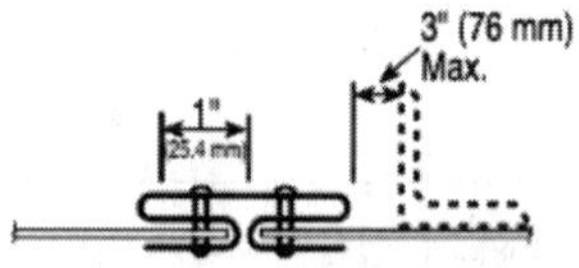

Double "S" Slip
T-8

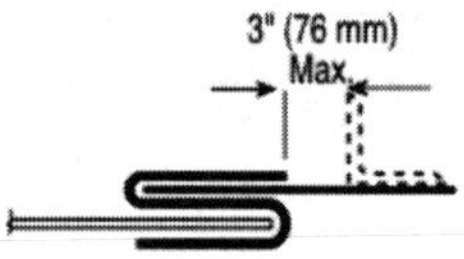

Plain "S" Slip
T-5

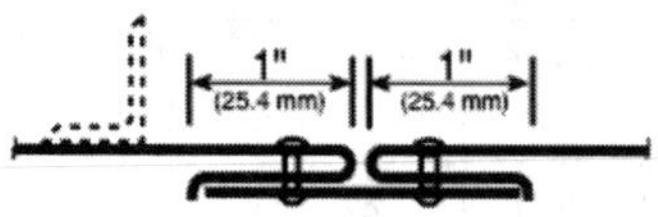

Inside Slip Joint
T-9

TABLE 6-2-J
Intermediate Reinforcement

Minimum Rigidity Class		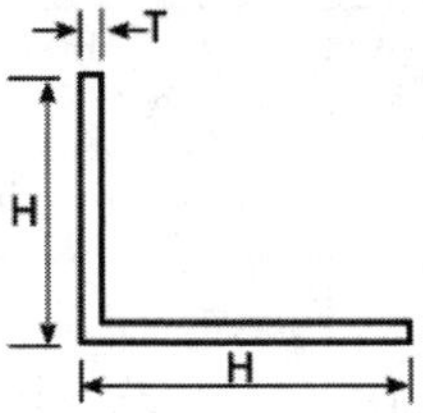 ANGLE		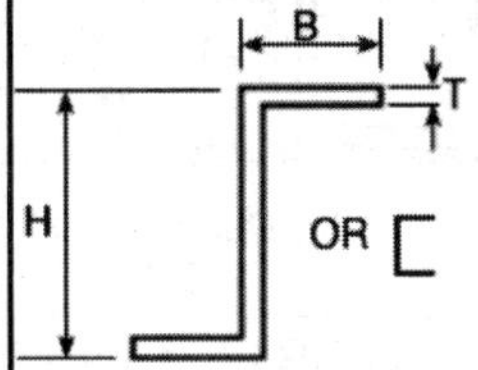 ZEE		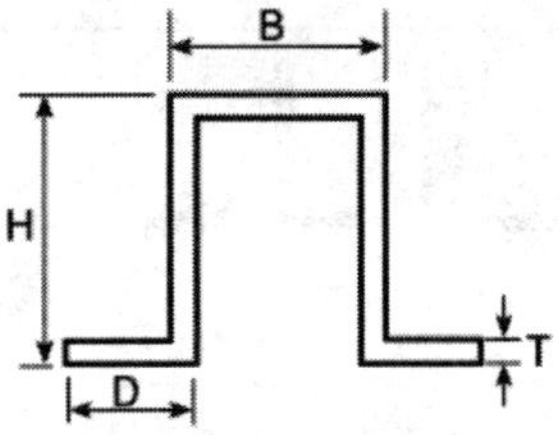 HAT SECTION		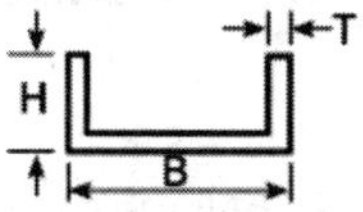 CHANNEL	
	E1*	H x T (MIN) (x 25.4 for mm)	WT/LF (x 1.488 for kg/m)	H x B x T (MIN) (x 25.4 for mm)	WT/LF (x 1.488 for kg/m)	H x B x D x T (x 25.4 for mm)	WT/LF (x 1.488 for kg/m)	H x B x D x T (MIN) (x 25.4 for mm)	WT/LF (x 1.488 for kg/m)
A	0.5	↑	↑	↑		↑		↑	
B	1.0			3/4 x 1/2 x 0.036" (20 ga.)	0.28				
C	2.5	1 x 0.047" (18 ga.) 1 x 0.058" (16 ga.) 3/4 x 1/8	0.34 0.44 0.59	3/4 x 1/2 x 0.047 (18 ga.)	0.36			3/4 x 3 x 0.047" (18 ga.)	0.81
D	5.0	1-1/4 x 0.047" (18 ga.) 1 x 1/8	 0.45 0.80	1 x 3/4 x 0.036" (20 ga.)	0.35			1-1/8 x 3-1/4 x 0.047" (18 ga.) 1 x 3 x 0.047" (18 ga.)	1.0
E	10	1-1/4 x 0.090 1-1/2 x 0.058" (16 ga.)	0.80 0.66	1 x 3/4 x 0.090 1-1/2 x 3/4 x 0.036" (20 ga.)	0.78 0.41		0.7	1 x 2 x 1/8	1.6
F	15	1-1/2 x 1/8**	1.23	1 x 3/4 x 1/8 1-1/2 x 3/4 x 0.047" (18 ga.)	1.03 0.54	1-1/2 x 3/4 x 5/8 x 0.036" (20 ga.) 1-1/2 x 1-1/2 x 3/4 x 0.036" (20 ga.)	0.7 1.1	1-1/4 x 3-1/2 x 0.058" (16 ga.) 1-1/8 x 3 x 1/8	1.32 2.1
G	25	1-1/2 x 3/16	0.96	1-1/2 x 3/4 x 1/8 2 x 1-1/8 x 0.036" (20 ga.)	1.23 0.6	1-1/2 x 3/4 x 5/8 x 0.058" (16 ga.) 1-1/2 x 1-1/2 x 3/4 x 0.047" (18 ga.) 2 x 1 x 3/4 x 0.036" (20 ga.)	0.82 1.1 0.90		
H	50	2 x 1/8	1.65	2 x 1-1/8 x 0.058" (16 ga.)	0.94	1-1/2 x 3/4 x 5/8 x 1/8 1-1/2 x 1-1/2 x 3/4 x 0.090 2 x 1 x 3/4 x 0.036" (20 ga.)	2.1 1.9 1.2	1.4 x 3	4.1
I	75	2 x 3/16	2.44	2 x 1-1/8 x .090	1.33	2 x 1 x 3/4 x .090 2-1/2 x 2 x 3/4 x 0.058" (16 ga.)	2.03 1.88	2 x 2 x 1/8 1.5 x 3	2.4 5.0
J	100	2 x 1/4 2-1/2 x 1/8	3.2 2.1	2 x 1-1/8 x 1/8 3 x 1-1/8 x 0.058" (16 ga.)	 1.74 1.2	2 x 1 x 3/4 x 1/8 2-1/2 x 2 x 3/4 x 0.090	2.63 2.67	1.6 x 4	5.4
K	150	2-1/2 x 3/16	3.1	3 x 1-1/8 x 0.090	1.64	2-1/2 x 2 x 3/4 x 1/8 3 x 1-1/2 x 3/4 x 0.058" (16 ga.)	3.57 2.0	Not given	
L	200	2-1/2 x 1/4	4.1	3 x 1-1/8 x 1/8	2.15	3 x 1-1/2 x 3/4 x 0.090	2.82		

Notes:
*Nominal El is number listed times 10^5.
** See tie rod options.

TABLE 6-2-K
Transverse Joint Replacement for Tables Nos. 6-2-G, 2-H, 2-L, 2-N, and 2-O

Minimum Rigidity Class		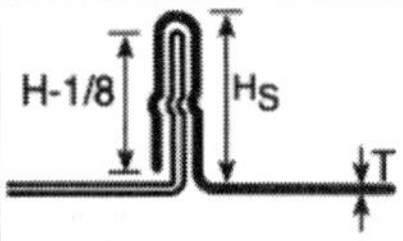 T-15 STANDING SEAM		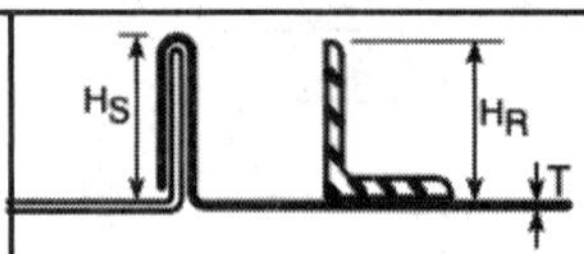 T-16 STANDING SEAM OR WELDED FLANGE REINFORCED 26 TO 22 GA. DUCT			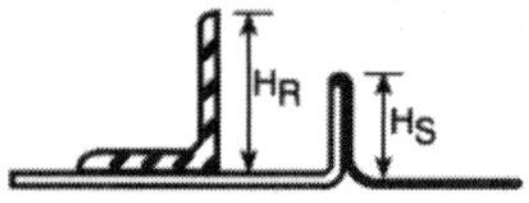 T-21a STANDING SEAM OR WELDED FLANGE REINFORCED 20 TO 16 GA. DUCT			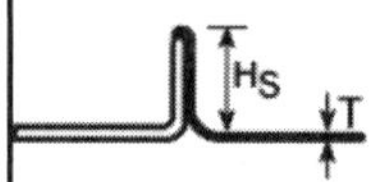 T-21 WELDED FLANGE	
	EI*	H_S x T (x 25.4 for mm)	WT/LF (x 1.488 for kg/m)	H_S	H x H x T (x 25.4 for mm)	WT/LF (x 1.488 for kg/m)	H_S	H x H x T (x 25.4 for mm)	WT/LF (x 1.488 for kg/m)	H_S x T (x 25.4 for mm)	WT/LF (x 1.488 for kg/m)
A	0.5	1/2 x 0.024" (24 ga.)	0.2		↑			↑		1/2 x 0.030" (22 ga.)	0.1
B	1.0	3/4 x 0.024" (24 ga.)	0.3							1/2 x 0.058" (16 ga.) 3/4 x 0.030" (22 ga.)	0.2 0.2
C	2.5	1 x 0.024" (24 ga.)	0.5							3/4 x 0.047" (18 ga.) 1 x 0.030 (22 ga.)	0.3
D	5.0	3/4 x 0.058" (16 ga.) 1 x 0.036" (20 ga.)	0.3 0.5	1"	1 x 1 x 0.058" (16 ga.)	1.0				1 x 0.047" (18 ga.) 1-1/4 x 0.030" (22 ga.)	0.4 0.3
E	10	1 x 0.058" (16 ga.) 1-1/2 x 0.024" (24 ga.)	0.7 0.7	1"	1 x 1 x 1/8	1.4	1"	1 x 1 x 0.058" (16 ga.)	1.0	1-1/4 x 0.047" (18 ga.) 1-1/2 x 0.030" (22 ga.)	0.5 0.4
F	15	1-1/2 x 0.036" (20 ga.)	0.7	1-1/2"	1-1/2 x 1-1/2 x 0.058" (16 ga.)	1.8	1-1/4"	1-1/4 x 1-1/4 x 0.058" (16 ga.)	1.7	1-1/4 x 0.058" (16 ga.) 1-1/2 x 0.036" (20 ga.)	0.6 0.4
G	25	1-1/2 x 0.047" (18 ga.)	0.8	1-1/2" 1-1/2"	1-1/2 x 1-1/2 x 1/8 2 x 2 x 0.058" (16 ga.)	2.0	1-1/2"	1-1/2 x 1-1/2 x 1/8	2.4	1-1/2 x 0.058" (16 ga.)	0.7
H	50	See T-16 and tie rod options		1-1/2"	2 x 2 x 1/8*	2.7	1-1/2" 1-1/2"	1-1/2 x 1-1/2 x 3/16 2 x 2 x 0.058" (16 ga.)	2.8 2.0	See T-21a and tie rod options Table No. 10-2-Q	
I	75				↑		1-1/2"	2 x 2 x 1/8*	2.7		
J	100			1-1/2"	2 x 2 x 3/16*	3.5	1-1/2"	2 x 2 x 3/16*	3.5		
K	150			1-1/2"	2-1/2 x 2-1/2 x 3/16*	4.1		↑			
L	200	↓		1-1/2"	2-1/2 x 2-1/2 x 1/4	5.3	1-1/2"	2-1/2 x 2-1/2 x 3/16*	4.1	↓	

Note:
*Nominal El is number listed times 10^5.

TABLE 6-2-K *continued*

Static Pressure (Pa)		Maximum Allowable Duct Width Dimension Inches (mm)		Maximum Reinforcement Spacing Interval Feet (mm)	
1/2" W.G.	(124)	20	(508)	10	(3,048)
1" W.G.	(249)	20	(508)	8	(2,438)
		14	(356)	10	(3,048)
2" W.G.	(498)	20	(508)	5	(1,524)
		18	(457)	8	(2,438)
3" W.G.	(764)	18	(457)	8	(2,438)
4" W.G.	(995)	16	(406)	5	(1,524)
		12	(305)	8	(2,438)
6" W.G.	(1493)	12	(305)	5	(1,524)

Note:
Where Table Nos. 6-2-G, H, L, M and N require Class A reinforcement, joint T-1 is acceptable as Class A within these parameters and without supplemental reinforcement from Table No. 6-2-J.

TABLE NO. 6-2-L Rectangular Duct Reinforcement

3" W.G. (746 Pa) Static Pos. or Neg. (x 25.4 for mm) Duct Dimension	0.058" (16 gauge) Minimum Class	0.047" (18 gauge) Maximum Spacing	0.036" (20 gauge) (x 304.8 for mm)	0.030" (22 gauge)	0.024" (24 gauge)
10" down	Not required	Not required	Not required	Not required	Not required
11" to 12"			↓	↓	A @ 8'
13" to 14"				A @ 8'	A @ 5'
15" to 16"		↓	A @ 8'	A @ 8'	A @ 5'
17" to 18"			A @ 8'	A @ 8'	A @ 5'
19" to 20"		B @ 10'	B @ 8'	A @ 5'	A @ 5'
21" to 22"		C @ 10'	B @ 8'	B @ 5'	B @ 5'
23" to 24"	↓	C @ 10'	B @ 5'	B @ 5'	B @ 5'
25" to 26"	D @ 10'	D @ 10'	C @ 5'	C @ 5'	C @ 5'
27" to 28"	D @ 10'	D @ 10'	C @ 5'	C @ 5'	C @ 4'
29" to 30"	D @ 10'	D @ 8'	C @ 5'	C @ 5'	C @ 4'
31" to 36"	E @ 8'	E @ 5'	E @ 5'	D @ 4'	D @ 4'
37" to 42"	E @ 5'	E @ 5'	E @ 5'	E @4'	E @ 3'
43" to 48"	G @ 5'	G @ 5'	F @ 4'	E @ 3'	E @ 2-1/2'
49" to 54"	H @ 5'	H @ 5'	G @ 3'	G @ 3'	E @ 2-1/2'
55" to 60"	H @ 5'	H @ 4'	G @ 3'	G @ 2-1/2'	G @ 2-1/2'
61" to 72"	I @ 4'	H @ 3'	H @ 4'	H @ 2-1/2'	F @ 2'*
73" to 84"	J @ 3'	J @ 3'	J @ 2-1/2'	I @ 2'	
85" to 96"	L @ 3'	K @ 2-1/2'	J @ 2'	G @2-1/2'*	
Over 96"		H @ 2-1/2'*			

Notes:
1. Reinforcement class refers to transverse joint or intermediate members having stiffness and other characteristics indicated in tables.
2. Joint spacing is unrestricted.
3. Reinforcement is evaluated for each duct side separately. When required on four sides, corners shall be tied. When required on two sides, corners shall be tied with rods or angles at the ends on 4 inch (995 Pa), 6 inch (1493 Pa), and 10 inch (2488 Pa) pressure classes.
4. The same gauge of metal shall be used on all four sides of a cross-section.

TABLE 6-2-O
Rectangular Duct Reinforcement

10" W.G. (2488 Pa) Static Positive (x 25.4 for mm) Duct Dimension	0.058" (16 gauge)	0.047" (18 gauge) Minimum Class	0.036" (20 gauge) (x 304.8 for mm)	0.030" (22 gauge) Maximum Spacing	0.024" (24 gauge)
8" down	Not required	Not required	Not required	Not required	Not required
9" to 10"	↓	↓	A @ 5	A @ 5'	A @ 4'
11" to 12"		A @ 8'	A @ 5'	A @ 5'	A @ 4'
13" to 14"	B @ 8'	B @ 8'	A @ 5'	A @ 4'	A @ 3'
15" to 16"	B @ 8'	B @ 5'	B @ 5'	B @ 4'	A @ 3'
17" to 18"	C @ 8'	C @ 5'	C @ 5'	B @ 4'	B @ 3'
19" to 20"	D @ 8'	C @ 5'	C @ 4'	B @ 3'	B @ 3'
21" to 22"	C @ 5'	C @ 5'	C @ 4'	C @ 3'	C @ 3'
23" to 24"	D @ 5'	D @ 5'	D @ 4'	C @ 3'	C @ 3'
25" to 26"	D @ 5'	D @ 5'	D @ 4'	D @ 3'	C @ 2-1/2'
27" to 28"	E @ 5'	E @ 5'	D @ 4'	D @ 3'	D @ 2-1/2'
29" to 30"	E @ 5'	E @ 4'	D @ 3'	D @ 3'	D @ 2-1/2'
31" to 36"	F @ 5'	F @ 4'	F @ 3'	E @ 2-1/2'	E @ 2'
37" to 42"	H @ 4'	G @ 3'	G @ 2-1/2'	F @ 2'	
43" to 48"	H @ 3'	H @ 2-1/2'	G @ 2'	G @ 2'	
49" to 54"	I @ 3'	H @ 2-1/2"	H @ 2'		
55" to 60"	J @ 3'	I @ 2-1/2'	I @ 2'		
61" to 72"	K @ 2-1/2'	K @ 2			
73" to 84"	H @ 2'				
85" to 96"	H @ 2'*				
Over 96"	H @ 2'*				

Notes:
1. Reinforcement class refers to transverse joint or intermediate members having stiffness and other characteristics indicated in tables.
2. Joint spacing is unrestricted.
3. Reinforcement is evaluated for each duct side separately. When required on four sides, corners shall be tied. When required on two sides, corners shall be tied with rods or angles at the ends on 4 inch (995 Pa), 6 inch (1493 Pa), and 10 inch (2488 Pa) pressure classes.
4. The same gauge of metal shall be used on all four sides of a cross-section.
5. *Plus tie rods at 48 inches (1,219 mm) maximum along joints and intermediates.
   Tie rod sizes:1/4 inch (9.5 mm) diameter for 36 inch (914 mm) length or less; 3/8 inch (9.5 mm) diameter for over 36 inches (914 mm). Space at even divisions of duct width.

TABLE 6-2-P-1
Rectangular Aluminum Duct

Adapted from 3" W.G. (746 Pa) or Lower Steel Construction Tables
Amend Tables 6-2-G, 6-2-H, and 6-2-L by using equivalent thicknesses from this table.

Galv. Steel Thickness, Inches	0.016″	0.019″	0.024″	0.030″	0.036″	0.047″	0.058″
(Gauge)	(28)	(26)	(24)	(22)	(20)	(18)	(16)
Min. Aluminum Equivalent*	0.023	0.027	0.035	0.043	0.052	0.068	0.084
Commercial Size	0.025	0.032	0.040	0.050	0.064	0.071	0.090

Aluminum weight: 171.1 pounds per cubic foot (2,738 kg/m³)

***Note:** The aluminum reference in the table is Alloy 3003-H14. If calculations are performed, the following allowable stresses shall be used: tension or compression, 10,000 psi (68.9 MPa); shear, 6,000 psi (41.4 MPa); bearing, 16,000 psi (110 MPa). Unstiffened flanges shall have stress reduced for the specified width-to-thickness ratio.

Table 6-2-P-2
Dimension Adjustments

inches x 25.4 = mm

Galvanized rigidity class	A	B	C	D	E	F	G	H	I	J	K	L
Aluminum dimension per galvanized class	C	D	E	F	H	H	I	K	**	**	**	**

**Calculate an I_X = 3 x that used for steel.

TABLE NO. 6-2-P-3
Reinforcements

Steel Angle Size, Inches	Code	Equivalent Aluminum*** Angle Size, Inches	Steel Bar	Aluminum Bar***
1 x 1 x 0.058″ (16 ga.)	C	1-1/4 x 1-1/4 x 1/8	1 x 1/8 1-1/4 x 3/16	1-1/2 x 1/8 or
1 x 1 x 1/8	D	1-1/2 x 1-1/2 x 1/8	1-1/2 x 1/8 2 x 3/16	1-1/2 x 3/8 or
1-1/4 x 1-1/4 x 1/8	E	1-3/4 x 1-3/4 x 1/8		
1-1/2 x 1-1/2 x 1/8	F	2 x 2 x 3/16 or 2-1/2 x 2-1/2 x 1/8		
1-1/2 x 1-1/2 x 3/16	G	2 x 2 x 1/4 or 2-1/2 x 2-1/2 x 1/8		
2 x 2 x 1/8	H	2-1/2 x 2-1/2 x 3/16		
2 x 2 x 3/16	I	2-1/2 x 2-1/2 x 5/16 or 3 x 3 x 1/4		
2 x 2 x 1/4	J	2-1/2 x 2-1/2 x 3/8 or 3 x 3 x 1/4		
2-1/2 x 2-1/2 x 3/16	K	3 x 3 x 3/8 or 3-1/2 x 3-1/2 x 1/4		

***Alloy 6061-T6 strength normally

Notes for Tables 6-2-P-1, P-2, and P-3:

1. For Tables 6-2-G, 6-2-H, and 6-2-L, use equivalent thickness from Table 6-2-P-1 for required thickness of aluminum stock.
2. For Tables 6-2-I and 6-2-K: 2a) A connector not using angles or bar stock shall have its thickness increased per Table 6-2-P-1 and its dimensions increased per Table 6-P-2. 2b) A connector using angles or bar stock shall have its aluminum thickness increased per Table 6-2-P-1 and shall use either aluminum stock or galvanized stock from Table 6-2-P-3.
3. Table 6-2-J: Galvanized steel members of dimensions given may be used, or aluminum members having thickness and dimension conforming to Table 6-2-P-3 may be used. Other suitable aluminum shapes having a moment of inertia three times that of steel may be used.
4. Button-punch snaplock seams shall not be used in fabricating aluminum rectangular duct.

Tie Rod Installations.

In Tables 6-2-G, H, L, M, N, and O, duct sizes over forty-eight (48) inches (1,219 mm) have a tie rod optional construction for positive pressures.

TABLE 6-2-Q
Intermediate Tie Rods at Reinforcement (Between Joints)

Duct Width	Table Reinforcement Grade	Reinforcement Grade for Tie Rod Option
49" to 96"	G, H, I, or J	1-1/2" x 1/2" x 1/8" angle with tie rod at center
73" to	K or L	1-1/2" x 1-1/2" x 3/16" angle with tie rod at center
97" up	H with tie rod	2" x 2" x 1/8" angle with tie rod at 60" maximum intervals

Tie Rods at Joints.

The direct attachment of the tie rod by brazing or welding to a vertical leg of the joint is allowed for joints T-15, T-20, T-22, and T-23.

The attachment of a tie rod to an angle that reinforces a joint is allowed for joints T-4, T-8, T-9, T-15, T-16, T-19, and T-21. The angle shall meet the requirements for tie-rodded intermediate reinforcement.

The attachment of two tie rods or tie straps to the duct wall by welding or bolting within one (1) inch (25 mm) of both sides of joints T-15, T-20, T-21 and T-23 is allowed.

Additional General Requirements.

Ties shall be attached in such a manner that they will not loosen or detach with normal operating pressure fluctuations in the duct. When ties are to be welded to the duct wall, the rod or strap shall have a 90 degree (1.57 rad) bend with 5/8 inch (15.9 mm) minimum extension beyond the bend. The extension of the rod shall be welded to the duct. The strap shall be welded to the duct at the 90 degree (1.57 rad) bend.

Holes made in the duct wall for tie rod passage shall be the minimum size.

Tie rods shall be galvanized steel of 1/4 inch (6.4 mm) minimum diameter if thirty-six (36) inch (914 mm) length or less and 3/8 inch (9.5 mm) minimum diameter for greater length.

Tie straps shall be one (1) inch by 1/8 inch (25.4 mm by 3.2 mm) galvanized steel minimum, and the smallest edge shall face the airflow direction.

Ties shall be spaced at even intervals of duct width not exceeding sixty (60) inches (1,524 mm).

The use of ties does not void the need to attach reinforcements to the duct wall. However, when ties occur outside of the duct, as on two-side reinforcements at four (4) inch w.g. (995 Pa) and over, the attachment within two (2) inches (50 mm) of the corner is not required.

When ties occur in two directions in the same vicinity, they shall either be prevented from contacting or shall be in permanent contact.

Ties may be used as components of the suspension system for ducts over ninety-six (96) inch (2,438 mm) width, provided that the hanger load is directly transmitted to a trapeze or duct reinforcement member beneath the duct.

The construction of ducts of widths greater than ninety-six (96) inches (2,438 mm) requires the use of tie rods on Grade H joints and intermediates (2" x 2" x 1/8") (51 mm by 51 mm by 3.2 mm) at intervals not exceeding sixty (60) inches (1,524 mm) in the transverse direction. For ten (10") w.g. (2,488 Pa), sixteen (16) gauge (1.47 mm) ducts with two (2) feet (610 mm) reinforcement spacing is required; eighteen (18) gauge (1.19 mm) duct is required at lower pressures with spacings of two (2) feet (610 mm) at six (6") w.g. (1,493 Pa); 2-1/2 feet (762 mm) at one (1") to four (4") w.g. (249 Pa to 995 Pa); and five (5) feet (1,524 mm) at 1/2" w.g (124 Pa).

APPENDIX A

UNIFORM MECHANICAL CODE STANDARD NO. 6-5
STANDARD FOR INSTALLATION OF FACTORY-MADE AIR DUCTS

Based on the 1989 Fibrous Glass Duct Construction Standards published by the Thermal Insulation Manufacturers Association, and the Guidelines for installing Flexible Ducts published by the Air Diffusion Council. See Section 602.3 of the *Uniform Mechanical Code.*

A.6.501 Scope.

These requirements are intended to supplement information contained in the manufacturer's installation instruction sheets included in each carton of material, which cover fabrication, closure methods, reinforcement, and hanging of factory-made rigid and flexible air ducts complying with UL 181, Standard for Factory-Made Air Ducts and Air Connectors. The standard is divided into two parts: Part A covering rigid ducts and Part B covering flexible ducts.

Part A— Rigid Ducts.

A.6.502 General.

The use of these ducts is governed by UMC Section 602.1, together with the restrictions contained in Section 604.2. For information regarding the fabrication of rectangular ducts from flat sheets of duct board, refer to the document upon which this standard is based and the manufacturer's instruction sheet.

A.6.503 Closure Systems.

(A) General. Closure systems are a vital element in the proper assembly of fibrous glass duct systems. They provide both the structural connection and sealing of seams and joints for air tightness. Only listed closure systems identified in the manufacturers' installation instructions are suitable for use with rigid fibrous glass duct systems. Listed closures include the following:

(1) Pressure-sensitive aluminum foil tapes.

(2) Heat-activated aluminum foil/scrim tapes.

(3) Mastic and glass fabric tape systems.

Listed closure systems may be used on listed duct board materials.

(B) Surface Preparation. In order to obtain satisfactory adhesion and bonding, the surface on which closures will be applied must be free from dust, dirt, oil, grease, moisture, and similar substances.

(C) Joint and Seam Preparation. Longitudinal seams are fabricated with a shiplap joint that is closed with the use of outward-clinching staples at two (2) inches (50 mm) on center through the stapling flap of the jacketing material. Transverse joints between two duct sections are prepared by joining two duct sections, pulling the staple flap over the adjoining section, and stapling as shown in Figure 6-5-1. When staple flaps are not present, crosstabs are used to hold seams and joints in position for reinforcing the closure system. Cross-tabs, made from eight (8) inch (203 mm) minimum lengths of closure tape, are to be equally spaced on each side of the joint. Cross-tabs are spaced on twelve (12) inch (305 mm) maximum centers, with at least one cross-tab per duct side (Figure 6-5-2). Cross-tabs may be placed either under or over the closure tape.

(D) Application of Pressure-Sensitive Aluminum Foil Tapes. Use minimum 2-1/2 inch wide (64 mm) listed pressure-sensitive tape. Position the tape along the edge of the flap in a manner that will allow one (1) inch minimum (25 mm) overlap on adjacent surfaces.

If tape has been stored at temperatures less than 50°F (10°C), it should be conditioned prior to use by placement in a warm environment in order to improve the initial tack.

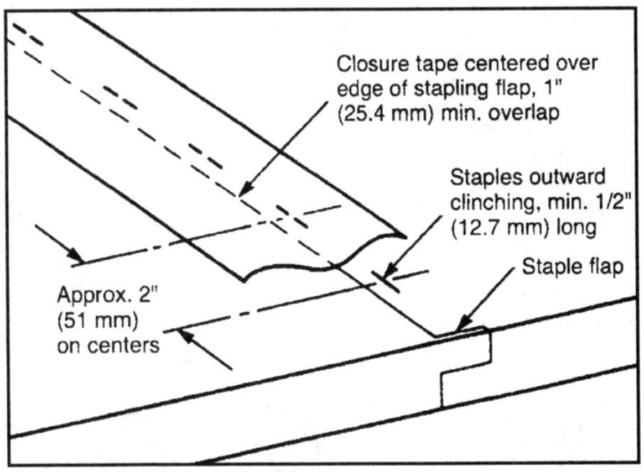

FIGURE 6-5-1 Tape Closure Joint with Staple Flap.

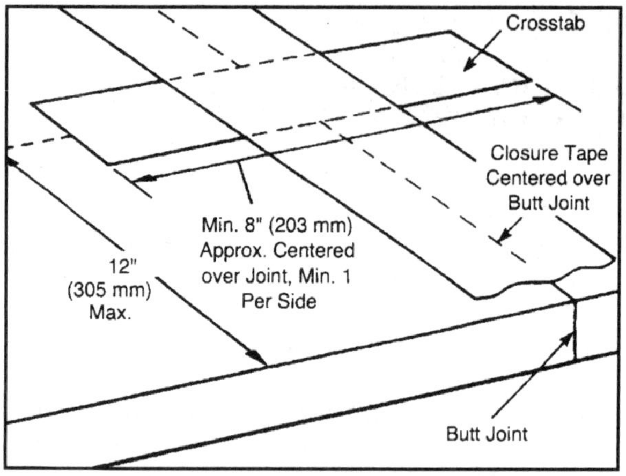

FIGURE 6-5-2 Tape Closure Joint Without Staple Flap.

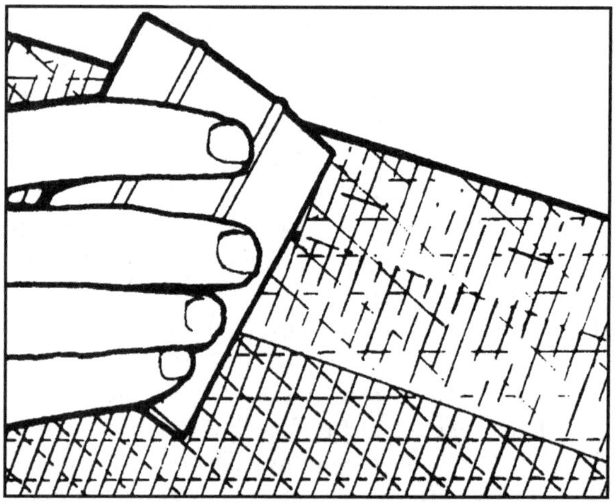

FIGURE 6-5-3 Pressure-Sensitive Tape.

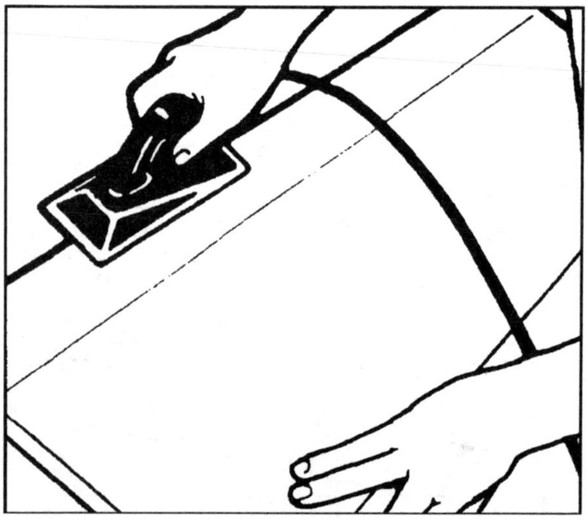

FIGURE 6-5-4 Heat-Activated Tape.

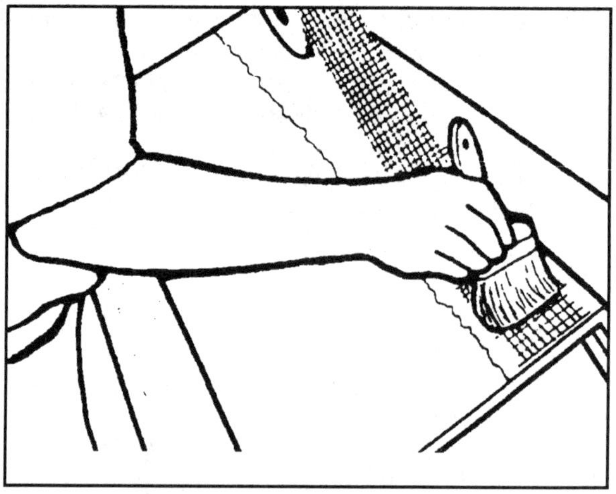

FIGURE 6-5-6 Glass Fabric into Mastic.

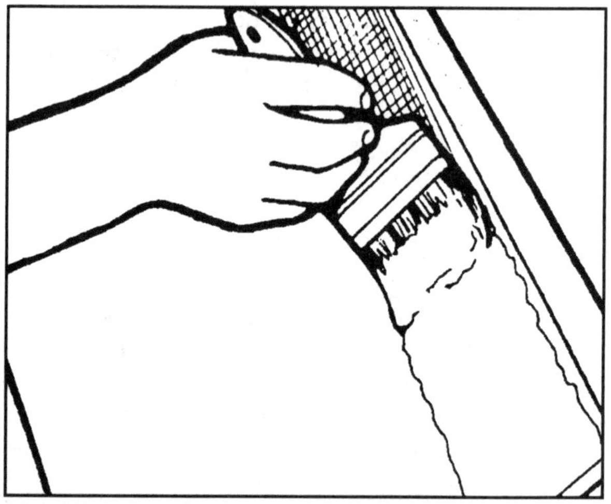

FIGURE 6-5-7 Second Coat of Mastic.

If the duct board has been located in an atmosphere of less than 50°F (10°C), the surfaces to be taped must be preheated to ensure a satisfactory bond of the tape. Using any suitable heating iron with the plate temperature set at 400°F (+25°F) (204°C + 14°C), preheat the area to be taped. Quickly position the tape on the preheated area and press in place. Pass the iron two or three times over the taped area using a rapid ironing motion (see Figure 6-5-4).

Rub tape firmly with a plastic squeegee (Figure 6-5-3) until the facing reinforcement shows through the tape.

When operating pressure is less than one (1) inch W.G. (249 Pa) (positive pressure), and sheet metal surfaces are cleaned in accordance with tape manufacturer's instructions, pressure-sensitive tape may be used to seal fibrous glass duct to sheet metal.

(E) Application of Heat-Activated Tapes. Position the 2-1/2 inch (64 mm) minimum tape along the edge of the flap in a manner that will allow a one (1) inch minimum (25.4 mm) overlap on adjacent surfaces. Using a suitable heating iron with a plate temperature between 550°F and 600°F (287°C and 316°C), pass the iron along the tape seam with sufficient pressure and dwell time to activate the adhesive (see Figure No. 6-5-4). A satisfactory bond has been achieved when the heat indicator dots have darkened.

Use a second pass of the iron to complete the bond by applying pressure to the front edge of the iron in a smearing action.

Allow all joints and seams to cool below 150°F (65°C) before any stress is applied. Avoid puncturing the tape at staple locations with excessive pressure from the iron.

(F) Application of Glass Fabric and Mastic Tape Systems. Apply a thin coat of mastic approximately 3-1/2 inches (89 mm) wide over the center of the joint seam. Firmly press the three (3) inch (76 mm) wide glass fabric into the mastic (see Figure 6-5-6).

Apply a second coat of mastic over the glass fabric, filling the scrim pattern as indicated in Figure 6-5-7.

Follow the manufacturer's label instructions on application rate, safety precautions, ventilation requirements, shelf life limitations, and minimum setup time required before stress can be applied to the joint or seam.

When connecting fibrous glass duct systems to sheet metal, as in connections to flanges or central air equipment, fasteners such as sheet metal screws and washers must be used to carry the mechanical load. Glass fabric and mastic can then be applied to seal the connections at these points. Usually, two widths of glass fabric will be required.

A.6.504 Reinforcement.

(A) General. Reinforcement is used to limit the deflection of the duct surface caused by internal static pressure loading. The maximum dimension of unreinforced ducts, which is a function of the system operating pressure, is shown on the manufacturer's label, affixed to each piece of duct board.

Duct systems made from Types 475 and 800 fibrous glass duct board may use one or more of the following reinforcement methods:

(1) Tie rod reinforcement.

(2) Channel reinforcement.

The reinforcement schedules contained in Tables 6-5-1, 6-5-2, and 6-5-4 are suitable for zero (0) inch to two (2) inch W.G. (0 Pa to 498 Pa) duct systems.

While some duct dimensions may not require any reinforcement in straight sections, certain fittings of the same dimension may require reinforcement. This requirement is noted wherever applicable in this standard and in Table 6-5-4. Fitting reinforcement is accomplished by the use of tie rods wherever possible and with channel reinforcement applied where tie rods cannot be used.

(B) Tie Rod Reinforcement—Positive Pressure.

(1) Tie Rods. Tie rods consist of straight lengths of twelve (12) gauge (2 mm) galvanized steel wire with flat steel volcano-type washers and any of three approved termination devices (see Figure 6-5-8). The 2-1/2 inch (64 mm) square shall be installed so turned edges face away from the duct board facing.

When butt joints are used instead of shiplap joints, tie rods are placed in rows located at three (3) inches (76 mm) on either side of the joint.

Alternate: A single tie rod reinforcement may be used if the butt joint is glued with an adhesive system documented by the duct board manufacturer.

(2) Location. In relation to transverse joints, the tie rods are spaced four (4) inches (102 mm) from the end of the female joint, as shown in Figure 6-5-9.

(3) Spacing. Tie rods are placed in rows of two (2) to five (5) rods, across the face of the duct, with row spaces varying from sixteen (16) inches to forty-eight (48) inches (406 mm to 1,219 mm); depending on duct size and operating pressure.

The spacing of tie rods is shown in Table 6-5-1.

When the sides of the duct exceed the dimensions that require reinforcement (see Table 6-5-1), horizontal tie rods are also installed per the schedule.

(4) Typical Tie Rod Reinforcement Examples, Positive Pressure Ducts. See Section 6.504(B)2 for placement of tie rods and sag supports in relation to joints. Sag support is required in ducts forty-eight (48) inches (1,219 mm) and greater in width.

The number of tie rods across the duct width shall be as required in the schedule contained in Table No. 6-5-1.

TABLE NO. 6-5-1

Tie-Rod System Reinforcement Schedule (Positive Pressure Only)

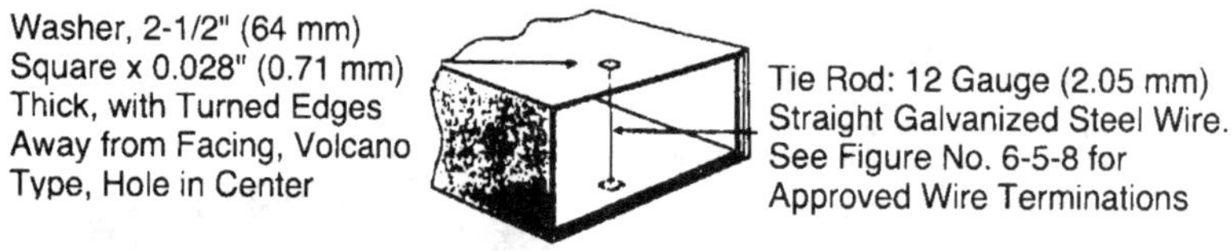

Positive Static Pressure	Maximum Inside Duct Dimension Inches	Type 475 Board			Type 800 Board		
		No. Rods Across Dimension	Maximum Longitudinal Spacing	No. Rods per 4 Ft. Section	No. Rods Across Dimension	Maximum Longitudinal Spacing	No. Rods Per 4 Ft. Section
(x 248.8 for Pa)	(x 25.4 for mm)		(x 25.4 for mm)			(x 25.4 for mm)	
0 through 1/2" W.G.	0-36		*			*	
	37-42	2	24"	4	2	48"	2
	43-48	↓		↓	↓		↓
	49-60	3		6	3	↓	3
	61-64	↓		↓	↓	24"	6
	65-80	4		8	4		8
	81-96	5	↓	10	5	↓	10
Over 1/2" through 1" W.G.	0-24		*			*	
	25-30	1	24"	2	1	48"	1
	31-32	↓		↓	↓	24"	2
	33-36	2		4	2		4
	37-48	↓		↓	↓		↓
	49-64	3		6	3		6
	65-80	4		8	4		8
	81-96	5	↓	10	5	↓	10
Over 1" through 2" W.G.	0-15		*			*	
	16-18	1	24"	2		*	
	19-24		↓	↓	1	24"	1
	25-32	↓	16"	3	↓	24"	2
	33-48	2		6	2		4
	49-60	3		9	3	↓	6
	61-64	↓		↓	↓	16"	9
	65-80	4		12	4		12
	81-96	5	↓	15	5	↓	15

*Straight ducts of these dimensions do not require reinforcement. However, some fittings of these dimensions may require reinforcement.

Notes: 1. Tie rods and washers must be no more than 16" (406 mm) on centers across duct dimension.

2. Ducts of 48" (1,219 mm) width and over require use of anti-sag devices. See Fig. 6-5-11A.

3. For duct dimensions over 96" (2,438 mm), maintain tie rod spacing on 16" (406 mm) centers across the duct dimension following longitudinal spacing for the design pressure.

used. These must have turned edges to prevent cutting into the duct board. In positive-pressure applications, wraparound channels need not be attached to the duct board except when required for sag control.

Table No. 6-5-2 gives longitudinal spacing, sheet metal gauge, channel height dimensions of reinforcement depending on the duct board type, maximum inside dimensions, and static pressure. This is a minimum reinforcement schedule applying to straight sections. This schedule will also apply to fittings, but a simplified schedule may also be used (see Table 6-5-3).

(2) Channels. Channels must be fabricated from galvanized sheet metal of the gauge shown in Table 6-5-2. Channels are three (3) inches (76 mm) wide, with a height (H) as shown in Table 6-5-2.

Each reinforcement may be fabricated from a continuous length of channel having three 90 degree(1.57 rad) bends and a fourth 90 degree (1.57 rad) corner that is fastened with bolts, screws, rivets, spot welds, or staples. Reinforcements may also be fabricated with two (2), three (3), or four (4) securely fastened corners.

TABLE NO. 6-5-2
Channel System Reinforcement Schedule

Static Pressure (x 248.8 for Pa)	Maximum Inside Duct Dimension (I.D.), in. (x 25.4 for mm)	Type 475 Board: Maximum longitudinal spacing (x 25.4 for mm)	Type 475 Board: Channel gauge	Type 475 Board: H dimension (see below) (x 25.4 for mm)	Type 800 Board: Maximum longitudinal spacing (x 25.4 for mm)	Type 800 Board: Channel gauge	Type 800 Board: H dimension (see below) (x 25.4 for mm)
0 - 1/2" W.G. negative	0-30		*			*	
	31-36	24"	22	1"	48"	22	1"
0 - 1/2" W.G. positive	0-36		*			*	
0 through 1/2" W.G. positive or negative	37-42	24"	22	1"	48"	22	1"
	43-48						↓
	49-60				↓		1-1/2"
	61-72				24"		1"
	73-84			↓			
	85-96	↓	↓	1-1/4"	↓	↓	↓
Over 1/2" through 1" W.G. positive or negative	0-24		*			*	
	25-30	24"	22	1"	48"	22	1"
	31-36				24"		
	37-42						
	43-48						
	49-60		↓			↓	
	61-72		18	↓		18	↓
	73-84			1-1/4"			1-1/4"
	85-96	↓	↓	↓	↓	↓	↓
Over 1" through 2" W.G. positive or negative	0-15		*			*	
	16-18	24"	22	1"		*	
	19-24	↓			24"	22	1"
	25-36	16"					↓
	37-48						1-1/4"
	49-60		↓		↓	↓	↓
	61-72		18	↓	16"	18	1"
	73-84			1-1/4"			1-1/4"
	85-96	↓	↓	1-1/2"	↓	↓	1-1/2"

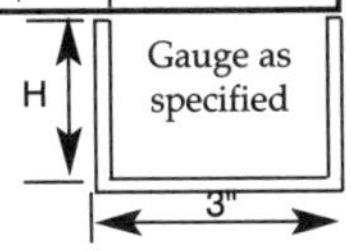

*Straight ducts of these dimensions do not require reinforcement. However, some fittings of these dimensions may require reinforcement.

Note: Ducts of 48" (1,219 mm) maximum width and over require use of anti-sag devices.

(3) Location of Channels. Channels are normally offset four (4) inches (102 mm) from the end of the duct section to facilitate installation of sag supports and the closure system.

Where wraparound channels without sag support are used, the channel is slipped over the closure tape after the tape is applied, centered over the female shiplap end for maximum support.

(4) Sag Support and Typical Channel Reinforcement. When the duct is forty-eight (48) inches (1219 mm) wide or greater, channels are secured to the top for sag support with No. 10 plated sheet metal screws and 2-1/2 inch (64 mm) square washers (see Detail A in Figure 6-5-12). The number of channels along the duct shall be in accordance with Table No. 6-5-2 .

(D) Negative-Pressure Channel Reinforcement.

(1) Locating Reinforcing Channels. In negative-pressure applications such as return-air ducts, the channel reinforcement is applied over the male shiplap. Special clips are installed inside the duct (see Figure 6-5-13) that will support both the male and female sides of the duct joint. The clips are fastened in place with two sheet metal screws.

(2) Clips. Clips are formed from 20 gauge (0.036 inch) (0.91 mm) galvanized steel, with turned edges as shown in Detail B.

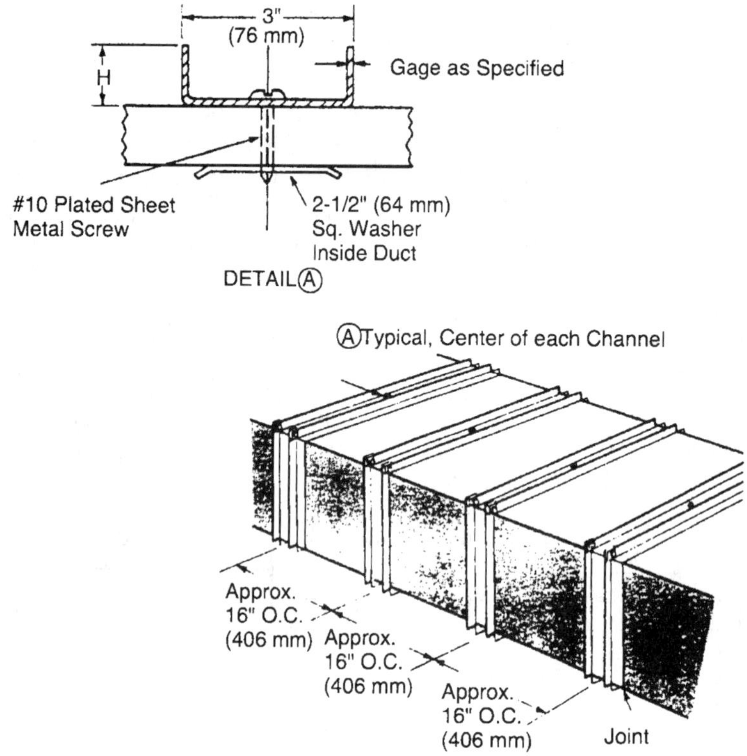

FIGURE 6-5-12 16-Inch (406 mm) Centers, 48-Inch (1219 mm) Duct Sections.

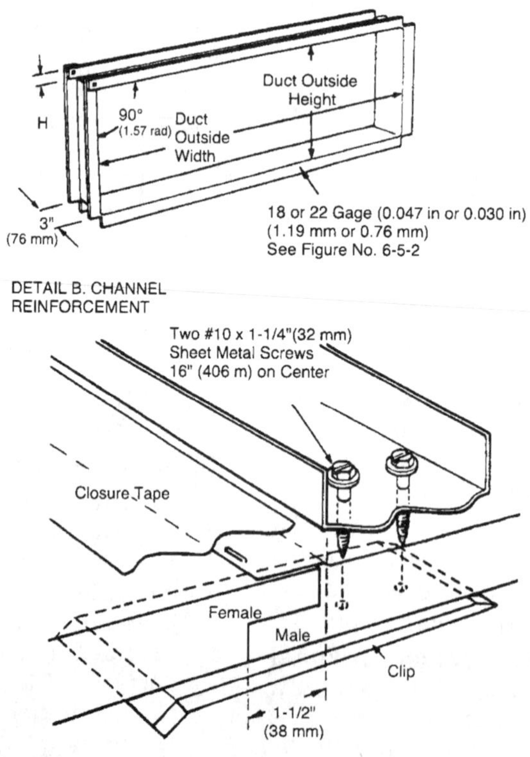

FIGURE 6-5-13 Channel Reinforcement for Negative Pressure Systems.

TABLE NO. 6-5-4
Partial Wraparound Reinforcement Schedule

Positive Static Pressure	Type 475 Board Maximum Inside Duct Dimension, Inches	Longitudinal spacing	Dimension L	No. of screws, each end	Type 800 Board Maximum Inside Duct Dimension, Inches	Longitudinal spacing	Dimension L	No. of screws, each end
(x 248.8 for Pa)	(x 25.4 for mm)				(x 25.4 for mm)			
0" through 1/2" W.G.	0-36	Not required			0-36	Not required		
	37-96	24"	4"	1	37-60	48"	4"	1
					61-96	24"	↓	↓
Over 1/2" through 1" W.G.	0-24	Not required			0-24	Not required		
	25-48	24"	4"	1	25-30	48"	4"	1
	49-64		7"	2	31-48	24"	↓	↓
	65-80		10"	3	49-64		7"	2
	81-96	↓	13"	4	65-80	↓	10"	3
					81-96		13"	4
Over 1" through 2" W.G.	0-15	Not required			0-18	Not required		
	16-24	24"	4"	1	19-24	24"	4"	1
	25-32	16"	↓	↓	25-32		7"	2
	33-48		7"	2	33-48		10"	3
	49-64		10"	3	49-60	↓	13"	4
	65-80		13"	4	61-64	16"	10"	3
	81-96	↓	16"	5	65-80		14"	↓
					81-96	↓	16"	5

Clips are spaced not more than sixteen (16) inches (406 mm) apart and not more than sixteen (16) inches (406 mm) from the sides of a duct. The number of clips required is shown in Table 6-5-3.

TABLE 6-5-3
Detail B—Clip, Negative-Pressure Systems
Fastener Requirements, Negative Pressure

Transverse Dimension in. x 25.4 = mm	Minimum Number of Clips or Washers per Reinforcing Member
16" - 32"	1
33" - 48"	2
49" - 64"	3
65" - 80"	4
81" - 96"	5

Note: The foregoing arrangements are important and, if not followed closely, may result in system failure.

(3) Additional Fasteners. When additional channels are required (as with twenty-four (24) inch and sixteen (16) inch [610 mm and 406 mm] spacing), they are attached to the duct with No. 10 plated sheet metal screws and 2-1/2 inch (64 mm) square washers (see Detail A of Figure No. 6-5-12).

The spacing of the fasteners through each reinforcement channel is the same as for joint clips (see Table 6-5-3). A typical layout for negative pressure ductwork is shown in Figure 6-5-14 and Detail C.

(E) Fitting Reinforcement—Positive-Pressure Systems.

(1) General. Fittings are reinforced with tie rods or channels based upon the duct dimensions using the criteria found in Tables 6-5-1 and 6-5-2.

Some fittings, branches, tees, or offsets may require reinforcement even though schedules for straight ducts of the same dimension may show reinforcement is not required.

(2) Partial Wraparound Reinforcement. Where reinforcement is required but cannot be fastened to opposite sides of a duct section or fitting, it is necessary to install formed sheet metal channels that partially wrap around a fibrous glass duct system fitting at the required location. (See Table 6-5-2 for appropriate gauge and profile.) In such cases, No. 10 by 1-1/4 inch (32 mm) plated sheet metal screws and 2-1/2 inch (64 mm) square washers, 0.020 inch (0.51 mm) (minimum) thick are used to attach the ends of the channels to the duct board. (See Detail D and Table 6-5-4.)

(3) Fitting Reinforcement—90 Degree (1.57 rad) Elbows. If neither dimension A nor B (see

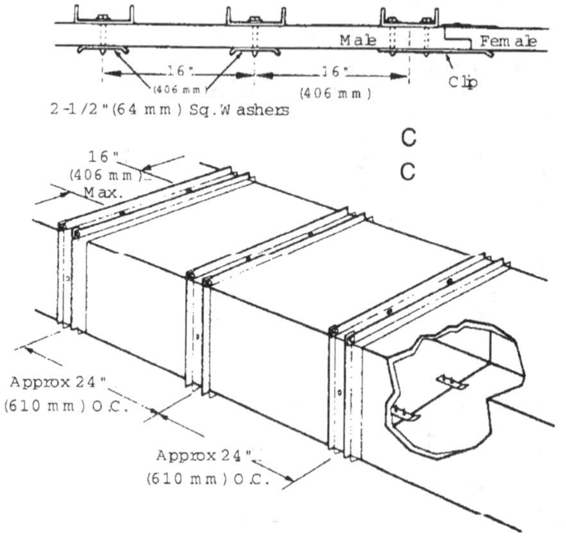

FIGURE 6-5-14 Channel Reinforcement on Twenty-four (24) Inch (610 mm) Centers, Forty-eight (48) Inch (1,219 mm) Duct Section.

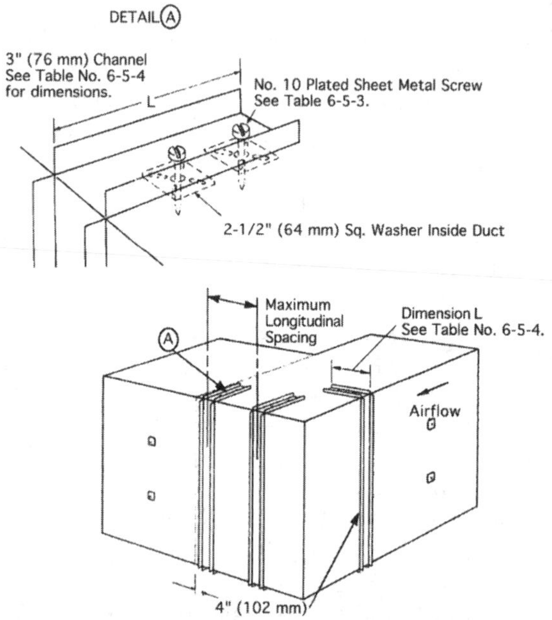

Total length of channel = outside duct dimension + twice L dim., Table No. 6-5-4

FIGURE 6-5-15 Partial Wraparound Reinforcement.

Figure No. 6-5-16) is greater than the maximum unreinforced duct dimension (see Table 6-5-1), but diagonal X-Y is greater than the maximum unreinforced duct dimension in accordance with Table 6-5-1, install tie-rod reinforcement at the midspan of the diagonal at point No. 1 in Figure 6-5-16.

If either A or B is greater than the maximum unreinforced duct dimension:

Reinforce in accordance with Table 6-5-1.

Reinforce four (4) inches (102 mm) upstream from female shiplap joints.

Reinforce at point No. 1 (Figure 6-5-16) where center lines intersect.

Note: Turning vanes in Figure No. 6-5-16 are omitted for clarity. Turning vanes do not replace reinforcement.

Heel and Throat Panels. If duct dimension H (see Figure 6-5-16) is less than the maximum unreinforced duct dimension from Table 6-5-1, but more than twenty-four (24) inches (610 mm), install sheet metal angle in accordance with Detail E below. (Angle may also be installed on inside of throat.)

When duct dimension H normally requires reinforcing, install sheet metal angle as shown at No. 2, Figure 6-5-17. Install tie rods through the angle on the upstream side, sixteen (16) inches (406 mm) on centers, in accordance with Table 6-5-1, with angle of length L from table below.

No. Tie Rods	1	2	3	4	5
Angle Length L, inches	4	20	36	52	68

Install three (3) inch (76 mm) channel reinforcement on heel panels as shown in Figure 6-5-17, spaced in accordance with Table 6-5-4 and fastened in accordance with Figure 6-5-15.

Note: Attachment of angles is best done after closure is completed. This requires sections to be short enough to allow the installer to reach inside to install the 2-1/2 inch (64 mm) square washers.

Note: For reinforcement of mitered elbows, refer to reinforcement standards for offsets.

(4) Branch Connections (Tees). Many branches may require reinforcement even though schedules for straight ducts of the same dimensions may show reinforcement is not required.

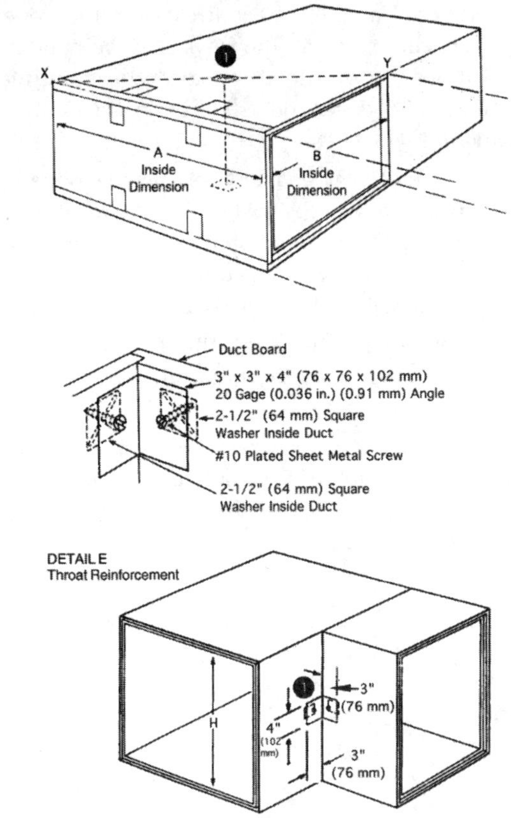

FIGURE 6-5-16 Tie Rod Reinforcement at Diagonal X-Y—Midspan.

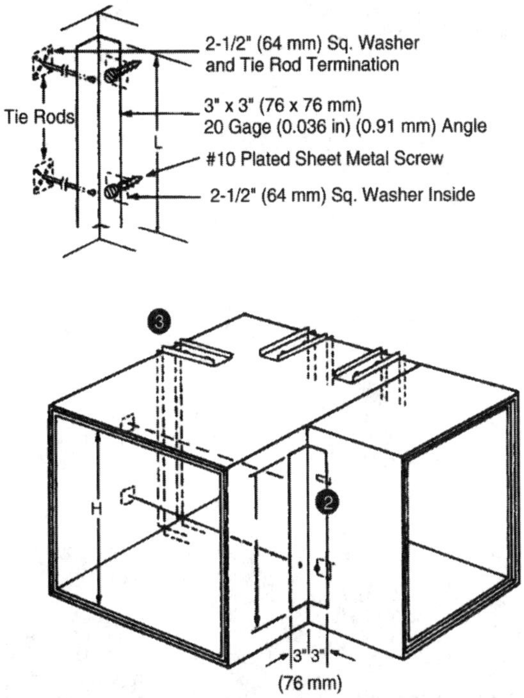

FIGURE 6-5-17 Sheet Metal Angle at Throat, Large 90-Degree (1.57 rad) Elbows.

Angled branches and positive takeoffs may be reinforced using the same methods as for tees. The amount and type of reinforcement depend on the dimensions of the takeoff and the system pressure (see Table 6-5-1).

If H is not greater than sixteen (16) inches (406 mm) and W (see Figure No. 6-5-18) is greater than twelve (12) inches (305 mm), install reinforcement in accordance with Detail E on the top of the branch where it intersects the trunk duct.

If H is greater than sixteen (16) inches (406 mm) and W is less than the maximum unreinforced duct dimension, reinforce the side of the branch as shown at No. 2, Figure 6-5-18.

Note: The maximum unreinforced dimension is thirty-six (36) inches (914 mm) at 1/2 inch W.G. (124 Pa), twenty-four (24) inches (610 mm) at one (1) inch W.G. (249 Pa), and sixteen (16) inches (406 mm) at two (2) inches W.G (498 Pa).

If H (Figure 6-5-18) is greater than sixteen (16) inches (406 mm) and W is greater than the maximum unreinforced duct dimension (see Table 6-5-1), install long angle clips in accordance with tie rod reinforcement through the branch and trunk ducts (similar to Figure 6-5-17), as required by Table 6-5-1. A combination of tie rods and partial wraparound reinforcement (see Figure 6-5-15) may be used.

If A (see Figure 6-5-19) is less than the maximum unreinforced duct dimension, but diagonals X-Y or Y-Z exceed the maximum allowable unreinforced duct dimension, install tie rods as shown at No. 2, four (4) inches (102 mm) from female shiplaps.

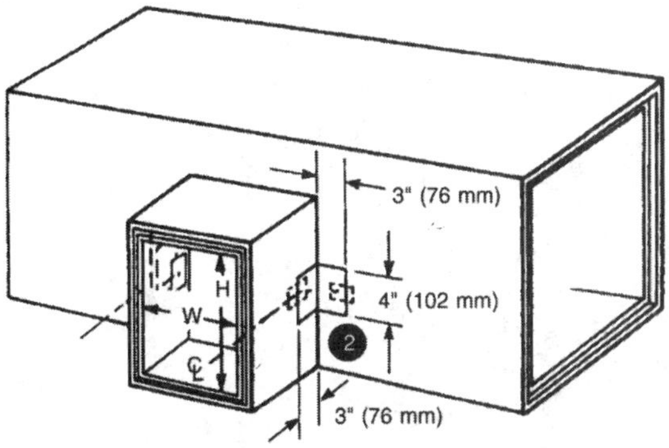

FIGURE 6-5-18 Sheet Metal Angle Reinforcement – Sides of Branch.

If A is greater than the maximum unreinforced duct dimension and B is greater than one-half the maximum unreinforced duct dimension, install tie rods four (4) inches (102 mm) from female shiplap joints as shown at No. 2 and also along branch centerlines and, additionally, spaced per Table 6-5-4 in the trunk duct.

Where a splitter damper interferes with tie rod reinforcement, wraparound channels must be used in their place.

Note: Turning vanes do not replace reinforcement.

(5) Offsets. Offsets require different types of reinforcement for the cheek panels (sides, as shown in Figure 6-5-20) and the heel and throat panels (top and bottom, per Figure 6-5-21).

If B is greater than the maximum unreinforced duct dimension and cheek panels have shiplap joints, reinforce in accordance with No. 1, four (4) inches (102 mm) from female shiplap (see Figure 6-5-20) with spacing in accordance with Table 6-5-1.

If the fitting uses butt joints, install the required spacing of tie rods three (3) inches (76 mm) on each side of the butt joint.

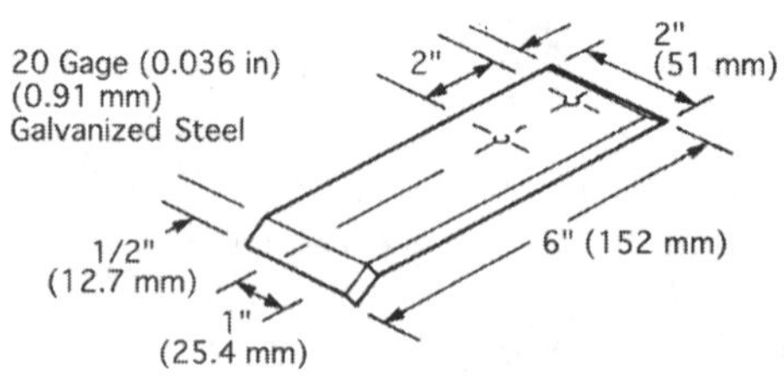

DETAIL A. CLIP, NEGATIVE PRESSURE REINFORCEMENT SYSTEMS

Where Table 6-5-1 requires more than one row of reinforcement, the rows shall be parallel to the edges of the duct, and the spacing between sets of tie rods shall be in accordance with Table 6-5-1.

The reinforcement for heel and toe joints is in accordance with Figure No. 6-5-21. Additional tie rods may be installed by using a longer plate at the heel, and also in the duct panel, spaced in accordance with Table No. 6-5-1.

The penetrations at the toe plate are grouped as close to the break in the metal as possible.

When dimension A requires more than one tie rod (see Table 6-5-1), sets of tie rods with metal angles shall be installed in rows parallel to the side of the duct.

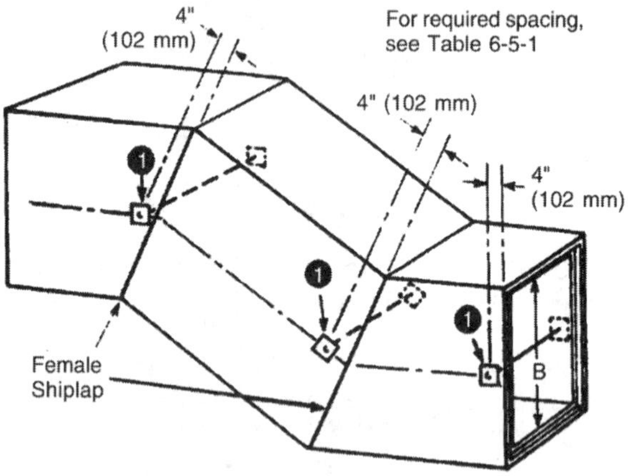

FIGURE 6-5-20 Offset Reinforcement Cheek Panels with Shiplap Joints.

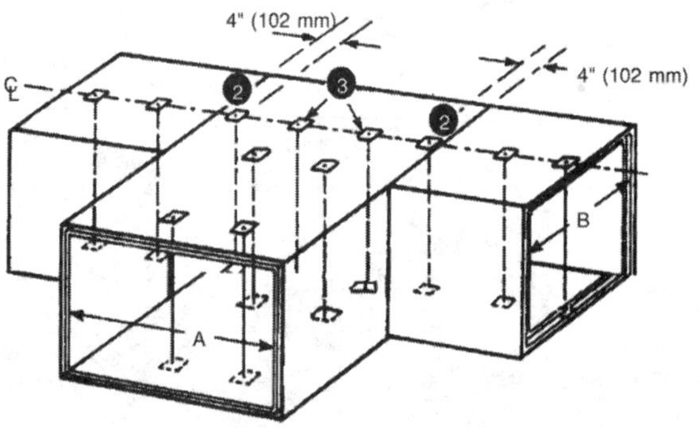

FIGURE 6-5-19 Tee Reinforcement—Normal Tie Rod Locations.

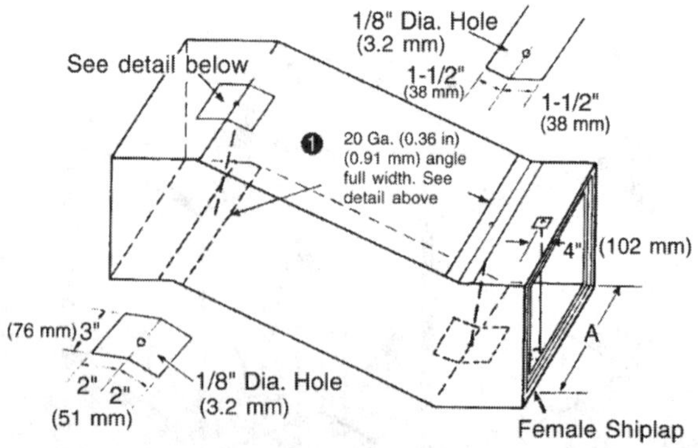

FIGURE 6-5-21 Offset Reinforcement with Sheet Metal Plates and Tie Rods.

Transitions are reinforced in the same manner with a tie rod extending from the flat side of the duct, through a plate or plates positioned on twelve (12), the heel joint of the fitting.

(6) Access Doors—Positive Pressure. If the access door width is not greater than the maximum longitudinal reinforcement spacing from Table 6-5-1, but interferes with reinforcement locations required by the table, install tie rods four (4) inches (102 mm) from both sides of the door opening in accordance with No. 1, Figure 6-5-22. Maximum reinforcement spacing must be in accordance with Table 6-5-1.

If access door height is greater than sixteen (16) inches (406 mm) and its width is greater than the maximum longitudinal reinforcement spacing shown in Table 6-5-1, install the frame for the access door inside the duct, securing it to the duct wall with screws and washers (see Figure 6-5-23).

Install tie rods near the vertical sides of the door frame (Figure 6-5-22) and also near the top and bottom sides of the frame, spaced as specified in Table 6-5-1 with a minimum of one tie rod on each side of the frame.

Note: Use channel reinforcement in place of tie rods between access door and fire damper where tie rods would interfere with damper access or operation.

The duct door is fabricated from one (1) inch (25 mm) duct board installed with one (1) x one (1) x one (1) inch (25 x 25 x 25 mm) twenty-two (22) gauge (0.76 mm) Z-channel framing around the access door. Doors are gasketed and secured in place with sheet metal screws, or they may be hinged with sash lock fasteners as shown in Detail G.

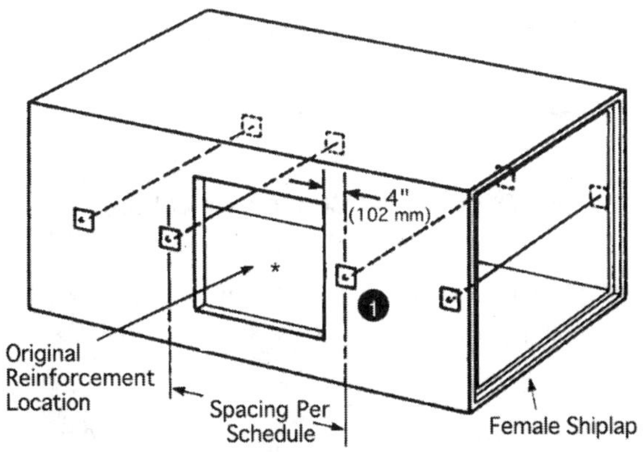

FIGURE 6-5-22 Access Door Location Interfering with Reinforcement.

A.6.505 Fire Dampers.

When fire dampers are required at penetrations of fire-rated walls, fibrous glass ducts must terminate at either side of such walls. Fire dampers shall be installed in accordance with the damper manufacturer's instructions and must be installed in a steel sleeve. The opening in the wall shall be large enough to allow for thermal expansion of the sleeve and to permit installation of 5/8 inch (16 mm) gypsum board between the sleeve and framing.

The sleeve must extend not less than three (3) inches (76 mm) beyond the face of the wall so that the duct can be slipped onto the sleeve. The duct is secured to the sleeve with screws and washers, as shown in Figure 6-5-24.

Sealing of fibrous glass duct board to the sheet metal sleeve must be made with glass fabric and mastic, except where operating pressure is less than one (1) inch W.G. (249 Pa) and the sheet metal surfaces are carefully cleaned, in that case pressure-sensitive aluminum foil tape may be used. See Section 6.503 for methods.

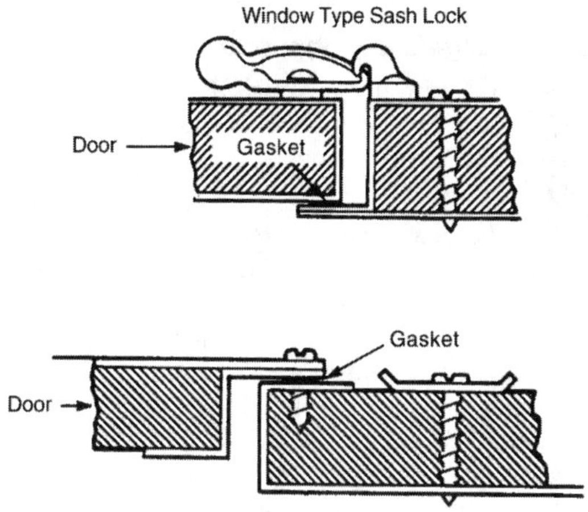

DETAIL G Window-Type Sash Lock, Flanged Door

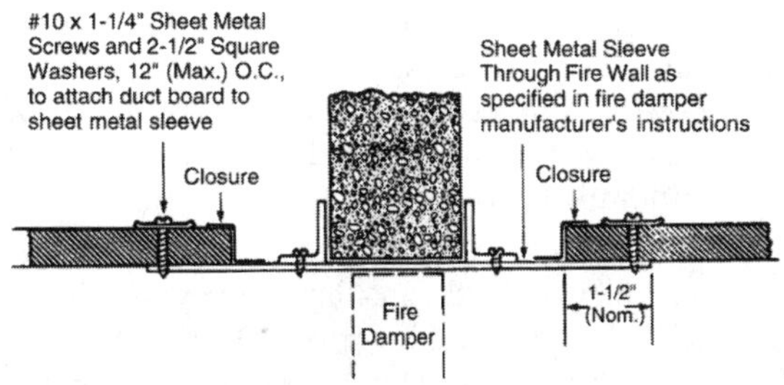

FIGURE 6-5-24 Attachment to Fire Damper Sleeves.

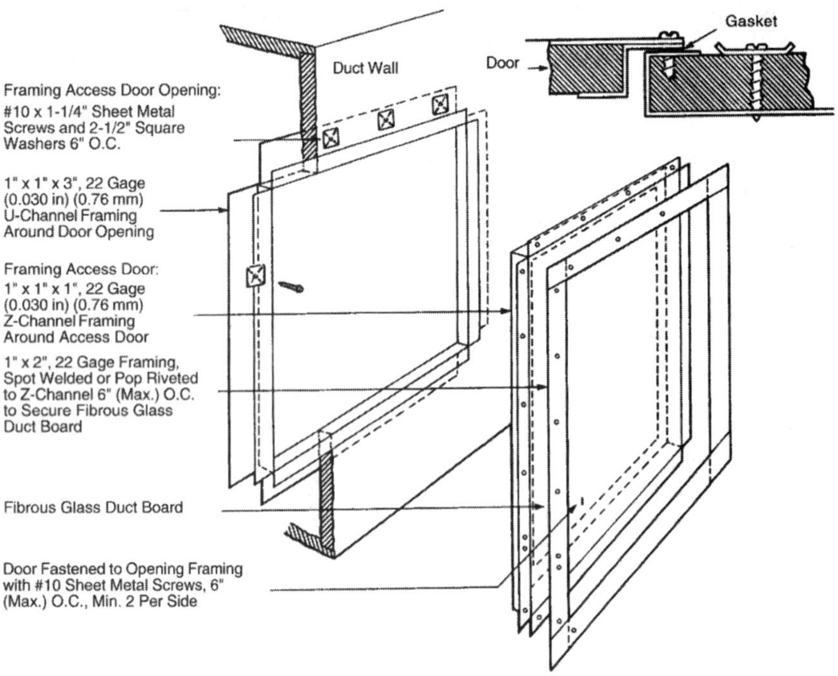

FIGURE 6-5-23 Duct Opening Frame.

A.6.506 Hangers and Supports.

(A) General. Fibrous glass ducts are light in weight, so that they can be supported with a minimum of hangers if care is taken as to placement of the supports.

The charts and examples illustrated in this section show that the hanger treatment and spacing required is dependent on duct dimensions. Trapeze-style channels, suspended by twelve (12) gauge (2.8 mm) (minimum) hanger wire (see Figure 6-5-25), are the preferred method of support. Channel gauge and profile vary with duct size, but in no case should the supporting channel be less than two (2) inches (51 mm) wide. Channels may also be suspended by means of metal rods of one (1) inch wide (25 mm) (minimum) galvanized steel straps.

When channel reinforcement members occur within maximum hanger spacing as shown in Table No. 6-5-5, sheet metal straps may be bolted to the channel reinforcement as shown in Figure No. 6-5-26. Support may also be made with twelve (12) gauge (2.8 mm) (minimum) wire.

(B) Hanger Design. Occasionally, hanger channels must be extended considerably beyond the duct sides so that the supports will clear other obstructions. The total extension (E + E′) of the supports beyond the duct sides (see Figure 6-5-27) governs the minimum dimensions (see Table 6-5-6) of the channel.

TABLE 6-5-5

Duct Size, Inches (x 25.4 for mm)	Maximum Hanger Spacing, Feet (x 304.8 for mm)
48" wide or greater	4
Less than 48" wide and less than 12" deep	6
Width between 24" and 48" and greater than 24" deep	6
Less than 48" wide and depth between 12" and 24"	8
Width 24" or less and depth greater than 12"	8

TABLE 6-5-6 Channel Selection

If Total Extension is Not Greater Than:	Min. Channel Gauge	Min. Channel Profile
6"	24	3" x 1.5"
18"	22	3" x 2"
30"	18	3" x 2"

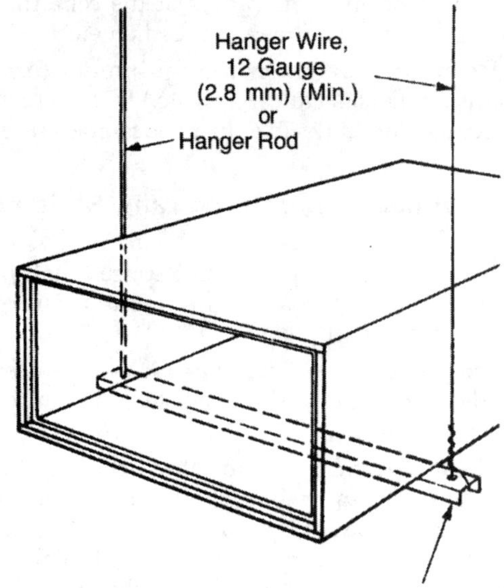

FIGURE 6-5-25 Attachment to Fire Damper Sleeves.

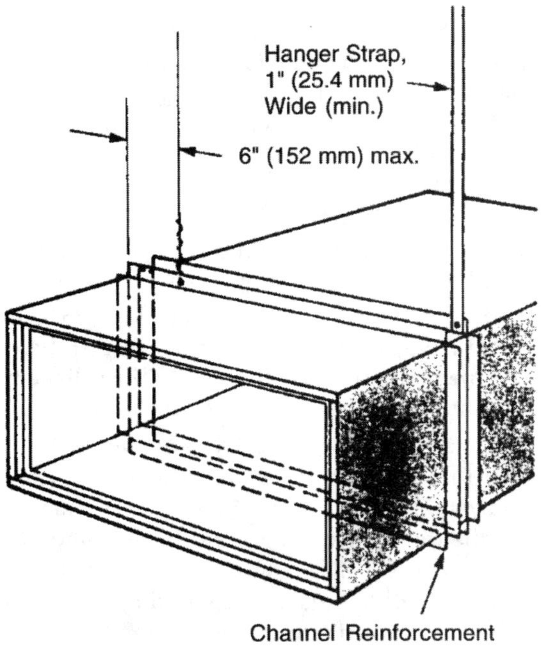

FIGURE 6-5-26 Hanging from Channel Reinforcement.

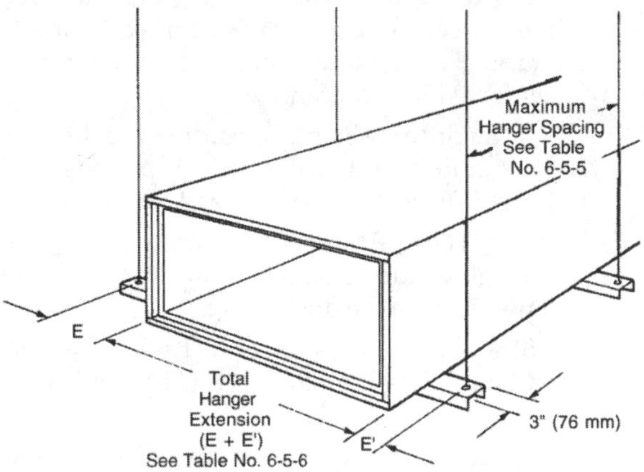

FIGURE 6-5-27 Hanger Spacing and Extension Using 3-Inch (76 mm) Wide Channels.

(C) Spacing. Hanger spacing per Table 6-5-5 or Figure 6-5-28 is based upon three (3) inch (76 mm) wide (minimum) channels.

For ducts not over forty-eight (48) inches wide by twenty-four (24) inches high (1,219 mm by 610 mm), two (2) inch wide by 1-1/2 inch channels (51 mm by 38 mm), spaced not more than four (4) feet (1,219 mm) apart may be installed. The total extension of the hanger supports shall not exceed six (6) inches (152 mm).

Hanger design and spacing for fibrous glass ducts is based upon extensive testing with loads exceeding twice the duct weight located between supports to ensure the integrity of the duct system. Recommended hanger spacing is shown in Figure 6-5-28. Caution should be taken with other types of hanger systems to ensure that excessive stress is not placed on the hanger or the fibrous glass duct system.

(D) Fittings. Proper support of duct fittings may require that additional hangers be installed.

For an elbow, hangers should be on each leg, within twelve (12) inches (305 mm) of the throat. If the width of the duct is greater than eighteen (18) inches (457 mm), an additional hanger must be installed (see Figure 6-5-29) so that it is approximately two-thirds of the diagonal distance from throat to heel.

Tees require support on the trunk as shown in Figure 6-5-30. If a tee runout hanger falls where the trunk duct is located, add hangers on either side of trunk. Do not exceed maximum hanger spacing.

Branch ducts are treated in a similar manner with hanger spacing on the trunk duct in accordance with Figure 6-5-28.

For diffuser drops, hangers are installed within three (3) inches (76 mm) of each side of the drop. The connection of the drop to the duct is reinforced in accordance with Figure 6-5-31.

If the drop assembly, including the diffuser, weighs more than twenty-five (25) pounds (11.3 kg), the diffuser must be separately supported.

(E) Hanging Rigid Round Duct. Preformed round fibrous glass duct should be hung so the hanger will not damage the duct facing.

Straps or saddles in contact with the duct must be not less than 5/8 inch (15.9 mm) wide. Avoid sharp edges and burrs.

Space hangers at a maximum of six (6) feet (1,829 mm) on center. Where practical, hangers should be located at circumferential joints. Provide hanger support at all fittings.

Part B — Flexible Ducts.

A.6.507 Suitable Installations.

(A) General.

(1) The routing and length of flexible duct, the number of bends, the number of degrees of each bend, and the amount of sag allowed between support joints will have serious effects on system performance due to the increased resistance each introduces. Use the minimum length of flexible duct to make connections. It is not recommended that excess lengths of ducts be installed, to allow for possible future relocations of air terminal devices.

(2) Avoid installations where exposure to direct sunlight can occur, e.g., turbine vents, skylights, canopy windows, etc. Prolonged exposure to sunlight will cause degradation of the vapor barrier. Direct exposure to UV light from a source lamp installed within the HVAC system will cause degradation of some inner core/liner materials.

(3) Terminal devices shall be supported independently of the flexible duct.

(4) Repair torn or damaged vapor barrier jackets with approved duct tape listed and labeled to Standard UL 181B. If the internal core is penetrated, replace the flexible duct or treat it as a connection.

(B) Installation.

(1) Install duct fully extended. Do not install in the compressed state or use excess lengths. This will noticeably increase friction losses.

(2) Avoid bending ducts across sharp corners or placing in incidental contact with metal fixtures, pipes, or conduits. The radius at the centerline shall not be less than one (1) duct diameter.

(3) Do not install near hot equipment (e.g., furnaces, boilers, steam pipes, etc.) that is above the recommended flexible duct use temperature.

A.6.508 Connecting, Joining, and Splicing Flexible Duct.

(1) All connections, joints, and splices shall be made in accordance with the manufacturers' installation instructions.

(2) Tapes, mastics, and nonmetallic clamps used with flexible air ducts shall be listed and labeled to Standard UL 181B.

(3) Sheet metal fittings to which the flexible ducts with plain ends are attached shall be beaded and have a minimum of two (2) inches (51 mm) collar length. Beads are optional for fittings when attaching metallic flexible ducts.

(4) Sheet metal sleeves used for joining two sections of flexible duct shall be a minimum of four (4) inches (102 mm) in length and beaded on each end. Beads are optional for sleeves when joining metallic flexible ducts.

(5) Flexible ducts secured with nonmetallic clamps shall be limited to six (6) inches w.g. (1,500 Pa) positive pressure.

Installation Instructions for Air Ducts with Plain Ends.

A.6.509 Supporting Flexible Duct.

(A) Flexible duct shall be supported at manufacturer's recommended intervals, but at no greater distance than five (5) feet (1,524mm). Maximum permissible sag is 1/2 inch per foot (42 mm/m) of spacing between supports.

A connection to rigid ducting or equipment shall be considered a support joint.

Long horizontal duct runs with sharp bends shall have additional supports before and after the bend, approximately one (1) duct diameter distance from the centerline of the bend.

(B) Hanger or saddle material in contact with the flexible duct shall be of sufficient width to prevent any restriction of the internal diameter of the duct when the weight of the supported section rests on the hanger or saddle material. In no case will the material contacting the flexible duct be less than 1-1/2 inches (38 mm) wide.

(C) Factory-installed suspension systems integral to the flexible duct are an acceptable alternative hanging method when manufacturers' recommended procedures are followed.

in. x 25.4 = mm
ft.. x 304.8 = mm

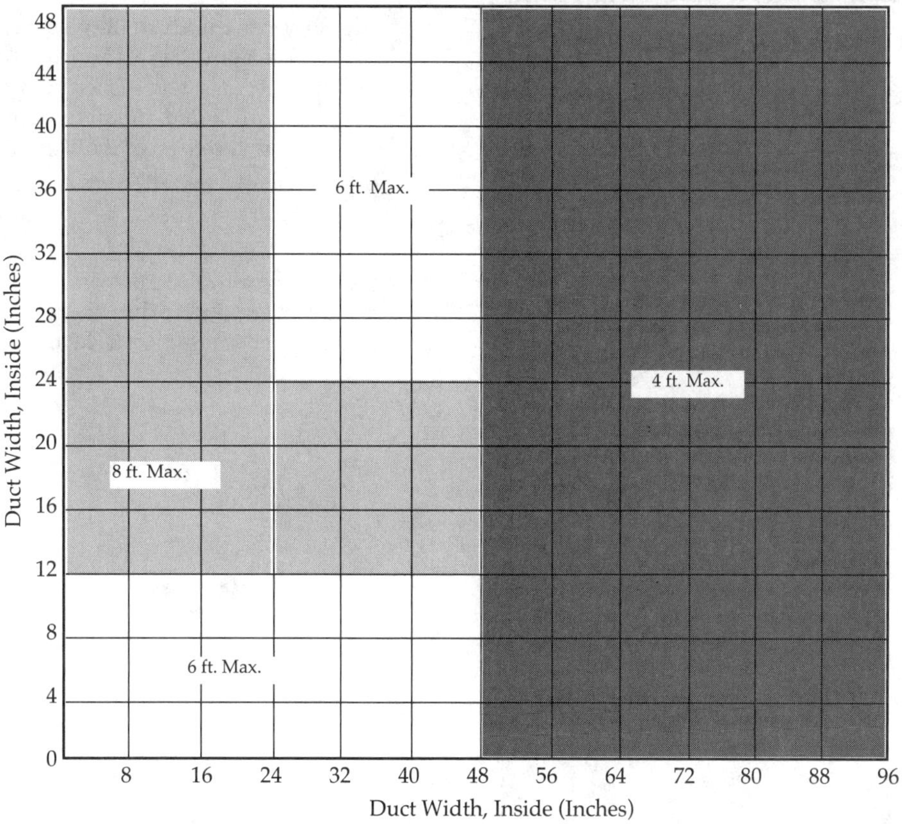

Figure 6-5-28 Minimum Hanger Spacing, Straight Duct, 3-Inch-Wide (76 mm) Channel.

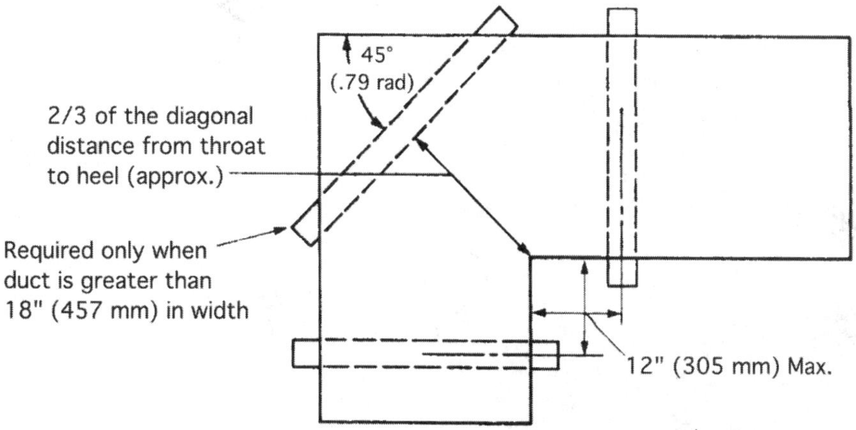

FIGURE 6-5-29 Elbow Support .

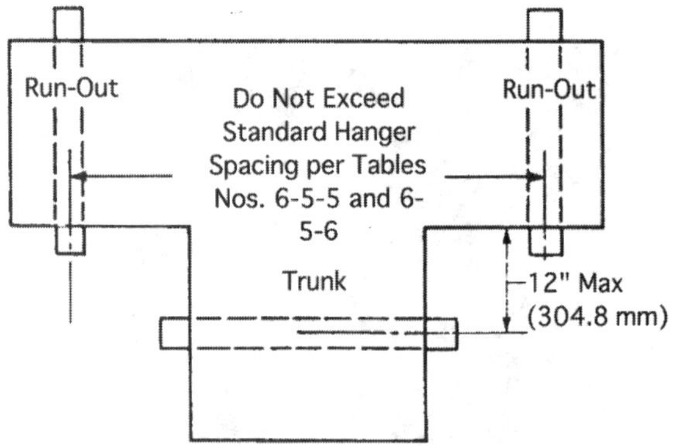

FIGURE 6-5-30 Tee Support.

(D) Flexible ducts may rest on ceiling joists or truss supports. Maximum spacing between supports shall not exceed the maximum spacing per manufacturers' installation instructions.

(E) Support the duct between a metal connection and a bend by allowing the duct to extend straight for a few inches before making the bend. This will avoid possible damage of the flexible duct by the edge of the sheet metal collar.

(F) Vertically installed duct shall be stabilized by support straps at a maximum of six (6) feet (1,829 mm) on center.

Note: Factory-made air ducts may not be used for vertical risers in air duct systems serving more than two stories. See Section 604.2 of this code.

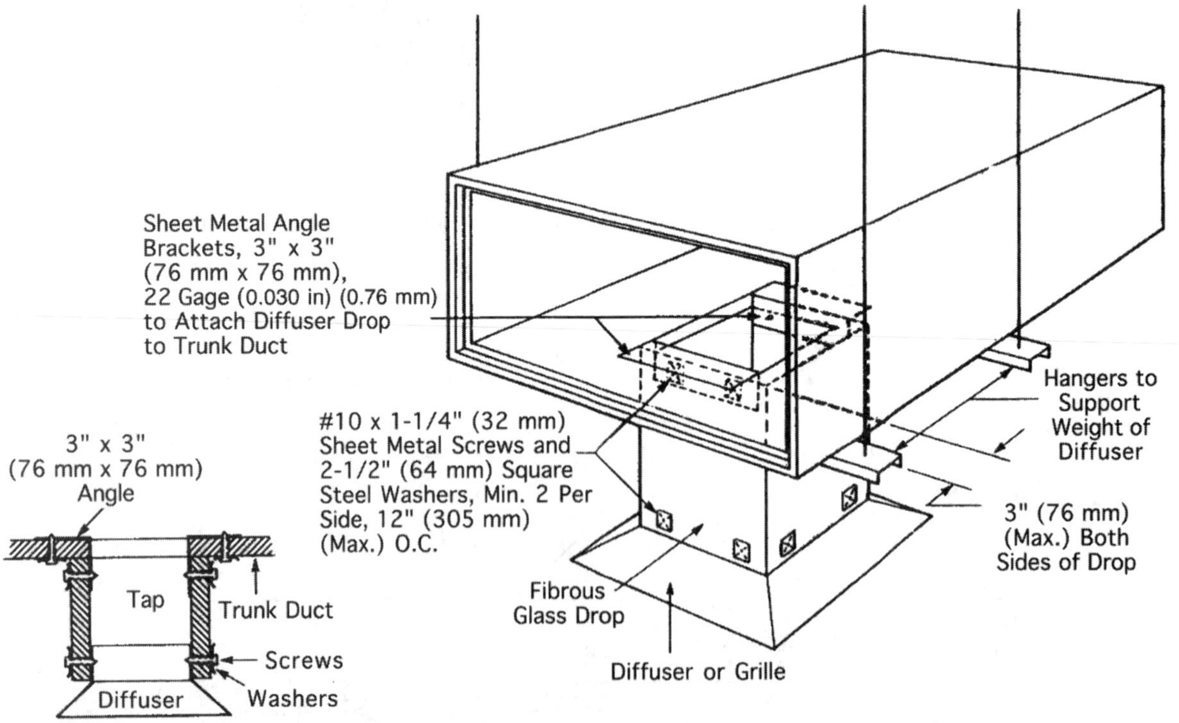

FIGURE 6-5-31 Diffuser Connection.

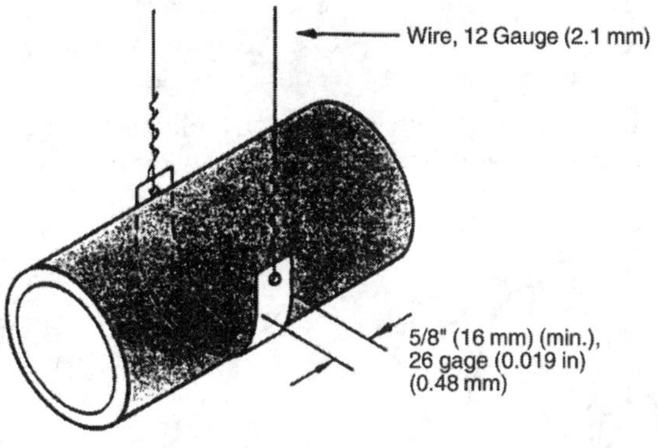

FIGURE 6-5-32 Hanging Round Duct with Saddle and 12 Gauge Wires.

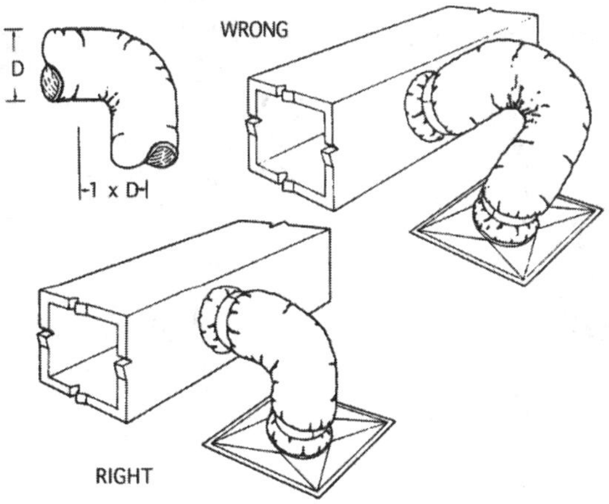

FIGURE 6-5-33 Extend Duct Fully.

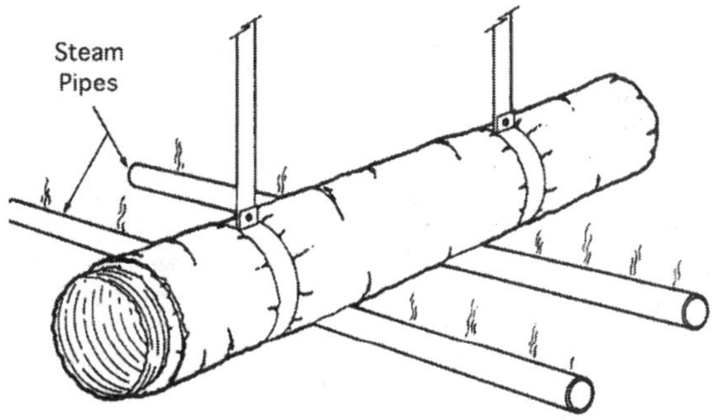

FIGURE 6-5-34 Wrong Method of Support.

The following figures contain the information in the previous edition Figures 6-5-35 and 6-5-36.

Connections

1. After desired length is determined, cut completely around and through duct with knife or scissors. Cut wire with wire cutters. Fold back jacket and insulation.

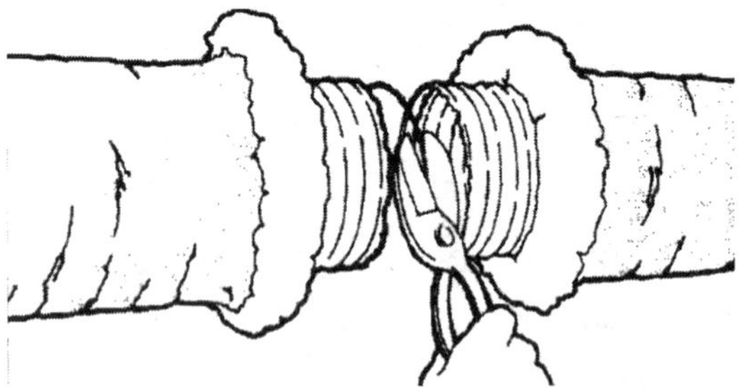

2. Slide at least 1" (25 mm) of core over fitting and past the bead. Seal core to collar with at least 2 wraps of duct tape. Secure connection with clamp placed over the core and tape and past the bead.

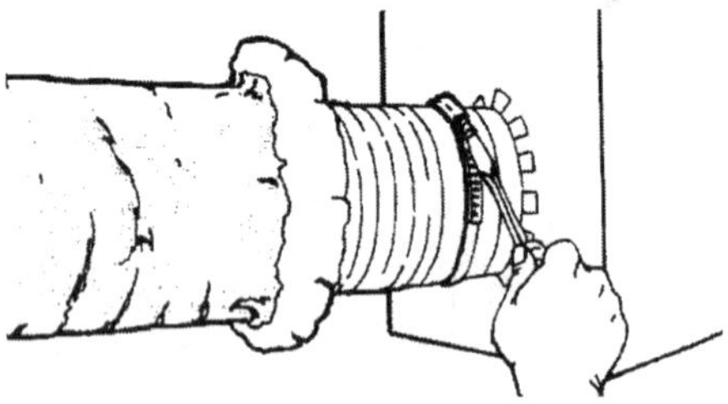

3. Pull jacket and insulation back over core. Tape jacket with at least 2 wraps of duct tape. A clamp may be used in place of or in combination with the duct tape.

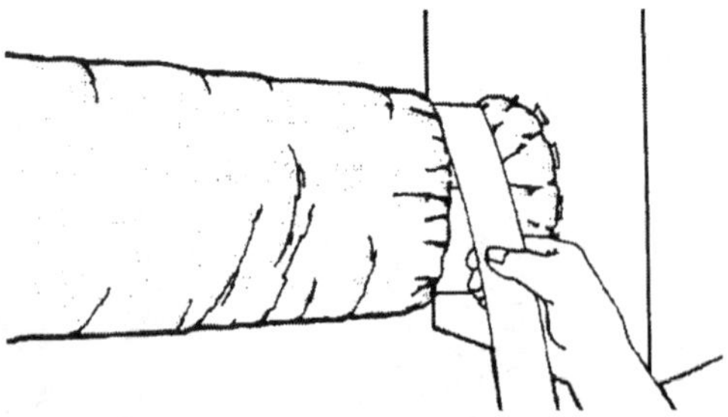

Splices

1. Fold back jacket and insulation from core. Butt two cores together on a 4" (100mm) length metal sleeve.

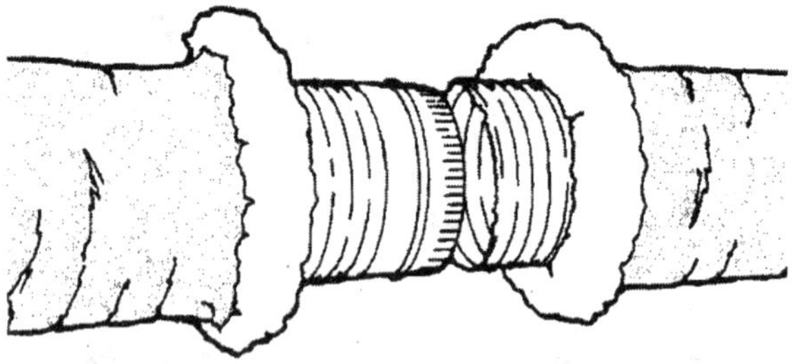

2. Tape cores together with at least 2 wraps of duct tape. Secure connection with 2 clamps placed over the taped core ends and past the beads.

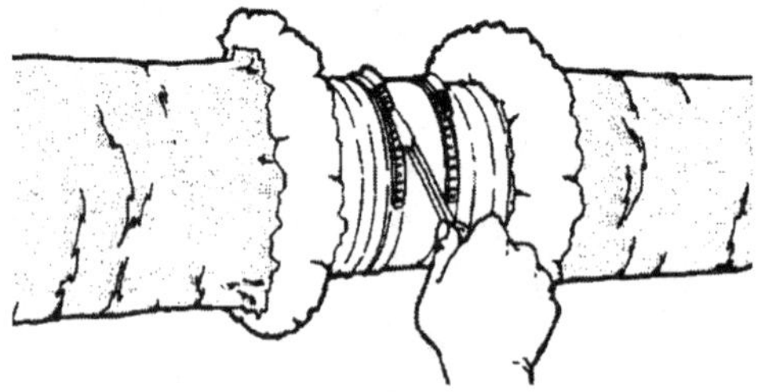

3. Pull jacket and insulation back over cores. Tape jackets together with at least 2 wraps of duct tape.

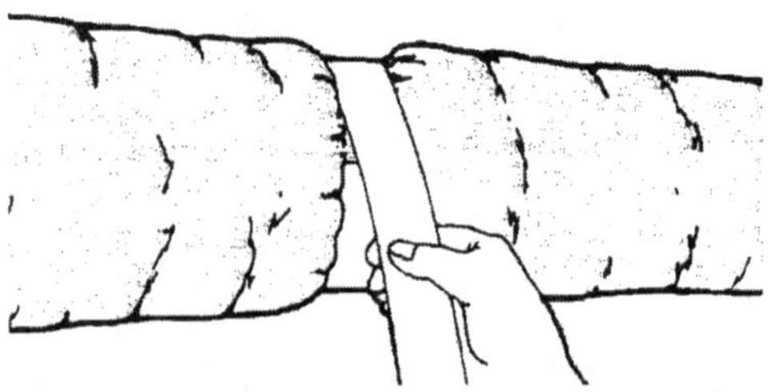

Note:

1. For uninsulated air ducts and air connectors, disregard references to insulation and jacket.
2. Use beaded sheet metal fittings and sleeves.
3. Use tapes listed and labeled in accordance with Standard UL 181B and marked "181B-FX".
4. Nonmetallic clamps shall be listed and labeled in accordance with Standard UL 181B and marked "181B-C". Use of nonmetallic clamps shall be limited to 6 in. w.g. (1500 Pa) positive pressure.

Alternate Installation Instructions for Air Ducts and Air Connectors – Nonmetallic with Plain Ends.

Connections and Splices

Step 1: After desired length is determined, cut completely around and through duct with knife or scissors. Cut wire with wire cutters. Pull back jacket and insulation from core.

Step 2: Apply mastic approximately 2" (50 mm) wide uniformly around the collar of the metal fitting or over the ends of a 4" (100 mm) metal sleeve. Reference data on mastic container for application rate, application thickness, cure times, and handling information.

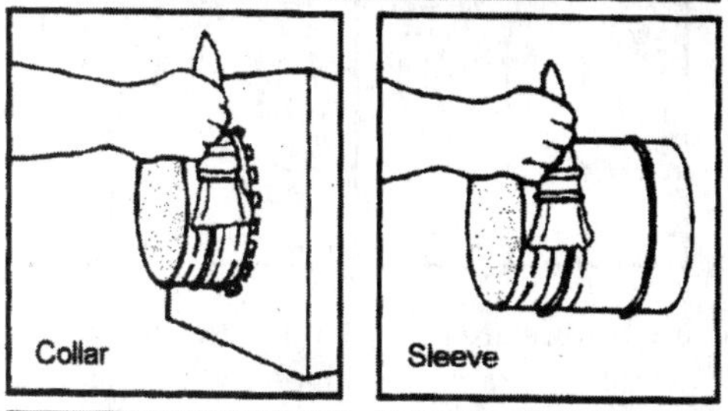

Step 3: Slide at least 2" (50 mm) of core over the fitting or sleeve ends and past the bead.

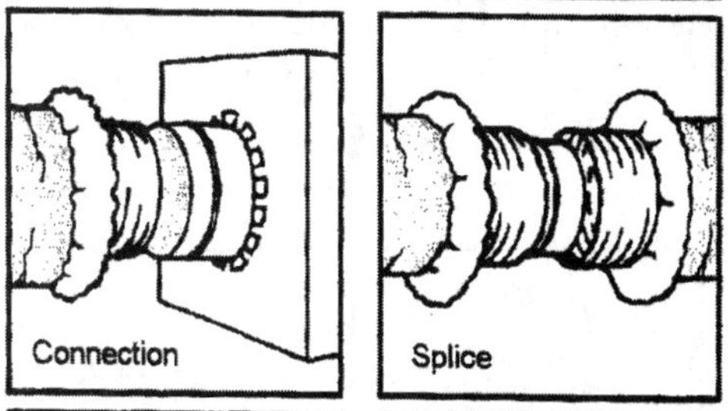

Step 4: Secure core to collar with a clamp applied past the bead. Secure cores to sleeve ends with 2 clamps applied past the beads.

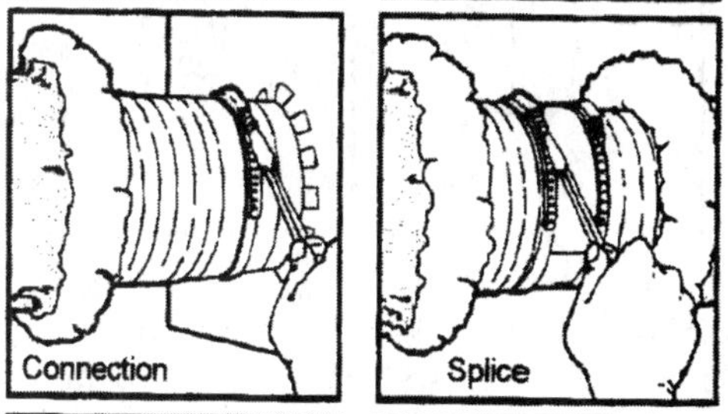

Step 5: Pull jacket and insulation back over core ends. Tape jacket(s) with at least 2 wraps of duct tape. A clamp may be used in place of or in combination with the duct tape.

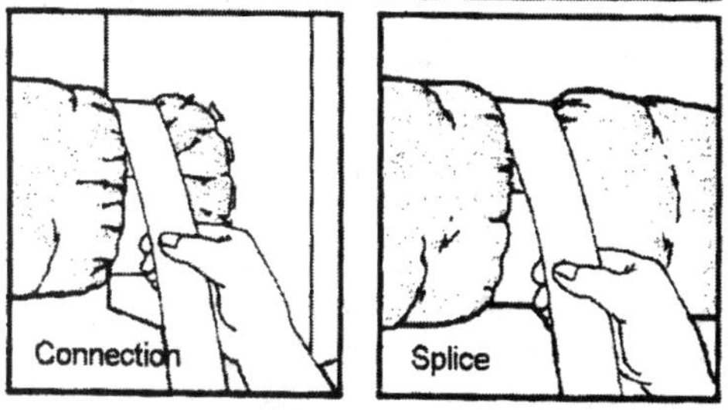

Notes:

1. For uninsulated air ducts and air connectors, disregard references to insulation and jacket.
2. Use beaded sheet metal fittings and sleeves.
3. Use mastics listed and labeled in accordance with Standard UL 181B and marked "181B-M" on container.
4. Use tapes listed and labeled in accordance with Standard UL 181B and marked "181B-FX".
5. Nonmetallic clamps shall be listed and labeled in accordance with standard UL 181B and marked "181BC". Use of nonmetallic clamps shall be limited to 6 in. w.g. (1,500 Pa) positive pressure.

Installation Instruction for Air Ducts and Air Connectors – Metallic with Plain Ends.

Connections and Splices

1. After cutting duct to desired length, fold back jacket and insulation exposing core. Trim core ends squarely using suitable metal shears. Determine optional sealing method (step 2 or 5) before proceeding.

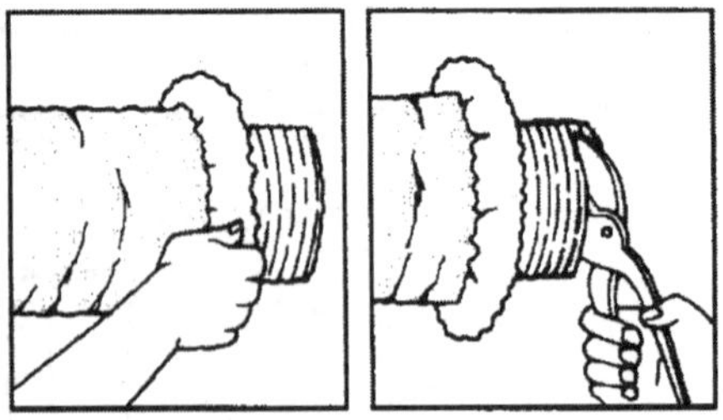

2. When mastics are required and for pressure 4" w.g. (1,000 Pa) and over, seal joint with mastic applied uniformly to the outside surface of collar/sleeve. (Disregard this step when not using mastics and proceed to step 3).

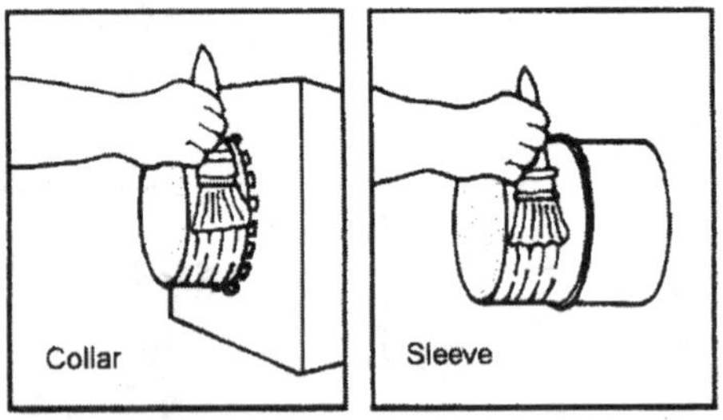

3. Slide at least 1" (25mm) of core over metal collar for attaching duct to take off or over ends of a 4" (100 mm) metal sleeve for splicing 2 lengths of duct.

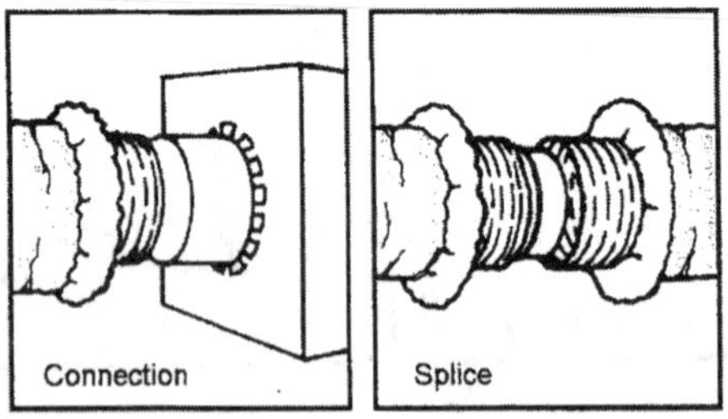

4. Secure to collar/sleeve using #8 sheet metal screws spaced equally around circumference. Use 3 screws for diameters under 12" (300 mm) and 5 screws for diameters 12" (300 mm) and over.

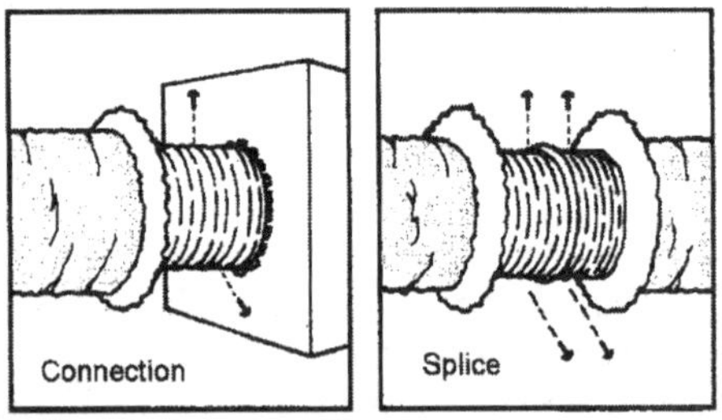

5. For pressure under 4" w.g. (1,000 Pa), seal joint using 2 wraps of duct tape applied over screw heads and spirally lapping tape to collar/sleeve. (Disregard this step when using mastics per Step 2.)

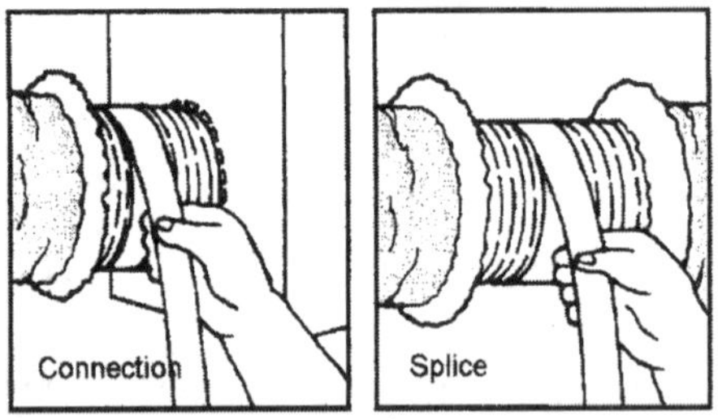

6. Pull jacket and insulation back over core. Tape jacket with 2 wraps of duct tape. A clamp may be used in place of or in combination with the duct tape.

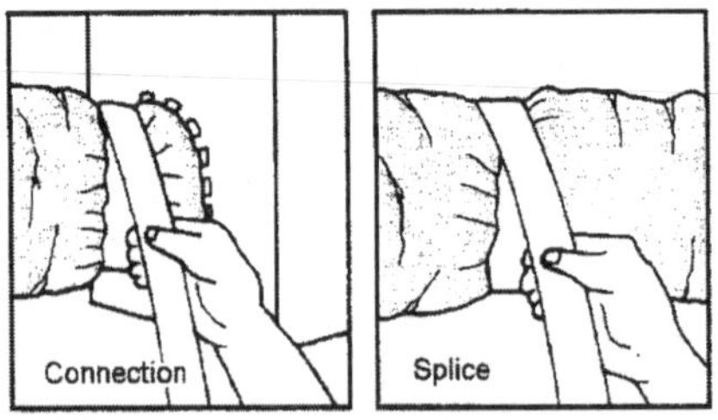

Notes:

1. For uninsulated air ducts and air connectors, disregard references to insulation and jacket.
2. Use mastics listed and labeled to Standard UL 181B and marked "181B-M" on container.
3. Use tapes listed and labeled to Standard UL 181B and marked "181B-FX".
4. Nonmetallic clamps shall be listed and labeled in accordance with Standard UL 181B and marked "181B-C".

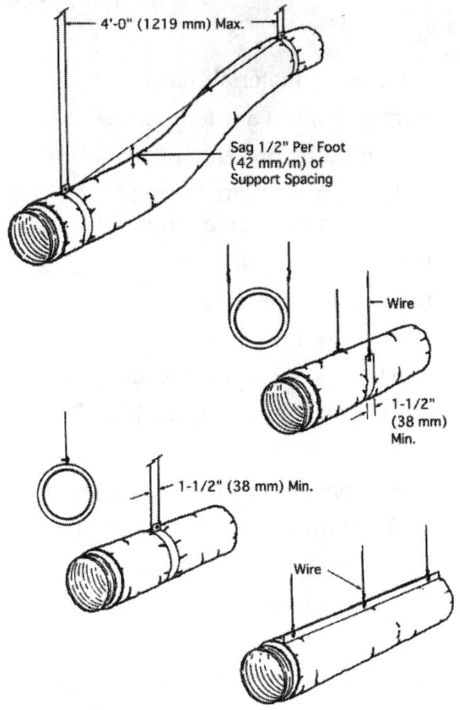

FIGURE 6-5-37.

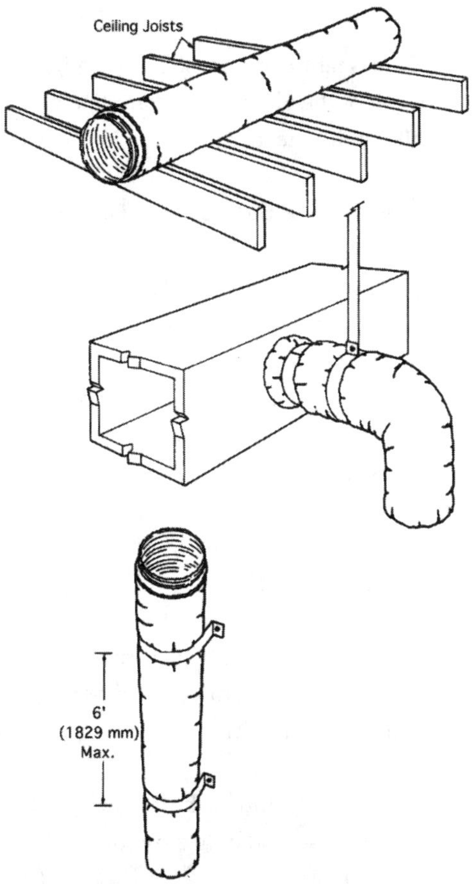

FIGURE 6-5-38 Proper Support.

A.6.510 Checklist.

The following checklist is provided for the benefit of the inspector, as well as the installer. It is designed so that the correct answer to all questions is yes.

General		Yes	No
___	Have all tears or punctures to facing material been repaired using the proper technique?	( )	( )
___	Are all sheet metal accessory items galvanized?	( )	( )
Product			
___	Is the UL label present? (Although each board is labeled, each section may not be, since there is only one label per sheet.)	( )	( )
Fabrication and Installation			
___	When metal parts are attached, are 2-1/2 inch (64 mm) minimum square steel washers used on 16-inch (406 mm) maximum centers?	( )	( )
___	When staples cannot be used, are 8 inch (203 mm) cross-tabs of approved closure being used in place of staples?	( )	( )
___	Are tab spacing requirements 12 inches, (305 mm) on center minimum one per side.	( )	( )
___	Are all system joints tight, free from bulges, with taped joints showing good workmanship?	( )	( )
___	Have offsets been installed so duct sections are not forced to bend around obstructions?	( )	( )

Fire Dampers—See Section A. 6.505 **Yes No**

___ Is sheet metal sleeve present? Is duct properly attached to sleeve with screws and washers 16 inches (406 mm) on center? (Fibrous glass ducts must not penetrate assemblies required to have a fire damper.) ( ) ( )

Access Doors

___ Is installation in accordance with Section A. 6.504(E)6? ( ) ( )

Grills, Diffusers, Registers—See Section A. 6.506

___ Is the extra weight of the item being separately supported and not dependent on the duct alone for support? (Exception: Registers not greater than 150 square inches (96.7×103 mm^2) in area may be attached to the duct with metal channel, without other support.) ( ) ( )

Connection to Units

___ Are sheet metal screws and washers used to secure the duct system to flange extensions? (Securing the duct to the unit flange without mechanical fasteners is sufficient.) ( ) ( )

Closures—See Section A. 6.503

___ Are all joints in the system properly sealed? ( ) ( )

___ Are closure materials for rigid duct systems of a listed type, as evidenced by presence of UL instruction sheet in duct board carton or imprinted UL 181A on the tape? ( ) ( )

Yes No

___ Are closure materials for flexible duct systems of a listed type, as evidenced by presence of UL instruction sheet in the duct carton and imprinted UL 181B on the tape, mastic, and nonmetallic clamps? ( ) ( )

___ Are there staples or cross-tabs, properly spaced, on circumferential joints? ( ) ( )

___ Are all pressure-sensitive tape closures rubbed down adequately, with staples or scrim infacing clearly visible through the tape? ( ) ( )

___ If heat-sealable tape closure was used, was it applied correctly, as evidenced by dot color change? ( ) ( )

___ If glass fabric and mastic are used, is the mesh of the glass fabric completely filled with mastic? ( ) ( )

Reinforcement—See Section A. 6.504

___ Is the reinforcement system in accordance with Section A. 6.504? ( ) ( )

___ Is the tie rod spacing correct according to duct span, board type, and static pressure? ( ) ( )

___ Are tie rod washers 2-1/2 inches (64 mm) square? ( ) ( )

___ Do tie rod washers have turned edges facing away from duct board so they will not cut into it? ( ) ( )

___ If tie rods reinforce a butt joint, are rods used on both sides of the butt joint? ( ) ( )

		Yes	No
___	Is wire termination one of those documented in Section 6.504(B)?	( )	( )
___	Are anti-sag devices used on ducts 48 inch (1,219 mm) span or greater to support top panel of ducts?	( )	( )
___	Are heels of tees, elbows, and end caps reinforced (formed sheet metal channel, tie rod, or combination)?	( )	( )
___	When formed sheet metal channel reinforcement is used, are sheet metal gauges, dimensions, and spacing correct?	( )	( )
___	For return ducts, are sheet metal channel reinforcements attached to ducts with screws and 2-1/2 inch (64 mm) square washers or 2 inch by 6 inch (51 x 152 mm) clips?	( )	( )

Hangers and Supports—See Sections A. 6.506 and A. 6.509

___	Are hangers installed in accordance with Section 6.506?	( )	( )
___	Are hanger designs in accordance with Table 6-5-6?	( )	( )
___	Are accessories that add weight to the duct system separately supported so as not to stress the system? (See Section A. 6.506(4).)	( )	( )
___	If formed sheet metal reinforcements are used as hangers, are attachments within 6 inches (152 mm) of duct sides?	( )	( )
___	Are all fittings supported by hangers?	( )	( )

APPENDIX B

PROCEDURES TO BE FOLLOWED TO PLACE GAS EQUIPMENT IN OPERATION

Note: Portions of this chapter have been reproduced in their entirety from Chapter 11 NFPA 54-2002 (ANSI Z223.1)

B.1.0 Adjusting the Burner Input.

B.1.1 Adjusting Input. The input shall be adjusted to the proper rate in accordance with the equipment manufacturer's instructions by changing the size of a fixed orifice, by changing the adjustment of an adjustable orifice, or by readjusting the gas pressure regulator outlet pressure (where a regulator is provided). Overfiring shall be prohibited.

B.1.2 High Altitude. Ratings of gas utilization equipment are based on sea level operation and shall not be changed for operation at elevations up to 2,000 feet (600 m). For operation at elevations above 2,000 feet (600 m), equipment ratings shall be reduced in accordance with one of the following methods:

(1) At the rate of 4 percent for each 1,000 feet (300 m) above sea level before selecting appropriately sized equipment.

(2) As permitted by the Authority Having Jurisdiction.

(3) Listed appliances derated in accordance with the terms of the listing shall be permitted.

B.2.0 Primary Air Adjustment.

The primary air for injection (Bunsen)-type burners shall be adjusted for proper flame characteristics in accordance with the manufacturers' instructions. After setting the primary air, the adjustment means shall be secured in position.

B.3.0 Safety Shutoff Devices.

Where a safety shutoff device is provided, it shall be checked for proper operation and adjustment in accordance with the manufacturer's instructions. Where the device does not function properly to turn off the gas supply in the event of pilot outage, it shall be properly serviced or replaced with new equipment.

B.4.0 Automatic Ignition.

Gas utilization equipment supplied with means for automatic ignition shall be checked for proper operation. If necessary, proper adjustments shall be made.

B.5.0 Protective Devices.

All protective devices furnished with the gas utilization equipment, such as a limit control, fan control to blower, temperature-and pressure-relief valve, low-water cutoff device, or manual operating features, shall be checked for proper operation.

B.6.0 Checking the Draft.

Vent-connected gas utilization equipment shall be operated for several minutes and checked to see that the combustion products are going up the chimney or gas vent properly by passing a lighted match or taper around the edge of the relief opening of the draft hood. Where the chimney or gas vent is drawing properly, the match flame will be drawn into the draft hood. Where not, the combustion products will tend to extinguish this flame. Where the combustion products are escaping from the relief opening of the draft hood, the equipment shall not be operated until proper adjustments or repairs are made to provide adequate draft through the chimney or gas vent.

B.7.0 Operating Instructions.

Operating instructions shall be furnished and shall be left in a prominent position near the equipment for the use of the consumer.

TABLE B-1-1
Gas Input to Burner in Cubic Feet per Hour
Size of Test Meter Dial

Seconds for One Revolution	1/2 ft.^{3}	1 ft.^{3}	2 ft.^{3}	5 ft.^{3}
10	180	360	720	1800
11	164	327	655	1636
12	150	300	600	1500
13	138	277	555	1385
14	129	257	514	1286
15	120	240	480	1200
16	112	225	450	1125
17	106	212	424	1059
18	100	200	400	1000
19	95	189	379	947
20	90	180	360	900
21	86	171	343	857
22	82	164	327	818

TABLE B-1-1 (continued)
Gas Input to Burner in Cubic Feet per Hour
Size of Test Meter Dial

Seconds for One Revolution	1/2 ft.^{3}	1 ft.^{3}	2 ft.^{3}	5 ft.^{3}
23	78	157	313	783
24	75	150	300	750
25	72	144	288	720
26	69	138	277	692
27	67	133	267	667
28	64	129	257	643
29	62	124	248	621
30	60	120	240	600
31	58	116	232	581
32	56	113	225	563
33	55	109	218	545
34	53	106	212	529
35	51	103	206	514
36	50	100	200	500
37	49	97	195	486
38	47	95	189	474
39	46	92	185	462
40	45	90	180	450
41	44	88	176	440
42	43	86	172	430
43	42	84	167	420
44	41	82	164	410
45	40	80	160	400
46	39	78	157	391
47	38	77	153	383
48	37	75	150	375
49	37	73	147	367
50	36	72	144	360
51	35	71	141	353
52	35	69	138	346
53	34	68	136	340
54	33	67	133	333
55	33	65	131	327
56	32	64	129	321
57	32	63	126	316
58	31	62	124	310
59	30	61	122	305
60	30	60	120	300
62	29	58	116	290

TABLE B-1-1 (continued)

Seconds for One Revolution	1/2 ft.^{3}	1 ft.^{3}	2 ft.^{3}	5 ft.^{3}
64	29	56	112	281
66	29	54	109	273
68	28	53	106	265
70	26	51	103	257
72	25	50	100	250
74	24	48	97	243
76	24	47	95	237
78	23	46	92	231
80	22	45	90	225
82	22	44	88	220
84	21	43	86	214
86	21	42	84	209
88	20	41	82	205
90	20	40	80	200
94	19	38	76	192
98	18	37	74	184
100	18	36	72	180
104	17	35	69	173
108	17	33	67	167
112	16	32	64	161
116	15	31	62	155
120	15	30	60	150
130	14	28	55	138
140	13	26	51	129
150	12	24	48	120
160	11	22	45	112
170	11	21	42	106
180	10	20	40	100

Note: To convert to Btu per hour, multiply by the Btu heating value of the gas used.

APPENDIX C

INSTALLATION AND TESTING OF OIL (LIQUID) FUEL-FIRED EQUIPMENT

C.1.0 General. Appendix C governs the installation, testing, or repair of oil or liquid fuel burners, oil-or liquid-fuel burning systems, oil-or liquid-fuel-burning equipment, and the oil-or-liquid fuel piping systems used in connection with buildings or structures and equipment within the property lines of the premises.

Definitions.

ANTI-FLOODING DEVICE – A primary safety control that causes the flow of oil or fuel to be shut off after a rise in oil or fuel level, or after receiving excess oil or fuel, and that operates before the hazardous discharge of oil or fuel can occur.

BURNER, AUTOMATICALLY IGNITED – A burner equipped so that main burner fuel may be turned on and ignited automatically.

BURNER, MANUALLY IGNITED – A burner equipped so that main burner fuel is turned on only by hand and ignited under supervision.

BURNER, MECHANICAL DRAFT TYPE – A burner that includes a power-driven fan, blower, or other mechanism as the primary means for supplying the air for combustion.

BURNER, NATURAL DRAFT TYPE – A burner that depends primarily on the natural draft created in the chimney or venting system to induce air required for combustion into the burner.

CONSTANT LEVEL VALVE – A device for maintaining within a reservoir a constant level of oil or fuel for delivery to a burner.

CONTROL LIMIT – An automatic safety control that is responsive to changes in fluid flow or level, pressure, or temperature, and that is normally set beyond the operating range for limiting the operation of the controlled equipment by shutting off the energy supply.

CONTROL SAFETY – Automatic interlock controls, including relays, switches, and other auxiliary equipment used in conjunction with them, to form a safety control system that is intended to prevent unsafe operation of the controlled equipment.

DRAFT BOOSTER – A power-operated fan, blower, or other device installed in the chimney connector to increase the natural draft developed in the connected chimney.

DRAFT REGULATOR, BAROMETRIC – A device built into a fuel-burning appliance or made part of a chimney connector or vent connector that functions to reduce excessive draft through an appliance to a desired value by admitting ambient air into the appliance chimney, chimney connector, vent, or vent connector.

FUEL – Natural, manufactured, or liquefied petroleum gas, or a mixture of these gases; all grades of fuel oil; wood; or any other combustible or flammable material or any mixture of combustible or flammable materials.

FUEL BURNER – A device used to convey the appropriate fuel into the combustion chamber zone in close proximity to its primary and secondary air supply to permit a stable controlled heat release compatible with the burner design, listing, and applicable approvals in a boiler, furnace, device or appliance. It includes, but is not limited to, burning oil or liquid fuel.

FUEL BURNER SYSTEM – The fuel burner and any conveyance system or piping system for the purpose of introducing the appropriate fuel into the combustion chamber zone.

FUEL-BURNING EQUIPMENT/APPLIANCE – An oil or fuel burner of any type including all oil or liquid fuel burners, oil, or liquid-fuel-fired-units, dual, or multifuel burners and heating and cooking appliances with their fuel burner system and with their tank or fuel storage system, piping system, vent connectors, vent flues, fans, blowers, valves, control devices, combustion air, wiring, controls, and related devices including all accessories and appurtenances for safe and proper operation of the appliance.

FUEL OIL – Any hydrocarbon oil as specified by ASTM D396, or the Canadian Government Specification Board, 3-GP-28, and having a flashpoint of not less than 100°F (38°C).

FUEL-PIPING SYSTEM – Any method of conveying liquid, vapor, steam, gases, or slurry from one point to another, including accessories, appurtenances, and equipment necessary for its proper operation.

INDIRECT-FIRED APPLIANCE – An oil- or fuel-burning appliance in which products of combustion (flue gasses) are not mixed in the appliance with the air or other medium being heated.

LABELED – Having attached a label, symbol, or other identifying mark of an organization acceptable to the Building Official and concerned with product evaluation that maintains periodic inspection of production of labeled equipment or materials and by whose labeling the manufacturer indicates compliance with appropriate standards or performance in a specified manner.

PREMIXING – A power burner in which all or nearly all of the air for combustion is mixed with the gas as primary air.

PUMP, OIL OR FUEL TRANSFER – An oil or fuel pump, automatically or manually operated, that transfers oil or fuel through continuous piping from a supply tank to an oil or fuel-burning appliance or to an auxiliary tank, and that is not designed to stop pumping automatically in case of total breakage of the oil or fuel supply line between the pump and the appliance.

TANK, AUXILIARY – A tank having a capacity of not over sixty (60) gallons listed for installation in the supply piping between a burner and its main fuel supply tank. It may be included as an integral part of an automatic pump or a transfer pump, or it may be a separate tank.

TANK, GRAVITY – A supply tank from which the oil or fuel is delivered directly to the burner by gravity.

TANK, INTEGRAL – A tank that is furnished by the manufacturer as an integral part of an oil- or fuel-burning appliance.

TANK STORAGE – A separate tank that is not connected to the oil- or fuel-burning appliance.

TANK SUPPLY – A separate tank connected directly or by a pump to the oil- or fuel-burning appliance.

TANK, VACUUM OR BAROMETRIC – A tank not exceeding five (5) gallons capacity that maintains a definite level of oil or fuel in a sump or similar receptacle by barometric feed. Fuel is delivered from the sump to the burner by gravity.

VALVE, MANUAL OIL, GAS, OR FUEL SHUTOFF – A manually operated valve in a fuel line for the purpose of turning on or completely shutting off the fuel supply to the burner.

VALVE, OIL, GAS, OR FUEL CONTROL – An automatically or manually operated device consisting essentially of a fuel valve for controlling the fuel supply to a burner.

C.2.0 Standards and Accepted Practices. The installation, testing, and repair of oil- or liquid-fuel-burning equipment systems must comply with this section, the appropriate standards listed in Chapter 17, and other information outlined in the UMC such as, but not limited to, combustion air, flue and breeching requirements, room clearances and dimensions, and control requirements.

C.3.0 Approval of Equipment. Oil-or liquid-fuel-burning equipment must be approved.

C.4.0 Placing Equipment in Operation. After completion of all installations, the installer shall test all safety and operating controls and venting before placing the burner in service. The correct input of liquid fuel must be determined and the fuel-to-air ratio set. Each oil or liquid fuel burner must be adjusted to its proper input according to the manufacturer's instructions. Overrating the burners or the appliance is prohibited. The input range must be appropriate to the appliance.

(1) For conversion burners installed in hot water (liquid) boilers or warm-air furnaces, the rate of flow of the oil or liquid fuel in btu/h must be adjusted to within plus or minus 5 percent of the design load, not to exceed the design rate of the appliance.

(2) For conversion burners installed in steam boilers, the oil or liquid fuel hourly input demand must be adjusted to meet the steam load requirements. The oil or liquid fuel input demand necessitated by an oversized boiler must be established and added to total input demand.

C.5.0 Pilot Operation. Igniter or pilot flames must be effective to ignite the oil or liquid fuel at the main burner or burners and must be adequately protected from drafts. Pilot flames must not become extinguished during the pilot cycle when the main burner or burners are turned on or off in a normal manner either manually or by automatic controls.

C.6.0 Burner Operation. In making tests to determine compliance with the requirements of this section, care must be exercised to prevent the accumulation of unburned liquid fuel in the appliance that might result in an explosion or fire.

(1) The flames from the burner must freely ignite the liquid fuel when operating at the lowest firing position.

(2) Burner flames must not flash back when the liquid fuel is turned on or off by an automatic control mechanism.

(3) Main burner flames must ignite freely from the pilot when the pilot flame is reduced to a minimum point that will actuate the pilot safety device.

(4) When ignition is made in a normal manner, the flame will not flash outside the appliance.

(5) Burners must not expel liquid fuel through air openings when operating at prevailing pressure.

(6) Burners must have a proper liquid fuel air mixture to ensure smooth ignition of the main burner.

C.7.0 Method of Test.

(1) The flue gas, venting, safety, and operating controls of the appliance must be checked by the installer to ensure their proper and safe operation.

(2) Method of test – atmospheric, induced draft, or fan assisted types.

The appliance must be allowed to operate until the stack temperature becomes stabilized, after which a sample of the undiluted flue products must be taken from the appliance flue outlet. The sample taken must be analyzed for carbon monoxide, carbon dioxide, and oxygen. Stack temperature must be noted.

Note:

Appliance designs incorporating induced-draft assemblies may require a flue gas sample to be taken ahead of the draft regulator or induced draft fan.

(3) Performance standards for atmospheric type must meet:

(a) Minimum of 75 percent efficiency as determined by flue gas analysis method at appliance flue outlet.

(b) Carbon monoxide concentration in flue gas not greater than 0.04 percent.

(c) Stack temperature not greater than 700°F (371°C), plus ambient.

(d) Carbon dioxide concentration between 8 percent and 13 percent.

(e) Oxygen concentration between 4 percent and 10 percent.

(f) Smoke test no higher than #2 for light oils or #4 for oils heavier than #4.

(g) Draft must be in accordance with the burner manufacturer's specifications.

(4) Performance standards for induced-draft or fan-assisted types:

(a) Minimum of 75 percent efficiency, as determined by flue gas analysis method at appliance flue outlet.

(b) Carbon monoxide concentration in flue gas not greater than 0.04 percent.

(c) Stack temperature not greater than 700°F (371°C), plus ambient

(d) Carbon dioxide concentration between 8 percent and 13 percent.

(e) Oxygen concentration between 4 percent and 10 percent.

(f) Smoke test no higher than #2 for light oils, or #4 for oils heavier than #4.

(g) Draft must be in accordance with the burner manufacturer's specifications.

Note:

Induced draft- and fan-assisted types of appliances may require a sample be taken after the induced-draft fan that may cause oxygen figures in excess of the limits stated. In such cases, safe liquid fuel combustion ratios must be maintained and be consistent with approvals and listings of the appliance.

(5) Method of test—power type

The appliance must be allowed to operate until the stack temperature becomes stabilized; after that, a sample of the undiluted flue products must be taken from the appliance flue outlet. The sample must be analyzed for carbon monoxide, carbon dioxide, and oxygen. Stack temperature must be noted.

(6) Performance standards for power type

(a) Minimum of 80 percent efficiency as determined by flue gas analysis method at the appliance flue outlet.

(b) Carbon monoxide concentration in the flue gas not greater than 0.04 percent.

(c) Stack temperature not greater than 700°F (371°C) plus ambient.

(d) Carbon dioxide concentration between 8 percent and 13 percent.

(e) Oxygen concentration between 4 percent and 10 percent.

(f) Smoke test no higher than #2 for light oils or #4 for oils heavier than #4.

(g) Draft must be in accordance with the burner manufacturer's specifications.

(7) After completion of the test of newly installed oil- or liquid-fuel-burner equipment as provided in this section, the installer shall file with the Authority Having Jurisdiction complete records of the test on a form approved by the Authority Having Jurisdiction. The tag stating the date of the test and the name of the installer must be attached to the appliance at the main valve.

(D) 2,500,000 to 12,499,999 Btu/h per burner (17.5 to 87.5 gallons per hour)

(1) One (1) approved manual shut-off valve lever handle.

(2) One (1) approved fuel oil filter, installed on the supply piping.

(3) Two (2) safety shutoff valves in series, with a combined flame failure response and valve closing time not to exceed five (5) seconds with strainer directly before the valves.

(4) Programmed electronic flame safeguard including proven low fire start, manual reset lockout, 100 percent shutoff (both pilot and main burner), and a separately supervised and proven pilot.

Note:

Flame-sensing systems utilizing a UV scanner must prove pilot and interrupt ignition spark prior to main burner valves being energized.

(5) Two (2) controls, one (1) operating and one (1) high limit, activated by temperature or pressure, as appropriate.

(6) All burners relying on mechanical means to provide air for combustion must have actual proof-of-air interlock device.

(7) Power burners must include proven prepurge of at least sixty (60) seconds at high fire damper settings. This prepurge must occur before every burner cycle, regardless of reason.

(8) Installations with dampered combustion air openings must prove damper open position before trial for burner ignition.

(9) Vent dampers and flue dampers must be properly interlocked to prevent burner ignition unless safely open.

(10) One (1) high oil or liquid fuel pressure interlock, reset from flame safeguard or manually.

(11) If hot water or steam, two (2) low water cutoffs.

(12) An atomizing medium proving switch.

(13) A low-oil-temperature switch for oil or liquid fuel requiring preheating.

(14) A high-oil-temperature interlock for oil or liquid fuel requiring preheating.

(15) A separate firing rate control valve.

(16) The burner oil pump must automatically not operate or rotate while the alternate fuel is firing.

(17) A pressure-relief valve must be provided between safety shutoff valves and between pump and safety valves if an integral valve is used with a pump.

(18) A separate relief device is required on each transfer pump.

(19) One (1) low-oil or liquid-fuel pressure interlock reset from flame safeguard or manually.

(20) Burners with automatic controls, prepurge, proof-of-closure, modulation, or postpurge shall not use relays external to the flame safeguard to accomplish these functions.

(E) 12,500,000 and greater Btu/h per burner inputs – these burners must conform with the requirements of the appropriate standards listed in Chapter 17 and the following (greater than 87.5 gallons per hour):

(1) One (1) approved manual shut-off valve lever handle.

(2) One (1) approved fuel oil filter, installed on the supply piping.

(3) Two safety shutoff valves in series, one with proof of closure, with a combined flame failure response and valve closing time not to exceed two (2) seconds with strainer directly before the valves.

(4) Programmed electronic flame safeguard including proven low fire start, manual reset lockout, 100 percent shutoff (both pilot and main burner), and a separately supervised and proven pilot.

Note:

Flame-sensing systems utilizing a UV scanner must prove pilot and interrupt ignition spark prior to main burner valves being energized.

(5) Two (2) controls, one (1) operating and one (1) high limit, activated by temperature or pressure, as appropriate.

(6) All burners relying on mechanical means to provide air for combustion must have actual proof-of-air interlock device.

(7) Power burners must include proven prepurge of at least sixty (60) seconds at high fire damper settings. This prepurge must occur before every burner cycle, regardless of reason.

(8) Installations with dampered combustion air openings must prove damper open position before trial for burner ignition.

(9) Vent dampers and flue dampers must be properly interlocked to prevent burner ignition unless safely open.

(10) One (1) high-oil or liquid-fuel pressure interlock; reset from flame safeguard or manually.

(11) A manual firing cock.

(12) If hot water or steam, two (2) low water cut-offs.

(13) An atomizing medium proving switch.

(14) A low-oil-temperature switch for oil or liquid fuel requiring preheating.

(15) A high-oil-temperature interlock for oil or liquid fuel requiring pre-heating.

(16) A separate firing rate control valve.

(17) The burner oil pump must automatically not operate or rotate while the alternate fuel is firing.

(18) A pressure-relief valve must be provided between safety shutoff valves and between pump and safety valves if an integral valve is used with a pump.

(19) A separate relief device is required on each transfer pump.

(20) One (1) low-oil or liquid-fuel pressure interlock, reset from flame safeguard or manually.

(21) Burners with automatic controls, prepurge, proof-of-closure, modulation, or postpurge shall not use relays external to the flame safeguard to accomplish these functions.

(F) Shutoff Valve. All oil or liquid-fuel burner installations must include a non-electric shutoff valve that is held open by a fusible link designed to close at 165°F (79°C), installed near the burner in the same room as the burner. This must prevent the flow of oil or liquid fuel to the burner through the supply pipe. A check valve is required in the return line if the tank is higher than the burner.

APPENDIX D

UNIT CONVERSION TABLES

SI SYMBOLS AND PREFIXES

BASE UNITS

Quantity	Unit	Symbol
Length	Meter	m
Mass	Kilogram	kg
Time	Second	s
Electric current	Ampere	A
Thermodynamic temperature	Kelvin	K
Amount of substance	Mole	mol
Luminous intensity	Candela	cd

SI SUPPLEMENTARY UNITS

Quantity	Unit	Symbol
Plane angle	Radian	rad
Solid angle	Steradian	sr

SI PREFIXES

Multiplication Factor		Prefix	Symbol
1 000 000 000 000 000 000	= 10^{18}	exa	E
1 000 000 000 000 000	= 10^{15}	peta	P
1 000 000 000 000	= 10^{12}	tera	T
1 000 000 000	= 10^{9}	giga	G
1 000 000	= 10^{6}	mega	M
1 000	= 10^{3}	kilo	k
100	= 10^{2}	hecto	h
10	= 10^{1}	deka	da
0.1	= 10^{-1}	deci	d
0.01	= 10^{-2}	centi	c
0.001	= 10^{-3}	milli	m
0.000 001	= 10^{-6}	micro	μ
0.000 000 001	= 10^{-9}	nano	n
0.000 000 000 001	= 10^{-12}	pico	p
0.000 000 000 000 001	= 10^{-15}	femto	f
0.000 000 000 000 000 001	= 10^{-18}	atto	a

SI DERIVED UNIT WITH SPECIAL NAMES			
Quantity	**Unit**	**Symbol**	**Formula**
Frequency (of a periodic phenomenon)	hertz	Hz	1/s
Force	newton	N	kg•m/s^2
Pressure, stress	pascal	Pa	N/m^2
Energy, work, quantity of heat	joule	J	N•m
Power, radiant flux	watt	W	J/s
Quantity of electricity, electric charge	coulomb	C	A•s
Electric potential, potential difference, electromotive force	volt	V	W/A
Capacitance	farad	F	C/V
Electric resistance	ohm	Ω	V/A
Conductance	siemens	S	A/V
Magnetic flux	weber	Wb	V•s
Magnetic flux density	tesla	T	Wb/m^2
Inductance	henry	H	Wb/A
Luminous flux	lumen	lm	cd•sr
Illuminance	lux	lx	lm/m^2
Activity (of radionuclides)	becquerel	Bq	1/s
Absorbed dose	gray	Gy	J/kg

CONVERSION FACTORS

To convert	To	Multiply by
LENGTH		
1 mile (U.S. statute)	km	1.609 347
1 yd	m	0.9144
1 ft	m	0.3048
	mm	304.8
1 in	mm	25.4
AREA		
1 mile^2 (U.S. statute)	km^2	2.589 99
1 acre (U.S. survey)	ha	0.404 687
	m^2	4046.87
1 yd^2	m^2	0.836 127
1 ft^2	m^2	0.092 903
1 in^2	mm^2	645.16
VOLUME, MODULUS OF SECTION		
1 acre ft	m^3	1233.49
1 yd^3	m^3	0.764 555
100 board ft	m^3	0.235 974
1 ft^3	m^3	0.028 316 8
	L (dm^3)	28.3168
1 in^3	mm^3	16 387.1
	mL (cm^3)	16.3871
1 barrel (42 U.S. gallons)	m^3	0.158 987

CONVERSION FACTORS (Continued)

To convert	To	Multiply by
(FLUID) CAPACITY		
1 gal (U.S. Liquid)**	L*	3.785 41
1 qt (U.S. Liquid)	mL	946.353
1 pt (U.S. Liquid)	mL	473.177
1 fl oz (U.S.)	mL	29.573541
1 gal (U.S. Liquid)	m^3	0.003 785
**1 gallon (UK) approx. 1.2 gal (U.S.), *1 liter = 0.001 cubic meters		
SECOND MOMENT OF AREA		
1 in^4	mm^4	416 231
	m^4	0.416 231 x 10^{-6}
PLANE ANGLE		
1° (degree)	rad	0.017 453 3
	mrad	17.4533
1' (minute)	urad	290.888
1" (second)	urad	4.848 14
VELOCITY, SPEED		
1 ft/s	m/s	0.3048
1 mile/h	km/h	1.609 344
	m/s	0.447 04
VOLUME RATE OF FLOW		
1 ft^3/s	m^3/s	0.028 316 8
1 ft^3/min	L/s	0.471 947
1 gal/min	L/s	0.063 090 2
1 gal/min	m^3/min	0.0038
1 gal/h	mL/s	1.051 50
1 million gal/d	L/s	43.8126
1 acre ft/s	m^3/s	1233.49
TEMPERATURE INTERVAL		
°F	°C	(°F-32)•(5/9)
	°K	Add 273.15 to °C
EQUIVALENT TEMPERATURE ($t_{°C} = T_K - 273.15$)		
t°F		9/5 $t_{°C}$ + 32
MASS		
1 ton (short***)	metric ton	0.907 185
	kg	907.185
1 Ib	kg	0.453 592
1 oz	g	28.3495
***1 long ton (2240 Ib)	kg	1016.05
MASS PER UNIT AREA		
1 Ib/ft^2	kg/m^2	4.882 43
1 oz/yd^2	g/m^2	33.9057
1 oz/ft^2	g/m^2	305.152

CONVERSION FACTORS (continued)

To convert	To	Multiply by
DENSITY (MASS PER UNIT VOLUME)		
1 lb/ft³	kg/m³	16.0185
1 lb/yd³	kg/m³	0.593 276
1 ton/yd³	t/m³	1.186 55
FORCE		
1 tonf (ton-force)	kN	8.896 44
1 kip (1,000 lbf)	kN	4.448 22
1 lbf (pound-force)	N	4.448 22
MOMENT OF FORCE, TORQUE		
1 lbf•ft	N•m	1.355 82
1 lbf•in	N•m	0.112 985
1 tonf•ft	kN•m	2.711 64
1 kip•ft	kN•m	1.355 82
FORCE PER UNIT LENGTH		
1 lbf/ft	N/m	14.5939
1 lbf/in	N/m	175.127
1 tonf/ft	kN/m	29.187 8
PRESSURE, STRESS, MODULUS OF ELASTICITY (Force per Unit Area) (1 Pa=1 N/m²)		
1 tonf/in²	MPa	13.7895
1 tonf/ft²	kPa	95.7605
1 kip/in²	MPa	6.894 76
1 lbf/in²	kPa	6.894 76
1 lbf/ft²	Pa	47.8803
Atmosphere	kPa	101.3250
1 inch mercury	kPa	3.376 85
1 foot (water column @32°F)	kPa	2.988 98
WORK, ENERGY, HEAT (1J = 1N•m = 1W•s)		
1 kWh (550 ft•lbf/s)	Mj	3.6
1 Btu (Int. Table)	kJ	1.055 06
	J	1055.06
1 ft•lbf	J	1.355 82
COEFFICIENT OF HEAT TRANSFER		
1 Btu/(ft²•h•°F)	W/(m²•K)	5.678 26
THERMAL CONDUCTIVITY		
1 Btu/(ft•h•°F)	W/(m•K)	1.730 73
ILLUMINANCE		
1 lm/ft² (footcandle)	lx (lux)	10.7639
LUMINANCE		
1 cd/ft²	cd/m²	10.7639
1 foot lambert	cd/m²	3.426 26
1 lambert	kcd/m²	3.183 01

Approximate Minimum Thickness (Inch) for Carbon Sheet Steel Corresponding to Manufacturer's Standard Gauge and Galvanized Sheet Gauge Numbers

inches x 25.4 = mm

Carbon Sheet Steel			Galvanized Sheet		
Manufacturer's Standard Gauge No.	Decimal and Nominal Thickness Equivalent (Inch)	Recommended Minimum Thickness Equivalent* (Inch)	Galvanized Sheet Gauge No.	Decimal and Nominal Thickness Equivalent (Inch)	Recommended Minimum Thickness Equivalent* (Inch)
8	0.1644	0.156	8	0.1681	0.159
9	0.1495	0.142	9	0.1532	0.144
10	0.1345	0.127	10	0.1382	0.129
11	0.1196	0.112	11	0.1233	0.114
12	0.1046	0.097	12	0.1084	0.099
13	0.0897	0.083	13	0.0934	0.084
14	0.0747	0.068	14	0.0785	0.070
15	0.0673	0.062	15	0.0710	0.065
16	0.0598	0.055	16	0.0635	0.058
17	0.0538	0.050	17	0.0575	0.053
18	0.0478	0.044	18	0.0516	0.047
19	0.0418	0.038	19	0.0456	0.041
20	0.0359	0.033	20	0.0396	0.036
21	0.0329	0.030	21	0.0366	0.033
22	0.0299	0.027	22	0.0336	0.030
23	0.0269	0.024	23	0.0306	0.027
24	0.0239	0.021	24	0.0276	0.024
25	0.0209	0.018	25	0.0247	0.021
26	0.0179	0.016	26	0.0217	0.019
27	0.0164	0.014	27	0.0202	0.017
28	0.0149	0.013	28	0.0187	0.016
			29	0.0172	0.014
			30	0.0157	0.013

*The thickness of the sheets set forth in the code correspond to the thickness shown under these columns. They are the approximate minimum thicknesses and are based on the following references:

Carbon Sheet Steel—Thickness 0.071 inch and over:
ASTM A 568-74, Table 3, Thickness Tolerances of Hot-Rolled Sheet (Carbon Steel).

Carbon Sheet Steel—Thickness less than 0.071 inch:
ASTM A 568-74, Table 23, Thickness Tolerances of Cold-Rolled Sheet (Carbon and High-Strength Low Alloy).

Galvanized Sheet Steel—All thicknesses:
ASTM A 525-79, Table 4, Thickness Tolerances of Hot-Dip Galvanized Sheet.

Minimum thickness is the difference between the thickness equivalent of each gauge and the maximum negative tolerance for the widest rolled width.

INDEX

– A –

– D –

– E –

– W –